Table of Contents

Detailed Table of Contents

When a firm initiates and invests into an information technology (IT) project, it is usually with the intention of realizing benefits in the informational, strategic, transactional, and infrastructure objective areas of its IT portfolio (Weill & Broadbent, 1998). From the project management perspective, it is critical to know how the project is performing from the viewpoint of scope, schedule, cost, and other constraints. Lewis (2008) reports that 70% of IT-related projects do not meet their objective. This work examines the use of a case-based reasoning decision support architecture that provides a collaborative intelligent agent system to aid in recommending the status of a project using color indicators (Red, Yellow, Green) derived from the progress and condition of the project-related constraints.

This chapter firstly reviews the importance of the Satisfiability Problem (SAT) for a wide range of applications, including applications in Operation Management such as planning. A review of methods nowadays employed by modern SAT-solvers is then presented. We then use Classical Planning as an illustrative example of how a significant problem can be translated into SAT. We point out important results and studies concerning reductions of planning into SAT, and explain how to construct a SAT instance which is satisfiable if and only if an instance of a bounded version of the classic blocks-world problem is solvable.

With the growing complexity of information technology (IT) projects, the management of these projects is proving to be a daunting task. The magnitude of this problem is underscored by the assertion that approximately 70% of IT projects fail to meet their objectives (Lewis, 2007). Computational intelligence (CI) is an area of research focused on developing intelligent systems to help with complex problems. Specifically, CI seeks to integrate techniques and methodologies to assist in problem domains in which information, data and perhaps even the problem itself are vague, approximate, and uncertain. It would seem that research aimed at leveraging the power of CI against IT project management problems is critical if IT project success rates are to be improved. This work examines the core CI technologies – fuzzy logic, neural networks, and genetic algorithms – and looks at current and potential future applications of these techniques to assist IT project managers.

Chapter 6
Piece-Mold-Machine Manufacturing Planning.. 105
O.J. Ibarra-Rojas, UANL, México
Y.A. Rios-Solis, UANL, México
O.L. Chacon-Mondragon, UANL, México

This chapter studies a manufacturing process of pieces. These pieces are produced with molds which are mounted on machines. The authors describe this process as an optimization problem using an integer linear programming formulation which integrates the most important features of the system, and determines the quantities of pieces to produce, including the allocation of molds to machines. The objective function is to maximize the weighted production since they seek to minimize the non-fulfilled demand. First they show that the addressed problem belongs to the NP-hard class. After observing that solving the problem in an exact way is time consuming, they propose a solution methodology based on an Iterated Local Search Algorithm. Through computational experimentation we make conclusions about the difficulty of the decisions determined in this manufacturing planning.

Chapter 7
Intelligent Simulation System for Supply Chain Event Management (SCEM) 118
Barin Nag, Towson University, USA
Haiying Qiao, SmartOps Company, Pittsburgh, USA
Dong-Qing Yao, Towson University, USA

In this chapter, the authors present an intelligent simulation system for supply chain event management for the purpose of designing and re-engineering the supply chain. The simulation framework mainly composes of component layer, process layer, intelligent execution layer, and output layer. The functional design of the layers is discussed with comments on the contribution of the simulation. Implementation issues are further addressed and an illustrative case study is reported.

Chapter 8
Mathematical Models for Optimizing the Global Mining Supply Chain 133
Bruno Santos Pimentel, Federal University of Minas Gerais, Brazil
Geraldo Robson Mateus, Federal University of Minas Gerais, Brazil
Franklin Assunção Almeida, Federal University of Minas Gerais, Brazil

The present chapter discusses the application of intelligent decision-support systems – mathematical programming-based in particular – to operations management within the mining industry. The underlying production and distribution planning and scheduling problems have often been addressed individually, in disregard of upstream and downstream operations. A supply chain approach to mining operations, however, requires an integrated perspective which takes into account mine, railway and port operations, as well as domestic and international supply stations served by appropriate logistics channels. Three main topics are discussed here: recent operations research developments in the mining industry; integrated approaches towards the development of decision-support systems to address a global mining supply chain; and possible solution approaches to the integrated problems. The main thread is oriented to mathematical programming approaches, but relevant applications of simulation and artificial intelligence techniques are also discussed.

The development of Information and Communication Technologies (ICT) allowed the enterprises to adopt an e-marketplace approach to Business to Business (B2B) applications. In particular, these kind of applications are demonstrating their capacity to provide real added value to manufacturing industries by allowing their global performance to increase. The implementation of the e-marketplace by firms is not considered an easy job because of the lack of automation: the human participation is still in all stages of the B2B process. The chapter proposes a three value added services: workflow design, Multi Agent System and negotiation approach. In particular, two negotiation, an auction and single round approaches with three customer behaviors are proposed. A simulation environment is developed in order to test the proposed approaches. The simulations have been conducted in several scenarios in order to highlight what is the best approach to perform.

Computer vision is nowadays a key factor in many manufacturing processes. Among all possible applications like quality control, assembly verification and component tracking, the robot guidance for pick and place operations can assume an important role in increasing the automation level of production lines. While 3D vision systems are now emerging as valid solutions in bin-picking applications, where objects are randomly placed inside a box, 2D vision systems are widely and successfully adopted when objects are placed on a conveyor belt and the robot manipulator can grasp the object by exploiting only the 2D information. On the other hand, there are many real-world applications where the 3rd dimension is required by the picking system. For example, the objects can differ in their height or they can be manually placed in front of the camera without any constraint on the distance between the object and the camera itself. Although a 3D vision system could represent a possible solution, 3D systems are more

This chapter describes the use of simulation systems for decision support in support of real operations, which is the most challenging application domain in the discipline of modeling and simulation. To this end, the systems must be integrated as services into the operational infrastructure. To support discovery, selection, and composition of services, they need to be annotated regarding technical, syntactic, semantic, pragmatic, dynamic, and conceptual categories. The systems themselves must be complete and validated. The data must be obtainable, preferably via common protocols shared with the operational infrastructure. Agents and automated forces must produce situation adequate behavior. If these requirements for simulation systems and their annotations are fulfilled, decision support simulation can contribute significantly to the situational awareness up to cognitive levels of the decision maker.

In many multi-stage manufacturing supply chains, transportation related costs are a significant portion of final product costs. It is often crucial for successful decision making approaches in multi-stage manufacturing supply chains to explicitly account for non-linear transportation costs. In this article, we have explored this problem by considering a Two-Stage Production-Transportation (TSPT). A two-stage supply chain that faces a deterministic stream of external demands for a single product is considered. A finite supply of raw materials, and finite production at stage one has been assumed. Items are manufactured at stage one and transported to stage two, where the storage capacity of the warehouses is limited. Packaging is completed at stage two (that is, value is added to each item, but no new items are created), and the finished goods inventories are stored which is used to meet the final demand of customers. During each period, the optimized production levels in stage one, as well as transportation levels between stage one and stage two and routing structure from the production plant to warehouses and then to customers, must be determined. The authors consider "different cost structures," for both manufacturing and transportation. This TSPT model with capacity constraint at both stages is optimized using Genetic Algorithms (GA) and the results obtained are compared with the results of other optimization techniques of complete enumeration, LINDO, and CPLEX.

Preface

The development and study of models, methods, and applications, of intelligent systems in operations is an area of research that is in rapid growth with increasing numbers of quality contributions. These contributions serve to develop and advance the body of knowledge in the field by the addition of new and improved intelligent methods, and by the use of models based on artificial intelligence techniques. Most of this research tends to be practical rather than theoretical, and extends existing applications or develops new applications in business and industry with benefits for all in terms of cost, productivity, and quality. This book presents a variety of current and original research in intelligent systems and methodologies that is directly relevant and is application oriented.

One of the key elements of the 21st century business environment is that business and industrial functions and operations are global in nature. Management decisions are often made on a global basis with consideration of global decision factors. Even when an operational decision is made locally, the impact of the decision can be global. Further, concerns about operational decision problems and the methods used to solve them vary from one country to another. However, the impact of the decision and the methods used might be effective in a different environment somewhere else in the world. This book presents work from many researchers from a number of countries around the work that are actively engaged in research on intelligent systems in operations. Some of this work is drawn from their experiences in their home countries but have impacts and applications in other countries as well. The roster of authors includes those who are resident in one country and are concerned with the problems there, but are really from another country and have an understanding of problems that span multiple countries.

The importance of the topic begins with the fact that an operation is at the core of every business and industrial process. A business exists for the single purpose of transforming inputs into outputs by some process that adds value to the product for its customers. The sale of the product to the customers generates revenue, and a profit that results from the differential of the revenue and the costs of the inputs. Even a non-profit organization operates on the same value added principle minus the generated profits. The concept of profitability of the organization depends on the productivity of the organization, and productivity is often defined as the quantity of output per unit of input. Productivity that is larger and more effective in terms of customer satisfaction leads to lower cost structures vis-à-vis the revenue pattern, and thus to greater profitability.

Another factor in the effectiveness of the operational business transformation process is the supply chain and the management of the supply chain. The supply chain may be defined as the sequence of supplies of input components and transformation processes from the initial items that may be derived from the ground in mining operations or from farming operations, all the way to the final products delivered to the customer or end user, in a manufacturing or service context. Whether complex or simple, the importance of the supply chain is that the supply chain cost structure dominates the process sequence, and significant advantages may be obtained by managing supply chain strategies and procedures compared to forced increases in revenue (Heizer & Render, 2007). The supply chain issue is addressed by several papers in the book.

The major question raised by the productivity and supply chain concerns as raised above has to do with decision issues. It was stated by Herbert Simon in a paper that won him a Nobel Prize in 1956 that decision making is about problem solving. In other words, a decision in a business process needs to be made only when there is a problem, i.e. a situation exists where the present output does not match the expected or desired output. Simon described problem solving in four phases labeled Intelligence, Design, Choice, and Implementation. The phases describe respectively gathering information, developing alternative solutions, selecting the best solution, and implementing, often in an iterative way (Turban & Volonino, 2009) . Simon's approach led to the description of problems in a spectrum ranging from structured and programmable to unstructured and solvable only by intelligent entities, with semi-structured problems in the middle. The goal of intelligent systems is to develop and implement automated and systemic solutions for the vast range of semi-structured problems requiring some level of intelligence in the solver, and preferably extending in the direction of unstructured problems.

Decision making in the business environment, as commonly performed by managers around the world, commonly involve people, equipment, targets, and uncertainties arising from market demands, the supply chain, performance characteristics, and the financial environment. It is quite common that the problem environment is less structured than one would want it to be. Conventional or programmable decision making is often inadequate as a computerized process. The field of operations management is one of the business decision areas having many problems designated as relatively unstructured because of uncertainties or other factors. The problems are important in the business context, and are in need of solution methodologies. Intelligent systems developed using concepts, tools, and methodologies, as used in Artificial Intelligence (AI) are transforming the world of Operations Management (OM) by the introduction of computerized and intelligent decision making tools, particularly in the areas of semi-structured decisions.

Intelligent systems are a collection of systems, methods, and technologies, derived from an established body of knowledge known as Artificial Intelligence, which is a way of replicating human intelligence and thinking in computer systems that are artificial or man-made. The applications of AI algorithms and intelligent systems embedded in various methods and devices are transforming operations around the world by increasing profitability and performance, by reducing the lead times for delivery, by making new manufactured and service products widely available at reduced costs and times to a larger number of customers, by improving performance, cost, and quality in the supply chain from the source to the customer, and by globalizing the supply systems and the customer base. This subject area is a bridge between the two traditional areas of Operations Management and Information Systems/Computer Science. The present focus is on the applications of Intelligent Systems in Operations Management (OM). This publication will present a variety of new research and new ideas in this context, as proposed systems, or as operational models, or as suggested and implemented methods in a variety of applications. Included are comparisons of modeling and application strategies in a global context. It is intended to cover all aspects of operations and the management of operations. Some of these functional areas are listed as Supply Chain Management, Manufacturing and Distribution with Logistics, Financial Services of all types, Healthcare and Medical Services, and all other manufacturing and service operations not mentioned above.

This book will provide knowledge and insights on present and future AI applications in OM from current research-oriented thinking on AI-based systems in the benefit of OM tools and decisions in terms of theoretical and empirical models, methods and their comparisons, and actual and proposed applications. It is intended to be inclusive with respect to all the diverse concerns of OM functions and issues, and will be global in its perspective. The diversity of content areas and applications is evident in the content. A brief look at the author biographies and affiliations shows the global perspective of the research presented. In summary, the book is not comprehensive in terms of AI methods in OM, as no book ever can be, but it presents a variety of research for thought in a variety of research and study areas, in a variety of applications, to support a variety of research.

The book will benefit professionals in industry seeking examples of applications, researchers looking for material that can be extended into further research that is as valuable as that found here, and graduate students in the field learning to do research and looking ideas and models as examples to build upon. The audience includes diverse application areas from industrial Operations Management to Business Process Management to Supply Chain Management, and also on the computational and algorithmic side from Computer Science to Information Science and Information Technology. In addition to covering all relevant functional areas as listed above, the book provides insights and executive support to executives in relevant industries tasked with decision making in these areas, and the management of expertise, methods, and knowledge, as it relates to these concepts.

Summary descriptions of the chapters are given below in alphabetical order with the first author. It is intentional that these are not in any sequence or progression to accommodate the various preferences of readers that have different outlooks and priorities:

Barnes, A.P. & Hammell, R.J. *Employing Intelligent Decision Systems to Aid in Information Technology Project Status Decisions.*

The initiation of an information technology (IT) project by a firm, and the investment decision, usually has an objective of realizing benefits in the informational, strategic, transactional, and infrastructural objective areas of its portfolio. In the project management perspective, the important concerns are the performance of the project from the viewpoints of scope, schedule, cost, and other constraints. It is common that IT projects fail to meet objectives. This paper presents a case-based reasoning decision support architecture to provide a collaborative intelligent agent system to aid in recommendations of project status using color indicators derived from the progress and condition of the project-related constraints.

Camarao, C., Galvao, M., &Viera, N. *SAT and Planning: An Overview*

This chapter studies the Satisfiability Problem (SAT), reviewing its importance in a wide range of applications. One application in Operations Management is planning. First, a review is presented of methods currently employed by modern SAT-solvers. Next, classical planning is used as an illustrative example to demonstrate the translation of a significant problem into SAT. Important results and studies are stated concerning reductions of planning into SAT. Finally, an explanation is given about the construction of a SAT instance which is satisfiable if and only if an instance an instance of a bounded version of the classic blocks-world problem is solvable.

Chen, W. & Decker, K.S. *Integrated Multi-Agent Coordination.*

The key topics of planning and scheduling from Operations Research and Multi-Agent Systems have typically focused at an abstract system level on the development of interaction protocols to be imposed on agents. The internal task structures of individual agents that affect the higher-level coordination behaviors have been less studied. Collaborative multi-agent planning addresses problems such as uncertainties in plan outcomes, anticipation of likely contingencies, and evaluations of the way agent actions achieve worth-oriented goals. This paper presents extensions and restrictions, called extended hierarchical task networks, to the traditional plan and schedule representations that allow the formal definition of an integrated multi-agent coordination problem. The chapter discusses open issues in multi-agent coordination, and proposes a general solution towards successful distributed goal achievement by analyzing the task structures of participating agents.

Dunkel, J., Bruns, R., & Pawlowski, O. *Complex Event Processing in Sensor-based Decision Support Systems.*

The chapter presents a reference architecture for sensor-based decision support systems, which enables the analysis and processing of complex event streams in real-time. The proposed architecture provides a conceptual basis for the development of flexible software frameworks that can be adapted to meet the needs of various applications. The architectural approach is based on semantically rich event models providing the different stages of the decision process. The approach is illustrated in the domain of road traffic management in high-capacity road networks. The need for this work arises from the high volume of continuously generated sensor events that decision support systems have to deal with. Conventional software architectures do not explicitly target the efficient processing of continuous event streams. The high volume of events, and the complex dependencies, implies that it is not possible to have a fixed or predefined process flow at the business level. Complex Event Processing (CEP) has been proposed recently as a general process model for event streams. CEP can provide mechanisms for computing high event volumes, but it does not define methodologies, models, and reference architectures that would make event processing mature software architecture.

Hammell II, R.J., Hoksbergen, J., Wood, J., & Christensen, M. *Computational Intelligence for Information Technology Project Management*

This chapter addresses the management of Information Technology (IT) projects, an area that has become difficult and challenging because of the growth in numbers and complexity of these projects. The magnitude of the problem is shown by the fact that over two-thirds of the IT projects fail to meet their objectives. This work uses Computational Intelligence (CI) methodology, an area of research focused on the development of intelligent systems to help with complex problems. CI is a process of integrating techniques and methodologies to assist in problem domains vagueness, approximations, and uncertainties, pervade the information, data, and even the problem itself. Leveraging the power of CI methods on IT project management is a logical approach to improving IT project success rates. Examined in this work are the core CI technologies of fuzzy logic, neural networks, and genetic algorithms, vis-à-vis current and potential future applications to assist IT project managers.

Ibarra-Rojas, O.J., Rios-Solis, Y.A., & Chacon-Mondragon, O.L. *Piece-mold-machine manufacturing planning*

This chapter is a study of a manufacturing process that puts together pieces produced on molds mounted on machines. To address the primary concern of minimizing the non-fulfilled demand, an integer linear programming formulation is used to integrate the most important features of the problem. The solution is used to determine the quantities of pieces to produce and the allocation of molds to machines. The objective function is designed to maximize production so that the non-fulfilled demand is minimized. The problem belongs to the NP-hard class and is not solvable in an exact way. A heuristic solution methodology is used based on an Iterated Local Search Algorithm. Computational experimentation is used determine the manufacturing planning decisions, and to make conclusions about the difficulty.

Nag, B.N., Qiao, H., & Yao, D. *Intelligent Simulation System for Supply Chain Event Management (SCEM)*

This chapter presents and discusses an intelligent simulation system for Supply Chain Event Management for the purpose of designing and re-engineering the supply chain. Supply Chain Event Management (SCEM) is a subtopic of supply chain management designed to analyze the risks associated with uncertain supply chains, and re-engineer risky supply chains that are not suitable for the usual analysis of supply chains. The simulation framework is discussed in a series of layers, consisting of a component layer, a process layer, an intelligent execution layer, and an output layer for the presentation of results. The functional design of the layers is discussed on the basis of a simulation experiment on a case study that is illustrative of the process. In addition, implementation issues are reported and discussed on the basis of this real case study.

Pimentel, B.S., Mateus, G.R., & Almeida, F.A. *Mathematical Models for Optimizing the Global Mining Supply Chain*

This chapter is a discussion on the application of intelligent decision support systems in the management of operations in the mining industry. An integrated perspective on the supply chain approach to mining operations takes into account mining, railway, and port operations, as well as the logistics channels that support domestic and international supply chain operations. The operations involve production and distribution planning, and scheduling problems, which have been addressed individually. Mathematical programming is the basis for this discussion. The primary topics discussed here are: recent developments in the use of operations research in the mining industry, integrated approaches towards the development of decision support systems to support global mining supply chains, and possible solution approaches to integrated supply chain problems. The focus is on mathematical programming formulations supported and solved by heuristic simulation and artificial intelligence solution techniques.

Renna, P. *Negotiation policies for e-procurement by Multi Agent Systems*

This chapter is about a special e-Business application known as e-Procurement, i.e. electronic procurement in the supply chain as commonly found in an e-marketplace approach in business-to-business (B2B) applications. B2B applications of e-marketplace follow the development of Information and Communication Technologies (ICT) that allowed enterprises to adopt the e-marketplace approach. These applications are providing added value to manufacturing operations in terms of increasing global performance. The e-marketplace implementation is still not easy because of the need of human participation in all stages of the B2B process. The value added services proposed in this chapter are: workflow design, Multi Agent System, and a negotiation approach. The negotiation approach is based on the auction environment, with simulations to test the proposed approaches. Results are presented for the simulations conducted in several scenarios to determine the best approach.

Sgarbi, M., Colla, V., & Bioli, G. *A 3D vision-based solution for product picking in industrial applications*

This chapter is about computer vision applications in robotic systems in industrial applications. Computer vision has become a key application factor in manufacturing processes, affecting quality control, assembly verification, and component tracking. Robotic systems play an important role in the automation of production lines. The common applications of 2D vision systems for placing objects on conveyor belts, where robot manipulators grasp the objects using only 2D information, is being replaced by 3D applications which enable valid solutions in bin-packing applications where objects are randomly placed inside a box. There are many real-world applications where the 3rd dimension is a requirement for successful operation. An example is where the objects differ in height, or are placed in front of a camera manually without any constraint on the distance between the object and the camera. The solution possibilities presented by a 3D system are balanced by the higher costs and complexities compared to 2D vision systems. This work presents a monocular system useful for picking applications by estimating the 3D position of a single marker attached to the target object in the assumption that the orientation of the object is approximately known.

Wautelet, Y., Achbany, Y., Lange, J., & Tran, V. *Developing a Collaborative Supply Chain Management Platform: a Service-Oriented Approach*

This chapter develops the concept of service-oriented computing and Service Oriented Architecture (SOA) that is becoming increasingly popular. It enables flexible designs and adaptable software systems that can be easily adapted to the demands of software customers. Supply Chain Management is a primary beneficiary of this concept. This paper is a proposal to apply ProDAOSS, a process for developing adaptable an open service systems. The application is directed towards an industrial case study in outbound logistics. ProDAOSS is conceived as a broader development methodology to cover the entire software development life cycle, and is intended as a plug-in for I-Tropos. Flexible business processes are typically generically modeled with different complimentary views at an analytical level. At first, an aggregate services view of the entire application package is offered. Then the services are split using an agent ontology, using the i* framework to represent it as an organization of agents. A dynamic view completes the documentation by offering the service realization paths. At the design stage, the service center architecture proposes a reference architectural pattern for services realization in ana adaptable and open manner. The work finally presents the implemented platform for a particular service to manage transport, so that a reader can realize how the developments have been achieved.

Acknowledgment

I would like to give a grateful note of thanks to every author who has contributed valuable research work, making the book that much more interesting and informative. State of the art research is always uncertain and risky. A research proposal, or a research plan, or even a commitment towards a research contribution, does not always result in one. One has to appreciate the work, and the commitment to work, of all the authors who have made contributions, often in a relatively short time. Each one was peer-reviewed. Some needed adjustments. In the end, all came out with flying colors.

At this time, I would also like to thank the reviewers, many of whom came from the list of authors. However, no reviewer had any idea of who the authors were. That is the honor code of reviewing, and the reviewers did a wonderful job, often returning the review in a very short time. I serve on the editorial boards of some journals and I really appreciate these fast reviews. I appreciate the time taken by the reviewers to do this.

One last comment refers to the global nature of the authors' affiliations, reflecting the global nature of the interests and research works represented in this book. Globalization of operations is the nature of the world today, and is of importance to me in terms of research interests. Globalization is reflected most effectively in this book with contributions from many countries in several continents.

Barin Nag
Towson University, USA

Chapter 1
Employing Intelligent Decision Systems to Aid in Information Technology Project Status Decisions

Anthony P. Barnes
Towson University, USA

Robert J. Hammell II
Towson University, USA

ABSTRACT

When a firm initiates and invests into an information technology (IT) project, it is usually with the intention of realizing benefits in the informational, strategic, transactional, and infrastructure objective areas of its IT portfolio (Weill & Broadbent, 1998). From the project management perspective, it is critical to know how the project is performing from the viewpoint of scope, schedule, cost, and other constraints. Lewis (2008) reports that 70% of IT-related projects do not meet their objective. This chapter examines the use of a case-based reasoning decision support architecture that provides a collaborative intelligent agent system to aid in recommending the status of a project using color indicators (Red, Yellow, Green) derived from the progress and condition of the project-related constraints.

INTRODUCTION

When a firm makes the decision to invest into an information technology project, the purpose may be to gain a competitive advantage, improve its position in the industry, or evolve its business strategy. Before the project is initiated, it is important for the firm to discuss risks, qualitative benefits, implementation costs, completion dates, and the staff required to successfully complete the project. It is advantageous of the firm to employ tools and techniques to help ensure the successful completion of the project. (Applegate, Austin & McFarlan, 2003)

Project management is the discipline that provides skills, tools, and techniques to guide the execution of a project (PMI, 2004). The definitive advantage that a firm gains through implementing project management techniques is a higher probability to have a customer (or the firm itself if it is

DOI: 10.4018/978-1-61520-605-6.ch001

performing an internal project) who is satisfied with the project's result (Gido & Clements, 2006). A project is defined as a temporary endeavor with the objective to produce a unique product, service, or result (Gido & Clements, 2006; PMI, 2004). The project objective is defined through its constraints of scope – the amount of work that will be completed to satisfy the customer; schedule – the amount of time needed to complete the scope through planned start and finish dates for project activities; and cost – the amount of money that the customer has budgeted for the completion and acceptance of project deliverables. (Gido & Clements, 2006; PMI, 2004)

A Project Manager assumes the lead role in planning, organizing, and controlling the tasks and resources that embody the project. Performing these duties includes defining the project objective; developing a plan to accomplish project objectives; securing proper resources to perform the project; assigning responsibility and delegating authority to project team members; and tracking actual and planned progress of the project. (Gido & Clements, 2006)

During the execution of a project, the status of completion of its tasks is pivotal in understanding if the project is progressing as planned. If a task is not completed on time, the impact could be a slippage in the project schedule, which can result in the need to obtain more resources to set the project back on the planned schedule. Acquiring more resources to complete a project may not only affect the schedule of the project (training time of newly acquired human resources), but also the scope (how much work can actually be completed before the delivery date) and cost (the amount of money needed to acquire more human resources and equipment). In all cases, the firm may inherit unplanned overhead that can have a negative impact on the strength of its information technology investment portfolio. (Weill, 1998)

Although there have been advancements in technology and modern processes, the ability to manage a large project has been problematic due to complexity. The intricacy associated with managing the complexity of information technology projects has led to reports that 70% of all IT-related projects do not meet their objectives (Lewis, 2008).

Therefore, it is important to correctly diagnose and maintain visibility of the status for large, complex projects (Burgess, 2003). Proper diagnosis of the project is a decision that requires input from other sources besides the project manager. In the natural project management environment, determining the status of a project also involves input from various lines of service and stakeholders -- the project team. Each of these lines of services and stakeholders serve as experts that possess knowledge and experiences that allow them to view the context of the project from a different point of view and give their own assessment of the project status. The input from these experts is invaluable due to their knowledge and experience in certain situational contexts. The decision of the project status is determined by a project manager that takes into account all of the available assessments. When a large number of decision variables are present, it becomes difficult for the human mind, in this case the project manager, to perform a successful decision analysis (Felsen, 1976).

To assist in the decision-making process, the project manager can employ decision support system (DSS) technology. In addition to DSS, the incorporation of intelligent agents can be a useful asset to aid in the goal of project status decision-making by providing the capability for knowledge to be applied to the problem without the imposition of human emotion. As a means to providing this capability, we will use an intelligent decision system called Recognition-primed Decision-making enabled Collaborative Agents for Simulating Teamwork (R-CAST) that was developed at the Pennsylvania State University (Fan 2005a).

In this work, we are concerned with using the R-CAST architecture to develop a collaborative agent system that will recommend the status of a project. More precisely, we aim to design, develop and demonstrate a system using the R-CAST architecture that can make a pertinent and correct recommendation of project status (Red, Yellow, or Green) after reasoning over a knowledge and experience knowledge base that is related to the context of a situation in a project management environment.

In the next section, background information is presented on DSS, case-based reasoning, and intelligent agents. After that, the R-CAST architecture is presented in high-level detail. The population of the necessary R-CAST components with project management knowledge is then outlined, followed by a description of our knowledge acquisition efforts and then a section discussing our ongoing work. We end with some suggestions regarding future work in this area.

BACKGROUND

In this section, we present background information on decision support systems, case-based reasoning, and intelligent agents to facilitate understating how they work together in the R-CAST system.

Decision Support Systems

Decision support system (DSS) technology is used as a tool to aid decision makers in complex decision-making and problem solving (Shim, 2002). A decision maker will discover that a DSS is a valuable tool that makes more information available for reasoning which allows the decision maker to make a well-educated decision.

During the late 1950s and early 1960s, efforts that were conducted in research areas such as theoretical studies of organization decision-making at Carnegie Institute of Technology, and the technical work conducted at MIT, provided the base from which DSS technology has evolved. In the middle of the 1970s, the concept of DSS developed into its own research area that has experienced significant advancements since that time period. In the early conceptual stages, DSS researchers concentrated on four areas of innovation:

i. tools and techniques to handle unstructured or semi-structured problems, which enhanced management science and operations research methodology;
ii. computer-based systems that promoted collaboration and interaction, which were constructed to aid managers and by characteristically more sophisticated than descriptive systems theory or traditional decision models;
iii. user-oriented systems, which outperformed MIS applications as a decision-making platform; and
iv. the division of data and models in computer applications, which created the expectation of the development of a foundation for efficient modeling. (Shim, 2002; Carlsson, 2002)

Over the past three decades, the evolution of DSS research has resulted in the birth of new concepts and views. For instance, new research has evolved the DSS into a role of assisting a specific type of decision-maker to solve a problem related to a specific domain, such as the development of a DSS for forest management (Varma, 2000), and the development of clinical DSS (Berner & Ball, 1998). DSS implementations have also made historical and current information about a firm, and other firms, available to strategic planners by giving them the capability to extract and format relevant business data and information (Martin, 1988). Overall, DSS research is a concentrated effort to improve the efficiency of the user's decision-making ability and the effectiveness of his decision. (Carlsson, 2002; Shim, 2002)

Case-Based Reasoning

From a systems approach, the manner in which data or information is structured in a system has a fundamental effect on how efficiently the system will perform. The structure of the information influences the method of how the knowledge is processed or reasoned to reach a solution (Watson, 1997). From the human cognitive perspective, people tend to have more confidence in someone that has experience in performing a task. Particularly when filling an employment vacancy, an employer wants assurance that a potential employee can perform a comparable task with a smaller probability of failure than one who has never performed the task. (Liebowitz, 1998) Whether it is from a systems or human cognitive approach, Case-based reasoning (CBR) is built on the idea of reusing previous experiences to solve new or existing problems (Liebowitz, 1998).

Origin of CBR

According to (Watson, 1997), the work of Schank and Abelson in 1997 is cited to be the origin of CBR, with a foundation based on cognitive science research. The early direction of CBR research conducted by the pair involved the use of *scripts* as the memory representation structure that provided the platform for humans to develop expectations and perform inferences from learned information stored in the brain (Watson, 1997). Experimentation results would later prove that a script could not adequately handle memory representation due to confusion that was present in scripts that had events with the same qualities. For instance, confusion would exist in the scripts with similar scenes such as a visit to a dentist's office and a doctor's office. (Watson, 1997)

Roger Schank continued to examine the influence that the memory of previous situations (cases) and situation patterns could have in the realm of problem solving and learning. Schank's ideas evolved through research performed by his graduate student group at Yale University; these efforts produced the first cognitive model for CBR and later the first CBR application. Janet Kolodner of Georgia Tech who developed the first CBR system called CYRUS in 1983 further developed Schank's concepts. Other CBR systems that were developed include CHEF, MEDIATOR, and PROTOS. (Liebowitz, 1998; Watson, 1997)

Watson (1997) offers that that the history of CBR is not extensive, in contrast to other decision support technologies such as rule-based system or neural networks, because of the straightforwardness of a CBR implementation.

Case-Based Reasoning Process

The CBR process is described as a recurring process where previous experiences are represented in the system in the form of a *case* (Liebowitz, 1998; Watson, 1997). The following steps explain the CBR process:

1. Retrieve the most similar case(s): relevant past cases are retrieved, or remembered, with the goal of retrieving the most useful cases.
2. Reuse the case(s) to attempt to solve the problem: the relevant cases are ranked and the instances that have the most potential for success are used.
3. Revise the proposed solution, if necessary: modifications may be made to an old solution to adapt it to a new problem situation.
4. Retain the new solution as a part of a new case: after testing and evaluation, the new case is stored into the case library. (Liebowitz, 1998; Watson, 1997)

Conceptually, the CBR process is similar to the manner in which humans solve problems. The process typically has human involvement, where the CBR tool is used as a case retrieval and reuse system, and the human user assumes the case revision (adaptation) role. (Watson, 1997)

Intelligent Agents

Intelligent agents are described as an application that performs a function that would normally require human knowledge, or intelligence, in place of a human user (Liebowitz, 1998). The foundation of intelligent agents is found in artificial intelligence (AI) research.

AI is a branch of computer science that aims to solve problems using methods that are related to the knowledge of a person and thus, not quantitative. AI research is focused on discovering ways of making machines superior in performing tasks that can normally only be completed through the application of human intelligence. AI aims to develop machines that use techniques to solve problems that are far too complex for a human by taking advantage of knowledge that is has acquired about a particular problem domain. (Liebowitz, 1998; Martin, 1988) Researchers have used agents to build systems that handle complex problems related to airport management, traffic guidance systems, and advanced robotics systems (Muller, 1996).

With its foundation laid in the 1970s and 1980s, the concept of intelligent agents evolves from AI research conducted in the areas of robotics, knowledge systems, planning, and learning techniques. As the intelligent agent research continued to progress, it began to be viewed as independent of AI. The reason for the partition is that researchers grouped systems and application that performed an intelligent, human task into its own class. (Liebowitz, 1998)

Adequately meeting the challenge of being able handle complex tasks requires the agent to possess basic characteristics. According to (Mueller, 1996), an agent should have the capacity to be: "*reactive*", having the ability to respond properly to unexpected events and changes in its operating environment; able to "*deliberate*", which gives the agent the ability to perform activities in a goal-oriented manner; be *efficient*, which allows them to solve tasks using their intelligence; "*interactive*", which allows the agent to manage positive and negative communication with other agents; and "*adaptable*", which allows the agent to be flexible in altering environment conditions.

Background Summary

This background section provided information regarding concepts that are used to shape the decision support system architecture used in our work. It should be noted that Carlsson (2002) mentions that the presence of intelligent agents is becoming apparent in DSS applications and that Shim (2002) points out how DSS is materializing into a technology that aids specific type of decision makers with specific types of problems. As an illustration of these two notions, we will now introduce the Recognition-primed Decision-making enabled Collaborative Agents for Simulating Teamwork (R-CAST) architecture.

RECOGNITION-PRIMED DECISION MAKING ENABLED COLLABORATIVE AGENTS SIMULATING TEAMWORK (R-CAST)

R-CAST is a decision aid that employs intelligent agent technology to enhance the quality of a decision making process. (Prindible, 2006) When used as a tool to support decision making, through experiments, R-CAST has shown that it has the ability to compensate for the negative influence that a human decision maker's cognitive capacity can have in the decision-making process (Fan, 2006). R-CAST has primarily been used as a decision aid in military training for battlefield scenarios to aid soldiers in their decision making process when multiple contexts, time pressure, and a human's cognitive capacity are factors that can degrade the quality of the decision (Fan 2005b, 2005c, 2006, 2007).

Figure 1. R-CAST components (adapted from (Prindible, 2006))

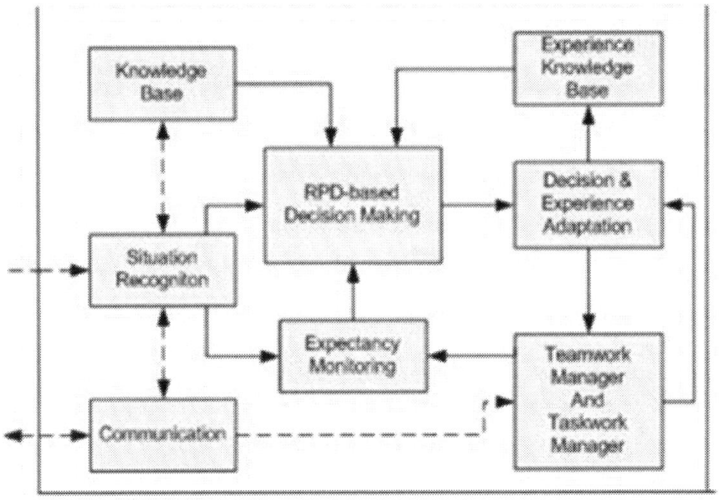

The cognitive aspect of the R-CAST architecture is based on Klien's Recognition-Primed Decision (RPD) Model, a naturalistic decision making model. Figure 1 provides an illustration of how the RPD Model, or "RPD-based Decision Making" component, interacts with the other major components of R-CAST. The R-CAST components are described as interchangeable and expendable depending on the intended use of the system (Prindible, 2006). To expand on that point as it relates to this research, the "Experience Base" component is used to hold the domain-specific knowledge; thus, changing the contents of this component changes the area of expertise of the system. Our efforts include performing knowledge acquisition activities aimed at populating the experience knowledge base with project management-specific expertise. Information about the knowledge acquisition activities conducted in this research is covered in the upcoming knowledge acquisition section; in this section we will explore the RPD decision-making process further. Detailed information concerning the other R-CAST components can be found in (Prindible, 2006).

RPD Decision Making Process

In this section, we will discuss the actions that comprise the RPD decision-making process in the R-CAST architecture. The RPD decision-making process component is based on Klien's RPD Model, which encapsulates the process of how a decision maker uses knowledge gained in prior situations (experiences) to decide what methods to use to solve a problem that relates to the current situation (Fan, 2005a).

The RPD Model has two phases: recognition and evaluation. As explained in (Fan, 2005a), the recognition phase of the decision making process deals with the agent identifying the courses of actions that are relevant to the current situation. In order to make this distinction, the agent becomes familiar with the current situation by developing an understanding of the decision context through analyzing the current state and behavior of the environment (situation awareness) (Fan, 2005a; Sarter, 1991). In the evaluation phase, the agent decides if a selected course of action will suffice by performing a mental simulation of the course of action in the environment with respect to the current situation context and constraints of the environment (Fan, 2005a; Prindible, 2006).

To translate to a human decision making process, first, a human decision maker will assess the problem and its domain; second, he will think of past events and the action he had to use to solve the problem; third, visualize and then choose the action to solve this particular problem; fourth, attempt to solve the problem; and finally, remember new steps that were taken, though successful or unsuccessful, when solving the problem.

Next we look further into the association that case-based reasoning has with the RPD Model.

RPD Model and Case-Based Reasoning (CBR)

When problem solving in R-CAST, an agent uses a set of experiences to test against the knowledge base in order to find a course of action that is suitable for the current situation. (Prindible, 2006) The set of experiences that are found in the experience knowledge base can be viewed as a set of cases, because they represent the structure of the information in the R-CAST system. This is equivalent to the manner that a case represents the structure of the information in a case-based reasoning system. In both instances the information within that structure (case information or experience knowledge) is pivotal to its relevance as a course of action for the current situation.

When an agent is unable to establish situation awareness in the recognition phase, RPD uses the methods of feature-matching and story-building. Of the two methods, feature-matching is first used, where an agent begins an examination of the situation by attempting to match the environmental cues from the domain with some of the features relevant to the domain. If the agent is deficient in experience or information, then the feature-matching method will fail and the story-building method is initiated. Story-building provides insight into a present situation by allowing a decision maker to view a sequence of events to realize the development of that situation. A logical relationship is formed using observed and accessible information about the situation to create a story that the decision-making agent will use to establish situation awareness. (Fan, 2005a, Prindible, 2006)

Once the agent achieves situation awareness, it attempts to identify a former experience to use in the current situation. The identification of an experience in the recognition phase corresponds with the first step of the CBR process where the case that is the most relevant to the problem is retrieved (Fan, 2005a; Leibowitz, 1998; Watson, 1997).

After the experience is retrieved, RPD performs a mental simulation of the relevant experiences and places them in order according to relevance. The agent uses the experience if the course of action seems suitable for the situation. This portion of the RPD process relates to the second step of the CBR process where the identified case is reused to solve the problem. Identically, the experience is also reused, as it is stored in the decision maker's experience knowledge base.

While the reused experience is being evaluated, RPD monitors the expectancy of that experience. The expectancy indicates what is likely to happen as recorded in the past experience. When expectancy is violated, the decision-making agent reiterates through the recognition phase to find an experience that is more applicable to the evolving situation (Fan, 2007). The relevant experiences are adapted to the evolving situation and the experiences with appropriate courses of action are used in the decision making process (Fan, 2006). This portion of the process corresponds to the third step of the CBR process where a proposed solution is revised and adapted to fit a new problem situation (Leibowitz, 1998; Watson, 1997)

The modifications of the cases and experiences are stored in the case library and experience knowledge base, respectively, thus corresponding to the fourth step of the CBR process. In both instances of an RPD experience and a case, the updated information will allow it to be a candidate for retrieval in the future for relevant situations. (Leibowitz, 1998; Prindible, 2006; Watson, 1997)

CONVERSION OF R-CAST TO OPERATE IN THE PROJECT MANAGEMENT WORLD

In this section we present our work to simulate a project management related environment using the R-CAST architecture. The steps taken to populate the necessary components of R-CAST with project management knowledge to accomplish this task is described in relation to guidance given in an R-CAST tutorial (Prindible, 2006) developed by researchers at The Pennsylvania State University. We will also discuss the knowledge acquisition activities performed, preliminary empirical results, and our plans to use R-CAST to design a collaborative agent system that will recommend the project status of an information technology project.

Project Management Simulation Example

As presented in (Barnes & Hammell, 2008), an example simulation was developed to test if R-CAST could reason over information related to a project management environment. The simulation was structured in relation to an example simulation presented in (Prindible, 2006), where the system was populated with knowledge and experience related to making decisions on the best driving route to take to "State College" when certain road situations were present. The scenario created in this research challenged the system to determine the project status (Red, Yellow, Green) of an information technology project based on its performance. The results of the experiment confirmed our hypothesis that R-CAST could select a course of action after reasoning over defined experiences and knowledge related to the project management domain. To provide the reader insight as to how the results were accomplished, a description of the parameters and components that were populated along with examples of the project management knowledge created for the simulation is presented in the following subsections. An explanation of the R-CAST syntax used in these sections can be found in the R-CAST tutorial (Prindible, 2006).

Active Knowledge Base (KB) Component

The active knowledge base component in the R-CAST architecture is the module where declarative knowledge is populated to form the agent's knowledge base. The KB structures the agent's knowledge on two levels of information dependency, *FactType* and *Fact*, along with optional constraints or *Rules*. The development of the *FactType*, *Fact*, and *Rule* knowledge is described next. (Prindible, 2006) Note that, unless otherwise specified, the remainder of the discussion in this section as well as that presented in the next section on the experience knowledge base component is from (Prindible, 2006).

FactType

In order to construct the agent's environment, the FactType variables are defined first. Facts cannot be defined and rules cannot be fired until these variables are populated to define the environment's structure. (Prindible, 2006) The following is an example of a FactType defined based on the project management simulation scenario:

(FactType cost (?status, ?costUtility))

For the FactType cost, the variables status and costUtility are used to define the project status where *status* denotes the color indicator (Red, Yellow, or Green) of the project's cost as it relates to the performance of the cost; the status of the cost is represented by the variable costUtility, with the values of "over" or "under". Thus, the FactType variable acts as the first level of abstraction, where the schema of the agent's environment is comprehensively formed when all of the FactType variables are considered. Importantly, this FactType is now able to assert facts into the knowledge base of the agent and perform other knowledge base functions.

Fact

A Fact is defined to assign an explicit value to a FactType variable. Therefore, a corresponding FactType variable has to be defined in order for a Fact values to be assigned. From this perspective, the Fact can be viewed as an instance of the FactType class, as indicated in the object oriented programming language paradigm. The following is an example of a Fact defined for the project management simulation:

(Fact daysVar (2))

In the knowledge base, the daysVar variable is defined to indicate the number of days that a project is behind schedule. In the above definition, the daysVar variable indicates that a project is 2 days behind schedule. In sum, the Fact variable serves as the second level of abstraction that allows declarative knowledge to be populated into the agent's knowledge base.

Rules

A Rule is defined to identify the restrictions that an agent has to follow when operating in a specified environment in R-CAST. Although optional depending on the situation, a rule can be defined to produce a series of implied facts when rules and facts are combined. The rule structure has the components of a rule name, which must be unique; an antecedent, the conditional statement; an antecedent-consequence separator (->), which denotes "therefore"; and a consequent state. In the project management simulation, the following rule is defined:

```
(Rule "determineHighPerfSched"
   # Rule def
      (daysVar ?daysVar)
            # Antecedent
      (< ?daysVar 3)
      ->    # Antecedent- conse
            quence separator
      (schedule green high)
            # Consequent state
)
```

In the determineHighPerfSched rule, an agent can imply that if the project is less than 3 days behind schedule then the performance of the schedule is 'high' and the status is 'green', as defined in the FactType schedule.

Once the knowledge base is defined, the experience knowledge base provides the repository of knowledge that the agent uses to recall experiences related to the environment. In the next section, the experience knowledge base and the knowledge created for the project management simulation is presented.

Experience Knowledge Base (EKB) Component

The experience knowledge base component in the R-CAST architecture is the module where sets of experiences that are related to the context of the decision are stored. These experiences are joined with the knowledge base to perform reasoning that will lead to a decision based on the situation. The overall purpose of the experience knowledge base is to model the decision making at an abstraction level that allows identification of the information requirements used during the decision-making process.

As facilitated by the RPD component, in the decision-making process the R-CAST agent will reason over the domain knowledge and experiences to prompt, or cue, a course of action relevant to the current situation and past experiences. The EKB is structured in a hierarchical manner in which there is an experience space of experiences and their cues (episodic memory), expectancies, anomalies, goals, actions, and results. Next, the reader is presented with the development of the experience space and the experiences used for the project management simulation.

ExperienceSpace Hierarchy

The agent, to organize experiences at a higher level, uses the experience space hierarchy. It is here that the recognition process of the RPD process is performed, where the degree of the match between the experience and the current situation is calculated. When a match is found, the agent is able to use other experiences found in the experience definitions.

The ExperienceSpace structure contains the following components: an exclusive ExperienceSpace definition that cannot be duplicated; a cue, the condition in the agent's KB that is used to draw out a previous experience; an expectancy, conditions in the current situation that the agent should look forward to; anomalies, a particular condition, if true, whose consequence will make the agent choose another experience; and experiences, which identify and link the current experience to other experiences or ExperienceSpaces. The following ExperienceSpace is defined in the project management simulation:

```
(ExperienceSpace getProjStatus
     (Cue (getProjStatus yes))
     (Expectancy (getProjStatus
           yes))
     (Anomaly (getProjStatus
           no))
     (Experience (statusGreen)
                 (statusYellow)
                 (statusRed))
)
```

When the getProjStatus variable is defined as 'yes', the getProjStatus ExperienceSpace becomes a working ExperienceSpace. Once this ExperienceSpace becomes active, the agent is given the options of the statusGreen, statusYellow, and statusRed experiences, which are related experiences, or nodes, of the getProjStatus ExperienceSpace.

The structure of an experience and how it is used in the project management simulation is explained in the next section.

Structure of an Experience

In a hierarchical structure, experiences are at the node or lowest level of each ExperienceSpace. An experience also uses cues, expectancies, anomalies and fuzzy logic in its calculations of finding a relevant experience for the current situation. In addition to the components that comprise the ExperienceSpace structure, an experience has a goal, that tells the goal of the course of action (COA); an action, which identifies a COA that is associated with a Plan defined in the Process Manager (component is further explained in the R-CAST tutorial); and a result, which conveys the consequence of the COA in the previous experience. The following is an experience that is defined in the project management simulation:

```
(Experience statusYellow
      (Cue (schedule yellow
           med)
             (cost green under))
      (Expectancy (project con
           tinue))
```

```
        (Anomaly (project halt))
        (Goal (project continue))
        (Action (projStatus Yel
            low))
        (Result success)
)
```

The experience statusYellow is recognized as a match when its Cues, defined as "(Cue (schedule yellow)(cost green under))", evaluates to be true where, FactType schedule is defined as 'yellow' because it is performing at a medium level; and the FactType cost is green and under, meaning that the cost of the project is under the budgeted cost thus it is green. In the statusYellow experience structure, the expectancy defined as "(project continue)" instructs the agent that it should expect the project to continue when the COA is performed. The anomaly component of the experience, defined as "project halt", instructs the agent that if the variable project is defined as halt, "(project halt)", then it should avoid the COA. The goal component, defined as "(project continue)" instructs the agent that the goal of the statusYellow experience is for the project to continue. The Action component of the experience defined as "(projStatus Yellow)", defines the name of the COA plan that is defined in the Process Manager. Finally, in the Result piece of the experience, the previous outcome of the statusYellow experience has been "success".

R-CAST Conversion Summary

In this section we described the format for the knowledge used for the project management simulation example. As presented in (Barnes & Hammell, 2008), preliminary results from the system demonstrated that R-CAST could select a course of action after reasoning over defined experiences and knowledge related to the project management domain.

In order to obtain a more detailed and in-depth representation of knowledge related to the project management environment, subject matter experts were interviewed. The following section describes the knowledge acquisition efforts.

KNOWLEDGE ACQUISITION ACTIVITIES – FIRST ITERATION

The first iteration of knowledge acquisition activities conducted in this research consisted of developing information technology project management scenarios and the use of the scenarios for interviews with subject matter experts (SMEs). In this section we will discuss the development of the project management related scenarios, the selection of the SMEs, the strategy used to conduct the interviews, and the results realized from the interviews during this first round of knowledge acquisition efforts. A second knowledge acquisition iteration, which focuses on incorporating fuzzy logic into the research, is discussed later.

Development of Project Management Scenarios

The method used to develop the scenarios for the knowledge acquisition activities was to create 3 examples that reflected each of the relevant project status indicators (Red, Yellow, Green). From this approach, the idea is that the SME would provide knowledge that fortifies or weakens the scenario's intent based on referencing a prior experiences. For instance, if the SME looks at the 'Red' scenario and thinks that it is actually 'Green', it is important to understand the SME's decision-making process and gain insight into the experience recalled to reach that determination. Overall, the goal of the project management scenarios is to acquire knowledge that can be transformed into the R-CAST system architecture.

Structure of Project Management Scenarios

Each project management scenario was structured to give the SME variables related to scope, schedule, and cost constraints. For example, as it relates to the schedule constraint, each scenario identified the date the project started, the expected completion date, and the current date in the schedule that the SME will need to take into account to make a decision of the project status. Each scenario also had an accompanying Gantt chart that was developed using the Microsoft Project tool to aid the SME in their decision-making process.

As mentioned in the preceding section, a scenario was developed to meet the expectations of each project status color indicator for a project. The choices are Red, where the project is at risk for not being completed before the expected completion date or as planned for the specific task; Yellow, where the project is at risk but has potential to become realigned with the planned schedule; and Green, where the project is on schedule and there are no immediate high impact risks that can lessen the quality of the project's performance.

Strategically, each of the scenarios and MS Project documents contained information that was vague and information that was purposely omitted to aid in the analysis of the SMEs decision-making process. The premise is that the expert will use the vague and missing information as a stimulus to aid in the identification of relevant factors necessary to determine the project status while referencing past experiences.

SME Selection Criteria

In this research, a SME experimental group of 10 professionals with experience in the field of project management was selected to participate in the knowledge acquisition activities. To establish a control within the group, each of the selected experts met the qualification of having the Project Management Professional (PMP) certification (PMI, 2008). The assumption made is that having the PMP certification means that the expert has been exposed to concepts related to project management and can use this knowledge to reach a decision on the status of an information technology project. However, the amount of experience that the SME has amassed was not a deciding factor in the selection process but was noted during the interview. If there are conflicts in rule development for the knowledge system, then the knowledge derived from the most experienced expert will be used in the system.

Knowledge Acquisition Interview Strategy

The goal of the knowledge acquisition activities was to use the interviews as a means to elicit the explicit, but most importantly the tacit, knowledge that a SME uses in determining the project status of an information technology project. Therefore, the research questions, scenario document, and MS Project documents were not pre-distributed to the experts. Since all of the experts used in the research met the qualification of being PMP certified, the assumption was made that they had access to the project management material that would allow them to prepare commonplace responses to the scenario questions. Also, to simulate an actual project management environment where issues may come about unexpectedly, a clear decision-making process can be observed.

During the interview, the SME answered a set of general survey questions directed to obtain background information on the their professional career. The questions ranged from inquiring about their years of experience in project management to their views on project management concepts. Next, the expert was presented the scenario and MS project documents for the first scenario. After observing and recording knowledge related to the decision-making process, the expert was provided with the scenario and MS Project documents for the next scenario.

As presented in (Barnes & Hammell, 2008), the following is an example of a scenario used during the knowledge acquisition interview:

Project Start Date: February 8, 2007
Completion Date: August 2007
Current Date: June 26, 2007
Resources: 1 project manager; 2 technical developers – 1 expert, 1 moderate expertise
Budget: $50, 000
Budget Spent: $20, 000

ABC Inc. is a company that loans money to students and gives them advice on when to pay their bill, the amount to pay in installments, and when to consolidate their loan. Project XYZ is a project in which the goal is to develop an application that can process the daily operating activities of ABC, Inc.

The project team has encountered a problem with the software that they were planning to use to implement the consolidation portion of the application. Due to incompatibility of versions of other COTS products used to develop the system, the project team will need 3 extra weeks to wait for the updated version of "Consolid8", the consolidation software, to be delivered and tested for functionality.

What is the status of the project?

For this scenario, the expected status decision for this project is Yellow. The scenario was developed with the intent to get the expert to recall past experiences related to budgeting, and resource availability.

As an example of vague information in the project scenario, the completion date is defined as August 2007. The purpose of this assignment is to lead the SME to question if it is at the beginning or at the end of the month of August, as this would affect the status decision of the expert.

Upon the completion of the 10 interviews, the acquired knowledge was reviewed and extrapolated to serve as the experiment results. This information is presented in the next section.

Preliminary Empirical Results

The expected decisions for the scenarios 1, 2, and 3 were Yellow, Red, and Green, respectively. As shown in Figure 2 (Barnes & Hammell, 2008), the experts did not unanimously choose the expected status decision for any of the scenarios. In the figure, the y-axis (Scenario #) values account for the 3 scenarios; and the x-axis (Expert's Classification) categorizes the expert's decision in respect to the Red, Yellow, and Green color indicators.

In the cases of Scenario #2 and Scenario #3, 70% of the experts concluded that the project status was Red and Green, respectively. (Note: Scenario #3 is the only scenario to have an entry in the 'Other' reflecting the instance where one SME did not provide a status decision.) However, in the case of Scenario #1, where the expected outcome was Yellow, half of the experts reached the conclusion that the project status decision was Red and only 40% of the experts reached the decision that the project was Yellow.

Interestingly, of the 50% of the experts who thought the status of Scenario #1 was 'Red', 80% of them have accumulated the most years of experience in the field of project management out of SME experimental group.

Overall, although none of the experts unanimously chose the expected status decision in any of the scenarios, we can deduce that a substantial majority did agree on the status in Scenarios #2 and #3. It is clear that Scenario #1, which was expected to yield the Yellow decision, is the most problematic. The reason for the disparity among the experts is not obvious. Maybe the scenario was not properly designed to depict a Yellow project.

Figure 2. SME scenario results

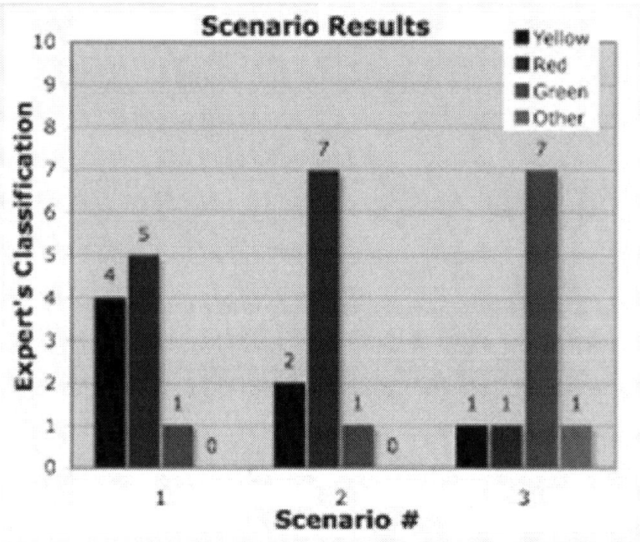

On the other hand, perhaps the inconsistency in the status decision indicates the precise problem that this research is hoping to overcome; that is, the uncertainty present when a project is "on the fence" between being mostly on track ('Green') or definitely in trouble ('Red') makes an accurate status decision difficult to obtain. Further investigation is necessary, and ongoing, with regard to the Yellow project status scenario.

The presence of ambiguity in the scenario decisions, especially in the case of the Yellow status decision, validates the efforts of this research. This ambiguity or uncertainty in project status decision-making is precisely the problem that we intend to reduce by populating the R-CAST system with project management domain knowledge. Overall, our position is that using the R-CAST system to aid in project status decision-making process will help a firm to avoid losing resources on failed projects, which consequently may hurt their information technology portfolio and their position in the marketplace (Weill & Broadbent, 1998).

INCORPORATING FUZZY LOGIC

One of the ultimate goals of this research is to integrate fuzzy logic into the R-CAST system to assist with the required complex decision-making. Background information on fuzzy logic and fuzzy systems is presented in this section.

Fuzzy Logic

Brief Background

Fuzzy logic was created from Lotfi Zadeh's idea that mathematics can provide a way to promote a relationship between language and human intelligence. By combining the concepts of crisp logic and the Lukasiewicz sets, Zadeh was able to use mathematics to define graded set membership. The idea is that it is easier to define some concepts by words rather than mathematics to construct better models of reality. Fuzzy logic and its expression in fuzzy sets supply such a discipline. (McNeill, Martin & Thro, 1994)

A crisp set is characterized by elements whose membership into the set is restricted to {0, 1}. However, in a fuzzy set, membership is not on a basis of 0 or 1, but rather on a 0 to 1 scale or interval [0, 1]. (Nguyen & Walker, 2000) Crisp language is described as ordinary language where words such as "No" and "Yes" are common. On the other hand, in fuzzy language, words such as "Almost" and "Slightly" are common thus indicating some level of ambiguity or vagueness. (McNeill & Thro, 1994; Nguyen & Walker, 2000)

Making decisions that contain elements of uncertainty or that require reasoning over values expressed in (perhaps vague) natural language has been shown to be difficult. In decision-making, establishing fuzzy sets is a tool that decision makers can use to handle uncertainty and vagueness. Fuzzy set membership values can be viewed as approximations, which gives fuzzy systems the ability to manage objects with imprecise values. (Ross, 1995)

Fuzzy Systems and Reasoning

Fuzzy systems began to be used by analysts and knowledge engineers in the expert system industry to aid in modeling complex business applications. Fuzzy modeling techniques provide a method that uses precise mathematical formulation to linguistically and approximately specify the relationships that are too complex or not well enough understood (Sudkamp & Hammell, 1998). Other forms of knowledge representation and natural computing paradigms, such as neural networks and genetic algorithms to model complexity, were considered before fuzzy systems. Fuzzy systems are viewed as a tool that will help advance the area of intelligent, information decision support systems through its ability to approximate systems that are complex and non-linear while employing a minimum amount of rules. (Cox, 1995)

A fuzzy reasoning system mixes knowledge representation in rules and fuzzy sets together to perform fuzzy, or approximate reasoning, on imprecise or ambiguous information. The fuzzy system uses fuzzy rules to encapsulate approximate relationships between the antecedent (input) and response (consequent) domains (Sudkamp & Hammell, 1994). A fuzzy rule restricts a region of space along the function surface and combines multiple regions to the same local space when multiple rules are implemented. Combining the multiple regions produces a composite region. The reasoning system produces an expected value of the solution variable by initiating the defuzzification process. The expected value, which represents the information contained in the output fuzzy set, is essential in most cases since many systems operate using numbers rather than fuzzy sets. (Cox, 1995)

Uncertainty, Fuzziness, and Project Management

In real-world problems, there are numerous forms of uncertainty that come into existence.

Uncertainty can crop up in a system where the objects that are defined in the system do not have boundaries that are detailed enough to specify a limit. Ambiguity, another kind of uncertainty, may also arise in real-world problems. (Nguyen & Walker, 2000)

As it relates to this research, examples of such vagueness have been shown in determining what constitutes a Red, Yellow, or Green project status. (Barnes & Hammell, 2008) When this uncertainty or vagueness is present, project managers and businesses are placed in the position where a strategic business decision cannot be reached as to what direction a project should continue, specifically when the project's status is Red or Yellow. By employing approximation techniques, a fuzzy set uses mathematics to determine membership into these classes as a matter of degree instead of pure certainty (Nguyen & Walker, 2000). In this research, efforts have been made to determine the membership of a Yellow fuzzy set. Before those efforts are discussed, it is important to explain how fuzzy logic will be used in this research to reduce cognitive dissonance and the aggregation methods that will be employed to determine the fuzzy set membership.

Using Fuzzy Logic to Reduce Cognitive Dissonance

The use of fuzzy logic can help in reducing cognitive dissonance by assigning a degree of truth to a set of crisp propositions acquired from a SME (Cox, 1995; Stanford, 2008). However, in acquiring these propositions, there is a variance in the way that the problem is articulated by the expert and the manner in which it is translated into code for the expert/decision support system. Employing fuzzy models condenses this variance, which is identified as cognitive dissonance. (Cox, 1995)

In the case of determining what constitutes a Yellow project, the crisp values acquired from the SME can be defined as a fuzzy set that specifies the range of values for the concept of a Yellow project as well as the degree of membership for any suspect characteristic value (Cox, 1995; Martin, 1988). This approximate definition of the associated values can be used as a mechanism to compensate for the vagueness in deciding what constitutes a Yellow project.

To explain further, as adapted from an example in (Cox, 1995), suppose a project manager that tells us how a project status is decided gives a statement:

When scope is on target but the schedule is behind, and cost is over budget, then the project status is Red.

This statement can be expressed in the following fuzzy rule:

```
If scope is on target
  and schedule is behind
  and cost is over budget
      then project status
       is Red
```

where *on target, behind, over budget,* and *Red* are fuzzy sets. A fuzzy rule exhibits more reasoning qualities than a crisp rule that is defined as a precise mathematical relation or expression. Unlike crisp expressions, a fuzzy rule does not depend on the value of the antecedent being true or false. With fuzzy rules, the degree to which the values of the antecedents match their corresponding fuzzy sets influences the membership strength the consequent value will have in its corresponding fuzzy set. That is, in this example the degree to which a specific value for 'schedule' matches the fuzzy set *behind*, somehow combined with the degree to which another specific value for 'cost' matches the fuzzy set *over budget*, will influence the degree of membership that 'project status' will have in the fuzzy set *Red*. (Cox, 1995) Note that *behind* and *over budget* are fuzzy sets over the input domains 'schedule' and 'cost', respectively, and that *Red* is a fuzzy set associated with the output domain 'project status'.

Fuzzy Aggregation

In a fuzzy reasoning system engine, fuzzy aggregation is the process that controls the correlation and management of solution fuzzy sets (Cox, 1995). The fuzzy aggregation mechanism cumulates uncertain and imprecise distributed information data to discover a measure of certain fuzzy information. (Malaviya & Peters, 1995)

Each solution variable has a corresponding fuzzy set region that is updated by each rule that identifies the solution variable in its consequent. Applying the fuzzy rules of the problem to the fuzzy set region of the solution variables contributes to the final shape and strength of the composite or solution fuzzy set (Cox, 1995). The rules of the system are classified as either unconditional fuzzy propositions or conditional fuzzy propositions. Conditional fuzzy propositions, which are most frequently used in expert and decision support systems, are relevant to this research as they have the traditional if-then syntax. Thus, aggregation methods used for this research will be focused on applicable operations to fuzzy sets with these conditional rules.

Fuzzy Aggregation Techniques

The following presents a brief description of two of the most common methods used to aggregate fuzzy sets. The presentation is taken from (Cox, 1995) and this reference should be consulted for a detailed explanation of the techniques along with examples.

Min/Max Aggregation

The min/max method ORs the correlated consequent fuzzy set with the contents of the solution variable's output fuzzy region. Using their common membership function in the process, the maximum of the consequent fuzzy set and the solution fuzzy set is recorded at every shared point.

To explain further, the correlation minimum (min), where the consequent fuzzy set is truncated at the truth of the premise, is taken for the consequent fuzzy set and aggregated to the output fuzzy region. The aggregation piece is performed by using a fuzzy union operation to aggregate the maximum (max) of the truth membership values into the current fuzzy region for the solution variable.

Additive Aggregation

In the additive aggregation method, the correlated consequent fuzzy set is added to the contents of the solution variable's output fuzzy region. While sharing a membership function in the process, the truth membership values of the consequent fuzzy set and the solution fuzzy set are added at each common point in the function.

In this method, the correlation product is used to produce the correlated consequent fuzzy set. The correlation product reduces the truth of the consequent by multiplying each truth membership by the premise truth. The fuzzy set is contracted so its maximum height is equivalent to the height of the predicate truth, thus preserving the shape of the fuzzy region. The additive method in this technique entails using the bounded sum operation for aggregation. Using this method makes it impossible for the composite membership value to surpass [1.0]. When added, if the sum of the solution fuzzy region and the correlated consequent region exceeds [1.0], then it is truncated at the [1.0] level.

KNOWLEDGE ACQUISITION ACTIVITIES – SECOND ITERATION

The second iteration of knowledge acquisition with the SMEs was concerned with incorporating fuzzy logic concepts into the research to assist with managing the uncertainty in determining the status of a Yellow project. The primary use of fuzzy sets in this research is related to the concept of dividing a domain into a group of overlapping classifications. Each of the classifications is a fuzzy set; together they represent a decomposition of the domain. An element in the domain has some grade of membership, from 0 to 1 inclusive, in each fuzzy set within its domain. The grade of membership is determined by a function called a membership function; the shape of the fuzzy sets used in the domain decomposition basically determines the membership function.

In Figure 3, the membership value is shown on the y-axis and the values contained within the domain "Temperature" are shown on the x-axis. The domain "Temperature" over the range 0 to 120 is decomposed into three fuzzy sets: Cold,

Figure 3. Example domain decompositions

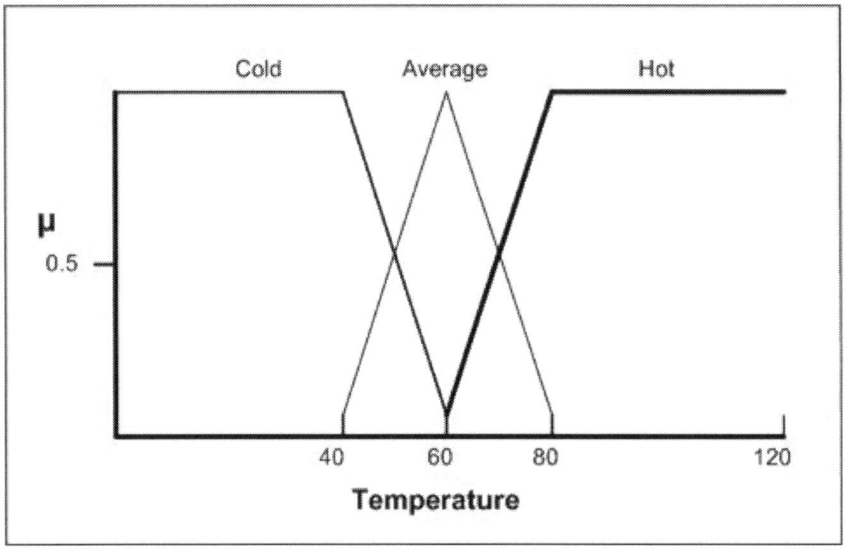

Average, and Hot. A "Temperature" value of 30 would have membership = 1 in fuzzy set Cold and membership = 0 in both the Average and Hot fuzzy sets. A "Temperature" value of 50 would have membership = 0.5 in both the Cold and Average fuzzy sets and membership = 0 in the Hot fuzzy set.

The use of fuzzy set concepts in the knowledge acquisition process was aimed at modeling, for the experts, that the categories of Green, Yellow, and Red could be interpreted as fuzzy sets over the cost/scope/schedule domains. This decomposition actually represents two key ideas. First, the change from one category to another is gradual. That is, there is a point where cost/scope/schedule is definitely in one of the categories (say, Green for instance) but as the value of the associated attribute pushes the status toward an adjacent category (in this example, Yellow) the membership in the original (Green) category gradually decreases (versus changing from 1 to 0 as a step function).

The second concept is related to trying to define the ranges in which an attribute could be classified into the three categories. Part of the uncertainty associated with judging whether a project is Yellow or not might be related to the difficulty in deciding where an attribute "suddenly" changes from Green to Yellow (and vice versa), or Yellow to Red (and vice versa). By modeling the domains of cost/scope/schedule as decompositions of overlapping fuzzy sets, it might be easier for the experts to provide, and agree upon, the ranges of the fuzzy sets (as opposed to the ranges of the crisp sets where there is no concept of gradual change).

Similar to the first iteration of interviews described in the previous section, one case with an expected outcome of Yellow was created for use in observing the SME's reasoning process. Also, various fuzzy exercises with the intent of identifying ranges for a Red, Yellow, and Green project with respect to the scope, schedule, and cost were developed. The remainder of this section will focus on the intent and results of the fuzzy exercises.

Fuzzy Logic Exercises

The fuzzy exercises were developed with the intent to use this information to model the condition of the scope, schedule, and cost constraints by

Figure 4. Knowledge acquisition exercise to elicit Over Budget fuzzy range values

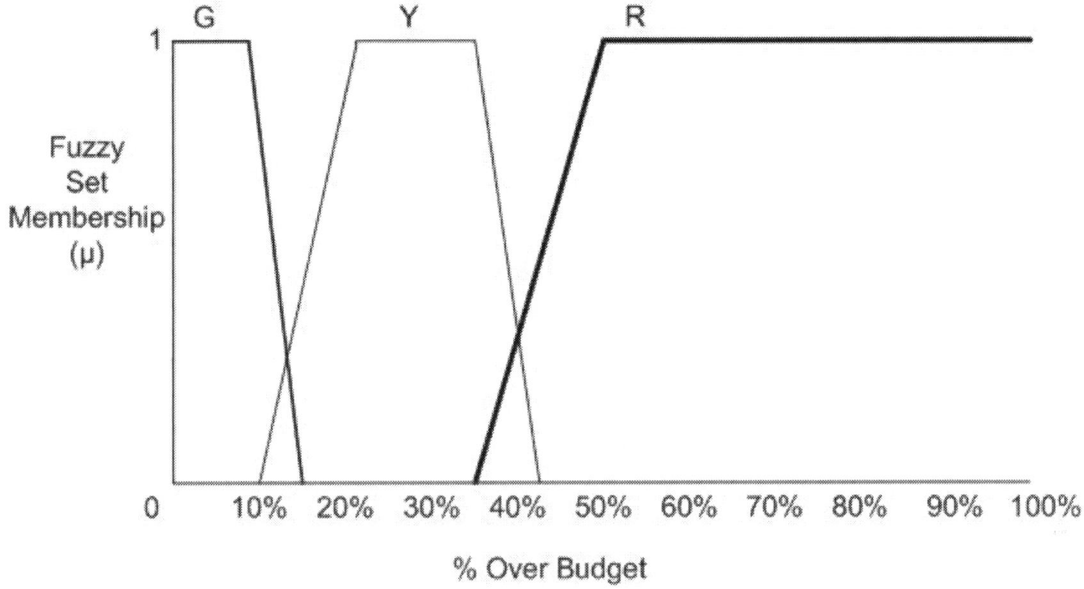

The figure above depicts a fuzzy set decomposition of the "over budget" domain.

The assumption used for the questions below is that the project has a fixed budget.

a) In general (at any point during the project), with respect to the percentage (%) over budget, what % range characterizes a GREEN project?

b) In this range, where is the project exclusively GREEN?

c) In general (at any point during the project), with respect to the percentage (%) over budget, what % range characterizes a YELLOW project?

d) In this range, where is the project exclusively YELLOW?

e) In general (at any point during the project), with respect to the percentage (%) over budget, what % range characterizes a RED project?

f) In this range, where is the project exclusively RED?

identifying crisp (McNeill, Martin & Thro, 1994) values that will be used to approximate fuzzy ranges. The idea is that modeling the condition of the constraints will provide an idea as to where the constraint can be classified within the Red, Yellow, and Green ranges at any point during a project. From the acquired SME knowledge, we will be able to infer fuzzy rules for the decision system (Hammell & Sudkamp, 1996).

Figure 4 depicts a fuzzy set decomposition of the "over budget" domain. The intent of this decomposition was to provide a starting point for the SMEs to reason about how to classify (that is, set the endpoints of the fuzzy sets to signify) a project as Green, Red, or Yellow when considering how much the project is over budget. The basic assumption is that a project is Green and Red if it is 0% and 100% over budget, respectively. In Figure 4, a project is considered Green if it is a member of the range of being 0% – 18% over budget; Yellow if it is in the 10% – 42% range; and Red if it is 35% - 100% range. The goal of the exercise questions in this example is to find out if the SME agrees with the assumed Red, Yel-

Table 1. Interview results: Fuzzy logic range values (2 SMEs)

Question	*SME #1*	*SME #2*	*Avg. Range*
A	0 – 10%	0 – 5%	0 – 7.5%
B	0 – 5%	0 – 4%	0 – 4.5%
C	5 – 30%	5 – 15%	5% - 22.5%
D	15 – 25%	13 – 15%	14 – 20%
E	25 – 100%	15 – 100%	20% - 100%
F	30 – 100%	15 – 100%	22.5 – 100%

low, and Green ranges that the researchers defined or if modifications to the ranges are necessary. In questions (a), (c), and (e), the SME is asked, respectively, to identify the range of a Green, Yellow, and Red project with respect to the cost being over budget. Questions (b), (d), and (f) request the SME to identify the subset of the indicated range where the membership into the set is 1. The information collected from the questions will be used for translation in the EKB of the R-CAST decision system to aid in determining the status of an information technology project.

The above example fuzzy exercise was only one of the four exercises that were conducted during the second iteration of knowledge acquisition activities to acquire fuzzy range data. Because all of the exercises produced similar results, only the results from the above Over Budget exercise will be discussed in detail as a representative sample.

Initial Fuzzy Logic Interview Results (% Over Budget Fuzzy Exercise)

The initial interviews using fuzzy logic concepts were conducted using two SMEs. The results indicated a significant difference in the assumptions made by the researchers and the range values provided by the SMEs. In addition to differing with the researchers, there was disparity between the 2 SMEs as well.

Table 1 presents the data collected from the SMEs that will be used to model a fuzzy set that represents the Red, Yellow, and Green range values for a project over budget. The '*Question*' column represents the letter of the question presented in the example in the previous section.

As indicated in Table 1, for question (a), SME #1 believed that a project could be considered Green until it is 10% over budget. Alternatively, the data collected from SME #2 were stricter, where a project could be considered Green only until it was 5% over budget. In question (b), the SMEs had a 1% variance in where they viewed the membership into the Green set as 1.

When focusing on the range of a Yellow status indicator question (c), it is clear that the SMEs agreed as to where the Yellow range should begin (5% over budget). On the other hand, there was a 15% difference between the 2 SMEs as to where the range should end. In question (d), SME #1 felt that a project could be considered Yellow if it is between 15 – 25% over budget. Again, the data collected from SME #2 was more rigid, only allowing for a 2% window (13 – 15%) as to when a project can have a membership of 1 in the Yellow range.

For question (e), where the range of a Red status indicator for a project that is over budget is considered, SME #1 believed that the range is between 25 – 100% over budget. SME #2 thought that the cost constraint should be Red if it lands in the 15% - 100% over budget range. In question (f), SME #1 felt that a membership of 1 in the Red range should span 30 – 100%. Interestingly, SME #2 thought that the entire Red range, which was provided by the SME in question (e), should be considered as having a membership of 1 into the fuzzy set. The reason that SME #2 provided was that there would not be enough money available after the project goes 15% over budget if it is a fixed budget as the exercise indicates.

As a result of the observable difference in the acquired range values, plans have been made by the researchers to interview more SMEs. The data attained from the interviews will be analyzed to determine if the disparity in the range values will decrease or increase in degree. Also, it is possible that the amount of experience of the SMEs may contribute to the disparity of the ranges. The relevance of this information will also be analyzed as more data is collected. As indicated in (Barnes & Hammell, 2008), the amount of experience of the SME will only be taken into consideration as a deciding factor in what information to use in an effort to arrive at a resolution to conflicting thoughts when developing rules for the decision system.

As a change in approach to the fuzzy exercise, different fuzzy ranges may be realized by removing the % ranges values from the x-axis of the figure that is used in the exercise (Figure 4) to acquire fuzzy range values. It is possible that by having the % values already on the scale, the SME may feel restricted and influenced to not deviate too far from what is indicated in the example. Removing the % ranges may encourage results that are more innovative. Thus, modifying the approach in the fuzzy exercises is under consideration for future knowledge acquisition interviews to attain fuzzy logic range values.

Similarly, a change in the method employed by the researchers to elicit fuzzy range values from the SME, may influence communication of different range values. A possible change to the interview approach could include taking time to educate the SME about fuzzy logic. Specifically, using examples of fuzzy sets that represent other domains to educate the SMEs may enhance their understanding of fuzzy logic and provide more clarity to the purpose of the interview when needed. However, this change in interview approach may be difficult to institute due to time constraints of the interviews. Further, the learning curve, a factor that will influence the amount of time needed for the interviews, of each of the SMEs in relation to comprehending fuzzy logic is unknown.

Another option for dealing with the disparity between the SMEs is to use the average of the values to set the ranges. While using the average of the range opinions may not be the optimal solution, we believe that it is a reasonable approach to use in the interim so that the construction of a "proof of concept" decision system can proceed. After additional SMEs are interviewed, other techniques will be investigated to aggregate the different range value opinions if necessary. One possible scheme is to use least squares regression analysis. This method can be used to find a representation that provides the best-fit and minimal error with respect to the data (Bretscher, 1995). Other techniques will also be considered.

ONGOING WORK

Ongoing efforts in this research primarily focus on designing collaborative agents with project management knowledge. Thus far in our approach, we have designed agents to manage each of the scope, schedule, and cost project constraints and a project manager (PM) agent that will provide a recommendation for the project after reasoning over the output from the three project constraint agents.

The next steps involve developing the knowledge and experience knowledge bases for the aforementioned agents using the knowledge acquired in the two knowledge acquisition iterations plus additional information as described below. The knowledge will be used to construct the appropriate FactType, Fact, and Rule variables for the project constraint agents and PM agent.

The PM agent will reason over the Red, Yellow, and Green recommendations from the project con-

Table 2. Assume that the constraints of scope, schedule, and cost are equally weighted. Which color indicator (Red – R; Yellow – Y; Green – G) would you use to indicate the project's status?

Scope: G, Schedule: G, Cost: G = ____	*Scope: Y, Schedule: G, Cost: R = ____*
Scope: G, Schedule: Y, Cost: G = ____	*Scope: Y, Schedule: R, Cost: G = ____*
Scope: G, Schedule: G, Cost: Y = ____	*Scope: Y, Schedule: G, Cost: G = ____*
Scope: G, Schedule: R, Cost: G = ____	*Scope: Y, Schedule: R, Cost: R = ____*
Scope: G, Schedule: G, Cost: R = ____	*Scope: R, Schedule: R, Cost: R = ____*
Scope: G, Schedule: Y, Cost: R = ____	*Scope: R, Schedule: G, Cost: R = ____*
Scope: G, Schedule: R, Cost: Y = ____	*Scope: R, Schedule: R, Cost: G = ____*
Scope: G, Schedule: Y, Cost: Y = ____	*Scope: R, Schedule: Y, Cost: R = ____*
Scope: G, Schedule: R, Cost: R = ____	*Scope: R, Schedule: R, Cost: Y = ____*
Scope: Y, Schedule: Y, Cost: Y = ____	*Scope: R, Schedule: G, Cost: Y = ____*
Scope: Y, Schedule: G, Cost: Y = ____	*Scope: R, Schedule: Y, Cost: G = ____*
Scope: Y, Schedule: Y, Cost: G = ____	*Scope: R, Schedule: G, Cost: G = ____*
Scope: Y, Schedule: R, Cost: Y = ____	*Scope: R, Schedule: Y, Cost: Y = ____*
Scope: Y, Schedule: Y, Cost: R = ____	

straint agents to recommend a project status based on the combination of the scope schedule and cost constraint variables. The scenario in Table 2 has been developed for use in upcoming knowledge acquisition interview to find out the overall status that our SME would give the project.

The intent of this scenario is to allow the SME to decide on the status of a project where all possible combinations of a Red, Yellow, or Green designation for each scope, schedule, and cost project constraint is present. This knowledge will serve as the foundation for the experience knowledge base of the PM agent. The recommendation that is made by the PM agent will provide the recommended status output from the R-CAST system that the human project manager can use aid in his project status decision for an information technology project.

FUTURE RESEARCH

Future work that we have planned includes the following:

- A set of scenarios to be used as "test cases" will be developed. A subset of the SMEs already used in the knowledge acquisition phase will be interviewed to get their decisions on the scenario's project status.
- Empirical testing of the R-CAST-based system developed in this work will be done by presenting it with the same scenarios developed in the previous step. The project status recommendation produced by the system will be compared to that of the SMEs. We are hopeful that this comparison will validate the efficacy of our system as a tool to aid project managers in their decision-making process.

After the usefulness of the system is illustrated as mentioned above, we hope that the system continues to evolve. A primary way this could be done is by making the knowledge base more complete as related to a complex project management environment. That is, other project constraints, variables, rules, facts and the like could be added to the system to make it even more helpful to project managers. The added knowledge could take different forms; for example, the goal could be to make the overall system more robust. On the other hand, perhaps the goal could be to make the system more specific to a particular project domain (construction projects versus information technology projects, for instance).

Another way future researchers could advance the system is to adapt it to use in other decision-making environments. Previous work by researchers at the Pennsylvania State University has illustrated that R-CAST can be used to construct a useful decision aid for military command and control teams in simulated military operations (Fan 2005b, 2005c, 2006, 2007). Our work should show that the architecture is also able to assist decision makers in the project management domain. Others could develop various knowledge and experience knowledge bases reflective of different decision-making environments to further examine R-CAST capabilities and construct decision aids for other domains.

CONCLUSION

When investing in a project it is important for a business to know if an IT project, that may affect their position in the marketplace, will be completed successfully. In the event where the project is complex, it may be beneficial to the project manager to use intelligent agents to aid in his project status decisions. In this chapter, we presented the R-CAST architecture as decision tool to provide a collaborative agent system that will recommend the status of a project. Thus far, we have shown that R-CAST can reason over information that is specific to the project management domain by defining relevant cases in the experience knowledge base and the project domain in the knowledge base.

The first iteration of knowledge acquisition activities demonstrated that ambiguity is present in the scenario where the expert had to decide that the status of a project is Yellow. From these findings, a second iteration of knowledge acquisition was performed that incorporated concepts from fuzzy logic to assist with handling the uncertainty associated with the status of a project when the outcome could fall on the borderline of two status indicators: Green-Yellow and Yellow-Red. While there was still some disparity associated with the range opinions from the experts, enough knowledge has been captured to allow the construction of a prototype decision support system to proceed. Ongoing work is aimed at designing collaborative agents with project management knowledge and developing the knowledge and experience knowledge bases for the agents. Future work will involve constructing a set of scenarios to be used as test cases and the subsequent empirical testing of the system to validate its efficacy.

REFERENCES

Applegate, L. M., Austin, R. D., & McFarlan, F. W. (2003). Corporate Information Strategy and Management Text and Cases: Sixth Edition. New Delhi, India: Tata McGraw-Hill Publishing Company Limited.

Barnes, A., & Hammell, R. J., II. (2008). Determining Information Technology Project Status using R-CAST. In *the Proceedings of CONISAR 2008 (v1),* Phoenix, AZ.

Berner, E. S., & Ball, M. J. (1998). Clinical Decision Support Systems: Theory and Practice. Berlin: Springer.

Bretscher, O. (1995). Linear Algebra With Applications, (3rd ed.). Upper Saddle River NJ: Prentice Hall.

Burgess, T., Byrne, K., & Kidd, C. (2003). Making Project Status Visible in Complex Aerospace Projects. *International Journal of Project Management, 21*(4), 251–259. doi:10.1016/S0263-7863(02)00022-4

Carlsson, C., & Turban, E. (2002). Introduction: DSS: Directions For The Next Decade. *Decision Support Systems, 33*(2), 105–110. doi:10.1016/S0167-9236(01)00137-3

Cox, E. (1995). Fuzzy Logic for Business and Industry. Rockland, MA: Charles River Media, Inc.

Fan, X. Sun, B, Sun, S., McNeese, M., Yen, J., & et al. (2006) RPD-Enabled Agents Teaming with Humans for Multi-Context Decision Making. In *Proceedings of the Fifth International Joint Conference on Autonomous Agents and Multi-Agent Systems (AAMAS'06)*, (pp. 34-41). Hokkaido, Japan: ACM.

Fan, X., Shuang, S., McNeese, M., & Yen, J. (2005b). Extending Recognition-Primed Decision Model For Human-Agent Collaboration. In *Proceedings of the Fourth International Joint Conference on Autonomous Agents and Multi Agent Systems*. (pp. 945-952).The Netherlands: ACM.

Fan, X., Shuang, S., Sun, B., Airy, G., McNeese, M., & Yen, J. (2005c). Collaborative RPD-Enabled Agents Assisting the Three-Block Challenge in Command and Control in Complex and Urban Terrain. In *Proceedings of 2005 Conference on Behavior Representation in Modeling and Simulation (BRIMS)*, (pp. 113-123), Universal City, CA.

Fan, X., Shuang, S., & Yen, J. (2005a). On Shared Situation Awareness for Supporting Human Decision-Making Teams. In *Proceedings of 2005 AAAI Spring Symposium on AI Technologies for Homeland Security*, (pp. 17-24). Stanford, CA: American Association for Artificial Intelligence.

Fan, X., & Yen, J. (2007). R-CAST: Integrating Team Intelligence for Human-Centered Teamwork. In *Proceedings of the Twenty-Second AAAI Conference on Artificial Intelligence (AAAI-07)*. Menlo Park, CA: Association for the Advancement of Artificial Intelligence (www.aai.org).

Felsen, J. (1976). Decision Making Under Uncertainty: An Artificial Intelligence Approach. New York: CDS Publishing Company.

Gido, J., & Clements, J. P. (2006). Successful Project Management, (3rd Ed.). Mason, OH: Thomson Higher Education.

Hammell, R. J. II, & Sudkamp, T. (1996). An Adaptive Hierarchical Fuzzy Model. *Expert Systems with Applications*, *11*(2), 125–136. doi:10.1016/0957-4174(96)00040-1

Lewis, B. (2008). The 70-percent failure. *InfoWord*. Retrieved August 2008, from http://www.infoworld.com/articles/op/xml/01/10/29/011029opsurvival.html

Liebowitz, J. (Ed.). (1998). The Handbook of Applied Expert Systems. Salem, MA: CRC Press.

Malaviya, A., & Peters, L. Extracting Meaningful Handwriting Features with Fuzzy Aggregation Method. In *Proceedings of the Third International Conference on Document Analysis and Recognition (ICDAR),* (pp 841-844).

Martin, J., & Oxman, S. (1988) Building Expert Systems: A Tutorial. Englewood Cliffs, NJ: Prentice Hall.

McNeill, F. M., & Thro, E. (1994). Fuzzy Logic: A Practical Approach. London: Academic Press, Inc.

Mueller, J. P. (1996). The Design of Intelligent Agents. Berlin: Springer.

Nguyen, H. T., & Walker, E. (2000). A First Course in Fuzzy Logic - Second Editon. Boca Raton, FL: Chapman & Hall/CRC Press LLC.

Prindible, M. (2006). An Introduction to R-CAST: Developing Simple Agents using Decision-making modeling and the R-CAST Architecture. *The Laboratory for Intelligent Agents, College of Information Science and Technology, The Pennsylvania State University*. Retrieved March 1, 2007, from http://agentlab.psu.edu/rcast/rcast_tutorial.pdf

Project Management Institute (PMI). (2004). Project Management Body of Knowledge. Newtown Square, PA: Project Management Institute, Incorporated.

Project Management Institute (PMI). (2008). The Project Management Professional (PMP) Credential. *Project Management Institute, Inc.* Retrieved August 2008, from http://www.pmi.org/CareerDevelopment/Pages/Obtaining-Credential.aspx

Ross, T. J. (1995). Fuzzy Logic with Engineering Applications. Boston: McGraw-Hill, Incorporated.

Sarter, N. B., & Woods, D. D. (1991). Situation Awareness: A Critical But Ill-Defined Phenomenon. *The International Journal of Aviation Psychology*, *1*(1), 45–57. doi:10.1207/s15327108ijap0101_4

Shim, J. P., Merrill, W., & Courtney, J. (2002). Past, Present and Future of Decision Support Technology. *Decision Support Systems*, *33*(2), 111–126. doi:10.1016/S0167-9236(01)00139-7

Stanford Encyclopedia of Philosophy. (2008). Search: Fuzzy Logic. *Metaphysics Research Lab, CSLI, Stanford University*. Retrieved August 22, 2008, from http://plato.stanford.edu/entries/logic-fuzzy

Sudkamp, T., & Hammell, R. J. II. (1994). Interpolation, Completion, and Learning Fuzzy Rules. *IEEE Transactions on Systems, Man, and Cybernetics*, *24*(2), 332–342. doi:10.1109/21.281432

Sudkamp, T., & Hammell, R. J. II. (1998). Scalability in Fuzzy Rule-based Learning. *Journal of Information Science*, *109*, 135–147. doi:10.1016/S0020-0255(98)00014-0

Varma, V. K., Ferguson, I., & Wild, I. (2000). Decision Support System for Sustainable Forest Management. *Forest Ecology and Management*, *128*(1-2), 49–55. doi:10.1016/S0378-1127(99)00271-6

Watson, I. (1997). Applying Case-based Reasoning: Techniques for Enterprise Systems. San Francisco, CA: Morgan Kaufmann Publishers, Inc.

Weill, P., & Broadbent, M. (1998). Leveraging the New Infrastructure: How Market Leaders Capitalize on Information Technology. Boston: Harvard Business School Press.

ADDITIONAL READING

Brenner, W., Wittig, H., & Zarnekow, R. (1998). Intelligent Software Agents: Foundations and Applications. Secaucus, New Jersey: Springer-Verlag New York, Inc.

Cleland, D. (1999). Project Management Third Edition: Strategic Design and Implementation. New York, New York: The McGraw-Hill Companies, USA

Cox, E. (1999). The Fuzzy System Handbook, Second Edition. San Diego, California: AP Professional.

Grady, R. B. (1992). Practical Software Metrics for Project Management and Process Improvement. Englewood Cliffs, New Jersey: Prentice Hall.

Hayes, J. R. (1962). *Human Data Processing Limits in Decision Making, Report No. ESD-TDR-62-48*. Air Force Systems Command, Electronics System Division, Bedford, Massachusetts.

Kerzner, H. (2006). Project Management: A Systems Approach to Planning, Scheduling, and Controlling. Hoboken, New Jersey: John Wiley & Sons, Inc.

Klir, G. J., & Yuan, B. (1995). Fuzzy Sets and Fuzzy Logic: Theory and Applications. Upper Saddle River, New Jersey: Prentice Hall PTR.

Klusch, M. (1999). Intelligent Information Agents: Agent-Based Information Discovery and Management on the Internet. Secaucus, New Jersey: Springer-Verlag New York, Inc.

Knapik, M., & Johnson, J. (1998). Developing intelligent agents for distributed systems: exploring architecture, technologies, & applications. New York, New York: McGraw-Hill, Inc.

Kosko, B. (1996). Fuzzy Engineering. Upper Saddle River, New Jersey: Prentice-Hall, Inc.

Leake, D. (Ed.). (1996). Case-Based Reasoning: Experiences, Lessons, and Future Directions. Menlo Park, California: AAAI Press/MIT Press.

Lucas, H. C. (1999). Information Technology and the Productivity Paradox. Oxford University Press.

McNeill, D., & Freiberger, P. (1994). Fuzzy Logic. New York: Simon & Schuster.

Murch, R., & Johnson, T. (1998). Intelligent Software Agents. Upper Saddle River, New Jersey: Prentic Hall PTR.

Padgham, L., & Winikoff, M. (2004). Developing Intelligent Agent Systems. Hoboken, New Jersey: John Wiley & Sons, Inc.

Schank, R. (1982). Dynamic Memory: A Theory of Learning in Computers and People. New York: Cambridge University Press.

Simon, H. (1955). A behavioral model of rational choice. *The Quarterly Journal of Economics*, *69*, 99–118. doi:10.2307/1884852

Snedaker, S., & Hoenig, N. (2005). How to Cheat at IT Project Management. Rockland, Massachusetts: Syngress Publishng, Inc.

Terano, T., & Asai, K. (1992). Fuzzy Systems Theory and its Applications. San Diego, California: Academic Press Professional, Inc.

Wang, L. X. (1994). Adaptive Fuzzy Systems and Control: Design and Stability Analysis. Upper Saddle River, New Jersey: Prentice-Hall, Inc.

Willcocks, L. (1994). Information Management: the evaluation of information systems investments. New York: Chapman & Hall.

Wooldridge, M. (1999). Intelligent Agents. In G. Weiss (Ed.), Multiagent Systems (pp. 1-44). Cambridge, Massachusetts: The MIT Press.

Yager, R. R., & Zadeh, L. A. (1992). An Introduction to fuzzy logic applications in intelligent systems. Boston, Massachusetts: Kluwer Academic.

Yen, J., & Langari, R. (1998). Fuzzy Logic: Intelligence, Control, and Information. Upper Saddle River, New Jersey: Prentice-Hall, Inc.

Zadeh, L. A. (1965). Fuzzy Sets. *Information and Control*, *8*, 338–353. doi:10.1016/S0019-9958(65)90241-X

Zadeh, L. A. (1973). Outline of a New Approach to the Analysis of Complex Systems and Decision Processes. *IEEE Transactions on Systems, Man, and Cybernetics*, *SMC-3*, 28–44.

Zimmerman, H. J. (1996). Fuzzy Set Theory. Norwell, Massachusetts: Kluwer Academic Publishers.

Chapter 2
SAT and Planning:
An Overview

Carlos Camarão
University of Minas Gerais, Brazil

Mateus Galvão
University of Minas Gerais, Brazil

Newton Vieira
University of Minas Gerais, Brazil

ABSTRACT

This chapter firstly reviews the importance of the Satisfiability Problem (SAT) for a wide range of applications, including applications in Operation Management such as planning. A review of methods nowadays employed by modern SAT-solvers is then presented. The authors then use Classical Planning as an illustrative example of how a significant problem can be translated into SAT. They point out important results and studies concerning reductions of planning into SAT, and explain how to construct a SAT instance which is satisfiable if and only if an instance of a bounded version of the classic blocks-world problem is solvable.

INTRODUCTION

Twenty years ago the use of formal logic as a basis for knowledge representation and processing was subject to severe critics, founded not only on problems of expressiveness, but also on problems of running time performance. Since that time the problem of expressivity has been object of intense research, which resulted in extensions and alternative formulations of logics intended to deal with important aspects of "intelligent" systems not dealt with by classical formal logic (the logic underlying purely mathematical reasoning) (Brachman & Levesque, 2004). From the point of view of running time performance, significant advancements have also been achieved (A. Robinson & Voronkov, 2001; Harmelen, Lifschitz, & Porter, 2008), even considering the undecidability of more expressive logics.

Although knowledge expression and processing in the area of artificial intelligence is still subject to intensive research, far from providing the necessary basis for its highest objectives (as the production of intelligent behavior), the last years have seen a growing use of logic based systems for solving many industrial level applications in several domains.

DOI: 10.4018/978-1-61520-605-6.ch002

In this chapter one of the simplest logics from the standpoint of expressivity will be considered: propositional logic. In spite of its simplicity, it is sufficient for the description and solution of an enormous amount of practical problems. Even considering that other logics are more adequate from a notational point of view, in order to have more concise descriptions for certain application classes, in many cases it is possible to produce automatically a translation to propositional logic (Cadoli & Schaerf, 2005; Navarro, 2007).

The justification for using a knowledge processor based on propositional logic, even in some situations where more expressive logics would be apparently more adequate (by having more concise sentences), is that the actual programs for propositional logic processing are so advanced that they allow the solution of many practical problem instances. They are typically not only more efficient than respective programs for other logics, but also than specific programs specially tailored to the application at hand (Gomes, Kautz, Sabharwal, & Selman, 2008).

The idea underlying the use of formal logic for problem solving is to obtain a solution for a problem instance from a given formal specification of the problem expressed in the language of that logic. There are two basic approaches for achieving this goal. In the approach based on *proof theory*, finding a solution consists of finding a proof for a formula α from a set of formulas Σ. For example, in the problems of software or hardware verification, Σ would be a formal description of an algorithm or a digital circuit and α a property to be verified. Another example: to verify the equivalence of two circuit descriptions given by two formulas α and β it is sufficient to establish the validity of the formula $\alpha \leftrightarrow \beta$ or, in other words, to prove that $\alpha \leftrightarrow \beta$ follows from the empty set of hypotheses. A huge number of other kinds of problems can be formulated as that of obtaining a proof of a formula (theorem) α from a set of formulas (axioms or hypotheses) Σ (Larry, 1984; Lifschitz, Morgenstern, & Plaisted, 2008).

The second approach is based on *model theory*. Here the goal is to find a model that satisfies a set of formulas Γ. In propositional logic, that corresponds to finding an assignment of truth values to the variables occurring in the formulas of Γ that makes all formulas true under the assignment. The problem of finding whether such assignment exists is what is called the *satisfiability problem* (SAT). Usually the algorithms for the satisfiability problem search, in a space of partial variable assignments, for a total assignment that satisfies the given set of formulas. An algorithm for SAT, as much as a proof procedure, can be used to find if a formula α is a logical consequence of a set of formulas Σ (which must be done, for example, to verify if a property follows from a set of specifications), as it is known that α follows from Σ if and only if $\Sigma \cup \{\neg\alpha\}$ *is not* satisfiable (is unsatisfiable). Consequently, to show that $\alpha \leftrightarrow \beta$ is valid (true for all possible assignments), it is sufficient to show that $\neg(\alpha \leftrightarrow \beta)$ is unsatisfiable. It is interesting to note that if the algorithm finds that $\Sigma \cup \{\neg\alpha\}$ (in the first case) or $\neg(\alpha \leftrightarrow \beta)$ (in the second case) is satisfiable, then the generated satisfying assignment gives a counter-example that helps at diagnosing software/circuit faults.

The SAT problem was the first to be proved NP-complete (Cook, 1971). The strength of this class of problems in general, and of SAT in particular, derives from the existence of many important practical problems that are NP-complete. As these problems are polynomial-time reducible to SAT, SAT algorithms with a good average-time performance are serious candidates to be applied at solving them. In fact, actually more and more problems are being tackled by means of modern SAT solvers, such as software (Chaki et al., 2003) and hardware (Velev & Bryant, 2001) verification, planning (Kautz & Selman, 1996), scheduling (Hoos, 2002) and many others. Actually, any NP-hard problem can eventually be solved through reduction or reformulation and use of some SAT solver sufficiently efficient to solve problem instances that really occur in practice. As already highlighted, it is remarkable that last years' advancements made it possible for some kinds of problems to be solved with performance comparable or even superior to that of solvers specially designed for them. For introductions to the theory of NP-completeness (and definitions of NP-complete and NP-hard problems), the reader may consult e.g. (Garey & Johnson, 1979; Sipser, 1997; Cormen, Leiserson, Rivest, & Stein, 2001).

The results obtained using SAT solvers have been so (surprisingly) good that some authors are foreseeing repercussions on the actual practice of algorithm design and analysis: the current excessive emphasis on the worst-case complexity analysis, that suggests the strict avoidance of non-polynomial performance, would not be appropriate any more (Gomes et al., 2008). Actually, such good results could be sensed early from the paper (D. G. Mitchell, Selman, & Levesque, 1992), which has shown that randomly generated difficult instances come from a relatively narrow zone of the space of possibilities. It was to be expected that for some classes of instances a specific approach to SAT solving could have a reasonable running time behavior, with a worst-case exponential behavior rarely or even never occurring.

What could have caused the excellent performance of modern SAT solvers to the point of opening the possibility of their use in the solution of very big instances that occur in industrial applications? Mitchell argues that the answer comes from three factors: better algorithms, improved implementation techniques and increased computer capacity (time and storage) (D. Mitchell, 2005). In particular, some algorithmic improvements (as, for example, heuristics and intelligent backtracking) are effective only with the computational resources recently available, which allow for efficient access to tens or hundreds of megabytes.

MODERN SAT SOLVING

Despite the worst-case exponential runtime of known SAT algorithms and despite the fact that SAT encodings generally substantially increase the size of the representation of a problem, modern SAT-solvers have been used effectively to solve very large problems, involving over a million variables. The approach of using SAT encodings and SAT solvers to solve large problems, from diverse areas, can be traced back to work on SAT published in the early nineties, but its use has increased significantly in the past few years, due to improvements in the performance of SAT solvers.

The work of (D. Mitchell, Selman, & Levesque, 1992) was pioneering in suggesting that, despite their worst-case exponential running time, SAT solvers could be used to solve quickly a large proportion of randomly generated SAT instances. Their experiments indicated that testing satisfiability (with a particular procedure, namely DPLL, see below) becomes less efficient, for any ratio of the number of clauses per number of variables, as the percentage of all formulas that are satisfiable approximates 50% (away from this "hard" area, satisfiability testing could be done efficiently).

In (Selman, Levesque, & Mitchell, 1992) a simple, incomplete, search method, called GSAT, was proposed, that solved instances in the hardest region effectively. GSAT's algorithm searches for a satisfying assignment of values to variables, starting from a random assignment and repeatedly changing the value assigned to a variable. A variable that leads to the largest increase in the number of satisfied clauses is chosen, until either a satisfying assignment is found or a fixed maximum number m of changes is reached. In the latter case the method is incomplete, since it can fail to find a satisfying assignment when m is reached.

Since 2002, competitions of SAT solvers are organized yearly as a satellite event at the international conference on Theory and Applications of

Satisfiability Testing (cf. http://www.satcompetition.org/, the international SAT Competitions web page). These have contributed to the improvement of implementations of many SAT solvers (e.g. (Silva & Sakallah, 1996; Moskewicz, Madigan, Zhao, Zhang, & Malik, 2001; Gold- berg & Novikov, 2007; Huang, 2007; Selman et al., 1992)) and to the availability of large sets of SAT benchmarks (e.g. (Hoos & Stützle, 2000)).

This chapter presents the main techniques used in modern SAT solvers and also a general idea of the performance of modern SAT-solvers, in particular with respect to the running time these solvers spend for solving SAT models of Operation Management problems cited in the previous chapter, considering inputs of various sizes.

TECHNIQUES FOR SOLVING SAT

Methods for SAT solving can be complete but also incomplete, according to whether the implemented function *sat:* $F \rightarrow Bool$ is a total or partial function. In other words: incomplete methods may fail to give an answer to some (hopefully few) formulas (problem instances). Complete methods are presented first, then incomplete ones. The motivation for the use of incomplete methods is that there are cases in which complete ones are not as efficient.

We use in the sequel common terms (like formula, clause, literal etc.) with their usual meaning (see e.g. (Garey & Johnson, 1979; Cormen et al., 2001)). In particular, a formula is a set of clauses and a clause is a set of literals. A literal is a variable or its negation. The negation of a literal α is denoted by $\bar{a}$. We use also the following functions:

tickle: (F, Set Lit) → F

unitProp: F → F

tickle (p,s) denotes the simplified formula obtained by removing from *p* all clauses with literals in *s* and deleting from all remaining clauses literals whose negations occur in *s*, and *unitProp(p)* is repeated tickling, until an empty clause is obtained or there are no more unit clauses.

$$tickle(p,s) = \left\{ x - \{\bar{k} \mid k \in s\} \mid x \in p \text{ and } x \cap s = \varnothing \right\}$$

$$unitProp(p) = \begin{cases} p & \text{if } \varnothing \in p \text{ or } s_1 = \varnothing \\ unitProp(tickle(p, s_1)) & \text{otherwise} \end{cases}$$

$$\text{where } s_1 = units(p)$$

$$units(p) = \left\{ k \mid \{k\} \in p \right\}$$

Complete Methods

Modern complete methods are still based, fundamentally, on the simple DPLL algorithm introduced in the early 1960's, which is based on selection and backtracking.

Martin Davis and Hilary Putnam (Davis & Putnam, 1960) defined in 1960 an algorithm, called DP, that used the resolution rule:

$$\frac{\{a\} \cup x \quad \{\bar{a}\} \cup y}{x \cup y}$$

as the basic rule for satisfiability testing.

In executions of programs based on an unrestricted use of the resolution rule, a large amount of useless consequents of large size can be generated, causing memory consumption to explode. In 1962, an alternative to DP was defined by Martin Davis, George Logemann and Donald Loveland (Davis, Logemann, & Loveland, 1962), usually called DPLL (the name DP is sometimes used to refer to DPLL, confusingly), that used a very simple modification of DP which replaced the use of the resolution rule by a select-and-backtrack (also called *splitting*) rule that avoids the often occurring exponential memory explosion problem of DP.

Figure 1. DPLL

$$dpll(p) = \begin{cases} \text{true} & \text{if } p = \varnothing \\ \text{false} & \text{if } \varnothing \in p \\ \text{otherwise, selecting any } k \text{ in a minimum size clause of } p\text{:} & \\ \begin{cases} \text{true} & \text{if } dpll(p_1) \\ dpll(p_2) & \text{otherwise} \end{cases} & \\ \quad \text{where} \quad p_1 = unitProp(p \cup \{k\}) & \\ \quad \qquad\qquad p_2 = unitProp(p \cup \{\overline{k}\}) & \end{cases}$$

DPLL is presented in Figure 1 in the form of a function *dpll:* $F \to Bool$.

The name DPLL is also used for variations of this original scheme that selects any literal occurring in a clause of minimum size, such as the one in which selection involves choosing any unassigned literal. The process of identifying unit clauses and performing unit clause propagation is sometimes referred to as *boolean constraint propagation* (BCP) (Nadel, 2002).

The pure literal rule—which removes clauses with pure literals, where a pure literal is one such that its negation does not occur anywhere —, usually included in DPLL, has been omitted here. Eliminating pure literals may be, however, worthwhile (cf. e.g. (Steedman,1995)).

The main efficiency improvements made in modern SAT solvers, namely clause learning and efficient unit propagation, introduce modifications to this basic DPLL algorithm, described next.

Selection and Backtracking by Clause Learning. The heuristics used for selecting a variable and its value, also referred to as the *decision strategy* or *branching heuristic*, varies significantly among SAT solvers. However, advances in the efficiency of modern SAT solvers can be attributed, to a large extent, to a technique called *clause learning*, that uses information obtained as a result of the selection process in order to guide further selection, by adding clauses to the set of clauses being tested.

The most successful learning scheme, used by modern solvers (e.g. GRASP(Silva & Sakallah, 1996), zChaff(Moskewicz et al., 2001), BerkMin(Goldberg & Novikov, 2007), siege (Huang, 2007)), is based on detecting conflicts (assignments that cause the formula to become un- satisfiable) and on adding a new clause, called a *conflict cause*, to the set of clauses being tested, in order to avoid the conflict to happen again.

A conflict clause is determined by using an *implication graph*, which records implications between variables that arise due to the assignment of values to variables performed during SAT solving.

An implication graph is a labeled directed acyclic graph (DAG) where each node corresponds to a selected or implied literal (implied by unit propagation) and each edge from node *a* to node *b* corresponds to an implication ($a \to b$), as a consequence of existing clauses and selected assignments (*e.g.* $a \to b$ is an edge if clause $\{\overline{a}, b, c\}$ exists and $\overline{c}$ has been selected or, else, is implied by unit propagation). More precisely, we have (cf. (Gomes et al., 2008)):

Definition 1: *An implication graph G = (V, E), at a given stage of a SAT algorithm A, is* *a directed acyclic graph with nodes V and edges E, defined inductively as follows:*

- *initially, let V contain the set of selected literals and E be empty;*
- *for each clause* $c = \{a_1, \ldots, a_n\}$ such that $\overline{a_1}, \ldots, \overline{a_n}$ *are nodes in G,* $n \geq 1$*, add node a to V and, for* $i = 1, \ldots, n$ and add edges $e_i = (\overline{a_i}, a_1)$ to G.

Example 1: *Let* $p = \{x_1, x_2, \ldots, x_6\}$, *where:*

$$x_1 = \{a_1, a_2\} \quad x_2 = \{a_1, a_3, a_7\}$$
$$x_4 = \{\overline{a_4}, a_5, a_8\} \; x_5 = \{\overline{a_4}, a_6, a_9\}$$

$$x_1 = \{a_1, a_2\} \quad x_2 = \{a_1, a_3, a_7\}$$
$$x_4 = \{\overline{a_4}, a_5, a_8\} \; x_5 = \{\overline{a_4}, a_6, a_9\}$$

$$x_3 = \{\overline{a_2}, \overline{a_3}, a_4\} \; x_2 = \{a_1, a_3, a_7\}$$
$$x_6 = \{\overline{a_5}, \overline{a_6}\} \quad x_5 = \{\overline{a_4}, a_6, a_9\}$$

$$x_3 = \{\overline{a_2}, \overline{a_3}, a_4\}$$
$$x_6 = \{\overline{a_5}, \overline{a_6}\}$$

and consider literals selected in the sequence $\overline{a_9}, \overline{a_8}, \overline{a_7}, \overline{a_1}$. We have then the implication graph in Figure 2.

A conflict occurs when both a and $\overline{a}$, called *conflicting literals*, occur in an implication graph (a is called a *conflicting variable*). An implication graph G contains a conflict if a and $\overline{a}$ are nodes of G, for some a, in which case G is said to contain a conflict graph (in other words, a conflict graph is a subgraph of G containing conflict literals a and $\overline{a}$, for some a).

Consider an implication graph $G = (V, E)$. Let a *cut* of a subset s of nodes of G, *cut (s)*, be the set of all edges going from nodes in s to nodes not in s, i.e. $cut(s) = \{e \mid e = (a, b) \in E, b \in s, a \notin s\}$.

Given an edge (a,b), a is called a source and b a target of e.

Any cut $C = cut(s)$ of a subset s of a conflict graph H separates H in two parts, called the reason and the conflict sides of H, the reason side including nodes from a selected literal to a source of an edge in C and the conflict side including nodes from a target of an edge in C to a conflicting literal. The sources of edges in C, plus conflicting literals, if they are in s, are *causes* of the conflict (at the conflict side) given by C, denoted by *causes* (C). The set (representing the disjunction) of all negations of causes is a *conflict clause of* C, denoted *conflictClause* (C).

For example, considering Example 1 we have (where s_{i+1} adds to s_i, for i = 0, 1, 2, the targets of edges in C_i, plus conflicting literals, if they are already targets of an edge in C_i):

$$s_0 = \{\overline{a_1}, \overline{a_7}, \overline{a_8}, \overline{a_9}\}$$
$$C_0 = cut(s_0) = \{x_1, x_2\}$$
$$causes(C_0) = \{\overline{a_1}, \overline{a_7}\}$$
$$conflictClause(C_0) = \{a_1, a_7\}$$
$$s_1 = s_0 \cup \{a_2, a_3\}$$
$$C_1 = cut(s_1) = \{x_3\}$$
$$causes(C_0) = \{a_2, a_3\}$$
$$s_2 = s_1 \cup \{a_4\} \quad conflictClause(C_1) = \{\overline{a_2}, \overline{a_3}\}$$
$$C_2 = cut(s_2) = \{x_4, x_5\}$$
$$causes(C_2) = \{a_4, \overline{a_8}, \overline{a_9}\}$$
$$s_3 = s_2 \cup \{a_5, a_6\} \quad conflictClause(C_2) = \{\overline{a_4}, a_8, a_9\}$$
$$C_2 = cut(s_2) = \{x_6\}$$
$$causes(C_2) = \{a_5, a_6\}$$
$$conflictClause(C_2) = \{\overline{a_5}, \overline{a_6}\}$$

In order to choose a conflict clause, in a clause learning scheme, each selected literal is usually associated with a number, from 1 upwards, called the *level* of the selected literal. The level of an implied literal is the maximum of the levels of all selected literals that are used in its implication. The level of a node of an implication graph is the level of its literal. Level zero is used to characterize unsatisfiability, when a conflict that does not depend on any selection is detected.

The backtracking scheme used in modern SAT solvers consists of changing the value assigned to one or more variables occurring in a conflict

Figure 2. Implication Graph

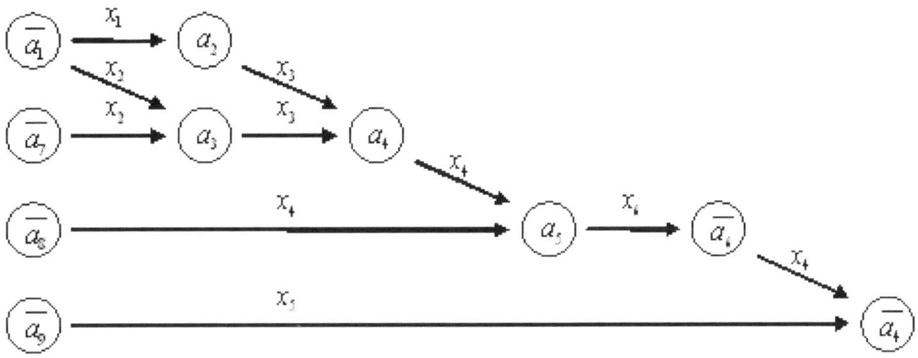

clause. A SAT solver uses some way — called a *clause learning scheme* — of obtaining a conflict clause. For example, Grasp (Silva & Sakallah, 1996) and zChaff (Moskewicz et al., 2001) use *FirstUIP*, which is a scheme that identifies a node in the implication graph that is closest to the conflict variable and is a unique implication point (UIP). Letting k and l_k be respectively the currently selected literal and its level, in a conflict graph H (k is the last selected literal and l_k is the highest level in H), a UIP of H is a node (literal) k' with level l_k such that every path from k to a conflicting literal must pass through k' (k itself is always a UIP) (Silva & Sakallah, 1996). Note that a conflict clause, the singleton $\{\bar{k}\}$ ($\{a_l\}$ in Example 1) thus always exists in a conflict graph. a_4 in our previous example is a *FirstUIP*.

A learning clause scheme enables distinct heuristics for performing and extending backtracking, in particular:

- *Backjumping* (also called *non-chronological backtracking*) is a term that generally refers to some way of performing backtracking by going directly to a previous point (level) in order to avoid irrelevant ones (Dechter & Frost, 1999). Several ways of performing *backjumping* have been used in SAT solvers.

Consider for example a situation in which n is any large positive number and the ordering of variables used for selecting values in variable assignment is:

$$a_1, b_1, b_2, \ldots, b_n, a_2, a_3$$

and consider clauses that include:

$$\{\ \{a_1, a_2, a_3\},\ \{a_1, \bar{a}_2, a_3\},\ \{b_1, b_2, \ldots, b_n\},\ \{a_1, a_2, \bar{a}_3\},\ \{a_1, \bar{a}_2, \bar{a}_3\},\ \{\bar{b}_1, \bar{b}_2, \ldots, \bar{b}_n\}\ \}$$

Any assignment ρ for which $\rho\ (a_l) = 0$ will lead to a contradiction, since any assignment of values to a2 and a3 will generate in this case a contradiction. If the selection scheme uses the variable ordering $[a_1, b_1, \ldots, b_n, a_2, a_3]$ and selection starts by assigning 0 to a_l, all 2^n assignments to b variables could be performed by the algorithm, all of them generating a contradiction. A naive backtracking strategy can take thus a very long time trying all such assignments before changing the assignment of a_l to 1. To avoid this, modern SAT solvers use backjumping. Backjumping identifies, at a failed choice point (i.e. when both assignments, of *true* and *false*, to a variable, in our example a_2, lead to a contradiction), the causes of the conflicts (consisting of previously selected or implied variables, in our example the singleton

$\{a_l\}$). In this case, backtracking changes the assignment of the variable with the greatest decision level (i.e. most recently selected). For example, Grasp and zchaff's backjumping heuristic allow backtracking to a lower decision level *d* when one of the branches that leads to a conflicting literal involves only literals at level *d* or lower with the exception of the literal with the current level (Gomes et al., 2008).

- SAT solvers can spend a lot of time in unproductive parts of a search tree. Many SAT solvers then randomly restart the search from scratch, with an empty assignment, taking into account the fact that searches will not be repeated due to a randomness that occurs in the selection process (Gomes, Selman, & Kautz, 1998).

Proposals to guarantee completeness in the presence of restarts, discussed in (Nadel, 2002), include: preserving all learned clauses, repeatedly increasing the number of conflicts between consecutive restarts, preserving an increasing number of learned clauses after each restart, and including a learned clause with the negations of all decision variables that participated in the conflict in order to avoid the search path that led to the conflict to be taken again.

- As the set of learned clauses increases, the ability to prune the search space increases but also does the cost of unit propagation. Heuristics that try to achieve a balance on the size of the set of learned clauses include associating a measure of activity for each variable, that is increased each time a variable (and also a clause, as in (En & Sorensson, 2003)) is involved in a conflict (Moskewicz et al., 2001). Periodically removing inactive clauses and the more common use of variable activity to guide the selection process are heuristics used in e.g. (En & Sorensson, 2003).

<u>Unit propagation with Watched Literals</u>. Efficient unit propagation is a key to current fast SAT solvers. It has been observed by profiling that around 90% of SAT solvers runtime is spent in unit propagation (Zhang, Madigan, Moskewicz, & Malik, 2001).

Almost all modern SAT solvers use an algorithm for unit propagation based on a simple technique of (so-called) *watched literals*, firstly used in the Chaff SAT solver (Moskewicz et al., 2001). The idea is to use, for each *active* (i.e. not yet satisfied) clause *c*, two literals k_1 and k_2 (called *watches* of *c*) such that $\overline{k_1}$ and $\overline{k_2}$ are not yet selected or implied (in other words, neither k_1 nor k_2 are *false*), and update information about watched literals whenever one of them, say k_2, is assigned to *false* (as a result of selection or implication, by unit propagation, of k_2), as follows:

1. k_1 is not implied; then there exists another literal k_3 in *c* that is not *false*, and k_1 and k_3 become the watched literals of *c*;
2. k_1 is implied (assigned to true); then $\overline{k_1}$ is assigned to *false* and unit propagation is done for k_1;

The watched literals technique is beneficial particularly because, upon backtracking, nothing has to be done or undone, since watched literals can stay the same.

The implementation of the watched literal scheme typically uses a *watch-list* for each literal *k*, that contains a list of clause references (numbers) for which *k* is a watch, and, for each clause reference (number), its pair of *watches*. When *k* is assigned to *false*, *k*'s watch list is scanned and, for each encountered clause *x*, *x*'s watches are updated.

Incomplete Methods

In principle, unsatisfiability could be searched but, in practice, most incomplete methods typically run with pre-set limits until a satisfying assignment is found or failure is reported, unsatisfiability being never reported. On problems of various domains, they run much faster than complete methods. Incomplete methods have been the subject of a lot of work since the early 1990's. References can be found in chapter 6 of the Handbook of Satisfiability, by Henry Kautz, Ashish Sabharwal and Bart Selman ((Kautz, Sabharwal, & Selman, 2008)), of which the material in this section is a condensed and adapted version.

The original work with incomplete local search methods was encouraged by experiments done with GSAT (Selman et al., 1992), which showed that global minima could be reached with local search strategies, in many cases much faster than with complete search methods. GSAT uses a very simple technique of choosing random assignments for all variables of a propositional formula p in cnf (conjunctive normal form, see e.g. (Prestwich, 2008)) and changing this assignment for the variable that leads to the greatest decrease in the number of unsatisfied clauses of p. GSAT is written in Figure 2 as a function *gsat: F → Bool*, overloaded for clarity and for parametrization on the number (t) of tries (number of ran- dom assignments chosen), with maximum value *maxTries*, and on the number (n) of flips (i.e. variable assignment changes), with maximum value *maxFlips* (*fVs* below abbreviates, say, *flippedVars*).

Experiments have demonstrated that GSAT usually proceeds quickly towards a truth assignment but in many cases it spends a lot of time flipping variables without changing the number of unsatisfied clauses, before finding another assignment in which this number does decrease. One of the most successful strategies to speed up this process has been the one used in Walksat (Selman, Kautz, & Cohen, 1996), which tries to select a variable to flip from an unsatisfied clause, chosen at random and, if that leads current satisfied clauses to become unsatisfied, either i) a random variable in C is flipped or ii) a variable in C with lowest *break-down* (the number of satisfied clauses that becomes unsatisfied, according to the current assignment) is chosen. The choice between i) and ii) is controlled by a so-called *noise* parameter ρ, which is a value between 0 and 1 that controls how often i) is chosen, instead of ii).

$$
\begin{aligned}
&gsat(p) = gsat(p,1)\\
&gsat(p,t) = \begin{cases} fail & \text{if } t > maxTries \\ gsat(p,\sigma,t,1,\varnothing) & \text{otherwise, } \sigma \text{ random assignment for } p \end{cases}\\
&gsat(p,\sigma,t,n,fVs) = \begin{cases} true & \text{if } \sigma \text{ makes } p \text{ satisfiable} \\ gsat(p,t+1) & \text{if } n > maxFlips \\ gsat(p,\sigma',t,n+1,fVs \cup \{v\}) & \text{otherwise} \\ \quad \text{where } v = flip(\sigma,p,fVs) & \\ \qquad \sigma' \text{ is } \sigma \text{ but with } \sigma(v) \text{ negated} & \end{cases}
\end{aligned}
$$

Equation (1)

walksat can be defined as *gsat* with only *flip* changed:

$flip(\sigma, p, vs) =$
  let x be an unsatisfied clause of p, chosen at random, in
   v if there exists v in x with break-down 0
   otherwise, where i/0 is chosen with probability $\rho \in [0,1]$,
    either i) a variable from x, chosen at random, or
      ii) a variable from x with smallest break-down

Equation (2)

Several other approaches for improving the performance of local search methods have been proposed, for which we refer interested readers to (Kautz et al., 2008).

PLANNING AS SAT

Due to significant improvements in the efficiency of satisfiability solvers in recent years, many scientific and industrial problems have been modeled as a translation into SAT (Chaki et al., 2003; Velev & Bryant, 2001; Kautz & Selman, 1996; Hoos, 2002). Since the 1990s, software for solving Planning and Scheduling problems of Operations Management, in particular, have experienced great advances with this approach

(Andrew & Baker, 1994; Kautz & Selman, 1996). This section gives a brief overview of the planning-as-SAT approach and explain the main issues involved in this approach.

Classical planning is a well-known problem in Artificial Intelligence and Operations Research. Planning is the problem of finding a sequence of actions that can be used for going from a initial to a final state, from given specifications of initial and final states and a set of action definitions, that specify changes of the state. Classical planning is its most basic form, where actions are deterministic.

Despite its importance in real world tasks, until the 1990s planners were not capable of solving real world instances, partly due to computer technology but mainly because of the algorithms used. During the 1990s dramatic advances occurred, with the introduction of three new methods: the GraphPlan algorithm (Blum & Furst, 1995), the compilation-to-SAT approach (first proposed by Kautz and Selman in (Kautz & Selman, 1992), extended by the same authors with improved encodings and test results in (Kautz & Selman, 1996) and extended in many ways in (Kautz, Mcallester, & Selman, 1996; Ernst, Millstein, & Weld, 1997)) and by new heuristic search based techniques (described in e.g. (Bonet & Geffner, 2001; Hoffmann & Nebel, 2001)).

The basic idea behind using SAT for solving problems of planning is to restrict planning to practically interesting cases where the sequence of actions to reach a final state from the initial state is bounded, by some length n. The problem can then be encoded so that a propositional formula for a given n is satisfiable if and only if there exists a plan of such length n; a different value of n must be searched for in the case of unsatisfiability.

To date, lots of different encoding schemes for compilation into SAT have been proposed (e.g. (Ernst et al., 1997; Xing, Chen, & Zhang, 2006; N. Robinson, Gretton, Pham, & Sattar, 2008)), but the basic idea behind them is to construct a SAT instance along the lines described by means of the classic "blocks world" example presented in Figure 3 (cf. e.g. (Kautz & Selman, 1992; Kautz, McAllester, & Selman, 1996)).

The blocks world problem consists of placing blocks, initially on a table (in Figure 3, blocks b_1, b_2, b_3 and table t), so that b_i becomes on top of b_{i+1} ($i = 1, 2$). Symbols $\neg$, $\wedge$, $\vee$, $\rightarrow$ and $\leftrightarrow$ are used in Figure 3 and in the sequel as the logical connectives of negation, conjunction, disjunction, implication and equivalence, respectively.

In the rather self-explanatory problem specification above, *init* marks the beginning of the specification of the initial state, *goal* of the final state and action of an *action* definition. Initial and final states are defined by propositions. In our example, a proposition consists of conjunctions of *predicate* applications (in general a proposition may use other logical connectives). Predicates are functions which return a boolean value, given elements in the domain of the problem instance (in our example, the domain of the problem instance consists of blocks b_i, $i = 1, 2, 3$, and table t). An action definition defines a predicate by means of:

1. a precondition: a proposition that must hold for the action to be taken. The definition of a precondition begins with the keyword *pre*.
2. an effect: a proposition that holds after the action is taken. The definition of an effect begins with the keyword *effect*.

A plan that solves the problem above in the smallest number of steps is the sequence of actions: [*move* (b_2, t, b_3), *move* (b_1,t, b_2)].

In order to translate a bounded version of the planning problem in terms of SAT, we have first to choose a limit n to the length of a plan to reach the final state. Algorithms for choosing values of n in order to improve the efficiency of the process of finding a solution to the planning problem (if it exists) — which is a critical step of the planning-as-SAT approach — are presented in e.g. (Rintanen, 2009). If no solution exists, a

Figure 3. Instance of the classic blocks-world planning problem

$$
\begin{array}{ll}
\mathit{init}: & \mathit{clear}(b_1) \wedge \mathit{clear}(b_2) \wedge \mathit{clear}(b_3) \wedge \\
 & \mathit{on}(b_1,t) \wedge \mathit{on}(b_2,t) \wedge \mathit{on}(b_3,t) \\
\mathit{goal}: & \mathit{on}(b_1,b_2) \wedge \mathit{on}(b_2,b_3) \\
\multicolumn{2}{l}{\mathit{action}\ (\mathit{move}(b,x,y),} \\
\quad \mathit{pre}: & \mathit{on}(b,x) \wedge \mathit{clear}(b) \wedge \mathit{clear}(y) \\
\quad \mathit{effect}: & \mathit{on}(b,y) \wedge \neg\mathit{on}(b,x) \wedge \mathit{clear}(x) \wedge \neg\mathit{clear}(y)) \\
\multicolumn{2}{l}{\mathit{action}\ (\mathit{moveToTable}(b,x),} \\
\quad \mathit{pre}: & \mathit{on}(b,x) \wedge \mathit{clear}(b) \\
\quad \mathit{effect}: & \mathit{on}(b,t) \wedge \neg\mathit{on}(b,x) \wedge \mathit{clear}(x))
\end{array}
$$

different value of n may be tried; if there is no n such that the planning problem is solvable with a plan of length n, this process iterates forever. The approach is thus only useful for problems for which a solution exists. For problems for which it is not known whether a solution exists and, furthermore, for problems for which the planning problem requires a formula that grows exponentially with the size of the problem instance, another approach should in general be used (see e.g. (Stockmeyer & Meyer, 1973)). However, in practice most planning problems have a solution — a plan — whose length grows polynomially with the size of the problem.

Once n is chosen, a propositional formula A is constructed that consists of a conjunction of "instantiations" of each action definition a, for each stage $k = 0,\dots,n-1$ and for each possible combination of arguments for a. An instantiation of an action a with precondition p and effect e is anequivalence $a(x_1,\dots,x_m,k) \leftrightarrow p_k^l \to e_{k+1}^l \wedge F_{a_k}$, where $x_1,\dots,x_m$ is sequence (possible combination of arguments of a, pre_k^l is an instantiation of the precondition for a for these arguments and state k, effect_{k+1}^l is an instantiation of the effects of a for these arguments and stage $k+1$, and F_{a_k} are so-called "frame conditions", which are implications that state that predicate instantiations that are not altered remain in the stage as they were in the previous state.

The propositional formula that corresponds to the bounded planning problem is a conjunction $S_0 \wedge G \wedge A \wedge E$, where:

- S_0 represents the initial state,
- G the final goal state,
- A is as described above, and
- E asserts that at most one action occurs at each stage.

In our example, for the chosen value of n = 3 these propositions are as follows (where c abbreviates clear):

$$
\begin{array}{lcl}
S_0 & = & c(b_1,0) \wedge c(b_2,0) \wedge c(b_3,0) \wedge \\
 & & \quad \mathit{on}(b_1,t,0) \wedge \mathit{on}(b_2,t,0) \wedge \mathit{on}(b_3,t,0) \\
G & = & \mathit{on}(b_1,b_2,3) \wedge \mathit{on}(b_2,b_3,3) \\
A & = & \text{combinations of } \mathit{move}(x,y,z,i) \text{ and } \mathit{moveToTable}(x,y,i), \\
 & & \quad \text{for } i = 0,\dots,n-1 \text{ and arguments } x,y,z, \text{ including e.g.} \\
 & & \quad \mathit{move}(b_2,t,b_3,0) \leftrightarrow \\
 & & \qquad (\mathit{on}(b_1,t,0) \wedge c(b_2,0) \wedge c(b_3,0) \to \\
 & & \qquad (\mathit{on}(b_2,b_3,1) \wedge \neg\mathit{on}(b_2,t,1) \wedge \\
 & & \qquad c(b_2,1) \wedge \neg c(b_3,1) \wedge (c(b_1,0) \leftrightarrow c(b_1,1)))) \\
E & = & \mathit{move}(x,y,z,i) \vee \mathit{moveToTable}(x,y,i) \vee \\
 & & \quad ((c(x,i) \leftrightarrow c(x,i+1)) \wedge (\mathit{on}(x,y,i) \leftrightarrow c(x,y,i+1))) \\
 & & \quad \text{for combinations of arguments of } \mathit{move}, \mathit{moveToTable}, \mathit{on}, c \\
 & & \quad \text{and for } i = 0,\dots,n\text{ - }1
\end{array}
$$

The presence of action definitions with 3 arguments (like *move* in the example above) or more leads in general to the generation of a proposition that is too big to be processed efficiently, which would make the approach impractical. However, there are techniques that may be used for reducing the number of parameters of action definitions and other techniques for reducing the number of variables and clauses of the generated proposition (cf. e.g. (Kautz, McAllester, & Selman, 1996)).

CONCLUSION

We have presented a brief overview of the problem of satisfiability of propositional formulas (SAT), which has many practical implications and is at the heart of the most important problem in theoretical computer science (the P versus NP question). There is a lot of research related to SAT, spanning from complexity analysis, heuristics, encodings of different problems into SAT, quantified boolean formulae, predicate logic and other kinds of logics, constraint programming etc. We have also given a gentle introduction to the approach of translating classical planning problems into SAT. The material aims at giving the reader a motivation for further studies in this vast and exciting field of research.

REFERENCES

Andrew, J. C., & Baker, A. B. (1994). Experimental results on the application of satisfiability algorithms to scheduling problems. In *Proc. of the 12th National Conference on Artificial Intelligence* (pp. 1092–1097).

Blum, A. L., & Furst, M. L. (1995). Fast planning through planning graph analysis. *Artificial Intelligence*, *90*, 1636–1642.

Bonet, B., & Geffner, H. (2001). Planning as heuristic search. *Artificial Intelligence*, *129*, 5–33. doi:10.1016/S0004-3702(01)00108-4

Brachman, R. J., & Levesque, H. J. (2004). Knowledge Representation and Reasoning. San Francisco: Morgan Kaufmann.

Cadoli, M., & Schaerf, A. (2005). Compiling problem specifications into SAT. *Artificial Intelligence*, *162*(1-2), 89–120. doi:10.1016/j.artint.2004.01.006

Chaki, S., et al. (2003). Modular verification of software components in C. In *Proc. of the 25th ACM/IEEE International Conference on Software Engineering,* (pp. 385–395).

Cook, S. A. (1971). The complexity of theorem-proving procedures. In *Proc. of the 3rd Annual ACM Symposium on the Theory of Computing,* (pp. 151–158).

Cormen, T. H., Leiserson, C. E., Rivest, R. L., & Stein, C. (2001). Introduction to Algorithms (Second ed.). Cambridge, MA: MIT Press.

Davis, M., Logemann, G., & Loveland, D. (1962). A machine program for theorem-proving. *Communications of the ACM*, *5*(7), 394–397. doi:10.1145/368273.368557

Davis, M., & Putnam, H. (1960). A Computing Procedure for Quantification Theory. *Journal of the ACM*, *7*(3), 201–215. doi:10.1145/321033.321034

Dechter, R., & Frost, D. (1999). Backtracking algorithms for constraint satisfaction problems (Tech. Rep.). Irvine, CA: Univ. of California, Irvine.

En, N., & Sörensson, N. (2003). An Extensible SAT-solver. In Sat (Vol. 2919, pp. 502–518). Berlin: Springer.

Ernst, M., Millstein, T., & Weld, D. (1997). Automatic SAT-Compilation of Planning Problems.In *Proc. of the 15th International Joint Conference on Artificial Intelligence (IJCAI-97),* (pp.1169–1176).

Garey, M., & Johnson, D. (1979). Computers and Intractability: A Guide to the Theory of NP-Completeness. San Francisco: Freeman.

Goldberg, E., & Novikov, Y. (2007). Berkmin: A fast and robust sat-solver. *Discrete Applied Mathematics*, *155*(12), 1549–1561. doi:10.1016/j.dam.2006.10.007

Gomes, C., Kautz, H., Sabharwal, A., & Selman, B. (2008). Satisfiability Solvers. In Handbook of Knowledge Representation (Foundations of Artificial Intelligence), (pp.89–134). New York: Elsevier B.V.

Gomes, C., Selman, B., & Kautz, H. (1998). Boosting combinatorial search through randomization. In *Proc. of the 15th National Conference on Artificial Intelligence (AAAI-98)* (pp. 431–437).

Hoffmann, J., & Nebel, B. (2001). The FF Planning System: Fast Plan Generation Through Heuristic Search. *Journal of Artificial Intelligence Research*, *14*, 253–302.

Hoos, H. (2002). An adaptive noise mechanism for WalkSAT. In *Proc. of the 18th National Conference on Artificial Intelligence (AAAI-2002),* (pp. 655–660).

Hoos, H., & Stützle, T. (2000). SATLIB: An Online Resource for Research on SAT. In *Proc. Of SAT 2000,* (pp. 283–292). Retrieved from http:/www.satlib.org

Huang, J. (2007). A Case for Simple SAT Solvers. In *Proc. 13th international conf. on principles and practice of constraint programming (cp'07), lncs 4741* (pp. 839–846). Berlin: Springer.

Kautz, H., Mcallester, D., & Selman, B. (1996). Encoding plans in propositional logic. In *Proc. of the 5th International Conference on the Principle of Knowledge Representation and Reasoning (KR'96),* (pp. 374–384).

Kautz, H., McAllester, D., & Selman, B. (1996). Encoding Plans in Propositional Logic. In *Proc. of the 5th International Conference on Principles of Knowledge Representation and Reasoning (KR'96)* (pp. 374-384).

Kautz, H., Sabharwal, A., & Selman, B. (2008). Incomplete algorithms. In Handbook of Satisfiability, (pp. 185–203). Amsterdam: IOS Press.

Kautz, H., & Selman, B. (1992). Planning as satisfiability. In *Proc. of the 10th European Conference on Artificial intelligence (ECAI'92),* (pp. 359–363). Hoboken, NJ: John Wiley & Sons, Inc.

Kautz, H., & Selman, B. (1996). Pushing the envelope: Planning, propositional logic, and stochastic search. In *Proc. of 13th the National Conference on Artificial Intelligence (AAAI'96),* (pp. 1194–1201).

Lifschitz, V., Morgenstern, L., & Plaisted, D. (2008). Knowledge Representation and Classical Logic. In Handbook of Knowledge Representation (Foundations of Artificial Intelligence), (pp. 3–88). New York: Elsevier B.V.

Mitchell, D. (2005). A SAT Solver Primer. *EATCS Bulletin*, *85*, 112–133.

Mitchell, D., Selman, B., & Levesque, H. (1992). Hard and Easy Distributions of SAT Problems. In Proc. of aaai (association for the advancement of artificial intelligence), 92, 459–465.

Mitchell, D. G., Selman, B., & Levesque, H. (1992). Hard and easy distributions of SAT solvers. In *Proc. of the 10th National Conference on Artificial Intelligence (AAAI-92)* (pp. 459–465).

Moskewicz, M., Madigan, C., Zhao, Y., Zhang, L., & Malik, S. (2001). Chaff: Engineering an Efficient SAT Solver. In *Proc. Design Automation Conference* (pp. 530–535).

Nadel, A. (2002). *Backtrack Search Algorithms for Propositional Logic Satisfiability: Review and Innovations*. Unpublished master's thesis, Univ. of Tel-Aviv, Israel.

Navarro, J. A. (2007). *Encoding and Solving Problems in Effectively Propositional Logic*. Unpublished doctoral dissertation, University of Manchester, UK.

Prestwich, S. (2008). CNF Encodings. In Handbook of Satisfiability, (pp. 75–97). Amsterdam: IOS Press.

Rintanen, J. (2009). Planning as SAT. In Handbook of Satisfiability, (Vol. 185, Frontiers in Artificial Intelligence and Applications, pp. 483–504). Amsterdam: IOS Press.

Robinson, A., & Voronkov, A. (Eds.). (2001). Handbook of Automated Reasoning (Vols. I & II). Amsterdam: The MIT Press.

Robinson, N., Gretton, C., Pham, D. N., & Sattar, A. (2008). A compact and efficient sat encoding for planning. Icaps (pp. 296-303).

Selman, B., Kautz, H., & Cohen, B. (1996). Local Search Strategies for Satisfiability Testing. In Cliques, Coloring, and Satisfiability: Second DIMACS Implementation Challenge, DIMACS Series in Discrete Mathematics and Theoretical Computer Science, (Vol. 26, pp. 521–532).

Selman, B., Levesque, H., & Mitchell, D. (1992). A new method for solving hard satisfiability problems. In Proc. of aaai 92 (pp. 440–446).

Silva, J. P. M., & Sakallah, K. (1996). Grasp — a new search algorithm for satisfiability. In *ICCAD '96: Proc. of the 1996 IEEE/ACM international conference on Computer-aided design,* (pp. 220–227).

Sipser, M. (1997). Introduction to the Theory of Computation. Boston: PWS Publishing Company.

Steedman, M. (1995). *Improvements to Propositional Satisfiability Search Algorithms*. Unpublished doctoral dissertation, Univ. of Pennsylvania.

Stockmeyer, L. J., & Meyer, A. R. (1973). Word problems requiring exponential time. In *Proc. of the 5th Annual ACM Symposium on the Theory of Computing* (pp. 1–9).

van Harmelen, F., Lifschitz, V., & Porter, B. (Eds.). (2008). Handbook of Knowledge Representation. New York: Elsevier.

Velev, M. N., & Bryant, R. E. (2001). Effective use of boolean satisfiability procedures in the formal verification of superscalar and VLIW microprocessors. In *Proc. of the 38th Conference on Design Automation,* (pp. 226–231).

Wos, L., Overbeek, R., Lusk, E., & Boyle, J. (1984). Automated reasoning: Introduction and applications. Englewood Cliffs, NJ: Prentice Hall.

Xing, Z., Chen, Y., & Zhang, W. (2006). MaxPlan: Optimal planning by decomposed satisfiability and backward reduction. In Proc. of the international planning competition (ipc), (pp. 53–56).

Zhang, L., Madigan, C., Moskewicz, M., & Malik, S. (2001). Efficient conflict driven learning in a boolean satisfiability solver. In *Proc. IEEE/ACM International Conference on Computer- Aided Design,* (pp. 279–285).

Chapter 3
Integrated Multi-Agent Coordination

Wei Chen
Intelligent Automation, Inc., USA

Keith S. Decker
University of Delaware, USA

ABSTRACT

Planning and scheduling have been a key topic in both Operations Research and Multi-Agent Systems. Most approaches are concentrated at an abstract system level on developing interaction protocols to be imposed on agents. There has been less concern about how the internal task structures of individual agents affect these higher-level coordination behaviors. Collaborative multi-agent planning addresses problems like uncertainty in plan outcomes, anticipating likely contingencies, and evaluating how agent actions achieve worth-oriented goals. This article presents extensions and restrictions, called extended hierarchical task networks (EHTN), to the traditional plan and schedule representations that allow the formal definition of an integrated multi-agent coordination problem. This chapter discusses open issues in multi-agent coordination (e.g. what to coordinate among agents, how much information to be exchanged, how to evaluate a planning approach) and proposes a general solution towards successful distributed goal achievement by analyzing the task structures of participating agents.

1. INTRODUCTION

Much work in multi-agent systems focuses on coordinating the activities of agents so that the end result approximates the solutions possible if one were to centralize the activities being carried out by these agents. Research into coordination has taken different forms: negotiation, scheduling, planning, organizational approaches, etc. *Multi-agent coordination*, defined as managing inter-dependencies between activities (Malone & Crowston, 1994), addresses the special issues arising from the relationships between multiple agents' tasks. We define *interdependency* as a relationship between agents' tasks, where the execution of one changes some performance-related characteristics associated with the other. This definition gives rise to the following questions: (1) how are the dependencies

DOI: 10.4018/978-1-61520-605-6.ch003

represented, and (2) how to manage these dependencies? Previous approaches, including GPGP (Generalized Partial Global Planning) (Decker & Li, 2000; Lesser, Decker & Wagner, 2004), have concentrated on high level solutions: keeping semi-coherence between agents' activities, representing the relationships between actions and goals, and developing coordination interaction protocols. For example, POMDP approaches (Boutilier, Dean & Hanks, 1999) have been influential in evaluating how system states will be affected by certain actions and thus make suitable decisions based on probabilistic analysis. A recent distributed POMDP approach (Nair, Pynadath & Yokoo, etc. 2003) has been focused on deriving joint policies to maximize certain rewards for a multi-agent group. However, there has been less concern about how an individual agent's internal task structures (i.e., hierarchically represented task network organization) and reasoning capabilities affect its coordination behaviors. In other words, our approach assumes that an intelligent agent is capable of managing its own activities (e.g., the general reasoning capabilities of AI planning) and it is the uncertainty in other agents' behaviors that affect its reasoning capabilities. Thus, it may be advantageous for the involving agents to coordinate on the estimates of their task executions so that they could remove the uncertainties by employing other agents' estimates. For example, Agent_A needs to finish a particular goal task consisting of both Task_A and Task_B and Task_B enables Task_A. Agent_A is able to carry out Task_A on its own, but Task_B can only be executed by Agent_B due to resource or capability constraints. Although Agent_A knows how long it will take for itself to finish Task_A, it cannot plan and schedule its own activities effectively since it is unknown how long it will take Agent_B to finish Task_B. Thus, before Agent_A starts its own task executions (keeping in mind that there may be other tasks for Agent_A), it is beneficial for Agent_A to ask Agent_B that what time Agent_B could be estimated to finish Task_B. Consequently, Agent_A can then plan and schedule its own activities with this estimated information. In particular, in order to understand the underlying assumptions, capabilities, and limitations of our approach, we need to formalize these linkages and relationships within the explicitly represented task structures.

Based on this definition and the (later) analysis of interdependency, we can give answers to the following two questions. (1) *Which applications required decentralized planning?* It is NOT about how many agents are present in a problem solving process, but whether interdependencies exist among the agents' activities, i.e. interdependencies demand decentralized planning. This article presents a set of sophisticated coordination mechanisms to manage those interdependencies. (2) *How can we evaluate Multiagent planning techniques?* Our answer lies in objective, quantitative analysis based on task features represented with extended hierarchical task networks (EHTN). (3) *What may be an effective solution to multi-agent problem solving?* We believe one of the efficient techniques to deal with the many problems inherent to the dynamics and uncertainties of multi-agent coordination problem is our extended set of coordination mechanisms including real-time plan structure alteration and associated coordination communication protocols. We argue that multi-agent problem solving is a comprehensive process, where the constituent reasoning capabilities (e.g. problem analysis, team formation, task assignment, planning and scheduling, task execution, etc.) are not separate, but tangled together as a whole forming an integrated multi-agent coordination problem. A couple of applications sharing this same research root will be introduced later in this chapter.

This article first focuses on the representation of dependencies using EHTNs (Chen, 2006; Chen & Decker 2006) and the restrictions we impose on them in practice to keep the planning problem decidable. This extension induces a second problem, one of resource scheduling. Previously, planning and scheduling were often studied separately in the field of artificial intelligence from multi-agent coordination, e.g., papers may analyze the use of certain market mechanisms to coordinate the use of certain resources. Such mechanisms have particular, sometimes guaranteed, performance characteristics. But most high level mechanism design work ignores viewpoints of individual agents wondering which market to participate in (planning) or how to set pricing expectations (related to scheduling of the agents' resources) or indeed whether to participate in this mechanism at all (unless simply forced to by the programmer) or how to react to dynamic situations (re-planning due to changed environmental variables, etc.). This article explains how planning, scheduling and execution are integrated to help agents improve their coordination.

Traditional HTNs (e.g., the effort by Erol, Hendler & Nau, 1994) enable goal-oriented programming, but are not expressive enough to represent worth-oriented goals, contingencies, or the uncertainties that arise when such plans are in fact distributed over multiple agents. Similar as briefly described earlier, one prerequisite input of a task, named TaskA, of Agent A is the result value of a task, named TaskB, of another agent, Agent B; in the absence of any additional information it is unlikely that TaskA in Agent A could be listed as a valid candidate plan step at all. Several approaches have tried to address parts of this problem. Collaborative planning (Grosz & Kraus, 1996) is proposed to coordinate multiple agents' activities by exploiting a revised and expanded version of SharedPlans to handle cases in which a single agent has only partial knowledge; but the task representation is still not expressive enough to analyze arbitrary coordination mechanisms. The modular structured approach (Nair, Pynadath & Yokoo, etc. 2003) uses schematic Bayesian network fragments and programming language constructs to represent actions, domain relations and plan space; although it provides some quantitative representation and time factors for the task structures, the characteristics of the tasks still remain vague. It has been stated that by annotating plan choices with conditions, a sophisticated coordination process may become more efficient (Weiss, 1999). The modern approach to general reasoning about worth-oriented goals, contingencies, and uncertainties is the POMDP approach (Boutilier, Dean & Hanks, 1999); but even Boutilier states that in practice the standard AI representations and algorithms survive because real "problems commonly possess structure... [therefore] specialized representations, and algorithms employing these representations, can achieve computational leverage by exploiting these various forms of structure." In another approach (Li, Durfee & Shin, 2003), execution resource constraints are defined to remove uncertainties about the plans of other agents and communicated among agents based on a convergence protocol; however, since the architecture is based on STRIPS, hierarchical structure information is lost (such structure is useful because often coordination only requires commitments at an intermediate level of abstraction).

In this chapter we present a brief introduction to formalized EHTNs to represent agent task features by annotating them like TÆMS (Task Analysis, Environment Modeling, and Simulation) (Decker & Li, 2000), an abstract modeling representation that represents interdependencies quantitatively, as shown in Section 2. Given EHTNs for representing tasks and environments, the remaining question is how to reason about the inter-dependencies between tasks. We show that these EHTNs generate a resource scheduling problem and recast our GPGP coordination mechanisms (Chen & Decker, 2006; Chen, Durfee & Dumas, 2009) within this formalism in Section 3. We introduce the schedule coordination problem and a corresponding solution in section 4 and 5. Implementation of previous research is sketched in Section 6. Finally, discussions will be presented for the implementation of the integrated coordination problem covering all reasoning and execution procedures in two selected domain applications: a performance-based solution for evaluating (team-based collaborative) planning approaches and the challenges posted by human involvement.

2. EXTENDING HIERARCHICAL TASK NETWORKS (EHTNS) TO REPRESENT COORDINATION PROBLEMS

Former HTN representations are goal-oriented, compared to EHTNs that are worth-oriented (Zlotkin & Rosenschein, 1996). This enables advanced reasoning capabilities, e.g. team performance assessment, real-time situational awareness, training and simulation based on projected plan results.

Following the definition of coordination (Malone & Crowston, 1994), interdependencies among multiple agents introduce the problems of contingency, uncertainty and worth-oriented goal tradeoffs. We concentrate on the agent behaviors of plan/subplan selection and action scheduling. We assume that if there is perfect information, the involved agents would behave in a cooperative way. A local scheduler cannot make appropriate schedules with unknown information about an agent's future tasks. Such information includes how the execution of some action or subplan at a particular time will affect the agent's performance, and what other non-local actions or subplans will change these effects. Thus a more expressive representation for agent task features is needed than classic precondition and effect. We use TÆMS (Decker & Li, 2000) as the source of this additional expressiveness, as it provides a way of precisely representing such local and non-local effects on agent performance.

Classical planning was oriented toward a notion of a plan as a (partially ordered) sequence of primitive actions. While developing the RETSINA HTN representation, Williamson, Decker & Sycara (1996) reviewed the more recent planning work and the push towards supporting a more sophisticated model of control flow, e.g., parallel execution, conditional branching, contingencies based on information gathering actions, etc. Thus we can think of classical plans with explicit control flow as a specialized case of contingency-filled plans based on information flows (and these, then, begin to look like a useful, tractable simplification of POMDP policies, as observed by Boutilier, Dean & Hanks (1999)). Similar to the way the explicit control flow in a classical plan is dictated by the plan step preconditions, sets of provisions attached to an action indicate when it is enabled for execution. Another important feature of the RETSINA HTN representation was that (unlike the effort by Erol, Hendler & Nau (1994)) a reduced task is not replaced with its subtasks, but instead the task network is given a tree-like structure. In the multi-agent context, this is important both for providing a way to send partial, abstract plans to other agents that omit some details, and to allow the specification of explicit task relationships such as enablement in complex cases.

There are two main features that TÆMS has that RETSINA HTNs do not have. First, a quantitative definition of a vector of measurable, utility-influencing characteristics, such as quality, cost, and task duration, and how these characteristics accumulate as actions are executed, called characteristic-accumulation-functions (CAFs). Secondly, explicit task relationships indicating how task progress affects primitive action characteristics elsewhere in the task structure. These extra features both move the planning problem from a goal-achievement view to a worth-oriented view, and more importantly provide a basic unit over which all coordination reasoning, no matter how complex, is ultimately concerned with (the task relationship).

We base our formal definition of the coordination problem on the work of Erol, Hendler & Nau (1994), and focus primarily on the differences. Note that we are only concentrating on an agent's (partially) local view—primitive or compound tasks at other agents are represented locally as non-local tasks (NLTs). The vocabulary L is a tuple <*V, C, P, F, T, N, NLT, I, O, A, CAF*>, where *V, C, P, F, T, N* have the same meanings as defined by Erol, Hendler & Nau (1994): *V* is an infinite set of variable symbols, *C* a finite set of constant symbols, *P* a finite set of predicate symbols, *F* a finite set of primitive task symbols, *T* is a finite set of compound task symbols, and *N* is an infinite set of symbols used for labeling tasks. *NLT* is a finite set of non-local task symbols, *I* is a finite set of input provision symbols, *O* a finite set of outcome symbols including "OK" and "FAIL", and A is a finite vector of domain-dependent task/action characteristics: $<A_1, A_2, ..., A_n>$, where A_i is the ith attribute of a task/action (e.g. quality, duration, cost, completeness, reputation, etc., but also in some work characteristics like duration-mean, duration-standard-deviation, etc). An instance of attribute A_i is shown as a_i (For example, the duration of a searching task at some certain search engine is 1 minute and the cost is $1.00.). *CAF* is a set of characteristic accumulation functions; a *CAF* specifies how the value of an attribute A_i of a parent task is derived from the cumulative values of its direct child tasks. Typically functions are max, min, sum, q_sum, q_sum_all as presented in certain relevant approaches (Graham & Decker, 2000; Lesser, Decker & Wagner, 2004). The cost of a task is the sum of the costs of its children; the duration of a task is the max child duration in parallel execution and sum in serial execution; tasks that only need one child completed ("OR" tasks) have max as quality-CAF, etc.

We modify Erol's representation to include explicitly named input provisions and outcomes (representing possible contingencies). A *Primitive Task* (informally, an action), is a syntactic construct of the form do[f(π_1, ... π_k) → (ω_1, . . . ω_l),A], where f ∈ F is a primitive task symbol, the π_i are unique provision symbols in *I*, and ω_i are unique outcome symbols in *O*. The function A: *A* → R, is a mapping from an attribute to the default value for that attribute. In some work, A is specified at the operator level (different for each outcome/contingency) (Wagner, Garvey & Lesser, 1997). A *Non-Local* Task, is a syntactic construct of the form NLTdo[nlt(π_1, ..., π_k) → (ω_1, . . . ω_l), A], where nlt ∈ *NLT* is a compound task symbol, and the other symbols as before, except that A may be partial or missing—this is one of the prime uncertainties associated with non-local tasks.

A Goal Task is a syntactic construct of the form achieve[l, φ], where l is a literal and φ is a Boolean formula constructed from characteristic vector constraints such as $a_i < c$. An instance of this Boolean formula could be ($a_{deadline}$ < 1:00PM) ^ (a_{cost} ≤ $50). A *Compound Task*, is a syntactic construct of the form perform[t(π_1, ..., π_k) → (ω_1, . . ., ω_l), *δ*], where t ∈ *T* is a compound task symbol, π and ω are unique provision and outcome symbols. The *δ*: A → *CAF* maps attributes to the CAFs that determine the value of the attribute from the attributes of the child tasks of the compound task.

Task networks are modified to explicitly include links from outcomes to provisions, and unlike some HTN formalisms the hierarchy from compound tasks to the associated task network is kept intact. This is extremely useful in multi-agent situations for creating abstractions of a task, or for making commitments to achieve a goal or perform a compound task without committing to a particular method. A *Task Network*, *T*, is a syntactic construct of the form [(n: α), (n_1: α_1)... (n_m: α_m), φ, L_{prov}, $L_{inherit}$, $L_{disinherit}$], where each task in the net has a unique label n and α is root task of *T* . The φ is a Boolean formula containing variable binding constraints, temporal ordering constraints, and truth constraints (such as the "protection intervals" that appear in POP-style planners). The *L* is a set of explicit links between provisions and outcomes as in Figure 1. (n_i, ω_i, n_j, π_j) ∈ L_{prov} links outcomes to input provisions of network siblings. (n, π_k, n_i, π_j) ∈ $L_{inherit}$ links the provisions of the parent non-primitive task to the provisions of its subtasks. (n_j, ω_j, n, π_k) ∈$L_{disinherit}$

Figure 1. Three kinds of links connecting provisions and outcomes of three tasks

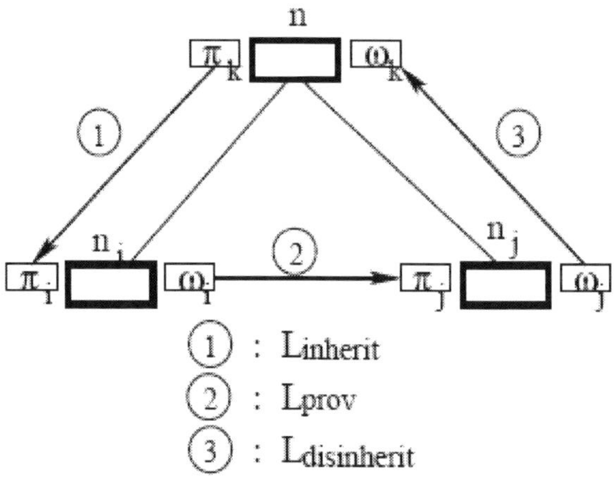

links the outcomes of the subtasks to the outcomes of the parent non-primitive task.

Erol's operator definition is modified to allow specification of each contingency separately. An *operator* is of the form *operator*(f(v_1, . . ., v_k), ω_i, l_1, . . . l_m) where the literals denote the primitive task's effects under the given outcome, or *perform*(nlt(v_1, . . ., v_k), ω_i, l_1, . . . l_m) where the literals denote the non-local task's effects under the given outcome. A *method* (α, d) still maps from a goal or compound task α to a task network *d*, representing one possible task reduction. A planning domain remains as a pair *D* = <Op, Me> where *Op* is a set of operators, and *Me* a set of methods, and a planning problem is still a triple *P* = <*d*, S_i, *D*>, where *D* is a planning domain, S_i is an initial state, and *d* is the task network we need to plan for (usually a goal task). Finally we slightly modify the definition of a primitive task network; it will contain only primitive tasks, non-local tasks, or non-primitive tasks that have been reduced to primitive or non-local tasks. All provisions must be linked to, and all outcomes have links from them. Based on the definition of *expressiveness* by Baadar (1996) and the similar ideas by Erol, Hendler & Nau (1994), we present the *expressiveness theorem* as follows:

Theorem 1. *L is strictly more expressive than the traditional HTNs.*

The expressive power of the extended hierarchical task networks (EHTNs) comes from the additional representation of information flow and control flow: the explicit definition of non-local tasks, input/outcome provisions and the computing ability for task characteristics. Without these new language constructs, *L* reduces to the equal expressiveness of traditional HTNs. The straightforward proof is omitted here.

3. GPGP MECHANISMS AND THE SCHEDULE COORDINATION PROBLEM

With the extended HTN representation, one can describe the management of the interdependencies among agents' activities using a set of GPGP coordination mechanisms. The key problem is that there exists uncertainty in the attribute vector values A of non-local tasks, which can lead to bad schedules, low efficiency, and resource misuse. We state a base assumption first: *Each agent is capable of reasoning locally about its schedule of activities and possible alternatives*. This assumption guarantees an appropriate local scheduling without uncertainty within task executions. A *schedule* S is a list of primitive tasks along with an earliest start time (EST) and latest finish time for each, indicating a specific execution path from a primitive task network T.

The scheduling problem is to select, from a set of possible execution paths, the "best" execution path (Graham & Decker, 2000). Two kinds of information are needed: (1) an estimate of the cumulative performance of the paths, and (2) a utility function. Recall that a vector of Characteristic Accumulation Functions (CAFs), δ, is defined to estimate characteristics of compound tasks, given scheduling information about primitive tasks. Recall that any task network *T* has a single named root, and that the attributes associated with that root node are defined purely recursively in the attributes of its subtasks. Thus for a primitive task network *T* that includes no non-local tasks we can simply compute the attributes associated with T given a schedule *S* that says what primitive tasks are executed and when. A *utility function*, *U*, is the reference from the user about the desirability of some set of attributes on a final plan and schedule, i.e. $U(S, T) \rightarrow R$. Thus, a local scheduler is defined as a function: $Schedule(T, U) = S$, that attempts to maximize $U(S, T)$. Solutions to this problem include the following: Garvey, Humphrey & Lesser (1993), Graham & Decker (2000), and Wagner, Garvey & Lesser (1997).

Besides the local scheduling problem there is a *schedule coordination problem* to provide information to the local scheduler, removing uncertainty and thus allowing construction of better schedules. The general idea is to find a way to fill the uncertain information in the uncoordinated task structures. If $A_i(T, S, n)$ is the value of attribute *i* in task network *T* and schedule *S* of named task *n*, then how to determine these values in the case of plans that contain non-local tasks? An example is like this: a subtask, SubTask1, in task network, T_A, of $Agent_A$ is *enabled* by a provision from a remote task in another agent and SubTask1 is uncertain about the duration of the *enabling* task. Our solution is that a specific GPGP coordination mechanism is applied and a result sent from the remote agent says that the enabling task will take 10 seconds to finish; a local scheduler for $Agent_A$ may now calculate the cumulative duration of SubTask1 based on the answer from the remote agent and the other local branches with known estimated durations.

Figure 2 shows the relationship and the information flow among the planner, the scheduler, and the coordination module in our agents' internal structure: The planner provides uncoordinated plans (with uncertainty) to the coordination module; the coordination module takes the uncoordinated plans as input, applies one or a combination of appropriate coordination mechanisms to the uncoordinated plans and produces coordinated plans that are the input to the scheduler; the scheduler uses the coordinated plans to make better schedules. The arrow from the scheduler to the coordination module indicates the latter takes advantage of the local scheduler's ability to evaluate/estimate the features of actions of the remote agents by asking "What-If" questions.

Generally speaking, there are three ways to coordinate multi-agent schedules: arrange the problem so as to avoid the need for coordination (by avoiding non-local tasks (NLTs), or using NLTs with static, known attributes), construct a special agent that functions as a centralized coordinator, or associate coordination mechanisms with each individual agent. All of these solutions have good and bad points and many possible realizations in different environments. GPGP (Generalized Partial Global Planning) (Chen & Decker, 2006; Decker & Li, 2000) is a domain-independent, extendible approach meant to encompass all of these alternative coordination mechanisms. GPGP views inter-agent relationships like NLTs as possible coordination points and then instantiates general coordination mechanisms at those

Figure 2. Relationships among the planner, the scheduler, and the coordination module

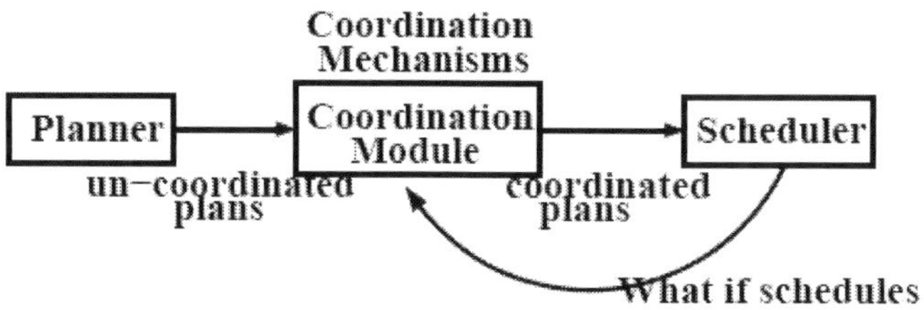

points. A GPGP mechanism is composed of two parts: (1) a set of protocols (represented by task structures) specific to the mechanism, and (2) a pattern-directed re-writing of the extended HTNs by analyzing the task structure, especially the NLTs, to acquire scheduling information.

The actual schedule coordination process has two steps: (1) locate the coordination points in the task structure, and (2) apply the coordination mechanisms at those points. The detection step is straightforward, except that the coordination point is the parent compound task T_i of a NLT (to enable structural rewriting during the application of a mechanism). The application of GPGP is formalized as $GPGP_{Apply}(T, T_i, M, role(T_i))$, where T is the task structure to be coordinated; T_i, found by $GPGP_{Detect}(T)$, is the compound task that contains NLT; $M = \{M_i \mid i = 1, 2, 3...\}$ is a set of GPGP coordination mechanisms; role determines the role of the agent (*enabler* or *enablee*). It was shown that $GPGP_{Apply}$ for a large set of mechanisms is deterministic through carefully defined processes (Chen & Decker, 2006). We present the formal representation of the application of one complex coordination mechanism in section 5.

> **Theorem 2.** *The GPGP processes for those coordination problems that can be represented with the extended HTNs are deterministic.*

The proof of this theorem is detailed in the Ph.D. dissertation by Wei Chen (2006) and we only present the brief idea here. A GPGP coordination mechanism is composed of two parts: the rewriting of the task structures and the coordination communication protocol. Based on the definition and explanation of these mechanisms in Section 5, it is easy to conclude that the coordination process is deterministic, i.e., once a coordination mechanism is selected and the coordination participators are ready, the coordination process is predetermined, and of course, not random. The detection of the coordination points has the same time complexity as traversing a tree - the hierarchical task network. The actual re-writing of the task structure is bounded based on which mechanism is chosen. The coordination communication is the only process that may introduce nondeterministic results because of uncontrollable factors, e.g., the reliability of the communication channel. The coordination protocols are implemented with time-outs, which mean that a "TIME-OUT" message will be collected in order to continue the coordination process, if a certain amount of time has passed after a coordination message is sent out. With this time-out mechanism, the largest communication time per message is also a fixed time period. Based on the above description, it is inductive the GPGP processes for those coordination problems that can be represented with our extended HTNs are deterministic.

4. COORDINATION MECHANISM SELECTION

There is an important observation that different mechanisms have correspondingly different performance characteristics, and that these can change dramatically in different environments (i.e., no one mechanism is best for all domains). Thus, it is a key for the agents to autonomously select the best coordination mechanisms based on some rules/strategies. If we did this, each agent would be able to learn over time to choose the best coordination mechanism in various situations according to its knowledge of its own capabilities, its belief about the other agents, and the dynamic environmental features. Since this process is also based on the planning and scheduling processes described in the previous two sections, we call this, the autonomous selection of appropriate coordination mechanisms under various environments, the coordination strategy problem, which is presented as Equation 1,

$$\textit{Given } k \textit{ nlts} \in NLT \textit{ in } E \ \& \ M, \textit{find } m_i \in M \textit{ for each } nlt_i, \textit{ that maximizes } EV(E, m_1, \ldots, m_k). \quad (1)$$

where E is a problem environment (including not just the formal planning problem P but also other characteristics explained later); the k nlts are the related coordination points in the current solution to that problem; M is our set of seventeen GPGP coordination mechanisms; m_i is the selected i^{th} mechanism in M to be applied at a particular coordination point, nlt_i; EV is a *performance EValuation function* under environment instance E and the coordination mechanisms $m_1, \ldots, m_k$ selected in response to the k nlts. The evaluation function provides feedback about whether a certain mechanism selection at a coordination point is good or not in various environment settings; $m_1, \ldots, m_k$ are also parameters of EV because the selection of mechanisms at different coordination points may be related.

Before we propose a general solution to the coordination strategy problem, a new term *coordination feature vector* needs to be defined first: a coordination feature vector consists of *task attributes* and *system attributes*. Task attributes were defined early in Section 2 as the task characteristic vector A: Quality, Duration, Cost, Reputation, etc. In contrast, system attributes represent characteristics of the application environment. There are two kinds of system attributes: *environment-specific attributes* and *situation-specific attributes*. Environment-specific attributes represent the physical backbones of agent coordination systems: the number of agents in the domain, whether communication is available or not, agent failure rate, communication bandwidth, etc. Situation-specific attributes represent the abstract features about the coordination situations: coordination task repeat rate, local load (how busy the agent is locally), interleaved NLTs, etc. The general steps to analyze the relationship between the GPGP coordination mechanisms and different environments are as follows (Chen & Decker, 2006):

1. Vectorize the coordination universe by filling in the information for the coordination feature vector;
2. Normalize the vector so that the vector elements can be formatted as input for learning algorithms;
3. Apply coordination mechanisms under various combinations of environmental values in the coordination framework to produce training data;
4. Apply learning over the training data to get the mappings, or rules, between environmental factors and the coordination mechanisms.

This general process, by solving the coordination strategy problem, proposes a practical method in turn to tackle the schedule coordination problem:

Given: (T, M)

T, an uncoordinated task network; M, the set of GPGP coordination mechanisms;

1. Apply the function $GPGP_{Detect}$ to T to get a set of *k* coordination points, CP = $\{nlt_i \mid i = 1, \ldots, k\}$;
2. While (CP != φ)

Figure 3. Re-writing of the task structure - demotion shift dependency

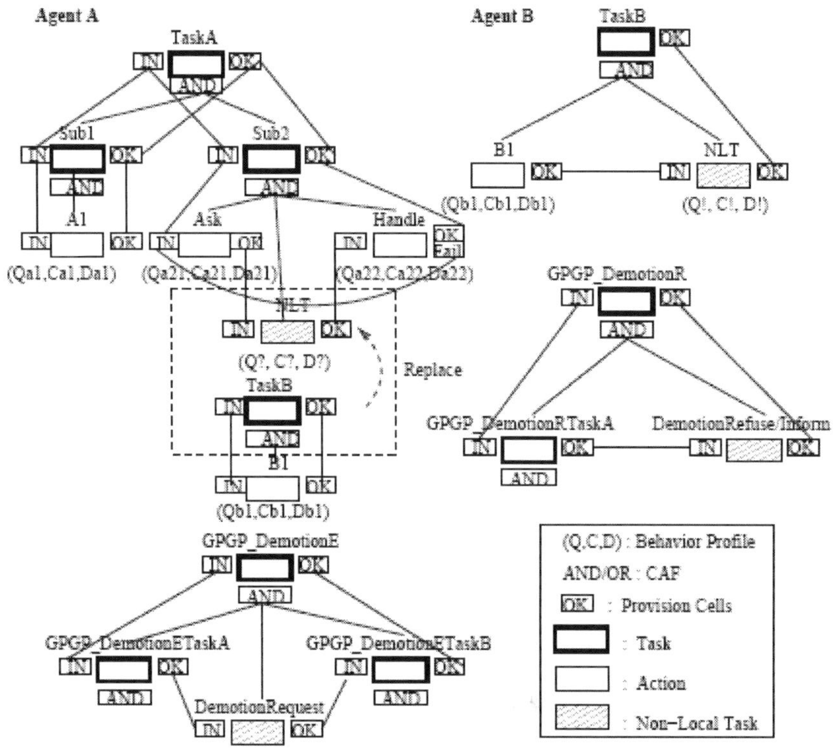

- Select nlt_i; CP ← CP − $\{nlt_i\}$;
- Select $m_j \in M$ according to Equation 1;

T ← $GPGP_{Apply}$ (T, nlt_i, M, role(nlt_i))

3. Generate better schedules (removing uncertainty): Schedule(T, U) = S;
4. Return S.

5. EXTENDED SET OF COORDINATION MECHANISMS

We introduced the GPGP coordination mechanisms in Section 3. Each GPGP coordination mechanism consists of a coordination protocol and a pattern-directed re-writing of the extended HTNs. Because of space constraints, we simply list these mechanisms and demonstrate the application of one particular mechanism— demotion shift dependency.

Seventeen coordination mechanisms have been catalogued for the *enable* relationship. They are: avoidance (with or without some sacrifice of quality), reservation schemes, simple predecessor side commitments (to do a task sometime, to do it by a deadline, to an earliest-start-time, to notify or send result directly when complete), simple successor-side commitments (to do a task with or without a specific Earliest Start Time), polling approaches (busy querying, timetabling, or constant headway), shifting task dependencies by learning or mobile code (promotion or demotion), various third-party mechanisms, or more complex multi-stage negotiation strategies. The inputs to these mechanisms are agents' task structures with structural information and annotations on the tasks. The result of the application of a mechanism may be either a structural alteration (i.e. removing or adding tasks) or an update of the annotations on the task structure, or both. Considering the

Figure 4. Coordination communication protocol - Demotion shift

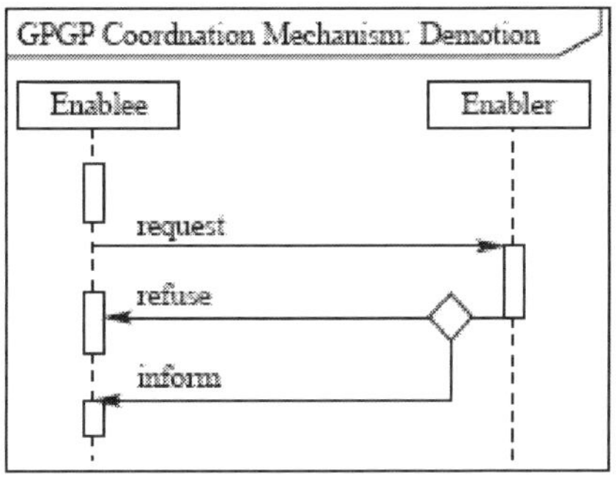

scheduling problem that a local agent may have no knowledge about the characteristics of non-local tasks, our approach removes this uncertainty and allows the local agent to make better scheduling decisions. Next we demonstrate the application of a particular mechanism—*Demotion Shift Dependency*; exploration of other mechanisms are in a paper by Chen (2006).

Under the assumption that AgentA's task, Sub2, is dependent on the execution of AgentB's TaskB, an example structure showing this interdependency is as in Figure 3; the coordination point for this dependency is the non-local task, NLT, within the task branch Sub2 for AgentA. TaskA and TaskB are the two agents' domain tasks; $GPGP_{DemotionE}$ and $GPGP_{DemotionR}$ are the coordination tasks initiated by the *Demotion Shift Dependency* mechanism; TaskB within the dashed box represents a modified mobile code sent from AgentB to AgentA. As shown in Figure 3, we name AgentB as Predecessor and AgentA as Successor. Demotion Shift gets the name because the task structure is transferred from the predecessor to the successor. Demotion shift dependency works as follows: AgentA detects the dependency relationship with AgentB's task, TaskB; AgentA sends a coordination request message to AgentB indicating task demotion; AgentB receives the message and sends back the task information and the modified object code (removing NLT) of TaskB; AgentA gets the returning message and dispatches it to its GPGP Module again; the GPGP module in AgentA unwraps the message and executes the object code to get the result and at the same time modifies the task structure so that the result is directed to the next execution task. The inserted task structures are shown in Figure 3 and the coordination protocol for Demotion is shown in Figure 4. The application of demotion shift dependency mechanism for the enablee agent, Agent A, is formally represented with EHTNs as follows:

Given $P = \langle d, I, Di \rangle$; suppose $d = [(n: \alpha), \ldots, (n_m: \alpha_m), (n_{nlt}: \alpha_{nlt}), \varphi, L_{prov}, L_{inherit}, L_{disinherit}]$, then

$GPGP(d, nlt, M_{demote}, r_{enablee}) = d'$, where

$d' = [\{(n_{T'}: T')\}, \varphi, L'_{prov} L'_{inherit}, L'_{disinherit}, \delta_A(T')]$,

$T' = T - \{NLT_{Sub2}\} + T'_{TaskB}$,

$L'_{prov} = L_{prov} - \{(n_{Ask}, OK, n_{NLT\text{-}Sub2}, IN)\} - \{(n_{NLT\text{-}Sub2}, OK, n_{Handle}, IN)\} + \{(n_{Ask}, OK, n_{TaskB}, IN)\} + \{(n_{TaskB}, OK, n_{Handle}, IN)\}$,

$L'_{inherit} = L_{inherit} + \{(n_{TaskB}, IN, n_{B1}, IN)\}$,

$L'_{disinherit} = L_{disinherit} + \{(n_{B1}, OK, n_{TaskB}, OK)\}$,

Figure 5. An EMS system information flow chart

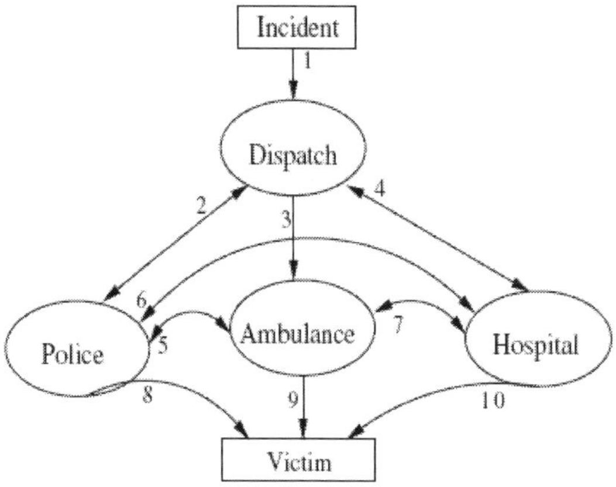

where φ is unchanged, and $T'_{TaskB} = T_{TaskB} - \{NLT_{TaskB}\}$ is the mobile code transferred from Agent B to Agent A to remove the NLT. For the enabler agent, Agent B, the coordination is to transfer modified mobile code of TaskB to Agent A as shown in Figure 3. Demotion Shift Dependency performs efficiently when the same tasks are requested for execution repeatedly. For example, in information gathering applications, very complex wrapper agent code can be demoted to the requester agent at the user side, especially when the wrapper is heavily loaded. *Promotion Shift Dependency*, where the object code transferred in the opposite direction, is also applicable in information gathering applications, for example, to upload tasks from a handheld agent or thin client to a large server.

6. IMPLEMENTATION

As introduced above in this article and as shown in the previous work (Chen & Decker, 2004; Chen & Decker, 2006), a novel set of seventeen GPGP coordination mechanisms have been implemented and applied following the general procedure in earlier sections to reveal the potential relationships between these mechanisms and the environmental characteristics. Experimentation has been carried out in a simulated Emergency Medical Service (EMS) system. The application of the extended set of coordination mechanisms in EMS scenarios will be introduced briefly as follows.

In practice an EMS response process includes patient flow, equipment flow and information flow as shown in Figure 5. A typical EMS process is as follows: A 911 call, activated by an incident, comes in and the 911 dispatch screens the call; the 911 dispatch contacts an appropriate response agency or multiple agencies based on the type of the incident; the dispatch of each agency broadcasts the incident to the response vehicles and sends appropriate vehicles to the incident site; the response vehicle finds its way to the incident site; proper treatment is provided and the incident is under control; if there is any victim who needs further medical care, the victim will be transported to a proper medical facility by an ambulance; if the incident changes so as to require additional care, other appropriate response vehicles will be dispatched to the site; all the response vehicles then find their ways back to their base stations or continue to respond to other emergency calls.

In EMS systems, events happen rapidly and unexpectedly; resources move accordingly in the same fashion; decision making is highly time-critical based on involved response organizations, institutions, and even geographical sites; doctors,

Figure 6. EMS coordination point one: between a dispatcher and a police agent

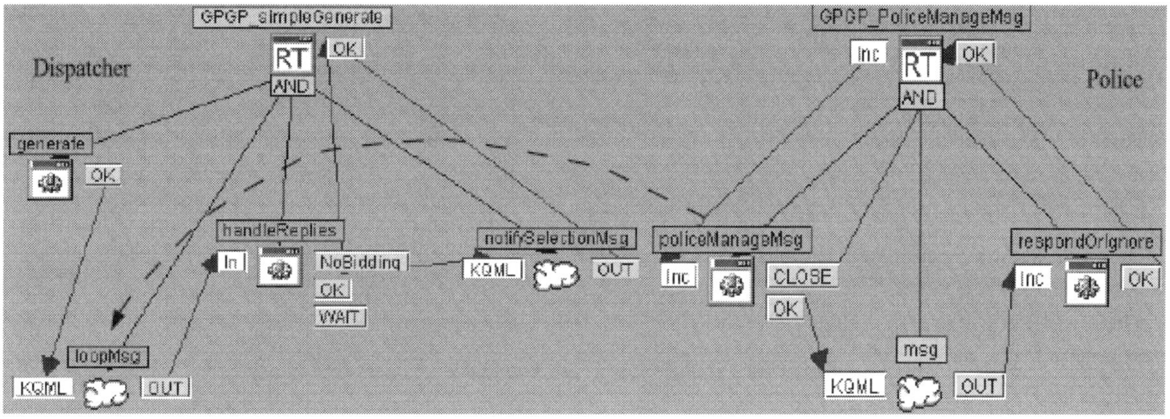

nurses, and paramedics are constantly on the move. Such high-velocity environments pose special computing and communication needs. Satisfying these needs can make a difference between efficient versus cumbersome processes, fully independent life versus permanent disability, and even life versus death. Information flow among the related entities needs to be highly effective; however, communication channels could be radically impaired based on certain environmental changes, such as traffic jams and power failures. Thus, effective coordination is a key to EMS systems.

Various interdependencies exist between the tasks of response agents, i.e., some tasks of one response agent depend on the completion of other tasks from a different response agent. We define three *coordination points* within a single emergency management procedure.

EMS *coordination point one* is as shown in Figure 6. When an incident happens, the dispatcher needs to deliver the incident message to certain response agents. Response agents will send a message back to the dispatcher indicating whether they are available to manage this incident. Based on response agents' replies, the dispatcher decides which agent should be selected in response to this incident. That is, the dispatcher cannot decide which agent to dispatch for an incident until reply messages arrive. Furthermore, the task of a response agent can not start until the notification of the selection of certain agents. Clearly the dependency is between the message-sending task of the dispatcher and the task of a response agent. Similarly, we define *coordination point two* as the interdependency between the tasks of a police agent and an ambulance, and *coordination point three* as the interdependency between the tasks of an ambulance and a hospital. Notably, the 'cloud' node represents a non-local effect, i.e., an estimate of a remote agent's task schedule realized by information exchange following certain coordination communication protocols. We omit the task illustration for coordination point two and three – please see a related article (Lesser, Decker & Wagner, 2004) for details.

The seventeen GPGP coordination mechanisms can be suitably applied to corresponding coordination points as shown in Table 1. A "Y" cell in the table shows that the mechanism (to the left of the row) can be applied to the corresponding coordination point (indicated by the coordination point number labeling the column. For example, the first mechanism, avoidance, can be suitably applied to coordination point two, but not to point one and three.

The experiment was set up as follows. There are two kinds of agents in the framework: static

Table 1. Application of extended GPGP mechanisms to the three coordination points in emergency medical service (EMS) framework

Index	Mechanisms	Point 1	Point 2	Point 3
1	***Avoidance***		Y	
2	***Sacrifice Avoidance***		Y	
3	***Reservation***		Y	Y
4	***Predecessor EST Commitment***		Y	
5	***Predecessor Deadline Commitment***		Y	
6	***Predecessor Notice at Start***		Y	
7	***Predecessor Sending Result***		Y	
8	***Successor Deadline Commitment***	Y	Y	Y
9	***Successor EST Commitment***	Y	Y	Y
10	***Demotion***		Y	
11	***Promotion***			
12	***Third Party Execution***	Y	Y	Y
13	***Third Party Coordinator***	Y	Y	
14	***Polling for Result***	Y	Y	Y
15	***Polling for Schedule***	Y	Y	Y
16	***Constant Headway / Timetabling***			Y
17	***Bidding***	Y	Y	Y

agents (e.g. hospitals, ambulance stations, police stations, dispatchers) and mobile agents (e.g., police cars and ambulances). Static agents are able to communicate with other agents; but they can not move around the local area map. Mobile agents are capable of both communicating and moving. The actual selection of each kind of response agents involves: three police agents, two ambulance agents, two hospitals, and one dispatcher. These facts are randomly generated and follow an exponential distribution: next incident arrival time (average arrival time, the mean value, is twenty seconds), service time for response agents (average service time for a police agent or ambulance agent to work at an incident location is two seconds), hospital service schedule time (the average schedule cycle, the arrival of a next victim not from this EMS framework, occupying the open beds, is ten seconds; the service time is four seconds). These facts are randomly generated as well but follow a uniform distribution: incident type (0: police call; 1: ambulance call), incident severity (0: victim is in low severity state and needs CPR or other basic first aid; 1: victim in high severity state, needs CPR or other basic first aid, and needs to be transported to a nearby hospital for further medical treatment.), incident location (uniformly selected from the twenty-six map locations in the University of Delaware local area map).

There are three main evaluation factors for EMS at a system level: response time for quality control, survival rate for effectiveness control, and coordination cost for efficiency control. These system evaluation metrics are defined as follows.

- Response Time = Call Evaluation + Caller Interrogation + Unit Dispatch Time +Launch Time + Routing Time.
- Survival Rate = The number of surviving victims / The number of all victims.

Figure 7. Experimental results for EMS coordination point one

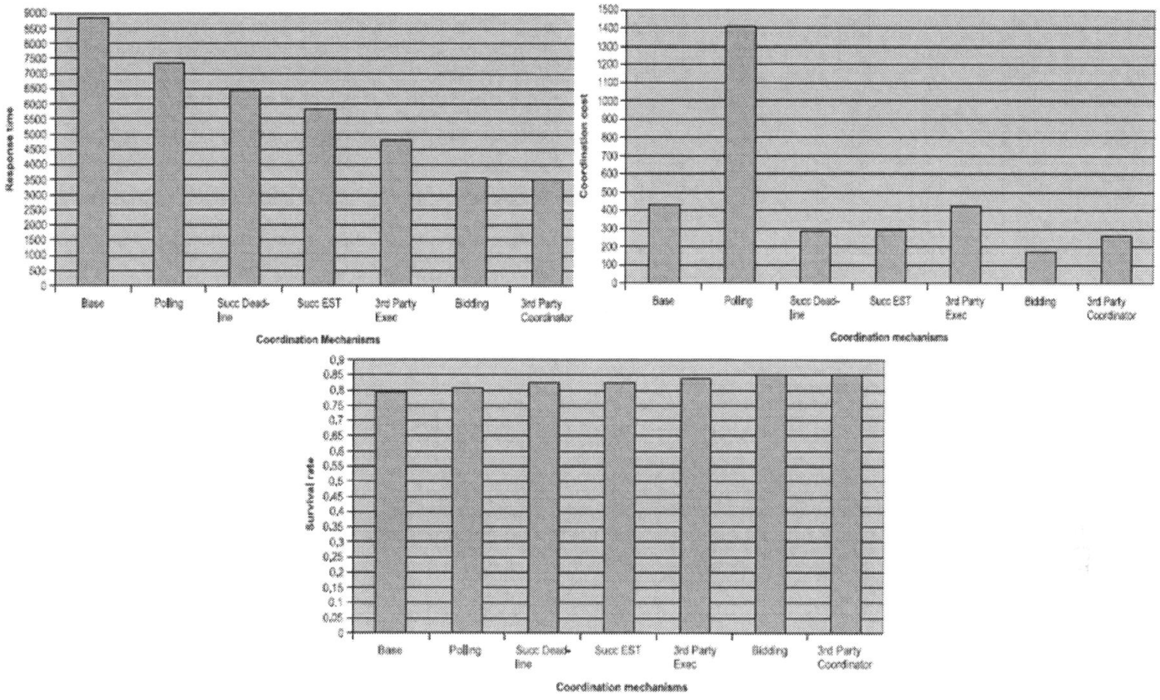

- Coordination Cost = Coordination Communication Time + Task Alteration Time.
- The experimental results for the three evaluation factors are shown in Figure 7. Notably, the "bidding" and "3rd party coordinator" mechanisms results in the shortest response time and the highest survival rate (due to the robust communication); the "polling" mechanism leads to the highest communication cost, but it is the most robust coordination mechanism thanks to the constant communication requests. Further experimentation results, e.g. the evaluation metrics in response to the other two coordination points, can be found in the Ph.D. dissertation by Chen (2006).
- This novel set of coordination mechanisms has been theoretically proven to be effective upon various environmental conditions in the EMS domain. Specifically, the applications of selected (by greedy strategy) coordination mechanisms to the three coordination points (i.e. interdependencies) within emergency medical service procedures improve the following three evaluation methods at the same time: reducing the average first response time to 38% of the base cases without coordination, increasing the average survival rate to 125% of the base cases, and reducing the total coordination cost (communication - message transfer amount cost and task modification cost) to 61% of the base cases (Chen, 2006; Chen & Decker, 2006).

General rules about applying which mechanisms in response to which particular environment have been presented, e.g. a couple of general rules are displayed below and more analytical results can be found in the Ph.D. dissertation by Chen (2006).

- **Rule 1** If (true), apply Avoidable and Sacrifice Avoidable mechanisms;
- **Rule 2** If (deadline is the user's main concern), apply Coordination by Reservation.

7. DISCUSSIONS

We did not delve into implementation details. Instead, we have discussed the key questions, e.g. what to coordinate among agents, how much information should be exchanged, and how to evaluate a planning approach, etc. The answers are summarized below. The task structures, the characteristics associated with task nodes, and the task achievement/progress status, represented using EHTNs, are the actual information to be coordinated following certain communication protocols; the amount of information to be exchanged among the agents depends on the relevance of that information to the target goals and the potentially different views held by those agents playing different roles; only those task structures (and status information) related to the designated dependent goals/sub-goals will be exchanged among involved agents. Performance based analysis could be used to evaluate whether such a planning approach is successful or not.

The Distributed Environment Centered Agent Framework, DECAF, is a suite of software that can be used to develop intelligent agents and multi-agent systems. A software agent developed by DECAF holds a particular architecture and is named as a DECAF agent. The internal structure of a DECAF agent with a GPGP (Coordination) Module is shown in Figure 8. These internal components work together to keep track of an agent's current status and to carry out its designated tasks for potential team goals: plan selection, scheduling, execution, information exchange, etc. The relationship among the GPGP_Module, the planner and the scheduler, as in Figure 8, is a concrete implementation of the model shown in Figure 2 and further explained by Chen and Decker (2006). Here we need to point out that the internal structure of a complex intelligent agent requires the integration of the underlying components as follows. A "plan file", representing general EHTN schemas, is created by a user (e.g., a system architect) in a pre-planning procedure; the "plan file" is then used by the planner component to generate a plan instance that may involve interdependencies with other agents; the interdependencies introduce uncertainties because one agent does not contain the state information of other agents; the GPGP module removes the uncertainty within the plans by communicating with related agents at a meta-level via coordination communication protocols and updating its task structures (e.g. task re-ordering, demotion shift, and task replacement); the subsequent coordinated plan is fed into the scheduler so that a better local schedule is produced given that the missing task characteristics about remote agents' schedules have been filled with appropriate estimate; the executor finally carries out the schedule by executing the designated tasks. We need to point out that the execution module will generate outgoing messages (to be sent to the agent itself) about execution results to affect the original plans and schedules — re-planning. As described above, all the reasoning components are inextricably linked to overall coordination behaviors; it is a promising approach to analyze how these reasoning components work together (involving other remote agents through interdependencies) to achieve mutual goals. The EHTN-based multi-agent coordination approach has been empirically proven to be effective in applications like Emergency Medical Services (EMS). We are also trying to apply the approach to other domain applications.

7.1 Look-Ahead Planning

We have stated earlier that the distributed nature of multi-agent systems cannot afford either centralized coordination or the distribution of constantly

Figure 8. DECAF (Distributed Environment Centered Agent Framework) agent architecture

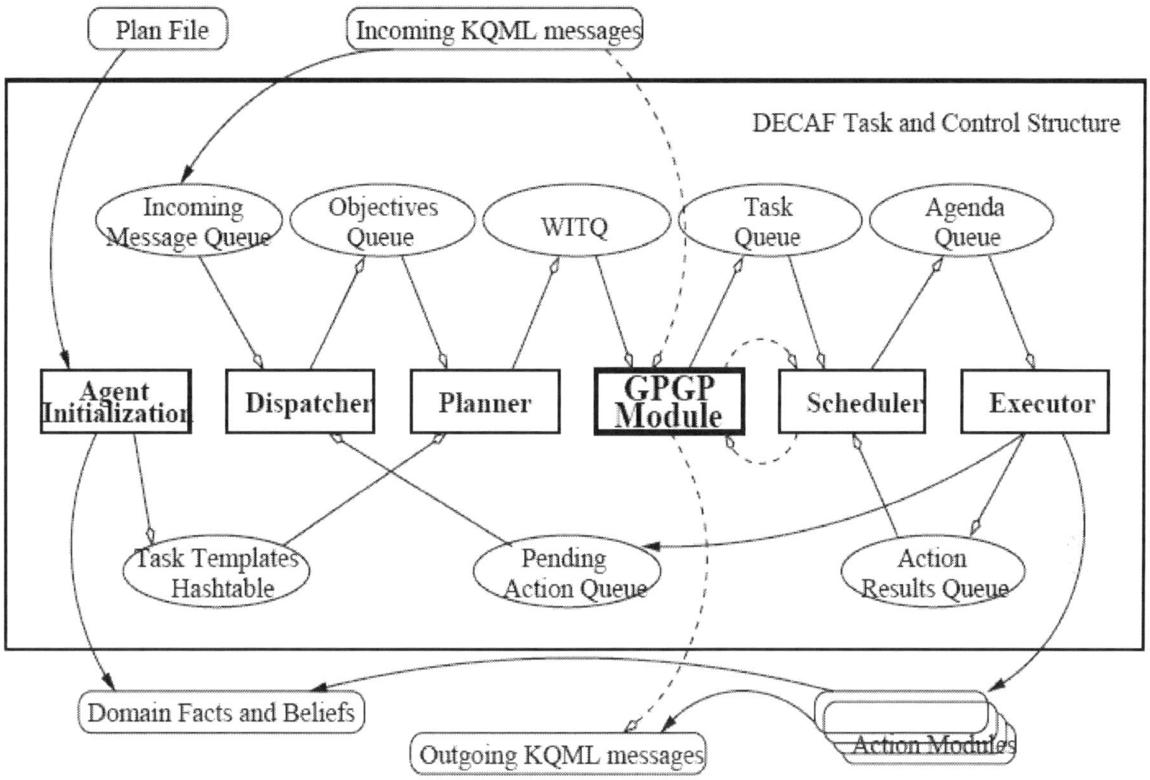

updated global view across all team participants. It is also difficult to coordinate agents' activities in the presence of uncertainty, contingencies, and worth-oriented goals. The base problem is that whatever plans and schedules generated at one time (e.g. the start of a goal task) may not reflect the updated situations at other times due to the aforementioned factors. Next a special planning strategy, look-ahead planning, will be discussed as an applicable solution to multi-agent coordination problems. Specifically, a performance analysis will be employed to assist look-ahead planning approach at both the short-term and long-term scales. [2]

Look ahead planning has been employed as one of the decision aids that constitute production control systems in the construction industry (Ballard, 1997). A master schedule is usually issued at the beginning of the construction phase (i.e., execution phase in agent research) specifying the entire project schedule ranging from the long term coordination to the terms of payments. However, such initial, total schedules cannot be accurately detailed too far into the future due to the lack of information about actual task durations and deliverables. Thus, short-term schedules are generated to coordinate and direct various trades and crews ahead several weeks into the future, called look-ahead schedules. The key assumption is that *thinking ahead accurately into the near future would be beneficial to the long-term goals.*

One feature of look ahead planning is its conceptual framework, which posits planning as a process of reducing uncertainty and maximizing throughput, counter-posing plan reliability to resource redundancy as alternative strategies for managing in the conditions of uncertain work flows (Ballard, 1997). The idea is to generate the best accuracy for short-term planning in order to achieve the overall long-term goals.

There are two kinds of scenarios concerning dependency: (1) there is no interdependency between the look-ahead plans and the future plans, and (2) interdependencies exist between the look-ahead plan and the later plans. In the former case, it is obvious that the achievements of look-ahead plans will not hinder the overall master plan for achieving the goal tasks. In the latter case, the coordination solution (applying appropriate coordination mechanisms to the resolve interdependencies) presented in early part of this paper could be employed to manage the interdependencies by removing the uncertainties, e.g., adjusting the intensity/priority of certain tasks to in the short run to achieve the ultimate production goals. This integrated coordination effort has been applied to a few ongoing projects as introduced below.

7.2 Domain Applications of the EHTN-Based Integrated Multi-Agent Coordination

In many research fields, problem solving can be simplified by an appropriate choice of knowledge representation. Representing the knowledge in one way may make the solution simple, while an unsuitable choice of representation may make the solution difficult and obscure. Hierarchical Task Networks (HTN) has been used for many practical planning systems. Earlier we discussed several aspects of traditional HTNs: that they are not expressive enough to represent worth-oriented goals, contingencies, or the uncertainties that arise when an agent's (or a robotic system's) plans are in fact distributed over multiple entities, e.g. a multi-human/robot team. Then, we introduced a more expressive representation, extended hierarchical task networks (EHTNs) and applied the EHTN-based approach to a collaborative domain. This approach is applicable to complicated distributed problem solving in collaborative environments and has the following features: unified representation of mission tasks for heterogeneous systems, the multi-agent based intelligence enabling extensive reasoning capabilities (e.g. multi-agent planning, scheduling, task assignment problems), reduced ad-hoc decision making to avoid errors upon dynamics and uncertainties (i.e., EHTNs regulate procedural processes with both manual and automatic task structure alteration that enables the simple and robust re-use of proven processes and technologies), scalability and collaboration (i.e., agents have different local views about themselves and provide information to cooperative peers when necessary). We will introduce two representative domain applications in this section: team performance and human agent collaboration. These two research areas share a similar root of knowledge representation using extended hierarchical task networks (EHTNs) with different advanced reasoning capabilities.

7.2.1 Team Performance

As mentioned earlier, look-ahead planning is a candidate solution to multi-agent coordination. It is intuitive that better collaboration leads to better performance and better performance reflects successful task planning. Whether dealing with team or individual team member's performance, the ability to accurately measure is a necessity — feedback, remediation, and ultimately learning, depend on it.

Cannon-Bowers and Salas (1998) advanced a set of requirements for performance measurement in team training systems, which have been further modified by other researchers. Their research sheds light on what and how to keep track of performance in order to evaluate planning approaches. First, performance measurement systems must consider multiple levels of measurements. In the simplest classification, performance can be measured at the individual and team level. Execution feedbacks for improving performance determine the level of analysis for measurement. Second, performance measurement systems should describe, evaluate, and diagnose performance. By addressing these issues, a more robust view of performance can be provided. The description of performance is conceptually simple, although in practice is often very difficult due to the complex dynamic nature of tasks and the conceptual frame in which they are described. The evaluation of performance is the comparison of observed behaviors to a model or standard and judgment concerning their relative values. Performance diagnosis is the determination of the underlying causes of both effective and ineffective performance. Performance diagnosis feeds into remediation, which is the next requirement. Third, performance measurement systems should provide a basis for remediation to support the task process. Measures should inform the agents with progress, provide results and direction for improving performance, and provide a basis for selecting future training. Performance measurement is vital to understanding and evaluating team behaviors; the degree to which performance measurement can meet this objective is capped by the ability of the assessment intelligence.

In summary, a set of suitable metrics for evaluating team performance could be: (1) the level of measurements (individual or team), (2) process content (how a task network is created as a goal task decomposition) and task outcome (task achievement and domain results), (3) the ability to describe, evaluate and diagnose performance, and (4) a remediation mechanism providing results and instructions to improve performance. EHTN-based approach can facilitate task content and outcome representation for team performance application, clarifies multiple levels of measurements (reflecting task progress measurement at different depths within a task network), display task progress information in a very intuitive way to the team performance (human) monitors, and provides subsequent mechanisms (e.g., alerting messages, response procedures) for remediation.

7.2.2 Human Agent Collaboration

As unmanned systems have been rushed into real world applications (e.g. anti-terrorist activities and fielded emergency response systems), it is necessary to incorporate humans and agents as peers for collaborated distributed problem solving (i.e., mutual goal achievement). However, human involvement is highly unpredictable compared with agents' computational capabilities (e.g., manmade utility based) for problem solving and decision making. A humans may reject a task assignment simply because "*I don't feel like to work today*", or "*I have a family emergency*", which is hard to model in computational systems. Intelligent software agents are more predictable—an agent cannot typically reject a task assignment if it is both capable (its capabilities covers the designated task) and available (there is no scheduling conflict between the designated task and its current schedule of activities). On the other hand, even though the availability of human expertise is less predictable, the actual expertise content (e.g., any domain specific expertise that requires humans' involvement) is generally more dependable, which is the reason why we want to reuse human expertise as much as possible. Based on the vision of human expertise re-use and retrieval in an on-demand fashion introduced by Wagner (2005), we believe the extended hierarchical task networks (EHTN) based integrated multi-agent coordination approach could be a viable solution, which is briefly introduced next.

The collaboration among *humans* and *artificial agents* is the kind of effort carried out by heterogeneous resources in pursuing mutual goals. The collaboration effort discussed in this article refers to the process for assisting decision-making at a meta-level (i.e., coordination level), instead of at a domain level. One sample scenario is the process of employing and managing a suitable team to solve a problem (e.g., the team formation of a group of language/cultural experts responding to an emergency) and the effort of enabling com-

Table 2.

Step 1	Initially the HAC system suggests widely diverse choices (both task plans and teams) to the user
Step 2	The user selects some suggestions of interest
Step 3	The user specifies further preferences/constraints
Step 4	The HAC system updates the suggestions accordingly and presents them to the user in the next round as variations on those narrower themes
Step 5	Go to Step 2 or exit upon convergence

munications (e.g., the international communication with a language barrier or the interactions among people with radically different cultural backgrounds). The process is not about the actual problem solving content (e.g., how a certain verbal expression in a foreign language should be understood correctly and suitably). Thus, our effort is general, not tied to any specific domain, and can be applied to potentially various scenarios. Particularly, the aforementioned human agent collaboration process addresses the following questions: How should a human collaborate with one or more software agents to solve problems that are beyond their individual capabilities? How do the human and agents converge on sufficient shared understanding of the problems that they are addressing, and on the ways in which each is contributing (and expecting the others to contribute) to solving the problems? How do they monitor individual and collective progress on solving problems, and re-define tasks or redirect efforts to improve collaborative performance? Etc.

Collaboration among *human activities* with possible agent assistance concentrates on the process of finding the suitable human expertise for the human users as an information requester. Human expertise needs to be retrieved upon requests in real time. A software agent may act as an assistant to facilitate the expertise retrieval with its computational power. Specifically, the collaboration among human activities addresses the following question: How can humans who are working on related tasks but in different contexts (physically separated, coming from different backgrounds, etc.) coordinate their activities so achieve their shared objectives efficiently and effectively? Etc.

The collaboration among humans and agents was named as *human agent collaboration* (HAC) and the collaboration among human activities was regarded as human to human activity collaboration (H2AC).

A problem specification about the human agent collaboration and our solution of employing suitable HAC process for solving the H2AC problems have been documented in an earlier approach (Chen, Durfee & Dumas, 2009). A general human agent collaboration process can be briefly illustrated in Table 2.

We believe that the human agent collaboration is effective in assisting human to human activities at the coordination meta-level and has shown that the EHTN-based, integrated multi-agent coordination technologies effectively support the notion of engaging human's biases, expertise and experiences and software agent's computational capabilities to achieve better decision-making.

8. CONCLUSION

The introduction of the EHTN-based integrated multi-agent coordination approach provides a general answer to the three reasons/questions about why coordination is necessary by Jennings (1996): (1) interdependencies are represented by the task structures—although we concentrate on ***enables*** in this article, other dependencies, e.g.

facilitates and ***hinders***, can also be represented, because the structural features of agents' tasks for various interdependencies are specified in the same way using EHTNs; (2) solution to resource constraints—using a new plan candidate (e.g. a sensing action) to discover the availability of the resources, or simply regarding the resource as a special task of an agent that enables related tasks by consuming (or producing) a certain amount of resources; (3) the partial view—the removal of uncertainty of agents' tasks provides different views from remote agents to local agents.

Most previous approaches of multi-agent coordination did not fully explore the low level (e.g., task structure feature-level), interleaved effects of planning, scheduling and execution on agent internal reasoning. This chapter projects an integrated multi-agent coordination problem and a general solution with an associated algorithm for other researchers' future explorations. Further discussions have been focused on key questions, e.g. what to coordinate (EHTN-based task structures and task achievement status), how much info to exchange among agents (only the information relevant to interdependencies), and how to evaluate a planning approach (using team performance objective metrics and adjustments with subjective metrics for human involvement). Finally we presented ongoing real world applications, to which our integrated multi-agent coordination approach has been applied. We believe the EHTN-based intelligence is effective and practical as a candidate solution to complicated real-world distributed collaborative problems.

Furthermore, we envision that a set of distributed collaborative problem solving applications can be addressed with this general human agent collaboration solution as long as they fall in the scope of the following problem: how to automatically form a team of suitable resources (human or artificial agent) from a loosely-coupled Internet of heterogeneous assets with adequate extensibility, generate a corresponding course of actions containing potential interdependencies for a given mission, monitor the mission progress, and adaptively update the course of actions based on resource, environmental and execution dynamics.

Still, this approach is restricted by the knowledge representation, i.e., applicable to those domains with clear, procedural, problem solving definitions. Also, this effort only provides a general solution for integrated coordination, but does not address how an action (the bottom level nodes in an EHTN structure) should be executed. For example, in the human agent collaboration domain, EHTN-based approach provides a solution to engage the most suitable team of human experts to manage a particular medical emergency. However, exactly how a specific action should be performed by a surgeon (the domain specific knowledge) is out of scope.

REFERENCES

Baader, F. (1996). A formal definition for the expressive power of terminological knowledge representation languages. *Journal of Logic and Computation*, *6*(1), 33–54. doi:10.1093/logcom/6.1.33

Ballard, G. (1997). Lookahead planning: the missing link in production control. In *proceedings 5th Annual Conference of the International Group for Lean Construction*, Griffith University, Gold Coast, Australia.

Boutilier, C., Dean, T., & Hanks, S. (1999). Decision-theoretic planning: Structural assumptions and computational leverage. *Journal of Artificial Intelligence Research*, *11*(1), 1–94.

Cannon-Bowers, J., & Salas, E. (1998). Team performance and training in complex environments: recent findings from applied research. *Journal of Current Directions in Psychological Science*, *7*, 83–87. doi:10.1111/1467-8721.ep10773005

Chen, W. (2006). *Designing an Extended Set of Coordination Mechanisms for Multi-Agent Systems*. Doctoral dissertation, Computer and Information Sciences, University of Delaware, Newark, DE.

Chen, W., & Decker, K. (2004). Managing multi-agent coordination, planning, and scheduling. In *Proceedings of the 3rd International Conference on Autonomous Agent and Multi-Agent Systems*, New York.

Chen, W., & Decker, K. (2006). Analyzing characteristics of task structures to develop GPGP coordination mechanisms. In *Proceedings of the 3rd International Conference on Autonomous Agent and Multi-Agent Systems* (pp. 662–669), Hakodate, Japan.

Chen, W., Durfee, E., & Dumas, M. (2009). Human Agent Collaboration in a Simulated Combat Medical Scenario. In *Proceedings of the International Symposium on Collaborative Technologies and Systems*, Baltimore, Maryland.

Decker, K., & Li, J. (2000). Coordinating mutually exclusive resources using GPGP. *Journal of Autonomous Agent and Multi-Agent Systems*, *3*(2), 133. doi:10.1023/A:1010074611407

Erol, K., Hendler, J., & Nau, D. (1994). HTN planning: Complexity and expressivity. In *Proceedings of the 12th National Conference on Artificial Intelligence* (pp. 1123-1128), Seattle, WA.

Garvey, A., Humphrey, M., & Lesser, V. (1993). Task interdependencies in design-to-time real-time scheduling. In *Proceedings of the 11th National Conference on Artificial Intelligence* (pp. 580–585), Washington, DC.

Graham, J., & Decker, K. (2000). Towards a distributed, environment-centered agent framework. In J. G. Carbonell & J. Siekmann (Ed.), Intelligent Agents VI, (LNAI-1757, pp. 290–304). Berlin: Springer Verlag.

Grosz, B., & Kraus, S. (1996). Collaborative plans for complex group action. *Journal of Artificial Intelligence*, *86*(2), 269–357. doi:10.1016/0004-3702(95)00103-4

Jennings, N. (1996). Coordination techniques for distributed artificial intelligence. In G. M. P. O'Hare & N. R. Jennings (Ed.), Foundations of Distributed Artificial Intelligence (pp. 187–210). Hoboken, NJ: Wiley.

Lesser, V., Decker, K., & Wagner, T. (2004). Evolution of the GPGP/TÆMS Domain-Independent Coordination Framework. *Journal of Autonomous Agents and Multi-Agent Systems*, *9*(1-2), 87–143. doi:10.1023/B:AGNT.0000019690.28073.04

Lesser, V., Wagner, T., & Raja, A. (2006). Modeling uncertainty and its implications to sophisticated control in TAEMS agents. *Journal of Autonomous Agent and Multi-Agent Systems*, *13*(3), 235–292. doi:10.1007/s10458-006-7669-2

Li, H., Durfee, E., & Shin, K. (2003). Multi-agent planning with internal execution resource constraints. In *Proceedings of the 2nd International Conference on Autonomous Agent and Multi-Agent Systems* (pp. 56-0567), Melbourne, Australia.

Malone, T., & Crowston, K. (1994). The interdisciplinary study of coordination. *Journal of ACM Computing Surveys*, *26*(1), 87–119. doi:10.1145/174666.174668

Nair, R., Pynadath, D., Yokoo, M., Tambe, M., & Marsella, S. (2003). Taming Decentralized POMDPs: Towards Efficient Policy Computation for Multiagent Settings. In *Proceedings of the 18th International Joint Conference on Artificial Intelligence* (pp. 705-711), Acapulco, Mexico.

Ngo, L., Haddawy, P., & Nguyen, H. (1998). A modular structured approach to conditional decision-theoretic planning. In Proceedings of Artificial Intelligence Planning Systems, (pp. 111–118), Menlo Park, CA.

Wagner, T. (2005). DARPA IPTO/IXO. We are not alone. In *Proceedings of the DARPATech* (pp. 75-80), Anaheim, CA.

Wagner, T., Garvey, A., & Lesser, V. (1997). Complex goal criteria and its application in design-to-criteria scheduling. In *Proceedings of the 4th National Conference on Artificial Intelligence* (pp. 294–301), Providence, Rhode Island.

Weiss, G. (1999). Multiagent Systems: A Modern Approach to DAI, (pp. 151–157). Cambridge, MA: MIT Press.

Williamson, M., Decker, K., & Sycara, K. (1996). Unified information and control flow in hierarchical task networks. In *Proceedings of AAAI-96 workshop on Theories of Planning, Action, and Control,* Portland, Oregon.

Zlotkin, G., & Rosenschein, J. (1996). Mechanisms for Automated Negotiation in State Oriented Domains. *Journal of Artificial Intelligence Research*, *5*, 163–238.

ENDNOTES

1 Notably, "avoidable", "sacrifice avoidable", and "coordination by reservation" are three of the seventeen GPGP coordination mechanisms developed in earlier work (Chen, 2006; Chen & Decker 2006).

2 The performance based planning assessment approach is applicable to self-interested agents' planning efforts given that both collaboration and self-interested coordination are based on objective performance analysis,

Chapter 4
Complex Event Processing in Sensor-Based Decision Support Systems

Jürgen Dunkel
Hannover University of Applied Sciences and Arts, Germany

Ralf Bruns
Hannover University of Applied Sciences and Arts, Germany

Oliver Pawlowski
Hannover University of Applied Sciences and Arts, Germany

ABSTRACT

Sensor-based decision support systems have to cope with a high volume of continuously generated sensor events. Conventional software architectures do not explicitly target the efficient processing of continuous event streams. Due to the high volume of events and their complex dependencies it is not possible to have a fixed or predefined process flow on the business level. Recently, Complex Event Processing (CEP) has been proposed as a general process model for event streams. Though CEP provides mechanisms for computing high volume of events, it does not define any methodologies, models and reference architectures, which would establish EDA as a mature software architecture. In this chapter the authors present a reference architecture for sensor-based decision support systems, which enables the analysis and processing of complex event streams in real-time. The proposed architecture provides a conceptual basis for development of flexible software frameworks that can be adapted to meet various applications needs. The authors' architectural approach is based on semantically rich event models providing the different stages of the decision process. They illustrate their approach in the domain of road traffic management for high-capacity road networks.

INTRODUCTION

Decision Support Systems (DSS) are information systems that provide assistance to humans involved in complex decision-making processes. The fundamental task for modern DSS is to help decision-makers in building up and exploring the implications of their judgements (French, 2000).

DOI: 10.4018/978-1-61520-605-6.ch004

Nowadays, many decision support systems are based on spatially distributed sensors.

Sensor systems have established themselves in many applications and systems with varying characteristics (Estrin et al., 2001). Possible application domains are extremely manifold, e.g. wild life observation (Mainwaring et al., 2002; Juang et al., 2002), glacier and ocean monitoring (Martinez et al., 2004; ARGO, 2009), monitoring vital signs of patients (Baldus et al., 2004), vehicle tracking (Palms, 2009), laboratory management (Dunkel & Bruns, 2008) or traffic management (Dunkel et al., 2008). Due to these different application areas the technical architectures of these systems differ widely as described in some more details in (Römer & Mattern, 2004). Generally, sensor networks are characterized by producing a high volume of fine-grained data, which is emitted by the sensors. Sensor-based decision support systems must deal with continuously arriving event streams (Luckham, 2002; Babu & Widom, 2001). Due to the high volume of events and their complex dependencies, no predefined workflow can be specified. A key issue of sensor networks is how to process sensor data streams and trigger appropriate business actions in downstream backend systems.

Current software architectures do not target event-based systems, because they are based on a process-oriented control flow, which is not sufficient for event-driven systems. In recent years, Event-Driven Architecture (EDA) has been proposed as a new general processing model for event streams (Luckham, 2002). The key concept is to use Complex Event Processing (CEP) as process model for event-driven decision support. Event streams emitted by sensors contain a high volume of different events, which must be transformed, classified, aggregated and evaluated to initiate appropriate domain actions.

Though CEP provides mechanisms for computing high volume of events, it does not define any methodologies, models and standards, which would establish EDA as a mature software architecture. Thus, one of the main issues is, identifying the general architectural components of an Event-Driven Architecture and to show, how CEP approaches can be integrated into enterprise applications. In particular, the main building blocks of EDA are strongly depending on the application domain.

In this chapter, we propose a general software architecture for sensor-based decision support systems based on Event-Driven Architecture and Complex Event Processing concepts. Our reference architecture enables the analysis and processing of complex event streams in real-time and is therefore well-suited for decision support in sensor-based systems. The proposed architecture provides a conceptual basis for development of flexible software frameworks that can be adapted to meet various applications needs. Our architectural approach is based on semantically rich event models providing the different stages of the decision process.

The remainder of the chapter is organized as follows. In the next section, the main concepts of Event-Driven Architecture and Complex Event Processing are introduced and the building blocks of EDA are identified. Subsequently, the state of the art in the area of Event-Driven Architecture and Complex Event Processing is discussed briefly. Afterwards, a reference software architecture for sensor-based decision support systems is derived. The approach is illustrated by means of the design of a traffic control system for managing a high-capacity road network. In the last section, we summarize the most significant features of our approach and provide an outlook on future lines of research.

FOUNDATIONS OF EVENT-DRIVEN ARCHITECTURE

Event-Driven Architecture (EDA) provides an architectural concept for dealing with continuous streams of events. In contrast to other approaches

Figure 1. Flow of events in EDA

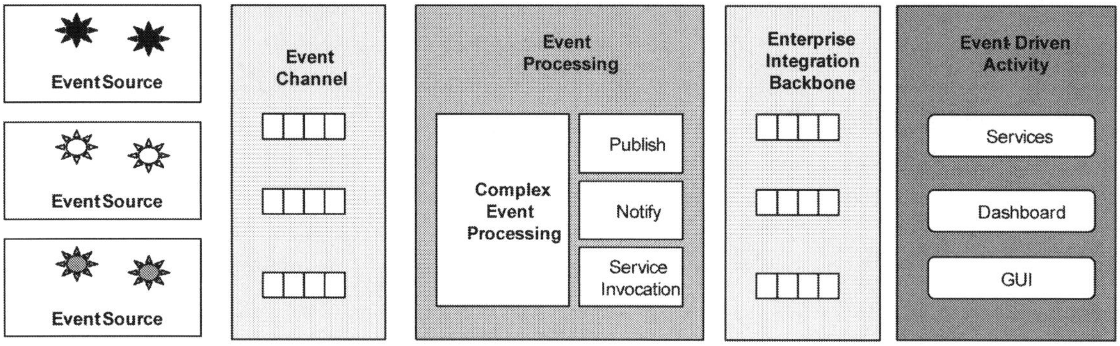

the control flow of EDA applications is essentially based on the processing of different kinds of events. Because EDA is focused on event processing, it is best-suited for event-driven business applications, e.g. logistic systems which track cargos by using RFID technology as well as for sensor network based applications

Events

Events can be considered as a change of a system state, which can be described by concepts of the business domain. The incidences that a number of cars have passed a sensor or that a traffic light has switched from green to red are possible events.

Each event instance must contain the necessary data and be passed in form of objects between the involved software components. An event description always contains general metadata: e.g. event ID, event type, event timestamp, and event occurrence number. Additionally, some event-specific information must be added, e.g. a *Sensor Event* in a traffic management system may contain traffic data like the number of cars passed and their velocity.

Usually, there are dependencies between events, e.g. Traffic Jam Event can only occur if previously High Traffic Events could be observed.

Event Flow

The general flow of events in an Event-Driven Architecture is illustrated in Figure 1.

- *Event Source*: Each event is generated by an event source. In a sensor network this might be a loop detector or a RFID reader. Each event must be described in a standard format, e.g. by an XML document, key value pairs or POJOs (Plain Old Java Object). The syntax of event descriptions should be well-defined by using a meta language, for instance XML Schema in case of XML.
- *Event Channel*: An event channel provides the infrastructure for the events to the event processing components. Typically, a message-oriented middleware (MOM) is used, which allows to send events in form of messages to all interested components. A MOM contains different message queues each dedicated for a certain event type. The communication complies with the publish/subscribe paradigm: events are forwarded to those components which have subscribed for the corresponding event type.
- *Event Processing*: In the event processing layer the continuously arriving stream of events is analyzed according to the Complex Event Processing (CEP) concept

(see below) to initiate appropriate actions. The main task of CEP is Event Pattern Matching to identify in a huge event cloud those patterns of events which are significant for a business domain. In traffic control systems millions of sensor-emitted events are analyzed to discover event patterns signifying upcoming traffic problems. Different reactions on matched patterns can be distinguished: usually, an event-handling service is invoked. Alternatively, events can just be published to other components or initiate notification of human operators.

- *Event-Driven Activities*: The real event handling is not provided by the CEP component but by the operational business applications and backend systems. They provide activities which implement the domain-specific event handling, e.g. services triggered by events, as well as dashboards or graphical user interfaces for visualizing events. For instance, traffic control systems provide services for rerouting the traffic or showing alert messages on control panels.
- *Enterprise Integration Backbone*: The Enterprise Integration Backbone (EIB) provides the infrastructure to connect the Event Processing Component with the Enterprise Backend Systems, e.g. information systems or legacy applications. The communication can be bidirectional: activities of the backend applications can be triggered by CEP events. Vice versa, a business application can generate a domain-specific event and propagate it to the CEP component using the EIB. Generally the EIB provides mechanisms for event transport, event subscription and for triggering event-driven activities implemented by the backend applications. It could be realized by a message-oriented middleware or an enterprise service bus (ESB).

Complex Event Processing

Complex Event Processing (CEP) is an event processing model introduced by David Luckham (2002) to cope with a huge number of events. Analyzing discrete events is most of the time meaningless. The main idea of CEP is that events are not independent from each other, but exhibit interdependencies and correlations. Therefore, event processing has to consider multiple events occurred over a longer historical period tying to match an event pattern that is meaningful for a certain application domain. Figure 2 shows the building blocks of a Complex Event Processing (CEP) system. Related to Figure 1 it refines the block "Event Processing". Mainly, three different blocks can be distinguished:

- *Event Metadata*: Automation of event processing requires a strong formalism based on metadata. At first, an *Event Model* must precisely specify all possible types of events and their interdependencies. An event model should provide a complete understanding of the different event types, its properties, constraints and dependencies. The event model is the base for the subsequent event processing. At second, *Event Processing Rules* can define correlations between events for detecting events patterns and determining corresponding actions. The rules consist of two different parts: event patterns specify a certain situation of events and event actions are executed when the event pattern is fulfilled, i.e. it matches. New events can be generated within the event action part. CEP relies completely on the event model, where all events used in the event processing rules must be defined. Event Patterns are *based on event types*, i.e. they define a sequence of event types that must be matched. Event sequences can be defined by some basic

Figure 2. Complex event processing components

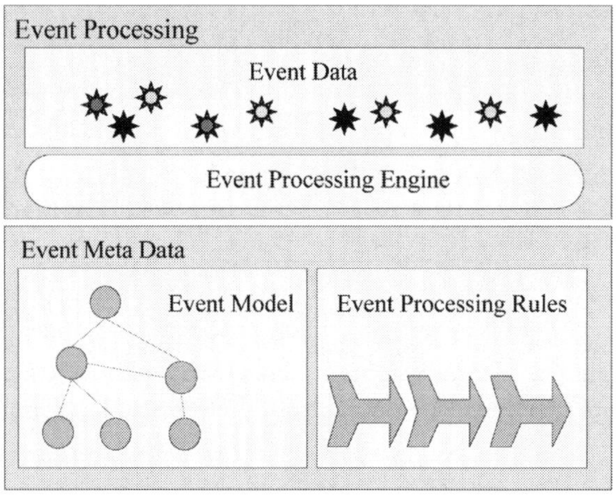

operators of an underlying algebra. Usually, event patterns are described by event processing languages (EPL) and executed by a corresponding rule engine. EPLs can be expressed as ECA (event condition rules) (Zimmer & Unland, 1999; Paton & Diaz, 1999) or as SQL-like continuous queries over event streams (Arasu et al., 2003).

- *Event Processing*: An event processing engine evaluates and executes event processing rules. The evaluation context is represented by the current instances of event data in the working memory. The event data manifests the occurred events and consists of a set of instances related to the event specification. Due to the continuously arriving data, the event processing engine has to cope with enormous data streams. Therefore, the pattern matching is characterized by continuous queries that are issued once and then run continuously over the data stream (Babu & Widom, 2001). To cope with the infinite number of data, sliding windows are used, i.e. a historical snapshot only considering the most recent set of events. It means that events have a certain lease time and are deleted when they expire. The appropriate lease times are event-type specific: the more abstract an event is, the longer its lease time is.

Pattern matching and event processing are performed by components called event processing agents (EPAs) (Luckham, 2002), which monitor the stream of occurred events analyzing it for event patterns. EPAs essentially perform the following tasks.

- *Event Filtering*: To cope with a huge number of events event processing agents have to filter out events that are of no importance. Filtering can be based on event type (e.g. *Traffic Data Event*) as well as on event content (e.g. a *Traffic Data Event* where the velocity is slower than 50% of the speed limit).
- *Content-based Routing*: The destination, whereto an event is routed, is determined by its content or its type.
- *Event Splitting and Event Aggregation*: Sometimes complex events must be split into different individual events for further processing (e.g. a *Traffic Data Event* containing different types of data like velocity, density, etc is split into single events for

each data type). Otherwise, if events are too fine-grained, multiple events can be aggregated to a single complex event, thus reducing the total number of considered events.

- *Event Transformation*: Additionally, event processing agents may transform the captured events for further processing. The simplest form of transformation is event translation which just transforms one event description format into another. Translation is required if the event sources are not capable of generating event objects in a standardized format. A more advanced step is content enrichment where domain specific knowledge is incorporated into the event object. To augment the event objects business data has to be accessed using services provided by the enterprise systems.
- *Synthesis of Complex Events*: Finally, new complex events can be synthesized from simple events. This step takes interdependencies and correlations between occurred events into account and provides the real power of complex event processing. For instance, technical sensor events must be synthesized into meaningful events, like a *Traffic Jam Event*.

Most of these tasks are well-known from the messaging patterns of Enterprise Application Integration (Fowler, 2002). In summary, EDA introduces a new software development paradigm. In contrast to classical software architectures, in an EDA the control flow is not predefined, but reacts dynamically and rule-based on event patterns.

STATE OF THE ART

Event-Driven Architecture (EDA) and Complex Event Processing (CEP) have received considerable attention in the research community recently.

Currently, a number of tools for developing EDA-based applications are available. Usually they provide an event processing language (EPL) and a corresponding processing engine. At the moment, no generally accepted standards exist for event definition, event pattern specification, or rule languages and engines. On the First Workshop on Event Processing, David Luckham identified as the main research challenges (Luckham, 2006), among others, the lack of

(a) higher level languages for events patterns and rules as well as standards for event hierarchies,
(b) reusable event processing agents,
(c) methods for the development of EDAs.

A key issue is the specification of a standardized language for the definition of event patterns and event rules, a so-called Event Processing Language (Schiefer et al., 2007). EPLs can be expressed by using general rule systems, like Jess (Jess 2009) and JBoss Rules (JBoss, 2009), which are not focused on event stream processing. They lack an inherent concept of describing temporal dependencies between events. The Rete algorithm used in general rule-based systems does not support temporal operators for defining temporal constraints. Recently, some work has been proposed for integrating temporal constraints into Rete (Berstel, 2002).

By contrast, the concept of time is already integrated in Event Query Languages like Esper (Esper 2009) or Coral8 (Coral8, 2009). EQL-based systems explicitly target the efficient processing of complex event streams in real time. Similar to databases, the response time depends mainly on the complexity of the event patterns, especially on how many event streams are joined.

Besides many research projects (Wu et al., 2006; Babcock et al., 2002, Wang & Liu, 2005) there are already some commercial products available, e.g. (Coral8, 2009, Tibco 2009, StreamBase, 2009).

Actually, there are no design patterns offering guidelines on how to build Event-Driven Architectures. In (Luckham, 2006), Event Processing Networks (EPN) consisting of several Event Processing Agents (EPA) have been introduced, which provide a general concept of event processing components. But an open question is, how Event-Driven Architectures can be structured by using such event processing agents. There is a lack of a construction kit containing standard event processing agents, which can be plugged together for developing EDA-based systems.

Experiences with EDA for sensor network applications are still very rare. In (Wu et al., 2006) first experiences with EDA for RFID are reported and an event processing language is proposed. So far, only few work exists that applies CEP on decision support systems (Dunkel et al., 2008).

EDA FOR SENSOR-BASED DECISION SUPPORT SYSTEMS

Event-Driven Architecture for Sensor-Based Systems

An EDA is well-suited for sensor-based networks, because the sensors produce a high volume of data: Common to most systems is that they rely on a huge number of sensors covering the area of interest. Each sensor is capable of monitoring its local environment within a relative small range. The data a sensor is capable to gather is very specialized and depends on the certain requirements of the particular domain. In most cases, wireless technologies are used to propagate the collected data, e.g. radio waves or inductive coupling (RFID – Radio Frequency Identification).

Other characteristics of sensor systems differ significantly. Especially the network topology and infrastructure can vary in many ways: in some systems sensors can communicate directly with each other; alternatively, base stations are used to provide the infrastructure for collecting and propagating sensor data. Regarding network topology, multi-hop ad-hoc networks can be distinguished from single-hop networks with a fixed star topology. Moreover, either the sensors are deployed spontaneously (e.g. by dropping from an aircraft) or their location is precisely determined.

Figure 3 shows the main EDA building blocks, which have been introduced in Figure 1 and Figure 2, applied to the sensor network context. Sensor networks are continuously emitting data. Consequently, they can be considered as event-driven systems and, as a result, they are particularly well suited for CEP.

Because sensor emitted data is fine-grained, single sensor events must be correlated for evaluating the current state of the physical environment. The interpretation and analysis of the continuously arriving event streams is executed by several *Event Processing Agents* (EPA), each of them dedicated to a certain task. Using CEP the EPAs identify those patterns of events which finally trigger an action in the backend systems. In general, the event processing steps executed by the EPAs corresponds to the decision-making process.

To demonstrate the approach, we refer to an EDA-based Decision Support Systems for traffic control (Dunkel et al., 2008). In high capacity road networks sensors installed in the roads emit events when cars are passing. The Decision Support System transforms the sensor events into more abstract and sophisticated domain events for evaluating the actual traffic situation and for initiating appropriate traffic control actions.

Structural Event Models

Obviously, events are the key concept of EDA and should therefore be defined precisely by a formal event model. A structural event model is semantically rich and should be based on a precise formalism, e.g. on ontologies. It is the indispensable basis to derive the software architecture of EDA-based systems.

Figure 3. Event-driven architecture applied to sensor networks

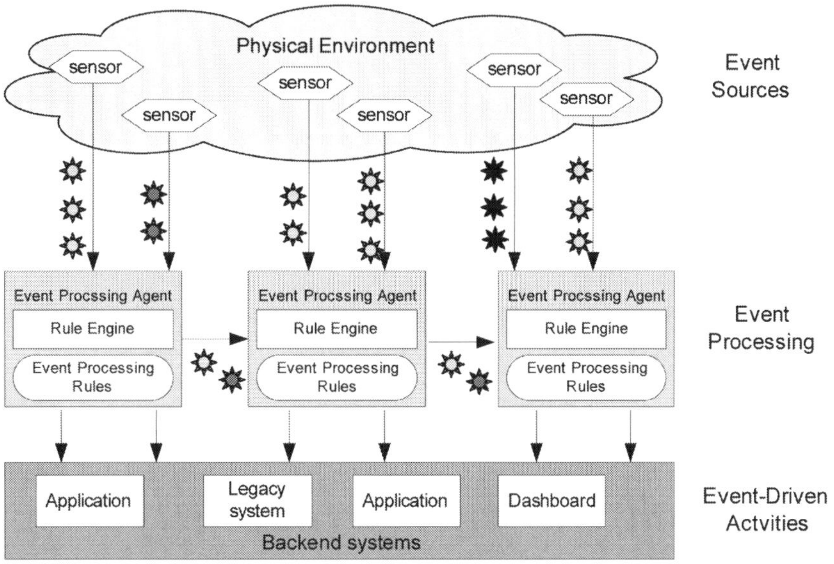

The key issue of a structural event model is the layered hierarchy of all kinds of event types that are significant for the application under investigation. The event hierarchy reflects the sequence of event processing steps: the system transforms raw sensor events into more abstract and sophisticated domain events. For sensor-based decision support systems we can derive a course-grained event hierarchy, as depicted in Figure 4.

- *Raw Sensor Events*: Raw sensor events are the physical events emitted by the sensors, e.g. loop detectors or RFID readers. Due to technical problems sensor data is often inconsistent: missed and duplicated readings, as well as unreliable readings must be compensated. For that reason, raw sensor events must be pre-processed and cleaned according to (Jeffrey et al., 2006) in order to detect and solve inconsistencies.
- *Domain Events*: The cleaned sensor data is too fine-grained and still uncorrelated. Sensor data tends to be strongly correlated in both time and space. Readings observed at one time instant are highly indicative of the readings observed at the next time instant. Applications are not interested in individual readings in time or individual devices in space, but rather in application-level concepts of temporal and spatial granules (Jeffrey et al., 2006).

Thus, domain events are derived by mapping raw sensor event data to domain concepts. For instance, domain events correlate data of sensors located in one road segment and evaluate average traffic data as traffic density and occupancy.

- *Status Events*: In a diagnosis and problem identification step the domain events are synthesized to status events that characterize the overall state of the physical environment monitored by the sensors. In traffic control systems those states are of interest that characterize a traffic problem. For example, a certain status event can identify high traffic density, traffic congestion or a blocked road.
- *Action Events*: Based on the status events, an action planning step is needed, which

Figure 4. Abstract event hierarchy for sensor-based decision support systems

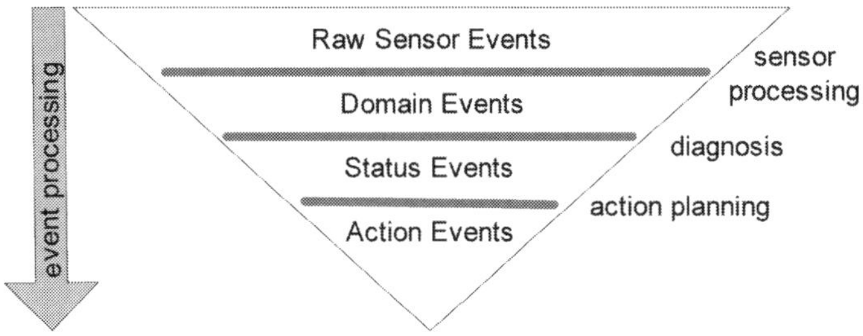

yields a sequence of actions suitable to a certain situation. Traffic rerouting to bypass a blocked road is an example of a feasible action. In EDA actions are mapped to special action events, which are propagated to enterprise backend systems for concrete event handling steps.

Note that the structural event model defines the properties of events, but it does not specify how they are processed. For a certain domain, the general structural event model proposed in Figure 4 must be refined by event subtypes and constraints.

Complex Event Processing for Decision-Support Systems

Complex Event Processing is responsible for processing streams of continuously arriving events, i.e. the operational or process behavior of EDA. The event hierarchy defined in the structural event model corresponds to the sequence of event processing steps. Event processing is nothing else than event transformation: raw sensor events are transformed into more abstract and sophisticated application-specific events for initiating appropriate control steps. The subsequent stages of event processing yield the basis for the software architecture. The event transformation steps are processed by corresponding event processing agents (EPA), which are connected to an event processing network (EPN) (Sharon & Etzion, 2008). The general EDA-based architecture for sensor-based decision support systems is shown in Figure 5. The different agents must be adapted to the refined domain-specific event hierarchy and processing steps.

- The *Sensor Processing Agent (SPA)* is responsible for cleaning the raw sensor events and for subsequent mapping to domain events. For instance, it relates technical sensor IDs and sensor data to road segments and traffic data. The normal traffic behavior can be calculated by aggregating data of a certain sensor over a longer period of time. Furthermore, in an abstraction step numeric values can be transformed into discrete qualitative values, e.g. car velocities are related to allowed speed limits and assigned to a qualitative category, e.g. 'SLOW', 'MEDIUM', 'FAST'. There may be both generic rules, applicable to all type of networks with a certain topology (congestions due to excess demand on trunks), as well as specific patterns (applicable to a set of sections with a particular location in the network).
- The *Diagnosis Agent (DA)* executes continuous queries on the stream of domain events to derive status events. In the traffic control scenario, the diagnosis agent applies adequate pattern matching rules on the incoming domain events to create status events, which signalize occurred

Figure 5. Abstract event processing network for decision support systems

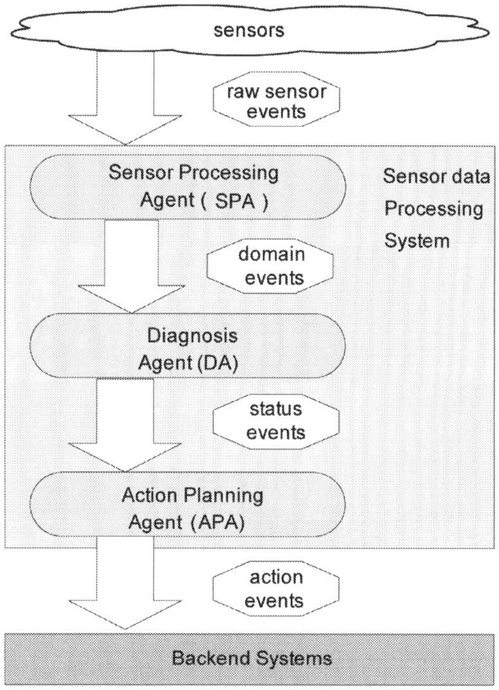

problems. A traffic problem can be understood as a state that deviates from normal traffic conditions. For example, congestions are characterized by the fact that the traffic demand exceeds the roads capacities. A problem event might be a subtype of the status event and represents a state that deviates from normal traffic conditions. Problems at certain sections, e.g. congestions and their importance are modeled in terms of traffic excess at those sections (comparing demand to capacity).

- The A*ction Planning Agent (APA)* generates appropriate action events to react on the actual state of the environment. The sequence of generated action events results from an action planning task. For example, warning signals or traffic rerouting to deal with traffic problems. Note that action events for different management areas may be in conflict: e.g. they may cause logical conflicts by deviating traffic to an already congested area. Context-sensitive priority rules must assure that the set of action events, which represent the recommendations to the operators at the traffic control center, are free of those conflicts at all times.

CASE STUDY: TRAFFIC MANAGEMENT

Scenario

In order to illustrate the usefulness of our approach we applied it to the problem of traffic management of the high-capacity road network. Figure 6 shows a fragment of the sensor network topology of a high-capacity road. Traffic data, registered by loop detectors, is received in the traffic control center, where traffic operators have to take decisions on what control actions to take in order to solve or minimize congestion. These actions include: (a) displaying messages on Variable Message Signal

Figure 6. Example of sensors positions in road sections

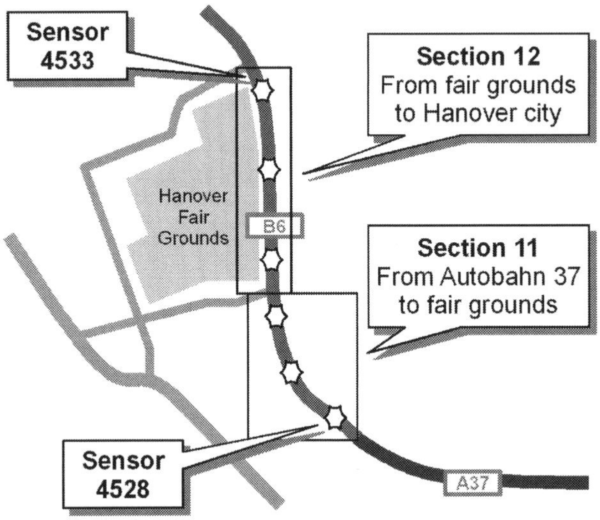

(VMS) panels installed above the road to warn drivers about traffic problems or recommend alternative routes and/or (b) contacting local authorities to send the responsible personnel to manage the situation.

Event Model

In the first step the general Structural Event Model of Figure 4 must be refined and adjusted to the traffic management scenario, i.e. specialized event subclasses must be defined, see Figure 7.

The RawSensorEvents are specialized into CarSensorEvent generated by the physical sensors located on the road. They contain event-specific information as a sensor ID (inherited from RawSensorEvent) as well as quantitative measures, e.g. velocity, density and the number of passed cars. Other sensor events are possible, like WeatherEvents containing data from a weather station.

A domain event is mapped to a TrafficEvent that aggregates information of sensors belonging to the same geographical area. The average velocity, traffic density and occupation are calculated using domain knowledge about roads capacities.

Diagnosis Agents (DA) apply different problem patterns on the traffic events to identify the actual traffic status. Each StatusEvent contains a location and is specialized to subtypes according to different problem cases. For example, a TrafficJamEvent informs about the cause, category and an AccidentEvent contains data about the location and severity of an accident.

The StatusEvents are transformed by the Action Planning Agent (APA) to ActionEvents including information about action plan proposals. Different action subtypes are possible, e.g. AlertEvent and ReroutingEvent.

Finally, the class AbstractEvent contains metadata that is common to all events types, i.e. the time of occurrence and an eventId. All event types inherit from this class to provide them with these concepts.

Event Processing Rules

The subsequent rules are examples of using the Esper engine and its continuous query language EQL (Esper, 2009) to implement event processing rules. The agents instantiate these rules with

Figure 7. Structural event model for the traffic management system

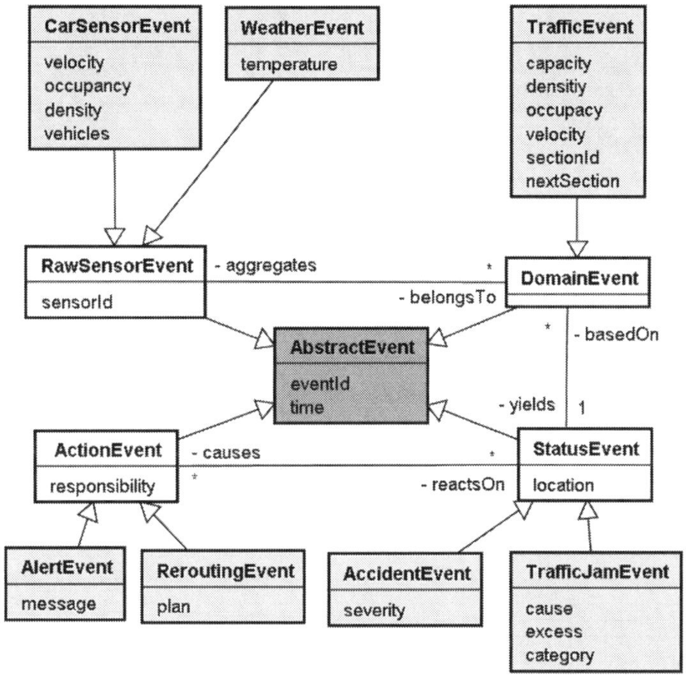

respect to their knowledge about the topology of their area of control and the expert knowledge about problem detection, diagnosis, and action planning for problem solving. There may be on the one hand generic rules, applicable to all type of networks with a certain topology (congestions due to excess demand on trunks), on the other side specific patterns (applicable to a set of sections with a specific location in the network). Problems in certain sections and their importance are modeled in terms of traffic excess in those sections (comparing demand to capacity).

The rule shown below contains the three main parts of EQL rules: (a) *insert into* creates a new event of a certain event type (here: TrafficEvent); (b) in the *select*-clause the fields of the new event are filled; and (c) source events are specified in the *from-clause*. In particular, that rule adds topology information about sections to the information obtained from all CarSensorEvents generated by sensors located in the section 'Section 11', i.e. 4528, 4529 and 4530. The section name (sectionId) and sensor data (speed, occupancy, density and vehicles) is specified and completed with domain knowledge, like the capacity of the section and topology information such as nextSection. A sliding window of length size 1 is used to achieve that each arriving CarSensorEvent generates a new aggregated TrafficEvent event.

There exist some similar rules which are only slightly different according to the structural properties of the road network. For instance, all sensor data belonging to the same road section should be aggregated. Such topology-dependent rules can be generated at compile-time from generic rule patterns based on topology knowledge, and are added automatically to the event processing engine.

Figure 8. System diagnosis pattern 'retention due to trunk incident'

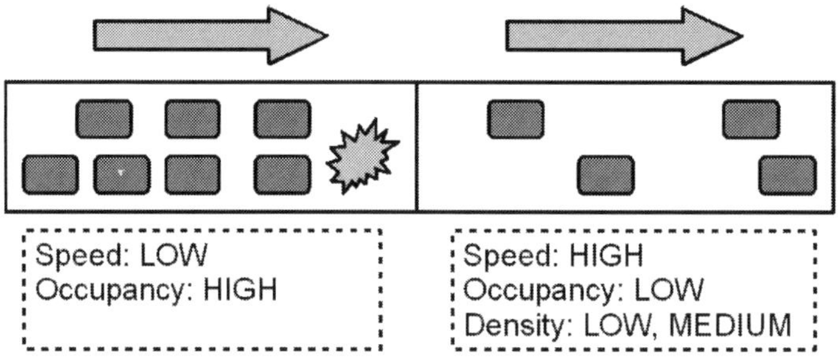


```
insert intoTrafficEventselect
'11' as sectionId,
     2820 ascapacity,
     '12' asnextSection,
     velocity asvelocity,
     occupancy asoccupancy,
     density asdensity,
   from CarSensorEvent (
     sensorId='4528'
     orsensorId='4529'
     orsensorId='4530').
win:length(1);
```

System diagnosis is based on a set of typical patterns that relate traffic conditions with problem types. Figure 8 presents one of those patterns: a retention caused by a trunk incident.

This is a generic pattern defined by the state of the two consecutive sections. This pattern matches when a section is characterized by low speed or high occupancy and the subsequent section shows low or medium density and either high speed or low occupancy.

The next rule implements this pattern with *Bef* and *Aft* referring to two consecutive sections. The abovementioned conditions are represented in the *where*-clause of the EQL rule. If the pattern matches a TrafficJamEvent is created where the location, cause and category are set with well-defined values and the excess is calculated as the difference between density and capacity.

```
insert into TrafficJamEvent
   select After.id as location,
      'retention because of in-
cident' as cause,
      After.density - After.ca-
pacity as excess,
      'problem' as category
   from TrafficEvent.win:time(30
seconds) After,
      TrafficEvent.win:time(30
seconds) Before
   where Before.nextSection =
After.sectionId and
      (Before.speed = 'LOW' or
Before.occupancy = 'HIGH') and
      (After.speed = 'HIGH' or
After.occupancy = 'LOW')
```

When two specific consecutive sections fulfill the pattern, a problem is identified and a new problem event is generated. As the agent knows the topology of the road network, in particular about consecutive sections, it can generate at runtime a set of rules like the following for each pair of consecutive sections. Rules expressing expert knowledge for local diagnosis and action planning can be represented in a similar fashion in EQL rules to define other Event Processing Agents.

CONCLUSION AND FUTURE RESEARCH

In this chapter, we suggested a general software architecture for sensor-based decision support systems. Our approach is based on Event-Driven Architecture and provides a flexible framework that can be adapted to different application scenarios. We distinguished two different stages in developing EDA-based sensor systems.

An event model specifies all kinds of event types occurring in the sensor network with their constraints and interdependencies. It should be based on a precise formalism like OWL ontologies. We presented a general event hierarchy for sensor-based systems which can be adjusted to different application domains. The Complex Event Processing defines event processing on base of the event model. The transformation between different event types is processed by Event Processing Agents applying event processing rules. Consequently, we have derived the basic structure of an event processing network corresponding with the event types of the decision support system. We defined some standard Event Processing Agents, which can serve as main building blocks for developing event-based decision support systems.

A case study of a real-world traffic management scenario illustrates how our approach can be refined in a certain application domain and provides a proof of concept. We presented that a real-world traffic management scenario can be effectively modeled and implemented within the EDA framework, thus greatly improving the real-time features of such systems.

For the future, we intend to develop further explicit architectural guidelines and design patterns to simplify the development of sensor-based applications. Basic guidelines and EDA patterns can be found in (Bruns & Dunkel, 2010). For this purpose we also attempt to apply model-driven software development for CEP. Current event processing languages are on a rather low level of abstraction and do not smoothly integrate semantically rich event models.

ACKNOWLEDGMENT

This work was supported by the European Community (European Regional Development Fund – ERDF) under Research Grant EFRE Nr. 2-221-2007-0042.

REFERENCES

W3C-The World Wide Web Consortium. (2004). *OWL Web Ontology Language Reference*. Retrieved on January 23, 2009 from www.w3.org/TR/owl-ref

Arasu, A., Babu, S., & Widom, J. (2003). CQL A language for continous queries over streams and relations. In 9th Intern. Conf. on Data Base Programming Languages (DBPL) (pp. 1-19).

ARGO. (2009). *Global Ocean Sensor Network*. Retrieved January 23, 2009, from www.argo.ucsd.edu

Babcock, B., Babu, S., Datar, M., Motwani, R., & Widom, J. J. (2002). Models and issues in data stream systems. In *Proc. of the 21st Symp. on Principles of Database Systems* (pp- 1-16).

Babu, S., & Widom, J. (2001). Continous queries over Streams. SIGMOD Record.

Baldus, H., Klabunde, K., & Muesch, G. (2004). Reliable Set-Up of Medical Body-Sensor Networks. In *Proc. EWSN 2004*, Berlin, Germany.

Barros, A., Decker, G., & Grosskopf, A. (2007). Complex Events in Business Processes. In W. Abramowicz (Ed.), *Proceedings of the 10th International Conference on Business Information Systems*, Poznan, Poland, (pp. 464-477). Berlin: Springer.

Berstel, B. (2002). Extending the RETE Algorithm for Event Management. In *Proc. of the 9th International Symposium on Temporal Representation and Reasoning,* (TIME'02).

Bruns, R., & Dunkel, J. (2010). Event-Driven Architecture - Softwarearchitektur für ereignisgesteuerte Geschäftsprozesse. Berlin: Springer.

Coral8. (2009). Retrieved January 23, 2009, from http://www.coral8.com/

Dunkel, J., & Bruns, R. (2008). Reference Architecture for event-driven RFID applications. In *2nd Intern. Workshop on RFID Technology (IWRT)* (pp. 129-135), Barcelona.

Dunkel, J., Fernández, A., Ortiz, R., & Ossowski, S. (2008). Event-Driven Architecture for Decision Support in Traffic Management Systems. In 11th Intern. IEEE Conf. on Intelligent Transportation Systems, (pp. 7-13).

Esper. ESPERTECH (2009). *Esper Reference Documentation*, Version 2.0.0. Tech. Rep. Retrieved January 23, 2009, from http://esper.codehaus.org

Estrin, D. Girod. L., Pottie, G. & Srivastava, M. (2001). Instrumenting the World with Wireless Sensor Networks, In *Intern. Conf. on Acoustics, Speech, and Signal Processing (ICASSP 2001)*, Salt Lake City.

Fowler, M. (2002). Enterprise Application Architecture. Reading, MA: Addison-Wesley.

French, S. (2000). Decision Analysis and Decision Support. Hoboken, NJ: John Wiley & Sons.

JBoss. (2009). *JBoss Rules*. Retrieved January 23, 2009, from http://www.jboss.com/products/rules

Jeffrey, S., Alonso, G., Franklin, M., Hong, W., & Widom, J. (2006). A pipelined framework for online cleaning of sensor data streams. In ICDE, (pp. 140-142).

Jess. (2009). *Jess the Rule Engine for the Java Platform*. Retrieved January 23, 2009, from http://herzberg.ca.sandia.gov/

Juang, P., Oki, H., Wang, Y., Martonosi, M., Peh, L. S., & Rubenstein, D. (2002). Energy-Efficient Computing for Wildlife Tracking: Design Tradeoffs and Early Experiences with ZebraNet. In *Proc. ASPLOS X*, San Jose, CA.

Luckham, D. (2002). Power of Events. Reading, MA: Addison-Wesley.

Luckham, D. (2006). A View of Current State of Event Processing. In *First Workshop on Event Processing*, New York. Retrieved January 23, 2009, http://complexevents.com/wp-content/uploads/2006/03/dluckham-workshop-03-2006.pdf

Mainwaring, A., Polastre, J., Szewczyk, R., Culler, D., & Anderson, J. (2002). Wireless Sensor Networks for Habitat Monitoring. In WSNA, Atlanta, GA.

Martinez, K., Ong, R., Hart, J. K., & Stefanov, J. (2004). GLACSWEB: A Sensor Web for Glaciers. In *Proc. EWSN*, Berlin, Germany.

Paton, N., & Díaz, O. (1999). Active Database Systems. *ACM Computing Surveys*, *31*(1), 63–103. doi:10.1145/311531.311623

Römer, K., & Matttern, F. (2004). The Design Space of Wireless Sensor Networks. *IEEE Wireless Communications*, *11*(6), 54–61. doi:10.1109/MWC.2004.1368897

Sandia. (2009). *Jess, the Rule Engine for the Java Platform.* Retrieved January 23, 2009, from http://herzberg.ca.sandia.gov/

Schiefer, J., Rozsnyai, S., Rauscher, C., & Saurer, G. (2007). Event-driven rules for sensing and responding to business situations. In *Proceedings of the 2007 Inaugural International Conference on Distributed Event-Based Systems,* (pp. 198-205).

Sharon, G., & Etzion, O. (2008). Event-processing network model and implementation. *IBM Systems Journal*, *47*(2), 321–334.

Streambase. (2009). *Streambase event processing engine*. Retrieved January 23, 2009, from http://www.streambase.com/

The 29 Palms Experiment (2009) *Tracking vehicles with a UAV-delivered sensor network*. Retrieved January 23, 2009, from http://robotics.eecs.berkeley.edu/~pister/29Palms0103/

Tibco. (2009). *Tibco BusinessEvents*. Retrieved January 23, 2009, from http://www.tibco.com/software/complex_event_processing/

Wang, F. & Liu, P. (2005). Temporal management of RFID data. *Very Large Data Bases (VLDB)*, 1128-1139.

Wu, E., Diao, Y., & Rizvi, S. (2006). High-performance complex event processing over streams. In *Proceedings of the 2006 ACM SIGMOD Intern. Conf. on Management of Data*, (pp. 407-418).

Zimmer, D., & Unland, R. (1999). On the semantics of complex events in active database management systems. In ICDE, (pp. 392-399).

Chapter 5
Computational Intelligence for Information Technology Project Management

Robert J. Hammell II
Towson University, USA

Julie Hoksbergen
Towson University, USA

James Wood
Towson University, USA

Mark Christensen
Towson University, USA

ABSTRACT

With the growing complexity of information technology (IT) projects, the management of these projects is proving to be a daunting task. The magnitude of this problem is underscored by the assertion that approximately 70% of IT projects fail to meet their objectives (Lewis, 2007). Computational intelligence (CI) is an area of research focused on developing intelligent systems to help with complex problems. Specifically, CI seeks to integrate techniques and methodologies to assist in problem domains in which information, data and perhaps even the problem itself are vague, approximate, and uncertain. It would seem that research aimed at leveraging the power of CI against IT project management problems is critical if IT project success rates are to be improved. This work examines the core CI technologies – fuzzy logic, neural networks, and genetic algorithms – and looks at current and potential future applications of these techniques to assist IT project managers.

INTRODUCTION

While the concept of project management has been around for centuries, the methodologies with which we are most familiar are not nearly that old. Tools developed during World War I and II, such as Gantt charts, PERT charts, and network diagrams laid the groundwork for the primary project management techniques we use today (Sisk, 1998). Even with

DOI: 10.4018/978-1-61520-605-6.ch005

these tools and the powerful information technologies at our disposal, successful performance of the multiple steps in project management is still a difficult task. The fact that approximately 70% of information technology (IT) projects fail to meet their objectives further emphasizes the difficulties throughout the life of a project (Lewis, 2007).

The sheer complexity of managing an IT project is a large part of the problem and has been cited as a primary cause in the failure of such projects (Daniels & LaMarsh, 2007). The IT industry is constantly changing which makes it difficult for users to envision and develop potential tools based on the newest technologies. Additionally, project management is still a relatively new science that is continuously evolving. Throughout the life cycle of an IT project, project leaders and their respective teams encounter many situations where they are faced with making important decisions which could determine the success or failure of the project. Often times many of these decisions are made with insufficient tools or resources. The lack of sophisticated IT tools directed specifically toward helping the project manager may lead to decreased performance in all aspects of project management. Problems are many, but examples include outdated or incomplete information for problem identification, planning and execution; basing decisions on past project experiences which may not apply to the current project; difficulty in monitoring the project due to inaccurate knowledge of the current project status; and perhaps even problems with capturing lessons learned from a project, regardless of whether it is a success or failure.

Computational intelligence (CI) is an area of research focused on developing intelligent systems to help with complex problems. Specifically, CI seeks to integrate techniques and methodologies to assist in problem domains in which information, data and perhaps even the problem itself are vague, approximate, and uncertain. The information technology project management environment is certainly one of these problem domains. With the ever increasing complexity of today's IT projects, the need for intelligent tools, processes, methods, and systems to assist in the management of these projects is also growing. While some work has been done in applying CI to IT project management problems, it appears that efforts in this area are still in their infancy. We submit that research aimed at leveraging the power of CI toward the problems in IT project management will be vital in the quest to improve IT project success rates.

The next section defines computational intelligence with respect to the work presented here. The core CI technologies that we will explore are also introduced and explained. A broad survey of current work related to applying these CI technologies to the field of IT project management is then given. We end with a section that describes potential future applications of CI to IT project management.

BACKGROUND

This section presents the background information related to computational intelligence necessary to understand the discussion regarding current and potential applications of computational intelligence constructs to the project management domain. A discussion of what computational intelligence means is provided first, and then each of its core technologies that are relevant to this work are presented and described. These technologies include fuzzy logic (FL), neural networks (NN), and genetic algorithms (GA). Finally, a summary of the background information is given along with some comments on two other computational intelligence technologies.

Computational Intelligence

Various definitions for computational intelligence (CI) appear throughout the literature. David Fogel (2008), President of the Institute of Electrical and Electronics Engineers (IEEE) Computational

Intelligence Society, provides the following on the Society website in his "Welcome from the President":

Welcome to the IEEE Computational Intelligence Society. This is your home for all computational and theoretical aspects of mimicking nature for problem solving. The core technologies within our scope include neural, fuzzy, and evolutionary computing, as well as hybrid intelligent systems that contain these and other related paradigms.

Venayagamoorthy (2009) says the initial definition for CI consisted of two of the technologies cited above – fuzzy logic (FL) and neural networks (NN) – but adds genetic algorithms (GA) in place of evolutionary computing (EC). He does say, however, that today's broader definition of CI is "the study of adaptive mechanisms to enable or facilitate intelligent behavior in complex, uncertain, and changing environments." With this definition, the relevant technologies would also include swarm intelligence and artificial immune systems. The book by Engelbrecht (2007) entitled "Computational Intelligence: An Introduction" agrees with this updated definition.

In (Magdalena, 2009), CI is equated to soft computing, where soft computing is defined as the core technologies mentioned by Fogel (FL, NN, EC) plus probabilistic reasoning. Magdalena portrays soft computing as "the design of intelligent systems which, in contract to classical hard computing techniques, are tolerant to imprecision, uncertainty, partial truth, and approximation" (p. 47).

Most of the definitions of CI include FL and NN without a doubt; many of the definitions are mixed about what else is contained in the CI field. Similarly, evolutionary computing (EC) has various slightly different definitions but many generally agree that it encompasses the areas of GA plus genetic programming, evolution strategies, evolutionary programming, and classifier systems (Eiben & Smith, 2003).

Regardless of the exact definition for CI that is used, general agreement exists that the goal is to build intelligent systems to help with complex problems in which information, data, and perhaps even the problem itself are vague, approximate, and uncertain. We submit that this line of reasoning places the project management domain squarely within the boundaries of the interests of CI. To focus the discussion in this work to a manageable and useful level, we will consider the core CI technologies of fuzzy logic (FL), neural networks (NN) and genetic algorithms (GA). Note that we are not passing judgment on what should and should not be considered under the CI umbrella, nor are we asserting what should or should not be considered part of evolutionary computing. The initial domain of CI included these technologies and we believe they have received the most attention from the field of project management in the past, and hold the most promise for assistance to information technology project managers in the immediate future.

Fuzzy Logic

Consider a banana. Is it still a banana if one bite is taken out of it? What about two? Or three? In the classical two-valued logic sense, when does the banana "suddenly" become a non-banana? The key word is "suddenly", because with conventional (crisp) logic there are only two classifications: the item is a banana or it is not a banana. In reality, the change of the object from banana to non-banana happens gradually; it transitions through degrees of "bananeness" and "non-bananeness". However, conventional set theory does not allow this type of realistic reasoning.

Fuzzy set theory provides the necessary *multivalued* framework for the above example. The transition from banana to non-banana is gradual, taking intermediate values in between. In 1965, Lotfi Zadeh wrote his famous paper formally defining multivalued, or "fuzzy" set theory (Zadeh, 1965). He extended the two-valued *indicator function* of traditional set theory to a multivalued *membership function*. The membership function is used to assign a grade of membership, ranging from 0 to 1, to each object in the fuzzy set. Zadeh formally defined fuzzy sets, their properties, and various mathematical operations on fuzzy sets. In a later paper (Zadeh, 1973) he introduced the concept of linguistic variables which have values that are linguistic in nature (i.e. speed = {slow, medium, fast}).

Fuzzy logic extends conventional Boolean (two-valued) logic so that it can handle truth values other than 0 (completely false) and 1 (completely true). That is, FL can work with values that indicate partial truth. For the example above, when the banana is half-eaten perhaps it is a banana with truth value equal to .5 or, equivalently, we can say the half-eaten object belongs to the set "banana" with a .5 membership value.

FL is built upon fuzzy sets and the basic concept is easy to grasp. In reality, we all really input, process, and output vague and imprecise information every day. If someone told you the world population right now was 6,667,073,197 would you really believe that number to be exact? Or would you interpret the value as an approximate (fuzzy) number? Suppose you are teaching your child to drive and are discussing rules for how to handle the approach to an intersection. Would you tell the child "If the light has been green for *30 seconds*, release the accelerator *75 feet* from the intersection"? Or would it be better to say "If the light has been green for a *long time*, let off the gas pedal as you *get near* the intersection"? The precision in the first rule makes it impossible to follow; the more vague, or fuzzy, second rule can be easily understood and applied.

One use of FL is to develop fuzzy inference systems; these systems provide the ability to perform approximate, or fuzzy, reasoning. Zadeh (1987) defines approximate reasoning as "the process or processes by which a possibly imprecise conclusion is deduced from a collection of imprecise statements." His idea of approximate reasoning uses FL which contains linguistic truth values (true, somewhat true, false, etc.) and approximate rules of inference. *Linguistic variables* are an important concept in fuzzy inference. Basically, a linguistic variable is used to approximately characterize the values of variables as well as their relationships. For example, numbers can be used to characterize a person's height, but using words instead might provide categories such as tall, quite tall, more or less tall, not very tall, more or less small, and so on. The imprecision introduced by using words may or may not be by choice. That is, the imprecision may be intentional based on not needing to be more precise. More often, however, the imprecision is dictated by the lack of a means to quantitatively specify the attributes of an object. (Zadeh, 1975)

Fuzzy rules of inference encapsulate the approximate relationships between the input and output, or in the terminology of rules, the antecedent and consequent, domains. A fuzzy rule with two antecedents has the form "If X is A and Y is B then Z is C" where A and B are fuzzy sets over the input domains U and V, respectively and C is a fuzzy set over the output domain W.

When using fuzzy sets in fuzzy inference, a domain is typically decomposed into overlapping fuzzy sets; each fuzzy set represents a classification. An element in the domain has some grade of membership, from 0 to 1 inclusive, in each fuzzy set in the domain. The membership function determines the grade of membership; the shape of the fuzzy sets determines the membership function. In Figure 1, an example domain is decomposed into the three fuzzy sets *slow*, *medium* and *fast*. The membership value is shown on the y-axis and the values contained within the domain are shown on the x-axis. A value of 1000 would have membership = 1 in the fuzzy set *medium* and membership = 0 in the fuzzy sets *slow* and *fast*. A value of 500 would have membership = 0.5 in both the *slow* and *medium* fuzzy sets and membership = 0 in the *fast* fuzzy set.

Figure 1. Example domain decomposition

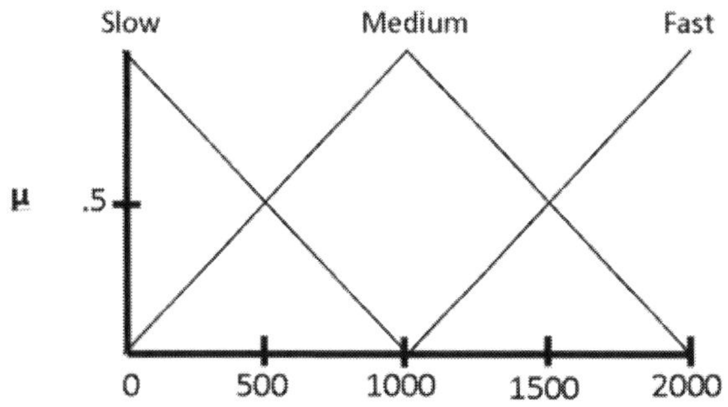

Note that the shape of the fuzzy sets do not have to be triangular as shown in Figure 1; they may be represented by many forms such as trapezoids or even beta, Gaussian, or other curves. Also, the fuzzy sets decomposing a domain do not have to overlap in a regular pattern as shown in Figure 1, nor does the sum have to be 1 for the membership values for all sets in which an element belongs.

Numerous approaches for inference using fuzzy rules appear in the literature, but most can be categorized as either generalizations of logical deduction or approximation theory. An extension of logical implication to fuzzy predicates is usually used for inference in fuzzy expert systems and databases. In the controls environment, the control surface of the system under consideration is approximated using fuzzy control rules. Thus, this form of inference is essentially functional approximation.

Advantages of fuzzy systems include their ability to handle imprecise, uncertain, and vague information; model complex non-linear systems which might be impossible to do mathematically; represent human decision making by handling vague data; and provide robustness due to being able to handle noisy and/or missing data (Yen & Langari, 1999). Historically, fuzzy rule bases have been obtained by knowledge acquisition from human experts; this is a highly heuristic and subjective process. However, recently this disadvantage has been overcome by the construction of fuzzy systems that can learn rules directly from numerical training data (Wang & Mendel, 1991; Sudkamp & Hammell, 1994). Fuzzy systems do have other potential disadvantages which include challenges related to how to determine the membership functions, correct number of rules, defuzzification techniques, and other system parameters (Yang et al., 2007).

Fuzzy logic algorithms have been implemented in many different technological areas. The earliest applications, and maybe still the most common, involve the area of control theory. FL controllers have been developed to control cement kilns (Larsen, 1980; Umbers, 1980), aircraft (Larkin, 1985), elevators (Fujitec, 1988), and automobile transmissions (Kasai & Morimoto, 1988), just to name a few. It has been applied to car engines that contain computers to stabilize the control of the car, used in software for interpreting handwriting on devices that have handwriting recognition capabilities, in cameras to control lighting, and for predicting how a target object may move (Bauer et al., 1996). It has been used in applications of case-based reasoning (Hanratty et al., 2007) data mining (Vincenti et al., 2005), and classification of chemical warfare agent simulants (Hammell & Schafer, 2007). FL has also been applied to many other fields, including pattern recognition, decision support, approximate reasoning, robotics, natural language and image understanding,

Figure 2. Example Multi-layer Neural Network

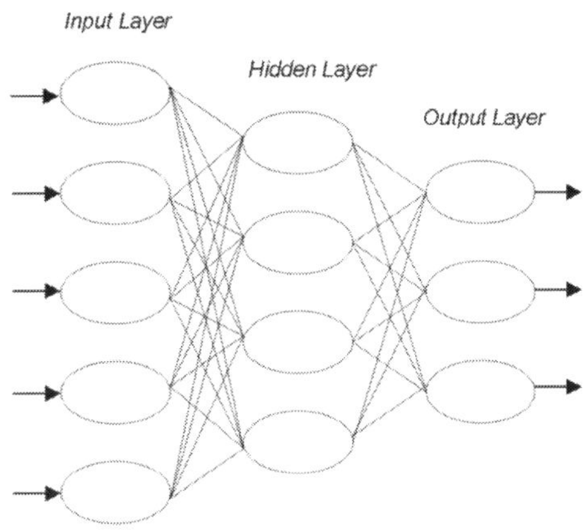

machine learning, and others (Zimmerman, 1985; Bezdek, 1993; Swartz & Klir, 1993).

Neural Networks

An artificial neural network (ANN), or simply called a neural network (NN), attempts to model the biological functions of the human brain (Zilouchian, 2001). A complex NN can be constructed with a large number of processors operating in parallel; each processor will have its own cache of knowledge and local memory. The general operation of a NN consists of a training phase followed by the operational phase.

The basic components of a neural network are the following:

- Weights: contained at each node and on each connection and are used to store the knowledge
- Thresholds: internal to each node as an offset to the activation of the output
- Activation function: at each node to perform some mathematical function to produce the output

A generic multi-layer, fully connected NN with one hidden layer is shown in Figure 2. The input layer receives the inputs from some external source. The hidden layer performs a summation operation from the outputs of the previous connected nodes. The activation function is then used to produce an output to go to the next connected node. The nodes at the output layer produce the final outputs for the system. (Samarasinghe, 2006; Zilouchian, 2001) The number of input nodes is determined by the number of arguments in the target function to be learned; the number of output nodes depends on the number of outputs required from the system.

The topology of a NN is defined by the number of layers of nodes and their connections. As mentioned, Figure 2 represents a NN with a single hidden layer and total connectivity. Note that there can be multiple hidden layers depending on the design of the NN system. Also, it is not required that every node at one layer be connected to every node at the next layer.

Weights on the connections and at the nodes are adjusted during the training phase. This is done by using some form of either supervised or unsupervised learning (Stergiou & Siganos, 1996). The training data is repeatedly presented to the system so that it "learns" the sample data by incrementally changing the weights. Training is continued through the required number of repetitions (epochs) until the output of the system matches the training data. The NN is then put into operation so that it may predict or approximate the output(s) for some new input(s) based on the underlying function which it has attempted to learn. (Zilouchian, 2001; Samarasinghe, 2006)

Neural networks are universal approximators, meaning they can approximate functions to any desired degree of accuracy. A main advantage is their ability to represent both linear and non-linear relationships and in their ability to learn these relationships directly from data. By analyzing complicated or imprecise data, neural networks can be used to extract patterns and detect trends. They are also well suited for handling problems with a large number of variables (input and/or output) and problems where the relationships between the variables are vague or hard to describe. A potential pitfall is "overfitting" or "overlearning" the training data wherein the NN models the training data exactly but does not accurately approximate the desired underlying function. Additionally, a large amount of training data might be required to correctly allow the learning of the target function and/or a large number of epochs might be required during the training phase. (Pomerleau, 1993) Determining the proper NN architecture for a given problem is also difficult, and the "black box" nature of NN systems makes human understanding of the underlying model represented by the connections and weights virtually impossible (Yang et al., 2007).

Neural networks have been used in a large variety of applications. They are especially good at identifying trends and recognizing patterns. They can be useful in function approximation, including modeling, predicting time series, and regression analysis; various classification problems; and filtering, compressing, and clustering data. Examples of specific applications include risk management, sales forecasting, customer research, marketing and credit evaluation, process control, medical applications (such as diagnosis) and voice, text, and visual recognition problems. (Stergiou & Siganos, 1996)

Genetic Algorithms

A genetic algorithm (GA) is used to solve problems by emulating the biological evolution process (Marczyk, 2004). While there is not agreement in the literature as to a universally accepted definition of a GA, most researchers and practitioners agree on the following common set of elements:

- population of *chromosomes*
- a selection mechanism based on some measure of *fitness*
- new offspring are created by a *crossover* scheme
- offspring are altered by a *mutation* process (Mitchell, 1998)

Figure 3 shows the outline of a representative GA. The first step in the algorithm is to generate a random initial population of chromosomes. Most commonly, the chromosomes are a bit string of 1s and 0s that encode a solution to the problem. However, arrays of numbers (integer or decimal) can also be used. In either case, each position in the string or array represents some portion of the solution.

Once the initial set of solutions is created, each of the solutions is evaluated according to some fitness function. It could be that an acceptable solution has been created at random and is contained in the initial population; this is not typically the case. However, the solutions need to be judged according to fitness anyway to select the "most fit" to reproduce.

Assuming that there are no acceptable solutions in the initial population, the third step is to select some number of potential solutions (now to be known as parents) to produce offspring. This set of parents will be the most "fit" of the existing "generation" of potential solutions. Once these parents are selected, they are combined

Figure 3. Sample genetic algorithm

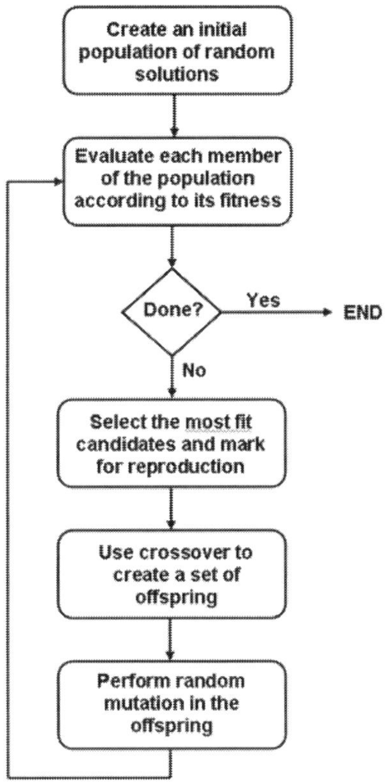

to generate some number of offspring using a crossover technique. Crossover simply allows the parents to swap pieces of their chromosomes (bit string or array positions) to produce offspring that are a combination of the parent's. There are various crossover techniques that determine the amount and placement of the positions that get swapped.

Once the parents produce the new generation of solutions, the offspring are allowed to mutate. Mutation causes small changes at single points along the chromosome. These changes occur randomly in one or more points in the bit string or array. The idea is to introduce some variation in the population.

After mutation, the process returns to step 2 – the evaluation of the current population. This process continues until one or more solutions meet some threshold as judged by the fitness function. The above algorithmic description is given in (Ladd, 1996; Marczyk, 2004); see these references for more details about the individual steps.

Genetic algorithms have some distinct advantages. They are useful for loosely defined problems and can rapidly find a "good" – perhaps not optimal – solution (Goldberg, 1989). They can easily exploit parallel exploration of the solution space due to the number of potential solutions generated at each pass. GAs are also well suited to solving problems with a very large solution space where exhaustive search is prohibitively time consuming. They do not get easily trapped by local minima in the solution space and they have the capability to manipulate many solution parameters at once. A GA also has an "open mind" about the problem; that is, they do not really know

Table 1. Comparison of computational intelligence technologies

Technology	Pros	Cons
Fuzzy Logic Based on fuzzy set theory	• Excellent for approximate reasoning • Can use linguistic variables • Handles imprecise, uncertain, vague, noisy, & missing information • Models linear and complex non-linear systems	• May need to get rules from a subject matter expert • Difficulty in determining optimal system parameters (membership functions, defuzzification technique, number of rules, etc.)
Neural Networks Models biological function of the human brain	• Universal approximators • Can learn directly from data • Good for extracting patterns and trends • Can handle large number of variables as well as vague, hard-to-describe relationships among variables • Models linear and non-linear relationships	• Overfitting to the data • May need lots training data • May need lots of training time • "Black box" nature of the model • Difficulty in determining the optimal architecture (number of hidden layers, pattern for node connectivity, etc.)
Genetic Algorithms Emulates biological evolution process	• Can quickly find a "good" solution • Works well with loosely defined problems • Can exploit parallelism in searching the solution space • Works well when there is a large solution space • Keeps "open mind" with respect to the solution	• May be hard to represent the solution in a bit string • Fitness function not easy to determine & evaluate • May evolve to suboptimal solution • Difficulty in selecting other system parameters (crossover and mutation, population size, etc.)

much about the problem but simply are looking for the best solution.(Marczyk, 2004)

GAs are not without some disadvantages, however. One that may have already been noticed is the potential difficulty in representing a solution by a bit string or array. Additionally, the construction of the fitness function by which to judge the solutions and compare them against one another is not a trivial task; in addition, the computation associated with repeatedly evaluating the fitness function may not be trivial. They may "evolve" to a suboptimal solution and may not work well when there is a single right answer to the problem. (Goldberg, 1989) Other considerations that can affect the performance of a GA include the population size, the rate of crossover and mutation, and how parents are selected. GAs can also suffer from premature convergence, in which a good fit is found early in the process leaving much of the search space (and potentially better solutions) unexplored. (Marczyk, 2004)

GAs have been used successfully in many practical applications. Examples taken from (Marczyk, 2004) include acoustics and signal processing (Tang et al., 1996), aircraft design (Obayashi et al., 2000), astronomy and astrophysics (Charbonneau, 1995), electrical engineering (Davidson, 1997), stock prediction (Mahfoud & Mani, 1996), game playing (Chellapilla & Fogel, 2001), geophysics (Sambridge & Gallagher, 1993), military and law enforcement (Kewley & Embrechts, 2002; Naik, 1996), and pattern recognition and data mining (Au, Chan & Yao, 2003; Rizki, Zmuda & Tamburino, 2002).

Background Summary and Additional Comments

This section presented several different viewpoints with respect to defining computational intelligence and the technologies it encompasses, as well as the working definition of CI used herein. The core technologies considered here – fuzzy logic, neural networks, and genetic algorithms – were then described in some detail. Table 1 provides a comparative summary of these technologies.

As mentioned at the beginning of this section, the literature includes various other technologies under the computational intelligence umbrella. Those described so far are used to represent the core CI technologies with respect to information technology project management applications. However, for the interested reader, we would also like to briefly introduce two other CI technologies with significant potential for the IT project management domain – Bayesian methods and swarm intelligence.

Bayesian methods area based on Bayesian probability; in this view, the concept of probability is seen as "a measure of a state of knowledge" (Hald, 1998). Bayesian probability follows Bayes Theorem, which illustrates the dependence of a conditional probability on its inverse as well as how evidence of observing some event alters the belief in a prior hypothesis (Howson & Urbach, 1993). Bayesian inference adheres to the premise that belief and uncertainty can be expressed as probabilities. As defined by David Hecherman (1999), a Bayesian network is "a graphical model for probabilistic relationships among a set of variables." Judea Pearl (1985), who coined the term "Bayesian network", identified three primary features of Bayesian networks: (1) the subjective nature of the inputs; (2) use of Bayes' conditioning for updating information; and (3) the distinction between causal and evidential modes of reasoning. Further introduction to Bayesian methods can be found in (Korb & Nicholson, 2004; Holmes & Jain, 2008); a glimpse into the potential for Bayesian networks with respect to modeling complex systems is presented in (Tian et al., 2003).

Swarm intelligence is a technique that falls under the broader category of "evolutionary computing" (Eiben & Smith, 2003). The key observation behind the concept of swarm intelligence is that relatively simple natural entities – such as bees, ants, termites, wasps, birds, fish, and the like – exhibit astounding collective capabilities. While the entities may have elementary individual behaviors, their interactions as a group lead to the emergence of intelligent behavior in the global sense. The main thrust is to use the idea that models related to self-organization of simple entities can be applied to develop intelligent systems (Bonabeau et al., 1999). A good introduction to swarm intelligence can be found in (Hinchey et al., 2007); further information on the topic is given by (Kennedy & Eberhart, 2001).

CURRENT CI APPLICATIONS FOR IT PROJECT MANAGEMENT

This section presents a broad survey of current work related to applying the core CI techniques described in the previous section to the field of project management with respect to information technology (IT) projects. While the examples presented here are certainly not an exhaustive treatment of IT project management-related applications of CI, they do serve as a representative sample of current work in the field.

Fuzzy Logic Applications

Criticality Prediction

One area in which FL has been applied to project management is with respect to performing criticality prediction. An example of determining criticality can be seen by considering the classical critical path method for highlighting activities that should be carefully monitored. Typically, critical tasks are those on the critical path(s) for the project; a critical path is usually defined as the path (or paths) through the project that is longest in terms of time and/or has activities with zero or negative slack (Gido & Clements, 2009). Once the critical path is identified, the activities along that path naturally receive the most attention. However, other factors related to the activities (such as cost, risk, and the like) might also affect their criticality. Jassbi et al. (2008) developed a fuzzy expert system to take into account various factors in addition to the duration of activities to predict their criticality. The fuzzy rules in the system were constructed from a bank of knowledge and experiences derived from subject matter experts. Although the fuzzy expert system was applied to a construction project case to assess its validity, the system could just as easily be used for IT projects. The results produced by the system were presented to the project manager and staff of the construction project; they agreed that the results were valid.

Another example of criticality prediction is related to predicting which software modules might be the most fault-prone within a software development project. Ebert and Baisch (1998) compared fuzzy classification with four other techniques – Pareto analysis, factor-based discriminant analysis, classification trees, and neural network approaches – and developed four different evaluation criteria to judge which was the best for predicting the likelihood of defects in software modules. The fuzzy classification technique provided the best results. Schenker and Khoshgoftaar (1998) used a fuzzy case-based reasoning system to identify fault-prone software modules. The system used fuzzy similarity measures (utilizing trapezoidal membership functions) to determine which known cases were most similar to some new case; software modules were then classified as either fault-prone or not fault-prone. Results were validated by classifying software modules from a military command, control and communication system. The results of the classification were compared to the actual classification that was done at a later phase in the project life cycle for the system. The fuzzy case-based reasoning approach achieved over 85% classification accuracy.

Software Quality Prediction

Predicting quality of software modules to assist with risk assessment is another instance of where FL has been applied. Liu et al. (2003) developed a fuzzy system to use various quantifiable metrics of software development to provide early warning about potential software quality issues. They assert that the use of fuzzy sets, fuzzy linguistic variables, and fuzzy inference rules helps handle "incomplete, inaccurate, and imprecise information" inherent in the software development environment. They claim that their system has the potential to detect software quality risk in the early stages of development. Seyyedi et al. (2005) use a fuzzy multi-agent system to measure software quality and maturity. They combine the Software Capability Maturity Model (SW-CMM), the Goal/Question/Metric method (GQM), and the Multi-level Fuzzy Inference model to measure software processes. A case study is presented to illustrate system performance, but the authors assert that their method is so different that it cannot be directly compared to existing methods.

Software Development Effort Estimation

The use of FL in estimating the effort required in software development projects is another area which has been investigated. Idri and Abran (2001) developed a set of fuzzy similarity measures to relate software project attributes when estimating software development effort for a new project. They state that many of the factors needed to judge similarity are not numerical but rather linguistic in nature, thereby making FL-based similarity measures an obvious choice. Their work describes preliminary empirical validation using the COCOMO'81 database but no results are presented based on effort estimation as related to an actual software development project.

Lopez-Martin et al. (2006) compare software development effort estimations obtained by linear regression with those from fuzzy logic-based systems. The Personal Software Process (PSP) model was used as a basis for their work; the PSP model is basically used to understand the goals, methods, and practices needed in small, personal software development efforts. Ninety-nine small programs developed by eighteen different programmers were used to construct the prediction systems. The systems were then tested on sixty programs developed by ten programmers, and the results were compared. In the majority of the comparisons the FL systems outperformed the linear regression systems.

Gu et al. (2006) apply FL techniques to estimate effort in the maintenance phase of a software project. Functional Size Measurement (FSM) techniques are used as the basis for estimating effort; specifically, the Common Software Measurement International Consortium Full Function Point (COSMIC-FFP) method is used. This method basically employs the functional perspective to measure the size of the software, versus using a technical perspective (such as lines of code). FL concepts are applied to the COSMIC-FFP method to produce a fuzzy regression decision-making model. The results from this new system are compared to existing linear and non-linear regression techniques for estimating software maintenance efforts. The results from the FL-enhanced model are shown to be better than either of the other two models.

Project Team Selection

Choosing a project team with the correct technical and cognitive skills can be difficult. The project team members are certainly "resources" within the context of the project, but the human aspects associated with choosing the right set of individuals is a much different resource allocation problem than what is typically addressed in the literature. While personality traits and team effectiveness have been well studied, attempts to model group dynamics in a way that maps individual skills and personalities with project requirements has proved to be a challenging task.

Shipley and Johnson (2009) describe a fuzzy approach for selecting teams that seeks to optimize a team's cognitive styles with the requirements of the project. They develop a model that uses FL to integrate aspects of each individual's personality and characteristics into the team selection process. Data on individuals were collected by using a cognitive style instrument (CSI) referenced from the literature. The CSI consisted of 24 questions spread over 8 scenarios. They then use four different team selection situations to illustrate the workings of the model. They conclude that their model is applicable to the project team selection process and could lead to improved project success rates.

Project Portfolio Management

Typically, there are more potential projects in an organization than available resources to complete them. As such, possible projects must be prioritized in some way so that those selected to go forward support the core business goals of the organization. A decision must be made based on probability of success of the project, potential risk, potential reward, possible market-based factors, and other tangible and intangible benefits.

Lin and Hsieh (2004) developed a fuzzy decision support system to assist with project selection and portfolio management. Their solution uses FL and goes through three basic steps: (1) alternative evaluation by using fuzzy weighted averages, (2) aggregation of the individual evaluation scores to produce a group result, and (3) selection of the optimal portfolio using fuzzy integer linear programming. They evaluate the model through an application for a real-world food corporation in Taiwan. The reported results illustrate a reduction in the time required to make a project portfolio selection decision and the ability of the system to deal with the uncertainty associated with the process. Additionally, the top managers in the company were satisfied with the approach and agreed with the system's recommendation.

Neural Network Applications

Software Error Estimation

The use of neural network techniques for estimating error rates in software development has been investigated. Zeng and Rine (2004) offer a NN-based system that uses "loosely-structured free text" describing software fix effort as input to train the system. They then use a NASA data repository to empirically evaluate the performance of the system. The results indicate good performance when estimating software development effort for projects similar to those used for training; the results are not as good when applied to development efforts in which a different development environment was used for the project.

Iwata et al. (2008) use NN to develop a model for estimating errors with respect to embedded software projects. They assert that, for embedded software projects, a relationship between the magnitude of the development and the number of resulting errors has not been investigated. They take nine previously proposed models that use regression analysis and consolidate them into 5 models based on their similarities, and then propose a new NN-based model. A comparison of experimentation results shows that the NN model is better at estimating errors in new projects than any of the other five models.

Software Cost Estimation

Tadayon (2005) proposes a NN approach for estimating software development cost. The Constructive Cost Model II (COCOMO II) algorithms are used as a basis for constructing the NN-based system. He takes the post architecture model from the COCOMO II suite of models and couples it with knowledge from software development subject matter experts, producing an adaptive NN system for software cost estimation. Unfortunately, no experimental results are reported to illustrate the efficacy of the model.

Software Risk Analysis

Neumann (2002) proposed a NN technique to provide software risk analysis. His system combined principal component analysis (PCA) with a NN to provide an objective analysis of potential software development risk. PCA is used to normalize the input data while the NN is used to determine and classify risk. He tested the method using a data set consisting of information about 390 software modules comprising a commercial imaging system. The combined PCA-NN system demonstrated significant improvement over using either of the two methods alone.

Software Process Improvement

Research by Haase (1998) addresses the question of how to decide what software process improvements to take once weaknesses have been indentified. A knowledge base of process quality data was constructed by collecting information from over 50 small software companies in Germany. A NN was used group similar companies into clusters and discover which techniques helped the most in improving software quality and maturity. Experimental results demonstrate that the NN approach is applicable to the problem of analyzing software process quality data.

Genetic Algorithm Applications

Project Scheduling

Scheduling the tasks of a project is often a complex effort, especially when there are constraints on the project resources. This problem is sometimes called the resource-constrained project scheduling problem (RCPSP). This complexity is exacerbated in a multi-project environment. Man et al. (2008) propose a genetic algorithm solution for scheduling project tasks in this multi-project resource-constrained situation. They combine a GA with the simulated annealing technique to produce an improved algorithm. The new approach takes on the advantages of each method and reduces the completion time for projects. Experimental results demonstrate that their solution is superior in terms of speed and optimality when compared to other heuristic algorithms from the literature.

Ge and Chang (2006) investigate the scheduling problem with respect to software development. They propose a framework for schedule generation that takes into account personnel capabilities when assigning tasks. Simulation results are used to show that the model produces optimal or near-optimal schedules and improves the ability of GA-based approaches to provide decision support to project managers.

Software Complexity Identification

In large software development projects, the more complex a software module is the more likely it is that the module will contain errors. Vivanco (2007) asserts that source code metrics can be used to classify complexity of software modules, but also says there are many such metrics which produce a very large search space for the problem. He uses a GA to perform feature selection; that is, the GA is used to identify the attributes that have the most discriminatory potential for correctly classifying a module's complexity. The GA is compared to a well-known metric suite for complexity classification that uses a particular subset of features, and to a method that uses all the available features. Initial results show that the GA exhibits better performance than either of the alternative methods.

Software Testing

The ability to discover and correct software errors during development, as opposed to later in the project life cycle, has the potential to save large amounts of time and money for software development projects. Berndt and Watkins (2005) explore GA-based algorithms targeted for high-volume, repeated testing especially with respect to complex, distributed systems. Their idea is to use a GA to provide automated test case generation for use in the long sequence testing approach. An automated vehicle simulation is being constructed to explore this methodology further.

Hybrid Applications

From the background discussion of FL, NN, and GA it is clear that each technique has distinct advantages and disadvantages. A promising field of research within CI is the investigation of combinations of these techniques to produce a new generation of intelligent systems. These *hybrid intelligent systems* seek to retain the advantages of the techniques while mitigating the effects of the disadvantages. This section presents examples of hybrid applications of FL, NN, and GA with respect to the IT project management domain.

Software Risk Analysis

With the association between software development projects and high project failure rates, risk analysis for such projects takes on increased importance. Yong et al. (2006) investigated combining NN, GA, and principal component analysis (PCA) within a system to assist with software project risk analysis. The PCA methodology is augmented with NN technology along the same lines as in (Nuemann, 2002) which was previously described in the NN applications section. Then, a GA was used to optimize the NN by suggesting the number of hidden layer units and the initial weights. Experimental analysis was done by making four comparisons from data collected for their research: predictions from a NN alone; predictions from a NN and PCA combination; predictions from a GA and NN combination; and predictions from the combination of all three methods. Results clearly show that the PCA-GA-NN combination exhibits the most accurate performance.

A later investigation by many of the same researchers (Hu et al., 2007) introduces the support vector machine (SVM) supervised learning method into the same problem space. This work compares software risk predictions from a NN and a GA-NN combination with those using the SVM methodology. While it appears that the researchers had high hopes for the SVM technique, the GA-NN combination demonstrated the best prediction performance of the three systems.

Software Quality Prediction

Yang et al. (2007) propose a FL-NN hybrid system for predicting software quality. They assert that the combination of these techniques is highly desirable since it provides the ability to handle objective data as well as the knowledge and experiences provided by experts. They say these types of information are most prevalent in the early stages of software development and early identification of potential software quality issues is the goal. Some preliminary experimentation with contrived data suggests that their system is applicable to the problem domain. However, they do admit that comparison of their system with other existing software quality prediction models is necessary to determine performance efficacy.

Software Maintenance

As software complexity increases there is a growing need for the ability to measure software maintainability in the early phases of the development life cycle to make it more manageable after deployment. Dahiya et al. (2007) propose enhancing a fuzzy maintainability metrics model with a GA to improve its stability. They define a model as stable "if given a small perturbation in inputs, output doesn't vary marginally." The GA is used to optimize the fuzzy system parameters to increase the stability of the system; specifically, the GA is used to adjust the fuzzy membership functions. They build a fuzzy model to measure maintainability by considering three software metrics: readability of source code, documentation quality, and understandability of the software. These metrics are obtained by standard methods used in the literature. They define a *condition number* to measure the stability; functions with a condition number closer to one are more stable. Experimental results show that the fuzzy system tuned by the GA significantly outperforms the fuzzy system without tuning. For example, in one test sequence the overall system condition number from the non-tuned fuzzy system was approximately 77 while the number dropped to about 2 when the fuzzy system was tuned by the GA.

Project Scheduling

The resource-constrained project scheduling problem (RCPSP) was mentioned in the GA applications section above. Work by Zhao et al. (2008) suggests combining a fuzzy critical chain project scheduling method with a GA to improve its performance. The identification of the critical chain (the RCPSP problem) is improved with the GA; the activity durations are represented by FL techniques. The authors demonstrate the application of the hybrid system on an example project with 30 activities, and assert that the method is applicable for project scheduling in which resources are constrained and uncertainty exists in the early project life cycle stages. However, no comparison is made with other scheduling methodologies nor are any case studies used to examine the system efficacy.

Potential CI Applications for IT Project Management

The observant reader will have already noticed that nearly all of the example applications presented in the previous section were related to improving the software development process. While software development projects are certainly among the most challenging in the IT project management domain, there are obviously other aspects of project management that could benefit from computational intelligence technologies. In this section we offer some suggestions on future work that could be undertaken to leverage the power of CI against other project management issues that would help the IT project manager and perhaps other types of project managers as well.

Project Status

While a project is ongoing, accurate information about the status of individual tasks in the project, as well as the overall project status, is crucial. It has been said that maintaining visibility of the correct project status, especially for large, complex projects, is of primary concern (Burgess et al., 2003). Following that premise, elsewhere herein is an article entitled "Employing Intelligent Decision Systems to Aid in Information Technology Project Status Decisions" which discusses this problem and presents ongoing work related to a proposed solution. The work basically describes a decision support system architecture that uses case-based reasoning coupled with intelligent agents. Also presented are ongoing and future efforts to incorporate fuzzy logic into the system to assist with the required complex decision-making.

The aforementioned work is only one example of how CI technologies could be used to assist with determining project status. It is likely that other approaches and uses of CI methods hold tremendous potential as well.

Earned Value Analysis

Another key parameter of project progress is that of earned value. Earned value is defined as the value, in monetary terms, of the work that has actually been performed (Gido & Clements, 2009). The earned value for a task is typically calculated by determining the completion percentage of the task, and then multiplying the total budgeted cost of the task by that percentage. This amount can then be compared against the actual amount that has been spent on the task to judge whether or not the work is proceeding in an efficient manner. Across the whole project, the cumulative earned value of all the tasks can be compared against the cumulative amount spent thus far on the entire project; it can also be compared against the cumulative amount that has been budgeted thus far for the project.

The determination of what percent of each task has been completed is usually a highly subjective process, which must sometimes be based on incomplete, vague, and uncertain information. An attempt to mitigate these problems is frequently accomplished by having an expert most closely associated with the task(s) be responsible for the assessment. Still, there are many factors (both political and technical) that can cause erroneous estimates to be made. CI-based systems hold much promise as decision aids for project managers in determining earned value.

Return on Investment Calculation

Return on Investment (ROI) is an important aspect of any project. ROI seeks to evaluate a potential project by comparing the investment cost to the expected monetary return once the project is completed; this return can be in terms of savings, increased profit, and the like. The timing of the expected gains from the project is also usually considered with respect to how long it will take to recover the project cost. ROI determination can be used in the project selection process to make the "go/no-go" decision about whether or not to even pursue the project. It can also be a factor in deciding on whether or not to terminate a project in the case where the cost, schedule, or scope constraints are not being met.

Given the uncertainty, ambiguity, and approximate nature of the ROI decision, it would certainly seem like a candidate for CI-based methods. CI technologies have already been used in other project environments, such as the health care industry for example (East, 2000). Assisting IT project ROI decision makers with CI systems could provide an excellent ROI itself.

Customer Satisfaction

Along with cost, scope, and schedule, customer satisfaction is considered by many to be a critical project constraint as well (Gido & Clements, 2009). The customer must be involved with the project throughout the life cycle and not just be asked at the end if he or she is satisfied. The process of gathering and interpreting the customer's opinions as to project satisfaction is inherently a vague, imprecise problem carried out with uncertain and approximate information. Again, CI-based systems would seem an appropriate approach in this area. One way might be to simply use fuzzy logic concepts in surveys to represent ranges of satisfaction (i.e., somewhat satisfied, very satisfied, neutral, somewhat dissatisfied, and the like); there would certainly be other, more sophisticated uses of CI to bring to bear on this problem.

Lack of PM Skills

In many failed projects, the lack of personnel with adequate project management skills is often cited as a primary reason for the failure. Indeed, the project manager usually shoulders much of the blame and is thought of as a "poor" (i.e., unskilled) project manager due to managing an unsuccessful project. It is often the case that many of the project team members may not have had the opportunity to obtain useful project management skills that could have helped avoid project failure.

Expert systems have been around for many years to assist decision makers and provide the knowledge of an "expert" to individuals and locations where no expert was available. In recent years, CI technologies have become an important aspect of such expert systems, particularly in the medical environment (Phuong, 1995). We feel that a large-scale effort to build a general project management expert system tailored for the IT industry has tremendous potential. While this would certainly not be a trivial effort it holds the promise for tremendous payoff, and could be a vital undertaking in the effort to reduce IT project failure.

Communication and Situational Awareness

Effective and timely communication is vital in any project. This communication must go back and forth from the project manager to the customer, members of the project team, company management, subcontractors, stakeholders, users, and anyone else involved with or impacted by the project.

The main idea of communication is to ensure that everyone has current and correct information. As such, it fosters a sort of "situational awareness" about all aspects of the project. Systems have been developed to assist with decisions needing up-to-date situational awareness (Fan et al., 2005), especially in the military environment (Hanratty et al., 2009). The development of such CI-enhanced systems aimed at IT project management seems a logical and essential step in assuring informed decision-making.

Document Management

Large amounts of documentation are produced in any reasonably sized project. There are certainly database-oriented types of document management and retrieval systems in use today. However, it would seem that adding intelligence to such a system could have nearly immeasurable potential. The use of CI technologies to improve searching and retrieval activities could help project managers and their teams in many aspects of their jobs, including communication and decision making.

In addition to simply adding intelligence to the storage and retrieval mechanisms, CI methods are applicable in other ways. For example, CI-applications to provide data mining actions might prove helpful in discovering unknown relationships in the stored information. Natural language processing, augmented by CI technologies, could have unlimited potential in providing information. Consider a system capable of analyzing the requirements document, schedule, risk assessment, and other initial documentation of a project. As other project documentation comes in over the life of the project such a system could analyze and compare all this information and potentially provide alerts and notifications of approaching risk, scope creep, schedule and resource violations, upcoming milestones, and many other items. This would be like having an unlimited number of assistants, continually analyzing and comparing the documents to provide information that could perhaps be overlooked by humans. Such a capability is precisely the goal of CI-related research.

CONCLUSION

In this paper we have presented a discussion about applying computational intelligence (CI) to the domain of information technology (IT) project management. The view of CI used in this work was defined, and the core CI technologies – fuzzy logic, neural networks, and genetic algorithms – were explained. A broad survey of current CI applications in the IT project management domain was presented. This was followed by a discussion of potential CI applications that we feel would be beneficial to IT project managers.

IT project management is a complex undertaking as evidenced by high IT project failure rates. It is clear that a lack of sophisticated tools directly aimed at helping project managers is a significant contributing factor in the lack of success suffered by many IT projects. Imprecise goals, vague information, approximate data regarding project status, and other aspects of reasoning under uncertainty plague the IT project manager throughout the project life cycle.

CI is an area of research focused on developing intelligent systems to help with complex problems. CI is particularly well suited to integrate techniques and methodologies to assist in problem domains in which information, data and perhaps even the problem itself are vague, approximate, and uncertain. The information technology project management environment is certainly one of these problem domains. We believe that leveraging the power of CI against the problems associated with IT project management will be instrumental in providing increased success rates for IT projects.

REFERENCES

Au, W., Chan, K., & Yao, X. (2003). A novel evolutionary data mining algorithm with applications to churn prediction. *IEEE Transactions on Evolutionary Computation*, *7*(6), 532–545. doi:10.1109/TEVC.2003.819264

Bauer, P., Nouak, S., & Winkler, R. (1996). Retrieved April 11, 2009, from http://www.esru.strath.ac.uk/Reference/concepts/fuzzy/fuzzy_appl.10.htm

Berndt, D., & Watkins, A. (2005). High Volume Software Testing using Genetic Algorithms. In *Proceedings of the 38th Hawaii International Conference on System Sciences (HICSS'05)*, (pp. 318b-318b), Big Island, HI.

Bezdek, J. C. (1993). Fuzzy Models – What are they, and why? *IEEE transactions on Fuzzy Systems*, *1*(1), 1–6.

Bonabeau, E., Dorigo, M., & Theraulaz, G. (1999). Swarm Intelligence: From Natural to Artificial Systems. New York: Oxford University Press.

Burgess, T., Byrne, K., & Kidd, C. (2003). Making Project Status Visible in Complex Aerospace Projects. *International Journal of Project Management*, *21*(4), 251–259. doi:10.1016/S0263-7863(02)00022-4

Charbonneau, P. (1995). Genetic algorithms in astronomy and astrophysics. *The Astrophysical Journal. Supplement Series*, *101*, 309–334. doi:10.1086/192242

Chellapilla, K., & Fogel, D. (2001). Evolving an expert checkers playing program without using human expertise. *IEEE Transactions on Evolutionary Computation*, *5*(4), 422–428. doi:10.1109/4235.942536

Dahiya, S., Chhabra, J., & Kumar, S. (2007). Use of Genetic Algorithm for Software Maintainability Metrics' Conditioning. In *Proceedings of the 15th International Conference on Advanced Computing and Communication (ADCOM'07)*, (pp. 87-92), Guwahati, India.

Daniels, C., & LaMarsh, W. (2007). Complexity as a Cause of Failure in Information Technology Project Management. In *Proceedings of the IEEE International Conference on System of Systems Engineering (SoSE '07)*. San Antonio, TX: IEEE.

Davidson, C. (1997). Creatures from primordial silicon. *New Scientist*, *156*(2108), 30–35.

East, T. (2000). *Return on Investment (ROI) Analysis in the Face of Uncertainty*. Retrieved March, 15, 2009, from http://www2.amia.org/pubs/symposia/D200476.PDF

Ebert, C., & Baisch, A. (1998). Industrial Applications of Criticality Predictions in Software Development. In *Proceedings of the Ninth International Symposium on Software Reliability Engineering*, (pp. 80-89), Paderborn, Germany.

Eiben, A. E., & Smith, J. E. (2003). Introduction to Evolutionary Computing. Heidelberg, Germany: Springer-Verlag.

Engelbrecht, A. (2007). Computational Intelligence: An Introduction (2nd Ed.). New York: John Wiley & Sons.

Fan, X., Shuang, S., & Yen, J. (2005a). On Shared Situation Awareness for Supporting Human Decision-Making Teams. In *Proceedings of 2005 AAAI Spring Symposium on AI Technologies for Homeland Security*, (pp. 17-24). Stanford, CA: American Association for Artificial Intelligence.

Fogel, D. (2008). Welcome from the President. *IEEE Computational Intelligence Society*. Retrieved April 2, 2009, from http://www.ieee-cis.org/about_cis/

Fujitec, F. (1988). FLEX-8000 series elevator group control system. Osada, Japan: Fujitec Co., Ltd.

Ge, Y., & Chang, C. (2006). Capability-based Project Scheduling with Genetic Algorithms. In *Proceedings of the International Conference on Computational Intelligence for Modelling Control and Automation, and International Conference on Intelligent Agents, Web Technologies, and Internet Commerce (CIMCA-IAWTIC'06)*, (pp. 161), Sydney, Australia.

Gibbs, W. (1996). Programming with primordial ooze. *Scientific American*, (October): 48–50. doi:10.1038/scientificamerican1096-48

Gido, J., & Clements, J. (2009). Successful Project Management (4th Ed.). Mason, OH: South-Western.

Gu, X., Song, G., & Xiao, L. (2006). Design of a Fuzzy Decision-making Model and Its Application to Software Functional Size Measurement. In *Proceedings of the International Conference on Computational Intelligence for Modelling Control and Automation, and International Conference on Intelligent Agents, Web Technologies, and Internet Commerce (CIMCA-IAWTIC'06)*, (pp. 199), Sydney, Australia.

Haase, V. (1998). Software Process Improvement Planning with Neural Networks. In *Proceedings of the 24th EUROMICRO Conference (EUROMICRO'98)*, (pp. 20808-20815), Vesteras, Sweden.

Hald, A. (1998). A History of Mathematical Statistics from 1750 – 1930. New York: Wiley.

Hammell, R. J., II, & Schafer, R. (2007). A Fuzzy Associative Memory for the Classification of Chemical Warfare Agent Simulants. In *Proceedings of the Annual Conference of the North American Fuzzy Information Processing Society (NAFIPS 2007)*, San Diego, CA.

Hanratty, T., Fan, X., & Hammell, R. J., II. (2009). Leveraging Visualization to Improve Sensemaking with a Computational RPD Model: A Military Perspective. In *Proceedings of the 9th Bi-Annual International Conference on Naturalistic Decision Making*, Covent Garden, London, accepted.

Hanratty, T., Hammell, R. J., II, Yen, J., & Fan, X. (2007). Utilizing Concept Maps to Improve Human-Agent Collaboration Within a Recognition-Primed Decision Model. In *Proceedings of the 2007/IEEE/WIC/ACM International Conference on Intelligent Agent Technology (IAT 2007)*. Silicon Valley, CA: IEEE.

Heckerman, D. (1999). A Tutorial on Learning with Bayesian Networks. In M. Jordan (ed.), Learning in Graphical Models (pp. 301-354). Cambridge, MA: MIT Press.

Hinchey, M., Sterritt, R., & Rouff, C. (2007). Swarms and Swarm Intelligence. *IEEE Computer*, *40*(3), 111–113.

Howson, C., & Urbach, P. (1993). Scientific Reasoning: The Bayesian Approach. Peru, IL: Open Court.

Hu, Y., Huang, J., Chen, J., Liu, M., & Xie, K. (2007). Software Project Risk Management Modeling with Neural Network and Support Vector Machine Approaches. In *Proceedings of the Third International Conference on Natural Computation (ICNC'07)*, (pp. 358-362), Haikou, China.

Idri, A., & Abran, A. (2001). A Fuzzy Logic Based Set of Measures for Software Project Similarity: Validation and Possible Improvements. In *Proceedings of the Seventh International Software Metrics Symposium (METRICS'01)*, (pp. 85-97), London.

Iwata, K., Nakashima, T., Anan, Y., & Ishii, N. (2008). Error Estimation Models Integrating Previous Models and Using Artificial Neural Networks for Embedded Software Development Projects. In *Proceedings of the 20th IEEE International Conference on Tools with Artificial Intelligence (ICTAI'08)*, (pp. 371-378). Dayton, OH.

Jassbi, J., Hafari, H., & Khanmohammadi, S. (2008). Fuzzy Expert System for Determining the Criticality of Activities in Mega Projects. In *Proceedings of the Eighth international Conference on Hybrid Intelligent Systems*, (pp. 246-251), Barcelona, Spain.

Kasai, Y., & Morimoto, Y. (1988). Electronically controlled continuously variable transmission. In *Proceedings of the International Congress on Transportation Electronics*, Dearborn, MI.

Keane, A. J., & Brown, S. M. (1996). The design of a satellite boom with enhanced vibration performance using genetic algorithm techniques. In I.C. Parmee, (Ed.), *Adaptive Computing in Engineering Design and Control '96 - Proceedings of the Second International Conference*, (pp. 107-113), University of Plymouth.

Kennedy, J., & Eberhart, R. (with Shi, Y.) (2001). Swarm Intelligence. San Diego, CA: Academic Press.

Kewley, R., & Embrechts, M. (2002). Computational military tactical planning system. *IEEE Transactions on Systems, Man and Cybernetics. Part C, Applications and Reviews*, *32*(2), 161–171. doi:10.1109/TSMCC.2002.801352

Ladd, S. (1996). Genetic Algorithms in C++. New York: M & T Books.

Larkin, L. I. (1985). A fuzzy logic controller for aircraft flight control. In M. Sugeno (ed.), Industrial Applications of Fuzzy Control (pp.87-104). Amsterdam: North Holland.

Larsen, P. M. (1980). Industrial application of fuzzy logic control. *International Journal of Man-Machine Studies*, *12*(1), 3–10. doi:10.1016/S0020-7373(80)80050-2

Lewis, B. (2007). *The 70-percent failure.* Retrieved August 15, 2008, from http://www.infoworld.com/articles/op/xml/01/10/29/011029opsurvival.html

Lin, C., & Hsieh, P. (2004). A Fuzzy Decision Support System for Strategic Portfolio Management. *Decision Support Systems*, *38*(3), 383–398. doi:10.1016/S0167-9236(03)00118-0

Liu, X., Kane, G., & Bambroo, M. (2003). An Intelligent Early Warning System for Software Quality Improvement and Project Management. In *Proceedings of the 15th IEEE International Conference on Tools with Artificial Intelligence (ICTAI '03)*, (pp. 32-38), Sacremento, CA.

Lopez-Martin, C., Yanex-Marquez, C., & Guitierrez-Tornes, A. (2006). Fuzzy Logic Systems for Software Development Effort Estimation Based Upon Clustering of Programs Segmented by Personal Practices. In *Proceedings of the Electronics, Robotics, and Automotive Mechanics Conference (CERMA'06)*, (pp. 367-372), Cuernavaca, Morelos, Mexico.

Magdalena, L. (2009, February). Soft Computing for Students and Society. *IEEE Computational Intelligence*, 47-50.

Mahfoud, S., & Mani, G. (1996). Financial forecasting using genetic algorithms. *Applied Artificial Intelligence*, *10*(6), 543–565. doi:10.1080/088395196118425

Man, Z., Wei, T., Xiang, L., & Lishan, K. (2008). Research on Multi-project Scheduling Problem Based on Hybrid Genetic Algorithm. In *Proceedings of the International Conference on Computer Science and Software Engineering (CSSE'08)*, (pp. 390-394), Wuhan, China.

Marczyk, A. (2004). *Genetic Algorithms and Evolutionary Computing*. Retrieved March 30, 2009, from http://www.talkorigins.org/faqs/genalg/genalg.html

Mitchell, M. (1998). An Introduction to Genetic Algorithms. Cambridge, MA: MIT Press.

Naik, G. (1996, January 16). Back to Darwin: In sunlight and cells, science seeks answers to high-tech puzzles. *The Wall Street Journal*, p. A1.

Neumann, D. (2002). An Enhanced Neural Network Technique for Software Risk Analysis. *IEEE Transactions on Software Engineering, 28*(9), 904–912. doi:10.1109/TSE.2002.1033229

Obayashi, S., Sasaki, D., Takeguchi, Y., & Hirose, N. (2000). Multiobjective evolutionary computation for supersonic wing-shape optimization. *IEEE Transactions on Evolutionary Computation*, *4*(2), 182–187. doi:10.1109/4235.850658

Pearl, J. (1985). Bayesian Networks: A Model of Self-Activated Memory for Evidential Reasoning. In *Proceedings of the 7th Conference of the Cognitive Science Society*, (pp. 329-334), Irvine, CA.

Phuong, N. (1995). Fuzzy Set Theory and Medical Expert Systems: Survey and Model. In G. Goos, J. Hartmanis, & J. Leeuwen (eds.), Lecture Notes in Computer Science: SOFSEM'95: Theory and Practice of Informatics (pp. 431-436). Berlin/ Heidelberg: Springer.

Pomerleau, D. (1993). Knowledge-based Training of Artificial Neural Networks for Autonomous Robot Driving. In J. Connell & S. Mahadevan (Eds.), Robot Learning (pp. 19-43). Boston: Kluwer Academic Publishing.

Rizki, M., Zmuda, M., & Tamburino, L. (2002). Evolving pattern recognition systems. *IEEE Transactions on Evolutionary Computation*, *6*(6), 594–609. doi:10.1109/TEVC.2002.806167

Samarasinghe, S. (2006). Neural Networks for Applied Sciences & Engineering. New York: Auerbach.

Sambridge, M., & Gallagher, K. (1993). Earthquake hypocenter location using genetic algorithms. *Bulletin of the Seismological Society of America*, *83*(5), 1467–1491.

Schenker, D., & Khoshgoftaar, R. (1998). The Application of Fuzzy Enhanced Case-Based Reasoning For Identifying Fault-Prone Modules. In *Proceedings of the Third International Symposium on High-Assurance Systems Engineering*, (pp. 90-97), Washington, DC.

Schwartz, D.G., & Klir, G.J. (1993). Fuzzy Logic: Concepts to constructs. *AI Expert*, 26-33.

Seyyedi, M., Teshnehlab, M., & Shams, F. (2005). Measuring Software Processes Performance on the Fuzzy Multi Agent Measurements. In *Proceedings of the International Conference on Information Technology: Coding and Computing (ITCC'05)*, (pp. 410-415), Las Vegas, NV.

Shipley, M., & Johnson, M. (2009). A Fuzzy Approach for Selecting Project Membership to Achieve Cognitive Style Goals. *European Journal of Operational Research*, *192*(3), 918–928. doi:10.1016/j.ejor.2007.10.006

Sisk, T. (1998). *The History of Project Management*. Retrieved March 15, 2009, from https://www.microsoft.com/downloads/details.aspx?FamilyID=c1f9b881-d879-4b54-b07b-55041685f15f&displaylang=en

Stergiou, C., & Siganos, D. (1996). Neural Networks. *SURPRISE 96 Journal*. Retrieved March 30, 2009, from http://www.doc.ic.ac.uk/~nd/surprise_96/journal/vol4/cs11/report.html

Sudkamp, T., & Hammell, R. J. II. (1994). Interpolation, Completions, and Learning Fuzzy Rules. *IEEE Transactions on Systems, Man, and Cybernetics*, *24*(2), 332–342. doi:10.1109/21.281432

Tadayon, N. (2005). Neural Network Approach for Software Cost Estimation. In *Proceedings of the International Conference on Information Technology: Coding and Computing (ITCC'05)*, (pp. 815-818), Las Vegas, NV.

Tang, K. S., Man, K. F., Kwong, S., & He, Q. (1996). Genetic algorithms and their applications. *IEEE Signal Processing Magazine*, *13*(6), 22–37. doi:10.1109/79.543973

Tian, F., Zhang, H., & Lu, Y. (2003). Research on Modeling with Dynamic Bayesian Networks. In *Proceedings of the IEEE International Conference on Web Intelligence (WI'03)*, (pp. 606-609), Halifax, Canada.

Umbers, I. G., & King, P. J. (1980). An analysis of human decision-making in cement kiln control and the implications for automation. *International Journal of Man-Machine Studies*, *12*(1), 11–23. doi:10.1016/S0020-7373(80)80051-4

Venayagamoorthy, G. (2009, February). A Successful Interdisciplinary Course on Computational Intelligence. *IEEE Computational Intelligence*, 14-23.

Vincenti, G., Hammell, R. J., II, & Trajkovski, G. (2005). Scouting for Imprecise Temporal Associations to Support Effectiveness of Drugs During Clinical Trials. In *Proceedings of the Annual Conference of the North American Fuzzy Information Processing Society (NAFIPS 2005)*, (pp. 22-25), Ann Arbor, MI.

Vivanco, R. (2007). Use of a Genetic Algorithm to Identify Source Code Metrics Which Improves Cognitive Complexity Predictive Models. In *Proceedings of the 15th International Conference on Program Comprehension (ICPC'07)*, (pp. 297-300), Banff, Alberta, Canada.

Wang, L., & Mendel, J. M. (1991). *Analysis and Design of Fuzzy Logic Controller*. Technical Report USC-SIPI-184, Signal and Image Processing Institute, University of Southern California, Los Angeles.

Yang, B., Yao, L., & Huang, H. (2007). Early Software Quality Prediction Based on a Fuzzy Neural Network Model. In *Proceedings of the Third International Conference on Natural Computation (ICNC'07)*, (pp. 760-764), Haikou, China.

Yen, J., & Langari, R. (1999). Fuzzy Logic: Intelligence, Control, and Information. Upper Saddle River, NJ: Prentice Hall.

Yong, H., Juhua, C., Zhenbang, R., Liu, M., & Kang, X. (2006). A Neural Networks Approach for Software Risk Analysis. In *Proceedings of the Sixth IEEE International Conference on Data Mining – Workshops (ICDMW'06)*, (pp. 722-725), Hong Kong.

Zadeh, L. A. (1965). Fuzzy sets. *Information and Control*, *8*, 338–353. doi:10.1016/S0019-9958(65)90241-X

Zadeh, L. A. (1973). Outline of a new approach to the analysis of complex systems and decision processes. *IEEE Transactions on Systems, Man, and Cybernetics*, *3*, 28–44.

Zadeh, L. A. (1975). The Concept of a Linguistic Variable - I. *Information Sciences*, *8*, 199–249. doi:10.1016/0020-0255(75)90036-5

Zadeh, L. A. 1987). A theory of approximate reasoning. In R. Yager, S. Orchinnikov, R. Tong, H. Nguyen (Eds.), Fuzzy Sets and Applications (pp 367-412). New York: John Wiley & Sons.

Zeng, H., & Rine, D. (2004). Estimation of Software Defects Fix Effort Using Neural Networks. In *Proceedings of the 28th Annual International Computer Software and Applications Conference (COMPSAC'04)*, (pp. 20-21). Hong Kong.

Zhao, Z., You, W., & Lv, Q. (2008). Applications of Fuzzy Critical Chain Method in Project Scheduling. *In Proceedings of the Fourth International Conference on Natural Computation (ICNC'08)*, (pp. 473-477). Jinan, China.

Zilouchian, A. (2001). Intelligent Control Systems Using Soft Computing Methodologies. Boca Raton, FL: CRC Press

Zimmerman, H. J. (1985). Fuzzy Set Theory and Its Applications. Boston: Kluwer-Nijhoff.

ADDITIONAL READING

Borgelt, K. (1998). Software Test Data Generation From A Genetic Algorithm. In C. Karr & L.M. Freeman (Eds.), Industrial Applications of Genetic Algorithms (pp. 49-68). Boca Raton, FL: CRC Press.

Cleland, D. (1999). Project Management Third Edition: Strategic Design and Implementation. New York, New York: The McGraw-Hill Companies, USA

Cox, E. (1999). The Fuzzy System Handbook, Second Edition. San Diego, California: AP Professional.

Dubois, D., & Prade, H. (1980). Fuzzy Sets and Systems. Orlando, FL: Academic Press.

Grady, R. B. (1992). Practical Software Metrics for Project Management and Process Improvement. Englewood Cliffs, New Jersey: Prentice Hall.

Holmes, D., & Jain, L. (Eds.). (2008) Innovations in Bayesian Networks: Theory and Applications. Berlin, Heidelberg: Springer-Verlag.

Jiang, H. (2006). Can the Genetic Algorithm Be a Good Tool for Software Engineering Searching Problems? *In Proceedings of the 30th Annual International Computer Software and Applications Conference (COMPSAC'06)*, (pp. 362-366). Chicago, IL.

Kerzner, H. (2006). Project Management: A Systems Approach to Planning, Scheduling, and Controlling. Hoboken, New Jersey: John Wiley & Sons, Inc.

Khoshgottaar, T. (2003). Software Engineering with Computational Intelligence. Norwell, MA: Kluwer Academic Publishers.

Klir, G. J., & Yuan, B. (1995). Fuzzy Sets and Fuzzy Logic: Theory and Applications. Upper Saddle River, New Jersey: Prentice Hall PTR.

Korb, K., & Nicholson, A. (2004). Bayesian Artificial Intelligence. Boca Raton, FL: CRC Press.

Liebowitz, J. (Ed.). (1998). The Handbook of Applied Expert Systems. Salem, MA: CRC Press.

Lin, J., & Yeh, P. (2001). Automatic Test Data Generation for Path Testing using GAs. *Information Sciences*, *131*(1-4), 47–64. doi:10.1016/S0020-0255(00)00093-1

Martin, J., & Oxman, S. (1988) Building Expert Systems: A Tutorial. Englewood Cliffs, New Jersey: Prentice Hall.

McNeill, D., & Freiberger, P. (1994). Fuzzy Logic. New York: Simon & Schuster.

McNeill, F. M., & Thro, E. (1994). Fuzzy Logic: A Practical Approach. London, England: Academic Press, Inc.

Mohanty, B. K., & Bhasker, B. (2005). Product Classification in the Internet Business – A Fuzzy Approach. *Decision Support Systems*, *38*(4), 611–619. doi:10.1016/j.dss.2003.10.002

Nguyen, H. T., & Walker, E. (2000). A First Course in Fuzzy Logic - Second Editon. Boca Raton, FL: Chapman & Hall/CRC Press LLC.

Nisar, M., Wang, J., & Elahi, M. (2008). Software Development Effort Estimation Using Fuzzy Logic – A Survey. *In Proceedings of the Fifth International Conference on Fuzzy Systems and Knowledge Discovery*, (pp. 421-427). Jinan, Shandong, China.

Pargas, R., Harold, M., & Peck, R. (1999). Test Data Generation Using Genetic Algorithms. *Software Testing . Verification and Reliability*, *9*, 263–282. doi:10.1002/(SICI)1099-1689(199912)9:4<263::AID-STVR190>3.0.CO;2-Y

Project Management Institute (PMI). (2004) Project Management Body of Knowledge. Newtown Square, Pennsylvania: Project Management Institute, Incorporated.

Shipley, M., Korvin, A., & Omer, K. (2001). A Fuzzy Logic-based Decision Model to Satisfy Goals for Successful Product/Service Introduction. *European Journal of Operational Research*, *135*(1), 209–219. doi:10.1016/S0377-2217-(00)00316-7

Wang, L., & Fu, X. (2005). Data Mining with Computational Intelligence. Berlin, Heidelberg: Springer-Verlag.

Watkins, A. (1995). The Automatic Generation of Software Test Data using Genetic Algorithms. *In Proceedings of the Fourth Software Quality Conference*, (pp. 300-309). Dundee, Scotland.

Watson, I. (1997). Applying Case-based Reasoning: Techniques for Enterprise Systems. San Francisco, California: Morgan Kaufmann Publishers, Inc.

Yager, R. R., & Zadeh, L. A. (1992). An Introduction to fuzzy logic applications in intelligent systems. Boston, Massachusetts: Kluwer Academic.

Chapter 6
Piece-Mold-Machine Manufacturing Planning

O. J. Ibarra-Rojas
UANL, México

Y. A. Rios-Solis
UANL, México

O. L. Chacon-Mondragon
UANL, México

ABSTRACT

This chapter studies a manufacturing process of pieces. These pieces are produced with molds which are mounted on machines. The authors describe this process as an optimization problem using an integer linear programming formulation which integrates the most important features of the system, and determines the quantities of pieces to produce, including the allocation of molds to machines. The objective function is to maximize the weighted production since the authors seek to minimize the non-fulfilled demand. First they show that the addressed problem belongs to the NP-hard class. After observing that solving the problem in an exact way is time consuming, they propose a solution methodology based on an Iterated Local Search Algorithm. Through computational experimentation they make conclusions about the difficulty of the decisions determined in this manufacturing planning.

INTRODUCTION

Automated planning is a branch of artificial intelligence which deals with the computation of the best strategies of action sequences, typically for execution by machines or robots (Russell & Norvig, 2003). In general, the automated planning solutions are complex, combinatory and have to be searched in a multidimensional space. Solutions usually resort to iterative trial and error processes commonly seen in artificial intelligence as the Iterated Local Search metaheuristic that we implement in this work.

More precisely, this chapter examines a real manufacturing system with molds and machines requirements for the production of pieces. These characteristics include, for example, plastic injection systems. Such an enterprise has a certain demand of pieces to fulfill. To produce a piece on a machine it is necessary to use a mold, and if it is decided to produce another type of piece in that machine with

DOI: 10.4018/978-1-61520-605-6.ch006

the same mold, a preprocessing time or setup time should be considered. This setup can reflect, for example, a change of color. Furthermore, each mold can make a certain number of pieces with a certain production rate. As for the type of pieces, molds changes involve a setup time (usually larger than for the pieces). In addition, there are limitations on the machines with molds with which to work, besides having a certain time availability. As a manufacturing process with these requirements increases (e.g., more types of pieces to produce, more molds, or more machines), one loses the ability to identify the type of pieces, and the quantities to produce as well as the optimal allocation of molds to machines. It is then important to implement formulations and solution methods to meet the planned objectives.

We consider the objective of maximizing the weighted fulfilled demand. Indeed, when a company fails to meet its demand, it is often forced to buy products to other companies so as not to lose customers. The objective function we propose deals with this issue since it can be viewed as minimizing the weighted non-fulfilled demand. Thus, the pieces to buy to other companies will be, less expensive.

With this chapter we show that the decisions of production flow (the type of pieces and the quantities to produce) is simple to operate, while the piece-mold assignments are mostly affecting the complexity of the problem. In Section 2, the characteristics of the system are presented together with some related scientific literature. To the best of our knowledge, there is not yet a model for this type of system. In Section 3, the problem is modeled by an integer lineal programming formulation covering most of the features present in these types of manufacturing processes. In Section 3.2 we prove that the studied problem belongs to the NP-hard class. Given the complexity of the problem and the computational time used to exactly solve large instances, we propose in Section 4 an Iterated Local Search algorithm to obtain "good" solutions in a short computational time. In Section 5 we present computational results obtained by the proposed methodology on random instances cases based on real data.

RELATED RESEARCH

Scheduling problems often consider setup times between the executions of different type of jobs (in our problem jobs can be considered as a type of piece). Moreover, there are cases where the setup times depend, besides the previous scheduled task, on the task to be executed. These are known as sequence dependent setup times (sij denotes the setup time between job i and job j). The latter is our case, where the change of type of piece to produce, induces a setup time. Scheduling problems with sequence-dependent setup times are, generally NP-hard (Brucker, 2001) and there exist several solution methods. However, these methodologies are focused on the single machine case (Eren, 2007; Eren & Güner, 2005; Sourd, 2006; Stecco, Cordeau, & Moretti, 2008). We can particularly cite Lou and Chu (2006) who worked in the minimization of the total tardiness scheduling problem on a single machine and sequence-dependent setup times. By using a branch and bound method (B&B) based on permutation schemes, they find exact solutions within reasonable time. This scheduling problem has the characteristic of being non preemptive which prevents a task to be interrupted during its execution. This assumption is not consistent with the characteristics of our problem, besides the use of a single machine.

Generally the articles dealing with the parallel machine case and setup times (Shim & Kim, 2006; Weng, Lu, & Ren, 2001) do not considered the use of an auxiliary equipment such as molds. Shim and Kim (2006) propose a B&B method for minimizing total tardiness where jobs are broken down into job subgroups that can be processed in parallel machines. A setup time is required between the productions of job subgroups. Additionally, it is possible to preempt the process of a job as in our problem. However, the processing time of the jobs are defined before hand, while in our problem a job-mold assignment is needed. In the problem we study, the molds play a vital role in the manufacturing planning since they are scarce resource and have to be assigned to a single machine at the time.

Chen and Powell (2003) propose to sequence job families on parallel machines where setup times appear when changing family type. In addition, each job can only be processed in a single machine. They use a column generation method to find optimal solutions for medium-scale instances (40 jobs, 8 families, and 6 machines) at a reasonable computational time. This work has applications when working with colors in the manufacture of plastic parts. Still, it differs from our work since in our case a product may have "multiple" families, which are defined by the molds that can be used to manufacture a piece type.

Chen and Wu (2006) study a scheduling problem on parallel machines with dedicated requirements for use of molds. In this work the production rates depend on the job and the machine which will execute it. In addition, there are setup times between jobs requiring different molds. Using heuristic methods based on tabu lists, they obtain solutions for large scale instances with the objective of minimizing total tardiness. Although, the work of Chen and Wu takes into account the requirement of molds, there is no possibility of producing a piece with a variety of molds, i.e., the pieces-molds assignment is not needed as it is in our problem.

Other formulations to represent related characteristics of our problem are based on job grouping (Chen, Ye, & Zhang, 2006; Haase & Kimms, 2000; Li, S., Li, G., & Zhang, 2004; Vilím, 2002). However, this type of problem does not considered a requirement for additional equipment such as molds.

In summary, to the best of our knowledge, the scheduling theory does not take into account the type of allocation decisions and the time restrictions as in the problem of our interest.

On the other hand, Jacobson et al. (1997) consider assembly problems of complex systems where it is needed to meet the demand of some products. They have a precedence order for the manufacture of the pieces and want to minimize the cost of the system. These authors develop some heuristics based on simulated annealing and tabu search to find feasible solutions for instances up to 100 pieces and 20 machines. However, in the assembly process a job can be executed with only one machine and they do not consider the existence of molds.

Our purpose is to formulate a new model incorporating all, or most of the key features of the real system we are interested in, such as allocation, production **flow**, and time constraint.

MANUFACTURING PLANNING

We focus in this chapter on a pieces-mold-machine planning problem present in some plastic injection manufacturing process. Figure 1 illustrates (left to right) a set of parts, a set of molds, and a set of machines. Arcs between the nodes of the sets represent the possible allocation of equipment. The characteristics of the equipment in the manufacturing system are: dedicated parallel machines, dedicated molds, and different production speeds for each pair (piece-mold). Notice that a piece can be produced by several molds, as well as the molds can be installed on different machines.

With Figure 2 we illustrate a manufacturing planning. In this figure, each bar represents a machine. The length of the bar represents the time limit while *(j, f)* represents the assignment of piece type *j* with mold *f*. Observe the use of a single mold at every moment in each machine. Moreover, the time limit for each machine is not exceeded. For instance, in machine 1 (*Ma1*) we have a setup time at instant 0 at which mold *f3* is installed. In the same machine, we find another

Figure 1. Representation of the manufacturing process

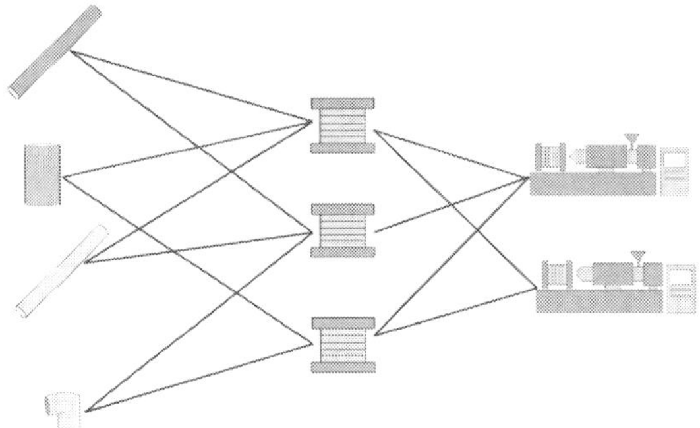

setup time for mold *f4*. It is noteworthy that with mold *f3*, pieces *j1* and *j5* are processed one after the other. The setup time between these two types of pieces is small, and it can be included in their processing time as we will see in section integer linear programming.

Integer Linear Programming

To avoid some complexity in the representation of the problem, we set that the jobs made with the same mold must be followed one after the other. This grouping idea has already yielded positive results (e.g., Chen & Wu, 2006) and is consistent with the reality: usually it is how companies wish to make their production. In fact, the company on which this study is based uses this simplification in practice. This simplification also implies the elimination of the jobs sequence-dependence setup times, which are now considered as a fixed time (as in Figure 1, Ma*1* and *f3*). By doing so, it is now possible to represent the other important characteristics observed in the real system as the equipment allocation, the production flow, and the time limit. Indeed, the main contribution of this chapter relies on the integration of most features of the studied system.

Let *J, F,* and *M* be the sets of parts, molds, and machines, respectively. Depending on the characteristics of production, let subset $J(f) \in J$ be the type of pieces compatible with mold $f \in F$. Let also subset $F(j) \in F$ contain the molds compatible with each piece type $j \in J$. Additionally, for each mold $f \in F$ there exists a subset $M(f) \in M$ of machines compatible with mold *f*. Finally, we add a set *Fm(k)* representing the molds compatibles with machine *k*.

The data required by the integer linear programming is described below.

d_j: Demand of piece type *j*.

v_{jf} : Production speed of piece type *j* using mold *f*.

w_j : Weight of piece type *j*. This weight can represent the price of this piece type in the market.

Figure 2. Five types of pieces produced by three machines and four molds

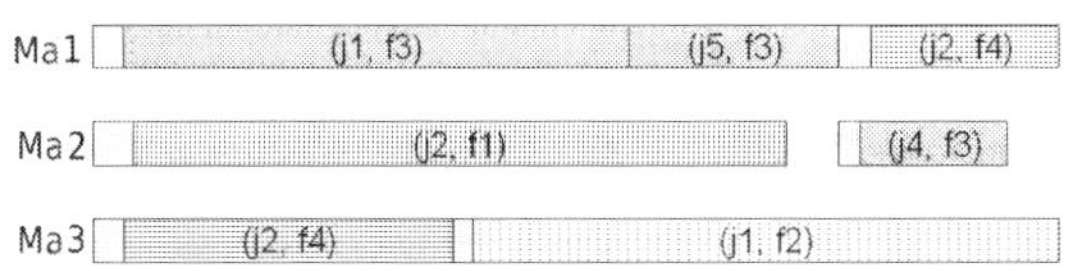

Figure 3. Instance of MKP followed by the PMM particular instance

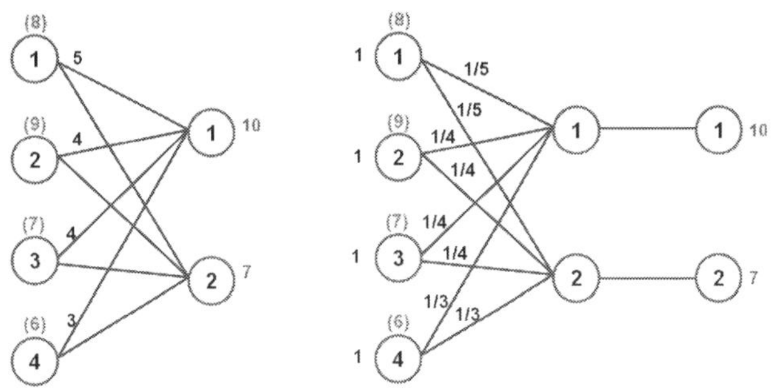

s_{jf}: Setup time of job (or piece type) j in mold f.
s_f: Time to put or remove mold f.
t_k: Available time on machine k.

For the model, it is necessary to include a set of binary variables z_{jf}, taking the value of 1 when the piece type j is produced with mold f (piece-mold arcs of Figure 1). Similarly, for each mold f is necessary another set of assignment binary variables y_{fk} taking the value of 1 when machine k is installed mold f (mold-machine arcs of Figure 3). Finally, variables x_{jfk} represent the amount of type piece j to be produced using mold f in machine k for all compatible combinations *(j, f, k)*.

A common problem for business is not complying with their production planning. For this reason, we consider as objective function the maximizing of the weighted fulfilled demand, i.e., minimizing the cost generated by the buying the not fulfilled demand. The objective is represented as

$$F(X) = \sum_{j \in J} w_j \sum_{f \in F(j)} \sum_{k \in M(f)} x_{jfk} .$$

System restrictions like time constraints, the setup times, and the number of molds, are described by the following expressions.

$$\sum_{f \in F(j)} \sum_{k \in M(f)} x_{jfk} \leq d_j \quad (4.1)$$

$$\sum_{k \in M(f)} y_{fk} \leq 1 \quad (4.2)$$

$$\sum_{k \in M(f)} x_{jfk} \leq z_{jf} d_j \quad (4.3)$$

$$\sum_{j \in J(f)} x_{jfk} \leq y_{fk} \sum_{j \in J(f)} d_j \quad (4.4)$$

$$\sum_{f \in Fm(k)} \left(\sum_{j \in J(f)} \left(\frac{x_{jfk}}{v_{jf}} + z_{jf} s_{jf} \right) + s_f y_{fk} \right) \leq t_k \quad (4.5)$$

$$x_{jfk} \in Z^+ \cup 0 \quad (4.6)$$

$$z_{jf} \in \{0,1\} \quad (4.7)$$

$$y_{fk} \in \{0,1\} \quad (4.8)$$

Restrictions (4.1) avoid exceeding the demand of piece type j. Equations (4.2) assign each mold f to at most one machine k. Expressions (4.3) indicate that the piece type j is produced with mold f, if piece type j is assigned to mold f, i.e., if $z_{jf}=1$. The set of restrictions (4.3) requires the value of $y_{fk}=1$, if exist production using mold f and machine k. Restrictions (4.5) are for not exceeding the available time at each machine. Finally, restrictions (4.6), (4.7), and (4.8) relate to the domain of the decision variables.

The optimization model of pieces-mold-machines manufacturing planning we are interested in (we name it PMM after Pieces-Mold-Machine) is determined by

max *F(X)*

subject to (4.1), (4.2), (4.3), (4.4), (4.5), (4.6), (4.7), and (4.8).

This model has assignment features: piece-mold and mold-machine assignments. However, in problem PMM one must also determine the production flow such to maximize the weighted sum of produced pieces. PMM problem differs from classical scheduling problem because permutations of jobs on each machine are not considered, the production has time limits on each machine, and the objective function is different from the ones used in the scheduling theory.

Computational Complexity of PMM

In this section we prove NP-hardness of the problem using a polynomial reduction from decision version of the multiple knapsack problem (denoted by MKP), which is NP-complete since it is a generalization of the classical knapsack problem (Karp, 1972).

Decision Version of Problem PMM (Named as dPMM)

INPUT: Let J, F, M, $J(f)$, $F(j)$, $M(f)$, and $Fm(k)$ the sets that define an instance of PMM. Let D, $W \in Z^{|J|}$ be the demand and weight vectors, respectively. Let V be a matrix of size $(|J| \times |F|)$ representing the production speeds; $S_1 \in R^{|F|}$ and S are matrices of size $(|J| \times |F|)$ representing the setup times; $T \in R^{|M|}$ corresponds to the availability of the machines; finally, K is a constant.

QUESTION: Is there any solution $x_{jfk} \in Z \cup \{0\}$, $y_{fk} \in \{0,1\}$, and $z_{jf} \in \{0,1\}$ such that $\sum_{j \in J} w_j \sum_{f \in F(j)} \sum_{k \in M(f)} x_{jfk} \geq K$, and satisfies (4.1), (4.2), (4.3), (4.4), (4.5), (4.6), (4.7), and (4.8)?

Decision Version of Multiple Knapsacks Problem(MKP)

INPUT: Let $(w_1', w_2', \dots, w_n') \in Z^n$ be the weights, $(p_1, p_2, \dots, p_n) \in Z^n$ be the benefits of choosing each article, Wi $(i \in \{1,2,\dots, m\})$ the capacity of each knapsack, and K' a constant.

QUESTION: Is there any solution $x'_{ij} \in \{0,1\}$ (x'_{ij} is 1 if item j is assigned to knapsack i, 0 otherwise) which do not exceeds the capacity of each knapsack, and satisfies $\sum_j \sum_i p_j x'_{ij} \geq K'$?

Now we establish the theorem for the complexity of the decision version of PMM.

Theorem 1: The decision version of the PMM problem is NP-complete.

Proof: The problem is clearly in NP. We reduce MKP to dPMM. For this, we build a polynomial reduction to obtain from a solution of an instance of MKP, a solution of a particular instance of dPMM.

Consider an instance of MKP. To obtain a particular instance of PMM one can do the following. Consider m machines corresponding to the knapsacks of the MKP instance. Define the piece type set and the mold set as $J=\{1,2,\dots, n\}$ and $F=\{1,2, \dots m\}$, respectively. We also define $F(j)=\{1,2,\dots, m\}$ for each piece type $j \in J$, $Fm(k)=\{k\}$ for any machine $k \in M$, $J(f)=\{1,2,\dots, n\}$, and $M(f)=\{f\}$ for each mold $f \in F$. Furthermore, we set $v_{jf}=1/w_j'$ for each combination of type piece $j \in J$ and mold $f \in F$. Likewise, we define the setup times as $s_{jf}=s_f=0$ for each piece type $j \in J$ and mold $f \in F$. We let $d_j=1$ and $w_j=p_j$ for each $j \in J$. Finally, we say $t_k=W_k$ for all machine $k \in M$.

By this reduction we define a particular instance for our PMM problem where the flow variables can only take the values {0,1} since the demand for each piece type is 1. Figure 3 illustrates an example with a MKP instance and its associated PMM particular instance. In this figure, the quantities in parentheses represent the benefits, the arc quantities are the weight in MKP, and the production speed in PMM, respectively.

The quantities of the right nodes of each network are the capacity of the knapsacks and machines, respectively.

In the particular PMM instance we only have the mold-machine assignment *(f,f)* for all $f \in F = M$. Then, our objective can be seen as

$$F(X) = \sum_{j \in J} w_j \sum_{i \in F} x'_{ij}$$

where *i* actually represents the pair of indices *(i, i)* for each $i \in F$. We can see that we have exactly the same objective as in MKP. We deduce the following lemma which does not require a proof.

Lemma 2: The objective values of MKP and dPMM problem are the same for each *X* feasible for both problems.

Now we demonstrated that the answer of the decision version of PMM is "YES" if and only if the answer to the decision version of MKP is also "YES." For this we made, without lose of generality, $K=K'$.

Step 1: If we have a solution of the particular dPMM instance, we do not need the information of Y and Z for the MKP feasibility. Let *X* be a solution of the dPMM problem such that the answer is "YES." As *X* is feasible in PMM, we have

$$\sum_{f \in Fm(k)} \sum_{j \in J(f)} \frac{x_{jfk}}{v_{jf}} \leq t_k \quad \text{for each } k \in M.$$

Since there are only mold-machines assignments *(f, f)* for each $f \in F$, we take the index i representing the pair of indices $i = (i, i)$, where $i \in F = M$. Then

$$\sum_j w'_j x'_{ij} \leq W_i \quad \text{for each } i \in \{1,2,..., m\},$$

which indicates the feasibility of this solution for MKP. By Lemma 2 we have that the answer to the question in the decision version of MKP is "YES."

Step 2: Let *X'* be a solution to a MKP instance such that the answer is "YES." Since *X'* is feasible for MKP, we have

$$\sum_{f \in Fm(k)} \sum_{j \in J(f)} \frac{x_{jfk}}{v_{jf}} \leq t_k \quad \text{for each } i \in M.$$

For the PMM instance we make $f=k=i$ (the mold-machine assignments *(f,k)* are indeed *(f,f)* for each $f \in F$) the above equation is equivalent to

$$\sum_{f \in Fm(k)} \sum_{j \in J(f)} \frac{x_{jfk}}{v_{jf}} \leq t_k \quad \text{for each } k \in M.$$

In the dPMM problem there are only mold-machines assignments *(f, f)* for each $f \in F$ and the demand for each piece is 1, then constraints (4.1) and (4.2) are redundant. Setting the piece-mold assignment as $z_{jf}=x_{jfk}$ for each piece $j \in J$ and mold $f \in F(j)$ satisfies constraint (4.3). Analogue to (4.3) setting the mold-machine assignment as

$$y_{fk} = \left\lceil \frac{\sum_{j \in J(f)} z_{jf}}{|J|} \right\rceil$$

for each $f=k \in F$ satisfies constraint (4.4). This indicates the feasibility of the solution for PMM. By Lemma 2 we have the answer for the decision version of PMM is "YES."

We have proved that MKP can be reduced to dPMM. Since MKP is NP-complete, then the decision version of PMM is NP-complete. Therefore, the optimization problem PMM is NP-hard.

SOLUTION METHODS

In this section we first make an experimental study to observe the behavior of an exact method (B&B). Through this study we conclude that the computational cost to obtain optimal solutions is time consuming. Therefore, we propose the use of a metaheuristic algorithm and obtain good quality

Figure 4. Experimental results using a B&B

	(|J|,|F|,|K|)	*dens(j,f)*	*dens(f,k)*	Average GAP
1	(50,30,5)	15%	60%	0%
2	(120,80,20)	5%	25%	4.71%
3	(120,80,20)	5%	60%	14.51%
4	(120,80,20)	15%	25%	13.42%
5	(120,80,20)	15%	60%	30.83%
6	(200,120,25)	5%	25%	5.59%
7	(200,120,25)	5%	60%	13.70%
8	(200,120,25)	15%	25%	15.47%
9	(200,120,25)	15%	60%	19.48%

solutions, and with this study we are able to make conclusions about the difficulty of the decisions in the manufacturing process of our interest.

Exact Solutions for PMM Using a B&B

We designed 9 types of problems with 10 instances each. We considered two density levels (percentage of possible assignments) for the piece-mold assignment and the mold-machine assignment. These density levels are denoted by *dens(j, f)* and *dens(f, k)*, respectively. The instances were generated, except for the weight of pieces, based on information provided by a real manufacturer of plastic products.

For solving the instances we use GAMS and CPLEX 9.0 (we kept the default settings). Figure 4 shows the sizes of the generated instances (column 1), the assignment densities (columns 2 and 3) and the average deviation (GAP) of the feasible solution from the best upper bound (column 4). This average deviation was obtained from solving the 10 instances of each size and by limiting the run time to an hour.

We observe that, for a medium size instance with low density level (row 2) we did not obtain optimal solutions. An increase in the density (row 3 and row 4) worsens the convergence of the B&B. Even letting the B&B run to up to three hours, the improvement in deviation from the best upper bound (GAP) is negligible.

We tried different alternatives to tackle PPM problem. For example, solving the continuous relaxation of the x_{jfk} variables does not improve the resolution time. The continuous relaxation of variables y_{fk} yields to quick solutions (476 seconds average time for row 5), but the problem losses its structure. Finally, eliminating the unique assignment restrictions of the molds does not improve the resolution time.

According to these experiments, even by relaxing some components of the integer program, we still face difficulties with the convergence of the B&B. This indicates us that it is not advisable to optimize one part of the problem and then the other. Such a methodology is not expected to obtain solutions with low computational cost as the optimization phase of a problem does not imply a good solution for the second.

Iterated Local Search

Given the complexity of the problem and the need of enterprises to find good and quick solutions without the cost of specialized optimization software, we propose the use of a metaheuristic to solve the PMM problem: the Iterated Local Search (ILS). ILS is used for scheduling tasks on machines (Alvarez & Tamarit, 1989), and it is not just applied as a solution algorithm, but as a component of more sophisticated procedures. The ILS proceeds as follows. Given an initial solution s^*, a change or disturbance is applied to s^*, this leads to an intermediate feasible solution s'. Then, a local search algorithm is run to find a solution $s^{*\prime}$. If this new solution passes an acceptance level, then we take solution $s^{*\prime}$ as current solution. We iterate until the algorithm is not able to find a better solution than the current one.

ILS is a simple and fast tool which can be applied and improved by increasing the quality of each one of its modules. When making these improvements, the interaction between intensification and diversification of the methodology is vital, and at the same time it is a challenging problem.

To determine an initial solution of reasonable quality, we implement a greedy algorithm based on priority lists. The idea is to program a combination of piece-mold with great benefit in the machine with more available time. For this, we define an order $(\succcurlyeq, J\times F)$. It is said that $(j_1, f_1)\succcurlyeq(j_2, f_2)$ when $w_{j1}v_{j1f1}\geq w_{j2}v_{j2f2}$. Similarly, we establish an order among the machines $(\succcurlyeq, K)$: $k_1\succcurlyeq k_2$ when $t_{k1}\geq t_{k2}$.

For the initial solution constructed by Algorithm 1 (presented below), a list of mold-piece combinations (*list(J, F)*) and a list of machines (*list(K)*) are ordered in a decreasing way (line 1). The algorithm makes iteration until it empties *list(J, F)* or consumes the time of the machines (line 2). During this iteration, the algorithm takes the first combination *(j,f)* of *list(J, F)* to assign it the highest priority machine with mold *f* (line 3, 4, and 5). Then, the production flow is computed depending on its demand and on the availability of the machine (lines 6, 7, and 8). Next, the algorithm updates the available time, *list(J, F)*, and the demand (lines 6, 8, and 10). Finally, it reorders the machine list (line 11).

```
Algorithm 1: Initial_solution
0. Initialize: x_jfk=0, z_ij=0, y_fk=0
1. Arrange (J, F) in decreasing
   order, arrange (K) in
   decreasing order
2. While (list(J, F) is not
   empty or t_k> 0 for a machine k)
3.   Take the first element (j,
     f) of list(J, F)
4.   Take first k compatible
     with f
5.   If(y_fk = 0)
6.     x_jfk ← min{ d_j,(t_k - s_f -
       s_jf)v_jf },   t_k ← t_k - s_f -
       s_jf - (x_jfk/v_jf), and y_fk = 1
7    Else
8.     x_jfk ← min{ d_j,(t_k - s_f -
       s_jf)v_jf } and t_k ← t_k - s_jf
       - (x_jfk/v_jf)
9.   End If
10.   d_j ← d_j - x_jfk, list(J,
     F)=list(J, F)\(j, f)
11.   Reorder list(K)
12. End While
```

For the ILS implementation, we use piece-mold and mold-machine assignment exchanges as local search methods. The operation *localsearch_mold* refers to molds exchange on different machines. While *localsearch_piece* refers to the exchange of pieces types in different molds. We propose a myopic algorithm which performs the first exchange which produces an improvement in the objective function.

```
Algorithm 2: localsearch_mold
0. Take an initial solution s
1. While (i < stop_criterion)
2.   For (f=0 to |F|-1)
3.    For (g=i+1 to |F|)
4.      Benefit ← 0.0, NewBenefit
        ← 0.0
5.      Benefit ← benefit(f)+
        benefit(g), NewBenefit ←
        newbenefit(f) +
        newbenefit(g)
6.      If (NewBenefit > Benefit)
7.       swap_molds (f, g)
8.      End If
```

Figure 5. Visualization of the 2 possible 3-exchanges of a given assignment

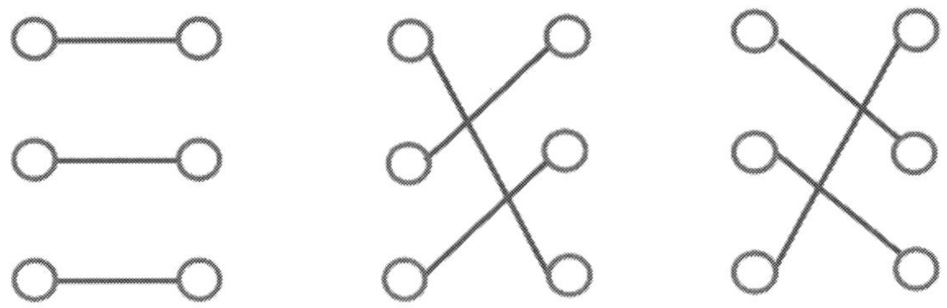

```
9.     End For
10.   End For
11. End While
```

Algorithm 2 is the operation *localsearch_mold*. It takes a couple of molds which can be exchanged. This algorithm estimates the benefits from the current solution *s* as well as the benefits generated by the mold exchange (line 5). Finally, it verifies if there is improvement in the solution (line 6) to accept or reject the operation. The algorithm for exchanging types of pieces (*localsearch_piece*) is similar.

The perturbation for the ILS should be an irreversible movement by local search. This is controlled by the kind of operations and the strength of the movement. With a "weak" perturbation few solutions could be explored, while a "strong" movement would almost do random exploration. We choose to implement a 3-exchange as shown in Figure 5. We implemented the perturbation of *k* mold-machine 3-exchanges, where *k* is a parameter of the algorithm.

The acceptance criterion is essential in the search conducted by the ILS. In this chapter, we implemented a greedy criterion, which accepts an operation only if there is some improvement in the objective function. A greedy algorithm may shed little diversity in the solutions. However, in our case, we obtain good results.

In preliminary results, the initial solutions built with Algorithm 1 were analyzed and compared with the optimal solutions of the small instances. It was noted the use of a great number of molds in the solutions generated with Algorithm 1, with this in mind, we propose a preliminary step *Rmv_molds(s, per)*, consisting in to remove a percentage (*per*) of the molds with little benefits in the cost. With this step we were able to generate higher profits using the ILS procedure. We use Bloodshed Dev-C++ version 4.9.9.2 for the heuristic.

```
Algorithm 3: ILS procedure
s* ← Initial_solution
s* ← Rmv_mold(s*, per)
s*← localsearch_piece(s*)
s*← localsearch_mold(s*)
While(no improvement)
   While(iter_no_improvement <
   limit)
      s*′← perturbation on s*
      s*′← localsearch_piece(s*′)
      s*′← localsearch_mold(s*′)
      s*← acceptation(s*, s*′)
   End While
End While
```

Algorithm 3 first constructs an initial solution *s** using Algorithm 1. Next, the algorithm applies procedure *Rmv_mold(s, per)* to the initial solution. Then, the local search methods are run. Finally, the algorithm iterates the ILS until no improvement is found.

Algorithm 3 is designed based on preliminary tests in which were carried different combinations of movements and perturbations. It was observed

Figure 6. Comparison results between GAMS and ILS for instances with low dens(j, f)

(|J|,|F|,|K|)	GAP(GAMS)	GAP(ILS)	Dem(GAMS)	Dem(ILS)
(50,30,5)	0%	19.42%	75.43%	72.59%
(120,80,20)	2.80%	10.48%	74.39%	73.17%
(200,120,25)	7.06%	10.52%	59.67%	58.99%

a far greater improvement using the exchange between molds compared with the exchange between pieces. For this reason, *localsearch_piece* it is implemented just to complement the exchange between molds.

Like any simple heuristic, Algorithm 3 can be improved by intensifying and improving each of its components. Still, the results we present in Section 5 show how such a simple algorithm finds good quality solutions when the instances have a low density in the piece-mold assignments as in the real case in which this study is based.

Computational Results

Ten instances were generated for three sizes of problems with low piece-mold assignment and high-density mold-machine assignment. The parameters were uniformly generated within the following ranges: $d_j \in [1000, 18000]$, $v_{jf} \in [120, 1000]$, $w_j \in [1, 10]$, $s_{jf} \in [0, 0]$, $s_f \in [.75, 1.15]$, and finally $t_k = 24$. As mentioned, the ranges of these parameters are based on real instances.

For the ILS is implementation we opted for four 3-exchange ($k = 4$) perturbation. It was observed, after an experimental analysis, that the percentage (*per*) of removed molds carried by the *Rmv_mold* function, depends on the size of the instance. Therefore, we empirically found the appropriated *per* for each size. For instances with 50 types of pieces, 5 machines, and 30 molds, we set *per= 4%* (2 molds). For instances with 120 pieces types, 80 machines, and 20 molds, *per= 19%*. Finally, for instances of 200 types of pieces, 120 machines, and 25 molds, *per = 21%*.

In Figure 6 we present the size of the instance (column 1), the relative average deviation from the best upper bound (GAP) obtained using GAMS and CPLEX 9.0 for one hour (column 2) and applying our ILS procedure (column 3). The ILS algorithm takes on average less than three minutes to run, even for large instances while GAMS was stopped after 1 hour. We also show in Figure 6 the average performance of the fulfilled demand obtained by using GAMS and ILS (columns 4 and 5, respectively). We observe a significant difference in the average relative

Figure 7. Comparison of demand performance for instances of size (120,80,20)

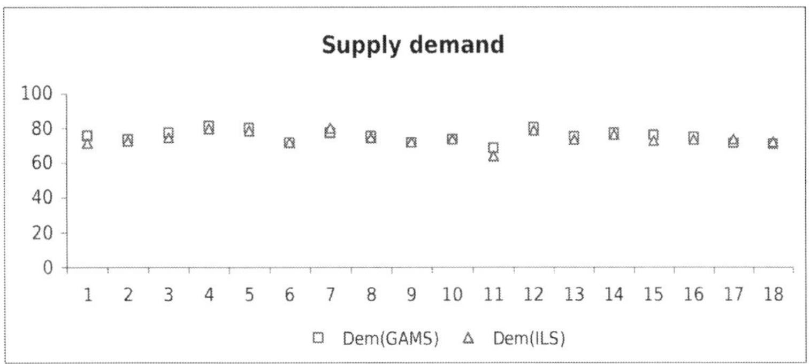

Figure 8. Results comparison between GAMS and ILS for problems with high dens(j, f)

(|J|,|F|,|K|)	GAP(GAMS)	GAP(ILS)	Dem(GAMS)	Dem(ILS)
(120,80,20)	29.72%	40.94%	80.48%	78.46%
(200,120,25)	25.90%	32.88%	70.42%	68.82%

deviation between the solution procedures. However, the fulfilled demand levels are similar. We also note that the larger the instance the smaller the quality differences between GAMS and the procedure using ILS.

Figure 7 shows the average percentage of the fulfilled demand for the instances of size (120,80,20), and illustrates the similarity of the level of fulfillment in both solution methods. This is because the production flow coincides in both solutions. Only in the piece-mold allocations of low priority relative to the benefit $w_j v_{jf}$, present significant differences.

We show in the following how the quality and speed of the proposed ILS algorithm but also of the B&B method are considerably affected when high-density piece-mold assignment is compared. In Figure 8, we can see the results of the one hour execution of GAMS and the outcome of the proposed Algorithm 3. For Algorithm 3, the running times are up to 16 minutes for the instances of 200 pieces types, 120 machines, and 25 molds. As in Figure 8, we notice that the larger the instance, the smaller the difference between the GAP of the solutions obtained by GAMS and by our ILS.

We make the following conclusions; the production flow can be reasonably obtained through the application mold-machine exchanges, being then the "easy" part of our PMM optimization problem. Furthermore, the piece-mold **assignment** presents difficulties not only in our solution algorithm, but also for the B&B algorithm. This assignment decision is the "complicated" part of the PMM problem.

CONCLUSION

We made a model of a real manufacturing system which integrates some essential features of the process. The objective is to fulfill the demand of pieces in a time limit through equipment (piece-mould-machine) assignment and calculating flow production considering the equipment constraints present in the manufacturing system. To the best of our knowledge, a model which implements these features has not yet been proposed in literature.

On the one hand, we proved the NP-Hardness of the PMM problem belongs. On the other hand, we generated good quality solutions through an Iterated Local Search procedure based on exchange movements between the mold-machines assignments.

The proposed resolution method allows us to meet a good percentage of the fulfilled demand. Moreover, we detected the great responsibility of the piece-mold assignment in the hardness of the PMM problem. Nevertheless, implementing piece-mold exchanges do not lead to better local searches and therefore to better algorithms. This is a field for future research.

REFERENCES

Alvarez, R., & Tamarit, J. M. (1989). Algoritmos heurísticos deterministas y aleatorios en secuenciación de proyectos con recursos limitados . *Qüestiió*, *13*(1), 173–191.

Brucker, P. (2001). Scheduling Algorithms. Secaucus, NJ: Springer-Verlag New York, Inc.

Chen, B., Ye, Y., & Zhang, J. (2006). Lot-sizing Scheduling with Batch Setup Times. *Journal of Scheduling*, *9*(3), 299–310. doi:10.1007/s10951-006-8265-7

Chen, J.-F., & Wu, T.-H. (2006). Total tardiness minimization on unrelated parallel machine scheduling with auxiliary equipment constraints. *Omega The International Journal of Management Science*, *34*(1), 81–89. doi:10.1016/j.omega.2004.07.023

Chen, Z.-L., & Powell, W. B. (2003). Exact algorithms for scheduling multiple families of jobs on parallel machines. *Naval Research Logistic*, *50*(7), 823–840. doi:10.1002/nav.10091

Eren, T. (2007). A Multicriteria Scheduling with Sequence-Dependent Setup Times. *Applied Mathematical Sciences*, *1*(58), 2883–2894.

Eren, T., & Güner, E. (2005). A bicriteria scheduling with sequence-dependent setup times. *Applied Mathematics and Computation*, *179*(1), 378–385. doi:10.1016/j.amc.2005.11.112

Haase, K., & Kimms, A. (2000). Lot Sizing and Scheduling with Sequence Dependent Setup Costs and Times and Efficient Rescheduling Opportunities. *International Journal of Production Economics*, *66*(2), 159–169. doi:10.1016/S0925-5273(99)00119-X

Jacobson, S., Johnson, A., Sullivan, K., Fleischer, M., & Kumar, A. (1997). Metaheuristic for a Flexible Assembly System Design Problems. *Journal of Heuristics*, *3*(2), 139–159. doi:10.1023/A:1009679413948

Karp, R. M. (1972). Reducibility Among Combinatorial Problems. In Complexity of Computer Computations, (85-103). New York: Plenum.

Li, S., Li, G., & Zhang, S. (2004). Minimizing makespan with release times on identical parallel batching machines. *Discrete Applied Mathematics*, *148*(1), 127–134. doi:10.1016/j.dam.2004.11.004

Lou, X., & Chu, F. (2006). A branch and bound algorithm of the single machine schedule with sequence dependent setup times for minimizing total tardiness. *Applied Mathematics and Computation*, *138*(1), 575–588.

Russell, S., & Norvig, P. (2003), Artificial Intelligence: A Modern Approach (2nd Ed.). Upper Saddle River, NJ: Prentice Hall.

Shim, S.-O., & Kim, Y.-D. (2006). A branch and bound algorithm for an identical parallel machine scheduling problem with a job splitting property. *Computers & Operations Research*, *35*(3), 863–875. doi:10.1016/j.cor.2006.04.006

Sourd, F. (2006). Dynasearch neighborhood for the earliness-tardiness scheduling problem with release dates and setup constraints. *Operations Research Letters*, *34*(5), 591–598. doi:10.1016/j.orl.2005.06.005

Stecco, G., Cordeau, J.-F., & Moretti, E. (2008). A branch-and-cut algorithm for a production scheduling problem with sequence-dependent and time-dependent setup times. *Computers & Operations Research*, *35*(8), 2635–2655.

Vilím, P. (2002). Batch Processing with Sequence Dependent Setup Times: New Results. *Lecture Notes in Computer Science*, *2470*, 153–167.

Weng, M. X., Lu, J., & Ren, H. (2001). Unrelated parallel machine scheduling with setup consideration and a total weighted completion time objective. *International Journal of Production Economics*, *70*(3), 215–226. doi:10.1016/S0925-5273(00)00066-9

Chapter 7
Intelligent Simulation System for Supply Chain Event Management (SCEM)

Barin Nag
Towson University, USA

Haiying Qiao
SmartOps Company, Pittsburgh, USA

Dong-Qing Yao
Towson University, USA

ABSTRACT

In this chapter, the authors present an intelligent simulation system for supply chain event management for the purpose of designing and re-engineering the supply chain. The simulation framework mainly composes of component layer, process layer, intelligent execution layer, and output layer. The functional design of the layers is discussed with comments on the contribution of the simulation. Implementation issues are further addressed and an illustrative case study is reported.

1. INTRODUCTION

Supply Chain Event Management (SCEM), sometimes referred to as Supply Chain Process Management (SCPM), is defined by APICS (APICS, 2005) as the application of a special type of software that simulates, controls, and responds to exceptions in a supply chain. SCEM can provide active visibility, and perform the functions such as real-time monitoring, measuring with key performance indicators (KPIs), notifying exceptions to decision makers, performing what-if simulation scenario analysis, and controlling unexpected events such as cancelled orders, supply disruption, and currency fluctuations, *etc*. The advantage of the application of SCEM is that connected processes can be monitored, even in a global context where the uncertainties are magnified, and that the processes can be modified if necessary (Christopher, 2005). In the global economy that is there today, the supply chain is global in nature and is becoming increasingly complex and dynamic. There are risks inherent in all stages of supply chain processes, from the supply side to the demand side (PRTM, 2005). As a result, global supply chain

DOI: 10.4018/978-1-61520-605-6.ch007

is vulnerable with respect to exceptional negative events, and can be easily disrupted by these negative events. The anticipation and avoidance by management of exceptional negative events becomes an important part of risk management in the global supply chain and of the management of the supply chain in general for effective performance.

2. APPLICATIONS OF INTELLIGENT SIMULATION SYSTEMS ON SCEM

The structure of SCEM as a working software solution is comprised of the following components: event management engine, alert manger, human decision maker, application, and data warehouse, and some other minor support functions (Otto, 2003). The key element known as the *event management engine* receives messages from the entire supply chain, and then proceeds to analyze these messages to make decisions based on predetermined rules and algorithms in a manner similar to an *expert system* (ES) as used in decision making and decision support. The ES is a well-known decision support method of using knowledge-based heuristic rules, typically with symbolic literal inputs, and needs no further introduction here. The analysis of the messages and the reporting of the anomalies to the ES is often are performed by intelligent agents that monitor the supply chain processes and issue alerts when unexpected events occur.

Uncertainties are inherent in a supply chain, and more so in a global supply chain. The supply chain uncertainty could be from demand side, supply side, or internal operational processes. Examples of the risks from demand side include demand forecast errors, canceled orders, order increase due to promotion, and demand decreases due to the introduction of new product. Examples of risks from supply side include supply disruptions, delivery delay, and quality issues. There are internal risks that include internal operations breakdowns, and quality issues.

In this paper, we will focus on the demand risk management because typically demand uncertainty accounts for 60% of total supply chain uncertainty. We demonstrate the construct of a simulation framework for SCEM in the electronics industry and present the implementation platform. It is well-known that it is typical for an electronics distributor to have a large number of products with rather short life cycles. For products with short life cycle or sometimes referred to as perishables, the costs of mismatch between supply and demand are very high because mismatch can lead to either lost sale or inventory leftover with little salvage value. In addition, the expected mismatch cost increases in demand variability (Cachon and Terwiesch, 2009). Typically, the customer demands for electronic products have large variability, and are thus hard to predict. In addition, when electronic products are supplied by vendors globally, the electronics distributor has very complex supply chain. The process of managing the supply chain and the different types of risks is a sizeable challenge to large electronic distributors (Raman and Ton, 2005).

Simulation plays an important role in practical applications of supply chain event management. However, the traditional simulation models focus on the static information and on modeling based on static information, and are thus unable to respond to risks that occur in real time. The failure of this type of simulation model is that these are designed without an intelligent component, and are therefore neither flexible nor reusable (Hung et al., 2004). Because of the dynamics of the hi-tech supply chain and the lack of intelligence, this type of simulation only can be used once, and has to be frequently redesigned (Yu et al., 2000).

In order to improve the model's effectiveness and efficiency, it is very important for a simulation framework to be designed for intelligence, thus the simulation model can be integrated with current information systems, such as data warehouse to analyze supply chain for different processes. Toward this end, in this chapter, we propose an intelligent business process simulation framework for hi-tech industries with products of very short life cycle. The framework focuses on first mapping various business processes, and then on analyzing different supply chain problems as needed. Although business problems vary dynamically, the business process is relatively stable. Therefore, mapping the business process is a basis for the business analysis (Xu et al., 2007). After the business process is mapped, it can be reused to analyze different business problems. In practice, simulation is frequently used to analyze business process because business process simulation allows the management to study the dynamics of the business with no or little risk (Barber et al., 2003; Greasley, 2003).

2.1. Simulation Framework

A global electronic company operates in many regions and the management processes vary from one region to another. Thus there is a need to simulate a variety of different processes. It is fortunate for this simulation that the basic components of the business processes in electronic industries are typically the same. It is only the attributes of these components that are different for specific processes. Therefore this project aims to build a generic intelligent supply chain simulation framework, the core of which is a process library to describe the process attributes in each of the different regions.

In order to make the simulation modeling system reusable and intelligent, our framework takes top-down approach. The first step is to abstract basic commonalities at the highest level. Afterwards, specific attributes of the components and processes are developed to populate the framework. Layered architecture can easily separate business logic, data access logic, and user interface, thus make the framework more flexible, and easily reused. As such, layered architecture is useful and popular to develop software and operations planning. For example, Lee and Hwang (2003) introduced five layered architecture and engineering approach to design agent-based system, and Westfechtel et al. (2001) proposed a layered architecture for a subset of SCM. Wang (1988) took advantage of a layered approach to plan manufacturing operations, in which the set of layers include a layer of facts, a layer of inference rules, and a layer of meta-rules.

At the highest level of abstraction, the simulation framework is made up of a multitude of entities and processes. The entities can be customers, products, warehouses, employees, and information, *e.g.* order information, forecast information, product information, or other classes of information. The processes are flows between these entities as guided by the business rules. Most Enterprise Resource Planning (ERP) systems in use today provide a backbone to integrate the business processes. In a similar way the simulation framework presented here provides only a generic description of these entities and processes because the set of business processes is relatively stable for an established company. The majority of functions then are customized in implementation stage when deployed to a specific business unit. However, the Continuous Process Improvement (CPI) initiative has been becoming more and more important. CPI provides data to support these initiative imperative in the dynamic business environment that exists today. In the complicated supply chain environment, it is difficult to predict the results of a process change, and even more difficult to quantify the change. A simulation model, which can model the interactions of different process, is a powerful tool that not only can provide the data support for such process improvement, but also provides insights into the interactions of these process. This is the key difference from an optimization model, designed to optimize static parameters but not the process itself. Thus, flexibility is another important requirement of a modeling system. The

Figure 1. The supply chain simulation framework

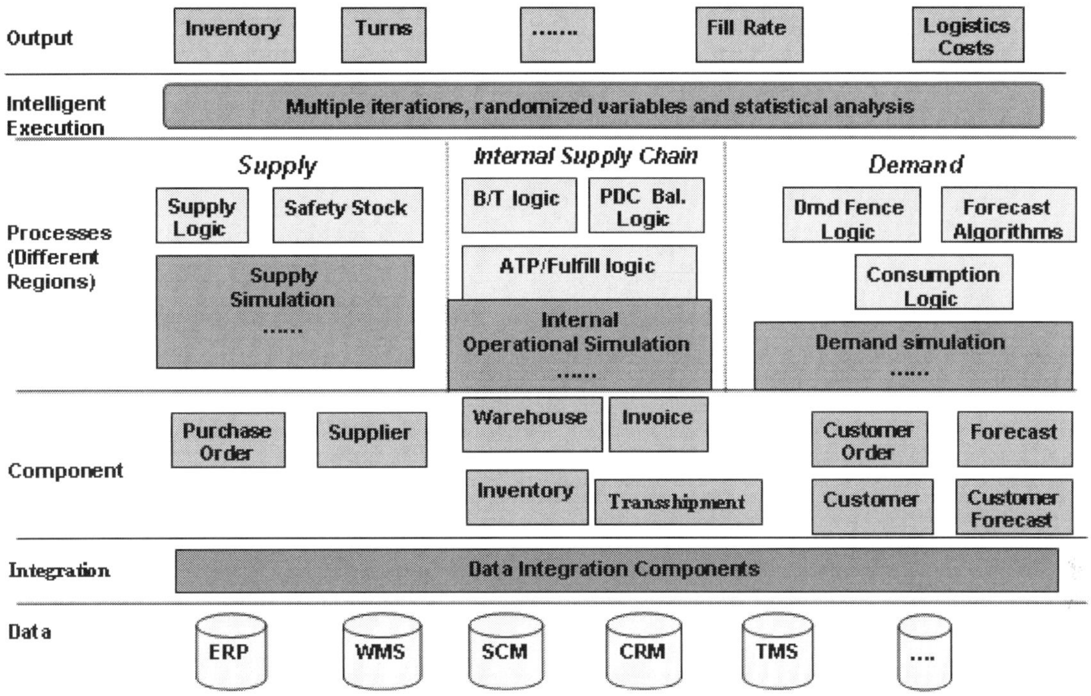

proposed system is designed to be flexible, and should be designed and implemented quickly to validate all kinds of process changes.

The proposed simulation framework for electronic distributors in a global system is shown below in **Figure 1.** The figure shows the primary layers of the design as discussed above. A description of the layers is given below. As the description shows, the layered architecture separates and identifies the parts of the system.

The Component Layer: This layer is used to describe business entities. All different processes will share these components in the layer. The components in this layer are reusable and configurable, and visible to model users. The key components in this layer include:

- *Orders*: orders include the purchase orders created by the buyer to the supplier, the sales orders to the customers, and internal orders.
- *Demand forecast*: customer forecast is an important component that includes buyer's forecasts and supplier's forecast.
- *Invoice*: invoices are documents that describe the goods/services and the terms of sales and are critical components of transactions and accounts.
- *Transshipment*: Transshipment means the shipments of items among different facilities in a supply chain moving from one to another. Transshipment takes advantage of risk pooling concept and can reduce the demand variability.
- *Logistics network*: The logistics network describes the physical movement of goods between components of the supply chain, and defines the structure of the supply chain describing vendors, manufacturers, and warehouses/distribution centers (DCs), etc.

It is seen that the component layer deals with the action components of the supply chain.

The Process Layer: This layer describes the business processes that integrate entities together to accomplish business functions. The processes include supply chain flows such as physical material flows, financial flows including procure-to-pay and order-to-cash cycles, and information flows. It also includes the logic and parameters used to trigger these process flows, as for example inventory replenishment policy, order promise logic, supply logic, and other such decisional characteristics. To mention a specific instance, the inventory replenishment policy may use a specific inventory policy, as for example a min-max inventory policy. The order promise logic defines the methods of finding the available-to-promise (ATP) quantities for different products. Whenever a supplier receives an order from a customer, it must determine how much uncommitted inventory is available to meet the demand. This determination is vital to providing quality customer service and a quick response to supply chain uncertainty. As shown in Xiong et al. (2006) and Jeong (2002), either heuristic rules or optimization techniques such as mixed-integer programming can be adopted to find the ATP quantity and date quoting. More details on the use of optimization techniques can be found in Zhao et al. (2005) and Chen et al. (2002). Consumption logic defines the way the forecasted demand is consumed by actual sales order demand. Typically a "window" is provided within which the forecast is consumed. Primary distribution centers (PDC) balance logic defines the way the inventory is balanced among the PDCs. Branch transfer (B/T) logic defines the way the transshipment moves among branches.

There are four different business process maturity levels in a supply chain. These are disconnected processes, internal integration, intra-company integration with limited external integration, and multi-enterprise integration (Heinrich and Simchi-Levi, 2005). Our generic simulation model aims to be on the maturity level three that works towards integrating business processes within a company with some limited collaborative forecasting for suppliers and customers in a sense that the point of sales (POS) data can be shared between suppliers and buyers. In a more comprehensive environment, this framework can be extended to integrate multiple enterprises in a global supply chain because different partners in a global supply chain can construct their business process simulation framework first, and then integrate them together with technologies such as service-oriented architecture (SOA). The overall process flow is as follows and shown in **Figure 2** describing the four phases of demand input, warehouse operations, demand fulfillment and replenishment.

The Intelligent Execution Layer: This layer is used to guide, control and monitor the simulation process. In this layer, the simulation parameters are set up, the iterations are determined, and different scenarios are simulated. It is also possible in this layer to incorporate different heuristic rules to choose simulation parameters to setup and handle various exceptions in the supply chain to make the decision support intelligent in the execution layer. For example, the optimization through simulation can be implemented in this layer by using heuristics libraries such as Genetic Algorithms, Tabu Search methods, and other approaches. Both Tabu Search and Genetic Algorithms are meta-heuristic algorithms used to solve the combinatorial optimization problems such as the Traveling Salesperson problem. Both algorithms are widely adopted in the advanced planning and scheduling mechanisms of the supply chain. For example, Brimberg et al. (2008) used the Tabu Search heuristic to examine the plant location problem to maximize the return-on-investment. As a Genetic Algorithm example, Tasan and Tunali (2008) recently conducted a survey of assembly line balancing optimization research using Genetic Algorithms. Because there are numerous simulation parameters, each of which can take up a wide range of values, Genetic Algorithms

Figure 2. Overall process flows

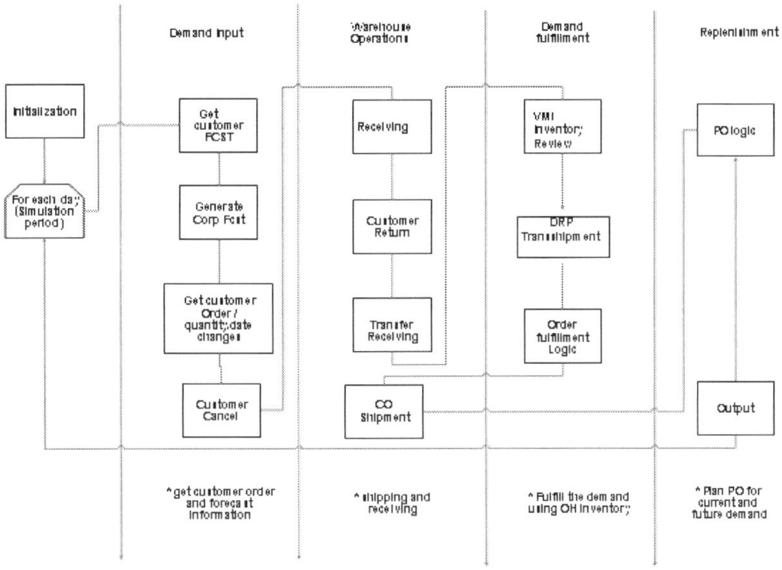

can be adopted to search optimal combinations of different parameters. In addition, heuristic rules are established to take care of some exceptions such as demand change, order cancelation, *etc.* If necessary, the execution layer will generate alerts for human interventions or approval. The conceptual execution is as shown in **Figure 3**.

All the logics can be performed by software agents. An example of agent activity is the monitoring by agents of the on-hand inventory level, and when it drops to a predetermined control level, the agent will trigger the placement of an order. An intelligent agent can also learn from experience to reduce the safety stock continuously and meet the customer service level at the same time, thus improving the performance. A special feature of the agents is the ability to monitor rapidly any exception and then take corrective actions, or send an alert for human intervention.

The Output Layer: This layer provides the metrics calculation in a dashboard format. In order to be effective in evaluating different options, calculations should be standardized and reused in the framework. This would also provide a significant advantage in comparing a variety of different scenarios in different regions. It is more appealing that the output layer can also include the feature of the graphic generation that is common in a number of simulation software products. Further, the output layer can respond to specific queries and make some online analytic processing (OLAP) to make it capable of performing daily business intelligence tasks.

The supply chain performance measures vary from one application to another as shown in Kleijnen and Smits (2003) and Hausman (2002). In professional business applications software, such as the Supply Chain Management module of Oracle E-business Suite, the key performance indicators include the inventory turns, on-time delivery, margin percentage, and planned utilization (Gerald et al., 2002). In addition, the Oracle advanced supply chain planning (ASCP) is integrated with the Business Intelligence Subsystem (BIS), and the performance measurements advocate continuous improvement throughout the supply chain. In the proposed simulation framework presented here, some of the performance metrics adopted are provided by the SCOR model of the Supply Chain Council (SCC, 2009). The SCOR

Figure 3. Conceptual execution

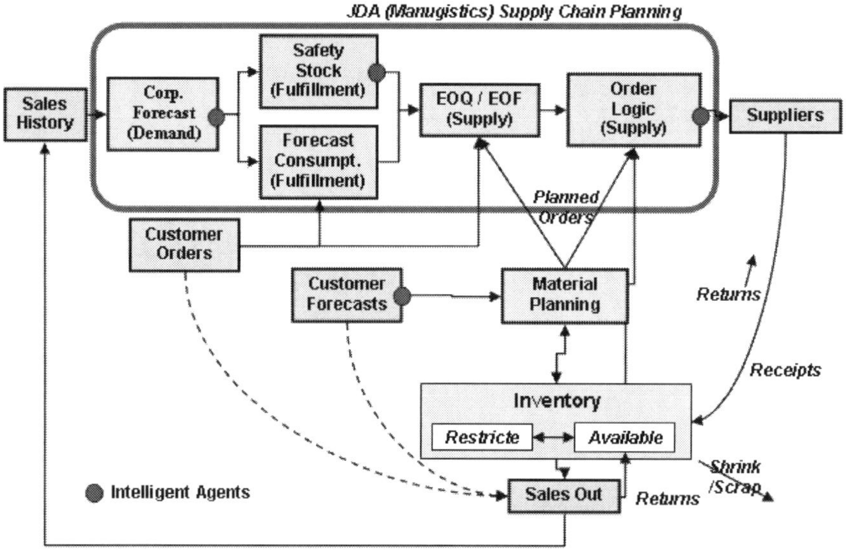

model is a process reference model for supply chain management that is based on five distinct business processes with four levels. It provides detailed supply chain performance measurement such as perfect order fulfillment, order fulfillment cycle time, and other metrics, for each level.

The **Figure 4** is the output illustration of the proposed simulation framework. The Figure 4 is a simple illustration of the structure of the simulation output of the proposed framework. The values and details of Figure 4 are shown later in **Figure 5** with the illustrative Case Study. It is shown with numbers and results derived from the execution of the case study. In this system, inventory turnover is the ratio of cost of goods sold to the average inventory level, an important measure of inventory velocity. The Customer Service Levels (CSLs) shown in the figure are used to measure the percentage of satisfied customer orders, while Available-to-Promise (ATP) refers to the uncommitted inventory that enables the company to make accurate promise to customer demands. Capable-to-promise (CTP) is another term that refers to the

Figure 4. Simulation outputs

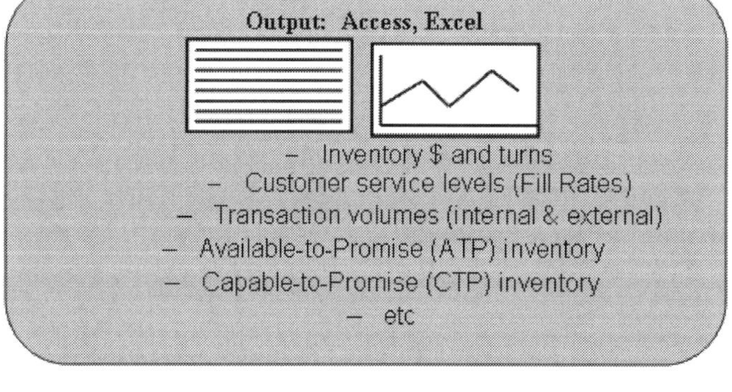

Table 1. Key implementation components

Language	JAVA
Development Tool	Eclipse
Random Distribution	Colt.jar
Input/Output	Access DB/BIRT/EXCEL
Forecast Algorithm	OpenForecast
Interface	Web pages/Tomcat

company's additional ability to determine the extra resources to meet unplanned demand.

2.2. Implementation Issues

The elements of implementation of SCEM simulation are software for various functions, web-based user interfaces, and computation algorithms. The key implementation components are presented in summary in **Table 1**.

As shown in the summary in **Table 1**, the simulation language chosen is JAVA because it is platform independent, and that makes the implementation more flexible. Another advantage is that most JAVA technologies are free for application developers and users and this makes the simulation framework cost efficient. To simplify the development process, the Eclipse development platform has been adopted because the Eclipse is open source that consist of extensible framework, tools, and runtimes to develop and manage simulation software across its life cycle (Eclipse, 2009). The core of the implementation for the supply chain event management is to generate random distributions of extreme and failure conditions for the framework, otherwise known as risk events. The system adopted in this project is Colt.jar. The Colt distribution provides an infrastructure for scientific computing in Java. Its library contains usable structures and algorithms for data analysis, Monte Carlo simulation, parallel and concurrent programming. The input/output design of the project, or the user interface, is a simple use of Excel spreadsheet, Access database, and the Eclipse-based open source reporting system BIRT. With BIRT, a variety of reports such as lists, charts, cross tabs, compound reports can be added to the application. The forecast algorithm is of great importance in the matching of supply and demand. The forecast algorithm used in this project is OpenForecast. It is a general purpose package with forecast models written in Java (OpenForecast, 2009). The web-based interface of the simulation project uses Web Pages/Tomcat. Tomcat is an implementation of Java Servlet and JavaServer Pages technologies.

With the progressive development of information system and technologies, increasing volumes of data become available. The sheer volume of data increases the challenges of business analysis. Traditional business analysis tools, such as analyses using Excel are still widely used and still have their values. However, these relatively simple tools are inadequate in managing the dynamics of business systems in the presence of very large volumes of data. Specially developed simulation and optimization models are often required to quantify and optimize the relationships and processes in the business domain. The next section addresses one key implementation issue, *viz.* data collection in the proposed intelligent simulation framework.

The data inputs to a model are critical success factors (CSFs) for the modeling process. To simplify the process and to reduce errors, the best practice is that the data collection and acquisition process be automated. In a large business, practical implementation of data acquisition is often a process of loading transaction data that has been transformed into a standard form into a Data Warehouse, otherwise known as a Very Large Data Base (VLDB). A Data Mart is defined as a focused subset of the Data Warehouse, where the focus is topical or based on functional use. A Data Mart is often useful in simplifying the data access for a particular application. A Data Mart can gather a variety of topical data from the data warehouse and provide a common and unified data access interface for the simulation application.

Operations Research (OR) mapping tools, such as JAVA hibernate, could be used to provide easy access to the database and provide the flexibility for late data structure changes. Even the data collection can be fully automated, with associated data cleansing. It is important to perform this last function as since iterative checking is still needed. There are several issues related to data collection and OR mapping tools stated as follows:

1. *Finding a fit distribution.* Fitting activities in the supply chain to certain distributions may oversimplify the process. It could be possible that no fitting distribution can be found. The assumption of a non-fitting distribution can be avoided by using a large sample of products and by using the actual historical data.
2. *Implementation using simulation software with object-oriented technology.* It has been mentioned earlier that the framework should be generic enough and high level enough so that it is easy to plug-in different process and component. Professional software such as Promodel can be used to help simulate some of the proposed supply chain simulation. Further, Promodel adopts intelligent evolutionary algorithm, such as genetic algorithms, to find optimal combinations of input parameters that can be integrated with proposed simulation framework.

In addition, the level of detail is an important decision parameter in a modeling process, *i.e.* the decision to decide what should be modeled and to what degree. Typically, this decision includes a large amount of prior analysis of the domain of the individual problem.

3. AN ILLUSTRATIVE CASE STUDY

In this section, a case study is presented as an illustration. The case is that of a leading electronics distributor in its efforts to improve its inventory performance and customer service by the use of the intelligent simulation model. The distributor's name is disguised because of the concerns of confidentiality. However, the reported business process framework and analysis is based on real data. The results reveal the advantages of the proposed simulation approach as presented earlier.

The essence of the simulation is to use the electronics distributor's sales data to make forecasts to match demand and supply, in a process of optimizing its supply chain operations. The intelligent simulation process makes extensive use of Intelligent Agents, also known simply as Agents. Agents, as described in Bodendorf and Zimmermann (2005), are software components working in the background that simulate intelligent human behaviors, perceptions, and actions. These agents can be used to manage and check system performance and response, and be used to optimize the performance of the overall system. The challenge of supply chain management is that a local decision, which could improve local performance, might have significant negative impact on the global system. The implied local optimization, as opposed to the desired global optimization, is equally a problem with human decision makers as it is with agent-based systems.

Manugistics is the name of the professional supply chain software, and also the name of the provider of the software, to be used for supply chain planning. Intelligent heuristic rules are set

Figure 5. Illustrative results for one product segment

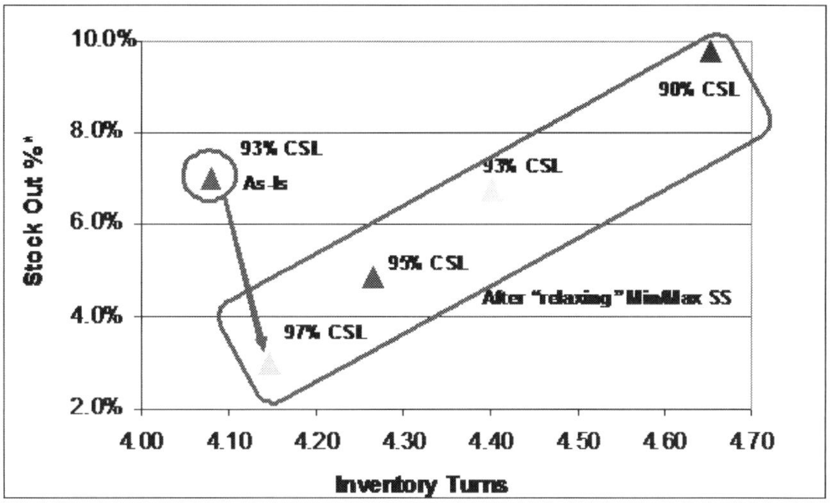

up for points of forecast demand, order logic, and for safety stock fulfillment. Human intervention is permitted, and sometimes required for reactions to supply chain situations as needed. The simulation problem was tested with one product segment including a number of different products. The results as presented in **Figure 5** are based on the product segment as currently using a min-max inventory policy with a customer service level of 93%.

In Manugistics, min/max safety stocks (SS) are the parameters used to calculate safety stock used for planning. If SS is higher than max SS then Max SS is used, while if SS is less than min SS then Min SS is used. A business may set those parameters based on judgment and experience. The challenge is determining whether these settings are good or not, and whether these human" intelligence" interventions can help improve the service level and inventory efficiency. This study investigates the factors related to the impact of relaxing these constraints, and questions about whether we need the extra human intelligence, or whether it is better to simply follow the statistics.

The example is based on data collected for a sample of 500 parts. In the simulation process the history is "rerun" based on the planning parameters as stated earlier. Some of the key inputs to the process are stated as follows:

1. Product information including SS, min/max SS parameters
2. Demand/forecast history
3. Supply chain infrastructure (warehouse etc.)

The simulation starts everyday with the customer placing or changing orders or sending customer demand information. At this time software agents begin work to evaluate customer demands, inventory positions, and the situation in the supply pipeline. The software agents use a set of rules that simulates the human behavior to decide whether to create, or to push or pull purchase orders. The safety stock parameters are the key parameters used to calculate the required safety stock at a particular location. In practice, the software agent compares the inventory position with the calculated safety stock and makes the above procurement decisions. The simulation first validates the model using existing system parameter settings to check whether it could match the actual historical data, and then it evaluates

Figure 6. Illustrative results for one product segment

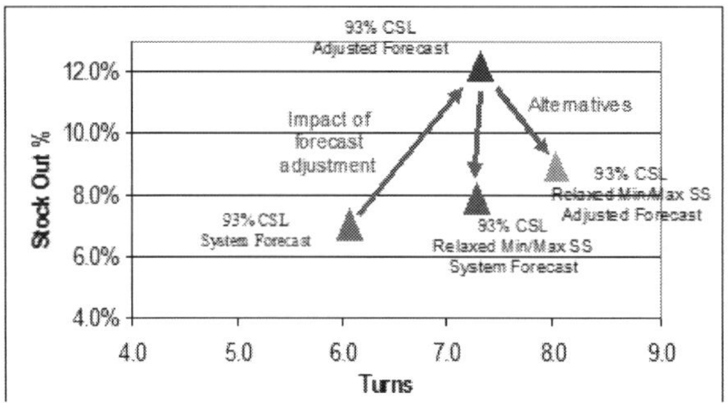

different options by relaxing various Min/Max safety stock constraints.

With the proposed intelligent simulation model, it is possible to improve the customer service level and to speed up the inventory turnovers at the same time. Figures 5 and 6 are included to show the illustrative results of this test simulation. The company used as an example has a 93% customer service level (CSL). The major concern with CSL in situations of uncertain demand is safety stock to avoid a costly stockout, *i.e.* the inability to satisfy customer demand because demand variability has drained the inventory. After relaxing the Min-Max Safety Stock conditions, as shown in Figure 5, we can see that the customer service level can be improved up to 97% and at the same time the inventory turns can be increased, thereby increasing revenue. Figure 6 shows the inventory turnover can be increased for the fixed customer service level of 93% by a simple procedure of adjusting forecasts and relaxing inventory replenishment parameters.

The other concern is about the supplier's order change event management. This can have negative impacts on the available inventory. The system background is that the suppliers prefer a stable demand pattern and tend to complain when there are a large number of change orders. This is a problem arising from the purchase order management systems and procedures used by Manugistics. The same order could be pushed or pulled a number of times because of several factors such as demand change, forecast netting, and other supply chain dynamics. In supply chain terminology, demand changes and forecast netting tend to be pull factors affecting placed orders, while push factors often involve production in the supply chain and other supply chain factors. The business team brainstormed more than a dozen of different ways to improve this situation and analyzed each one systematically. The simulation model evaluated each of them and their respective impacts on inventory.

In practice, any PO change with a reduction in quantity could result in an inventory increase. The challenge is not only to find the way to reduce the PO changes in number and quantity, but also to minimize the impact on inventory resulting from the PO changes. In a manner similar to the previous case described, the six months of demand information is captured and the demand changes tracked. The demand changes are the biggest drivers of PO changes. This change information is encoded into the software agents managing the system. The software agent managing the system has additional rules coded to make decisions about whether to change PO or otherwise.

Figure 7 shows the different inventory impacts of different settings of two different rules, and illustrates the differences between "pushing" inventory from an inventory perspective versus a "pull" representing customer demand. Push $ is the threshold of amount used to determine whether to push the order or not to do so. If the

Figure 7. Reduction in total change orders

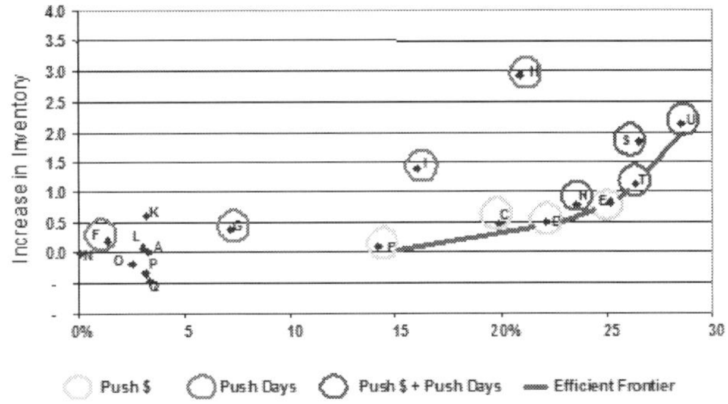

order is less than the Push $, then the decision is not to push out. If the push out days count is less than push days, then the decision is not to push. It is clear that these decisions have a direct and immediate impact on inventory. The study is about the impact of the modeling and decision method on inventory. The diagram shows that with about 0.5% inventory increase, it is possible to reduce PO changes by more than 20%.

The interesting finding of the simulation model is that it does not support some of the most promising proposals. One obvious suggestion is that it is not recommended to pull or push orders within certain frozen windows, typically about 3 weeks, because of the dynamics of the supply chain, and resulting from a number of small orders are created due to this frozen window. Another suggestion disapproved by the simulation study is to pull the planning order by certain number of weeks. The planning orders are the internal orders placed for customer forecasts. Because of the dynamics of inventory planning parameters, these planning orders can suddenly change ASAP orders, and result in many PO changes. The concept is built around a pull factor in the required date of these planned orders that will reduce the sudden changes in ASAP orders. After the simulation study, it was found that the order changes actually increased and inventory increased significantly. The reason later found is that even when we pull in the required

Figure 8. Set proper trans-shipment threshold to minimize total cost

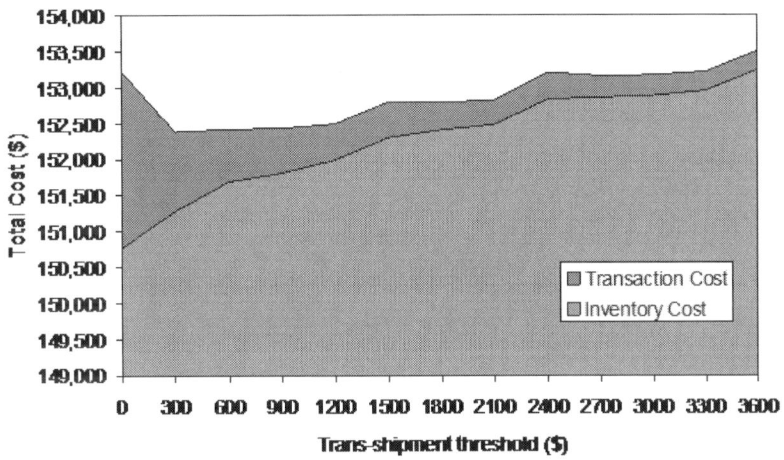

date of these orders, the demand change dynamics are still existing, and sometimes even worse. Eventually, it is needed to handle more forecasts and forecast changes and bring inventory much earlier than needed.

The prototype system was also used to search for the proper transshipment thresholds with the objective to minimize total inventory cost. Thus, the decision levels generated by the prototype could be used to balance the warehouse inventories, and to validate the network strategy for transshipment. **Figure 8** reveals that total inventory cost reaches minimum when transshipment cost is between $300 and $1200. Therefore, a transshipment strategy can be used for cost reduction if the transshipment cost is within this range.

4. CONCLUSION

Presented in this chapter is an intelligent simulation framework for supply chain event management. Supply chain event management is a technique and a framework used to analyze and evaluate dynamic supply chains which have large variability and high risk environments in terms of satisfying customer demands. A four-layer process model is proposed and the implementation technologies are also addressed and discussed. An actual case is presented, analyzed, and discussed. This case is one of a large electronics distributor, typical of the uncertain demand supply chain situation.

An event-based simulation model has been presented to analyze and improve the performance of the supply chain. The primary concern in the performance of the model is the inventory performance of the electronics distributor in the presence of uncertainty and supply chain events in adequate supply of the customer demand. The presented model has been tested to evaluate the inventory performance for a segment of product for an electronics distributor, and the results show the effectiveness of the proposed simulation framework.

The study shows that the simulation model is effective and that it achieves the desired results of higher customer satisfaction with lower inventory levels. The combination of enhanced customer satisfaction with lower inventory levels is a way to achieve higher profitability. Thus, it becomes a highly desired situation for a business. In other words, the contribution of the chapter is a new system to analyze a business supply chain by methods of simulation of events, and develop a framework for lower inventory, better management of uncertain supply chain activity, and the achievement of a higher profitability.

There are several possible future extensions to the research. For example, the simulation model can be further extended to include other business intelligent tools to make it more intelligent, For example, data mining can be adopted to find hidden patterns or exceptions from the data available in the data warehouse, so that the system can respond quickly to any supply chain risk or market change.

REFERENCES

APICS – The Association for Operations Management. (2005). *The Certified Supply Chain Professional Learning Systems*. APICS – Advancing Productivity, Innovation, and Competitive Success, Chicago, IL. Retrieved from http://www.apics.org/

Barber, K. D., Dewhurst, F. W., Burns, R. L. D. H., & Rogers, J. B. B. (2003). Business-process modeling and simulation for manufacturing management: A practical way forward. *Business Process Management Journal*, *9*(4), 527–542. doi:10.1108/14637150310484544

Bodendorf, F., & Zimmermann, R. (2005). Proactive Supply Chain Event Management with Agent Technology. *International Journal of Electronic Commerce*, *9*(4), 57–89.

Brimberg, J., Hansen, P., Laporte, G., Mladenovic, N., & Urosevic, D. (2008). The maximum return-on-investment plant location problem with market share. *The Journal of the Operational Research Society*, *59*(3), 399–406. doi:10.1057/palgrave.jors.2602349

Cachon, G., & Terwiesch, C. (2009). Matching Supply with Demand: An Introduction to Operations Mangement, (2nd Ed.). New York: McGraw-Hill.

Chen, C., Zhao, Z., & Ball, M. O. (2002). A model for batch advanced available-to-promise. *Production and Operations Management*, *11*(4), 424–440.

Eclipse. (2009). Retrieved from http://www.eclipse.org/

Gerald, B., King, N., & Natchek, D. (2002). Oracle E-Business Suite Manufacturing and Supply Chain management. New York: McGraw-Hill/Osborne.

Greasley, A. (2003). Using business-process simulation within a business-process reengineering approach. *Business Process Management Journal*, *9*(4), 408–420. doi:10.1108/14637150310484481

Hausman, W. H. (2002). Supply Chain Performance Metrics: The Practices of Supply Chain Management. Boston: Kluwer Academic Publishers.

Heinrich, C., & Simchi-Levi, D. (2005). Do IT Investments Really Pay Off. *Supply Chain Management Review*, *9*(4), 22–28.

Hung, W. Y., Kucherenko, S., Samsatli, N. J., & Shan, N. (2004). A Flexible and Generic Approach to Dynamic Modeling of Supply Chains. *The Journal of the Operational Research Society*, *55*, 801–813. doi:10.1057/palgrave.jors.2601740

Jeong, B., Sim, S., Jeong, H., & Kim, S. (2002). An available-to-promise system for TFT LCD manufacturing in the supply chain. *Computers & Industrial Engineering, 43*(1,2), 191-212.

Kleijnen, J. P. C., & Smits, M. T. (2003). Performance Metrics in Supply Chain Management. *The Journal of the Operational Research Society*, *54*, 507–514. doi:10.1057/palgrave.jors.2601539

Lee, S. K., & Hwang, C. S. (2003). Architecture layers and engineering approach for agent-based system. *Information and Software Technology*, *45*(13), 889–898. doi:10.1016/S0950-5849-(03)00057-0

Martin, C. (2005). Logistics and Supply Chain Management, (3rd Ed.). Englewood Cliffs, NJ: Prentice Hall.

OpenForecast. (2009). Retrieved from http://openforecast.sourceforge.net/

Otto, A. (2003). Supply Chain Event Management: Three Perspectives. *The International Journal of Logistics Management*, *14*(2), 1–13. doi:10.1108/09574090310806567

PRTM – Pittiglio. Rabin, Todd, & McGrath. (2005). *Managing Complexity in Global Supply Chain*. PRTM's 2005 Supply Chain Complexity Survey. Retrieved from http://www.prtm.com/

Raman, A., & Ton, Z. (2005). Operational Execution at Arrow Electronics. Harvard Business Case, Harvard University, Cambridge, MA.

Supply Chain Council. (2009). *SCOR Model (Version 9.0) of Supply Chain Council (SCC)*. Retrieved from http://www.supply-chain.org/

Tasan, O. S., & Semra, T. (2008). A review of the current applications of genetic algorithms in assembly line balancing. *Journal of Intelligent Manufacturing*, *19*(1), 49–69. doi:10.1007/s10845-007-0045-5

Wang, H. (1988). A Layered Architecture for Manufacturing Operations Planning. *Computers & Industrial Engineering*, *14*(2), 201–210. doi:10.1016/0360-8352(88)90029-0

Westfechtel, B., Munch, B. P., & Conradi, R. (2001). A layered architecture for uniform version management. *IEEE Transactions on Software Engineering*, *27*(12), 1111–1133. doi:10.1109/32.988710

Xiong, M. H., Tor, S. B., Bhatnagar, R., Khoo, L. P., & Venkat, S. (2006). A DSS approach to managing customer enquiries for SMEs at the customer enquiry stage. *International Journal of Production Economics*, *103*(1), 316–331. doi:10.1016/j.ijpe.2005.08.008

Xu, K., Liu, L., & Wu, C. (2007). A three-layered method for business processes discovery and its application in manufacturing industry. *Computers in Industry*, *58*(3), 265–278. doi:10.1016/j.compind.2006.06.001

Yu, B., Harding, J. A., & Popplewell, K. (2000). A Reusable Enterprise Model. *International Journal of Operations & Production Management*, *20*(1), 50–69. doi:10.1108/01443570010301083

Zhao, Z., Ball, M. O., & Kotake, M. (2005). Optimization-Based Available-To-Promise with Multi-Stage Resource Availability. *Annals of Operations Research*, *135*(1), 65–85. doi:10.1007/s10479-005-6235-7

Chapter 8
Mathematical Models for Optimizing the Global Mining Supply Chain

Bruno Santos Pimentel
Federal University of Minas Gerais, Brazil

Geraldo Robson Mateus
Federal University of Minas Gerais, Brazil

Franklin Assunção Almeida
Federal University of Minas Gerais, Brazil

ABSTRACT

The present chapter discusses the application of intelligent decision-support systems – mathematical programming-based in particular – to operations management within the mining industry. The underlying production and distribution planning and scheduling problems have often been addressed individually, in disregard of upstream and downstream operations. A supply chain approach to mining operations, however, requires an integrated perspective which takes into account mine, railway and port operations, as well as domestic and international supply stations served by appropriate logistics channels. Three main topics are discussed here: recent operations research developments in the mining industry; integrated approaches towards the development of decision-support systems to address a global mining supply chain; and possible solution approaches to the integrated problems. The main thread is oriented to mathematical programming approaches, but relevant applications of simulation and artificial intelligence techniques are also discussed.

INTRODUCTION

The present decade has shown extreme variations in ore prices and demand. In a growing economy environment, when ore prices and demand are high, mining companies tend to invest in establishing new ventures, expanding capacity, or consolidating the market. The high ore prices improve the feasibility of lower grade operations and geographically distant ore producers, especially if product prices are significantly higher than freight costs. Customers tend to prioritize quantity over quality and there is

DOI: 10.4018/978-1-61520-605-6.ch008

Figure 1. A global mining supply chain

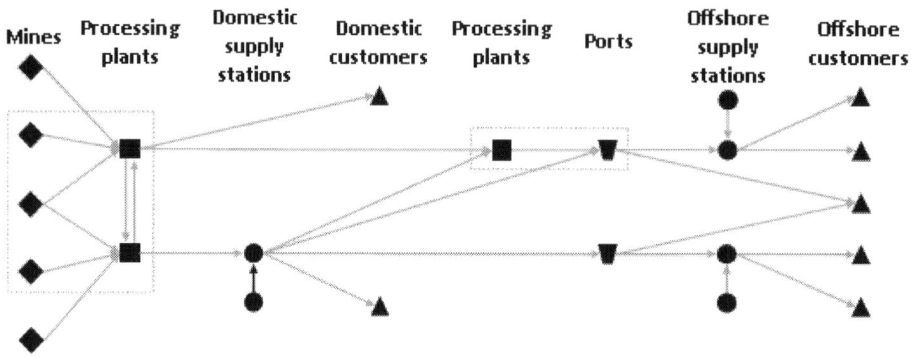

a high pressure on the availability of production and distribution systems. However, in an unstable or shrinking economy, the situation is quite the opposite. Demand and prices drop, thus forcing mining companies to reduce production levels and shutting down expensive or lower grade operations. Customers tend to prioritize quality over quantity and there is a high pressure on reducing production and distribution costs. Such dynamic and uncertain scenario has driven base metals mining companies to pursue higher efficiency levels in their global operations. This can be achieved by adequately modeling and planning operations in an integrated framework, such as the global supply chain.

A global mining supply chain can be defined as network of integrated facilities designed to convey ore products from mines and processing plants, through a variety of modals including roads, ducts, railways, rivers and oceans, to metal processing customers, which can be located at significant geographic distances. Figure 1 depicts a general global mining supply chain.

Although there has been significant effort in developing intelligent decision-support systems for the mining industry, these are usually focused in tasks such as mine development, hauling equipment dispatch and railway scheduling. An approach to the global mining supply chain, on the other hand, requires an integrated perspective which must take into account mine, railway and port operations, as well as domestic and international supply stations served by appropriate logistics channels. There is a whole growing stream of research aiming at the integration of strategic, tactical and operational decisions in supply chain planning (Melo et al., 2009) that could be exploited to develop intelligent systems to assist operations management in an integrated mining industry. Such developments, nevertheless, do come with an additional degree of complexity in both mathematical modeling and computational requirements.

The aim of this Chapter is thus threefold: firstly, to present recent developments of intelligent systems in the mining industry by discussing problems, models, applications and challenges to implementation; secondly, to discuss an integrated approach towards the development of decision-support systems to tackle the global mining supply chain problems; and thirdly, to discuss possible solution approaches to the problems addressed. Although the iron ore mining industry is used as a case study, there would be no significant loss of generality if different base metals industries were considered. Due to its scale and geographical range, the iron ore mining industry includes many of the operational aspects relevant to modern commodity production and distribution systems. It is important to notice, however, that the scope of this Chapter does not include an extensive analysis of supply chain theory. The most relevant supply chain concepts, mainly those related to operations design, planning and optimization, are addressed in a way to support the functional and inter-temporal integration that spans today's large-scale mining operations. The computational and mathematical systems that can be developed to solve the related problems comprise the main focus of the discussions.

BACKGROUND

Traditionally, a greater amount of effort has been directed towards the optimization of mine development and its long-term production plan. However, in such a continuously consolidating industry, it is common to observe complex integrated operations involving not only open-pit and underground mining, but also all logistics operations concerned with hauling equipment dispatch, cargo composition, railroad transportation planning and traffic control, and port handling, blending and loading. Operations Research approaches to mining industry problems can provide management support throughout that functional dimension, as well as at strategic, tactical and operational decision levels. Depending on the scenario and on the problem size, different solution approaches may be required, especially when there is a choice between mathematical programming and simulation approaches. Within the mining industry, where large financial and material flows are involved, there are clearly several improvement opportunities.

An Overview of Mining Operations

Mining is an extremely capital intensive industry and, as such, demands high efficiency with regard to operation management decisions. There are today two important trends that have been influencing the decision-making process in large-scale mining operations: the continuous ore grade decrease in mines throughout the world, which impacts product quality and processing costs; and the ever increasing pressure over operational costs. Managerial decisions must then be quantitatively evaluated in order to account for those issues and still attain the expected results.

The first major and critical decision in mining operations is the establishment of a new mining venture, since it often requires the highest levels of capital investment. Traditional evaluation of mining projects includes drilling and sampling, generating a representative orebody model, deciding mining and processing methods, assessing capital and operating costs and developing a technical and financial life-of-mine plan. After establishing a feasible mining operation, another major decision involves determining the mine design and development plan. Proper mine design must account for access opening, orebody detailing, open pit (or underground tunnel) preparation, among other strategic decisions. In the mine development plan, detailed orebody data is used to build a discrete block model that allows scheduling ore removal according to the net present value, expected grade, processing plant requirements and the terrain's structural constraints.

Ore extraction is performed by a fleet of heavy-duty, off-road trucks and shovels, which remove the run-of-mine (ROM) from the pits and take it to appropriate locations – valuable ore to milling and screening facilities and valueless material to waste stockpiles. Since the operation of hauling equipment is rather costly, appropriate allocation and dispatching of trucks and shovels to load and discharge points is of utmost importance (See Figure 2).

Milling and screening facilities are employed to reduce ROM size according to the requirements of the processing plants. Besides size, those requirements may also include grade considerations, in which case the ROM may need to be blended and homogenized. Blending may be achieved by appropriately feeding the milling and screening facilities according to the characteristics of each mining work bench and thus allocating hauling equipment accordingly. Another way of achiev-

Figure 2. The mine pit and milling subsystem

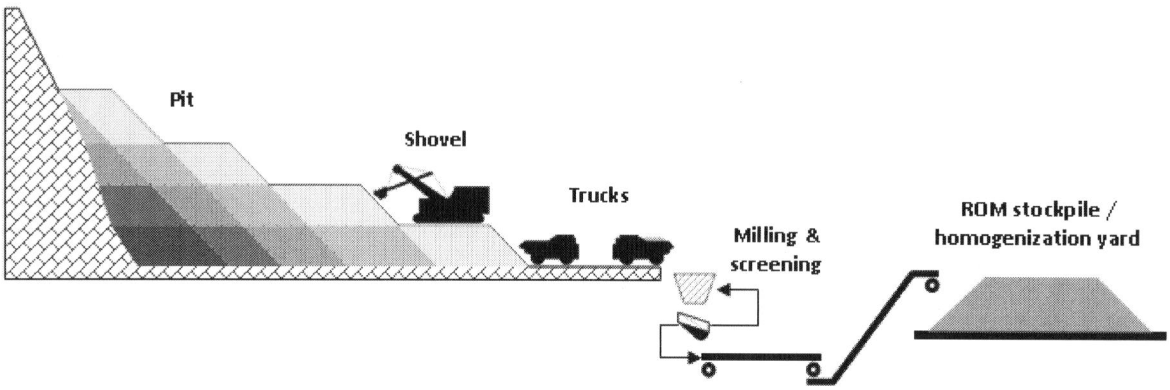

ing proper blending requires a homogenizing yard, where different ROM are properly mixed in stockpiles so that physical and chemical properties can be homogenized. Conveyor belts are used to transport ROM from the milling facilities to the homogenizing yard and from there to the processing plant feed.

Processing plants enhance ore product quality by means of classification (grinding and screening) and concentration (for example, flotation and magnetic separation for iron ore) processes. Processing plant operations are aimed at maximizing mass efficiency, which depends not only on the characteristics of the ROM offered by the mine, but also on adequate planning and control. In the iron ore industry, pelletizing plants can also be employed to process ultra-fine materials generated at the processing plants. Although such ultra-fine materials are also sold as end products, they cannot be used to feed a blast furnace directly and must, therefore, be agglomerated into pellets. Pelletizing plants can be located at the mine complex or at the port (or even at both). The pellets produced from pellet-feed ore and other agglomerating materials is also transferred to the mine or port product stockyards and from there to steelmaking customers.

In the product stockyard, ore products are stored according to commercial programs and logistics (railways, roads, rivers or sea) availability. A set of stackers form stockpiles and a set of reclaimers retrieve those stockpiles to compose specific ore product shipments. Again, a set of conveyor belts, or even trucks and shovels of smaller size, can be used to accomplish lot loading (See Figure 3).

Railroads are often used to transport ore products to the port, where exportation takes

Figure 3. The mine's product stockyard and railroad interface

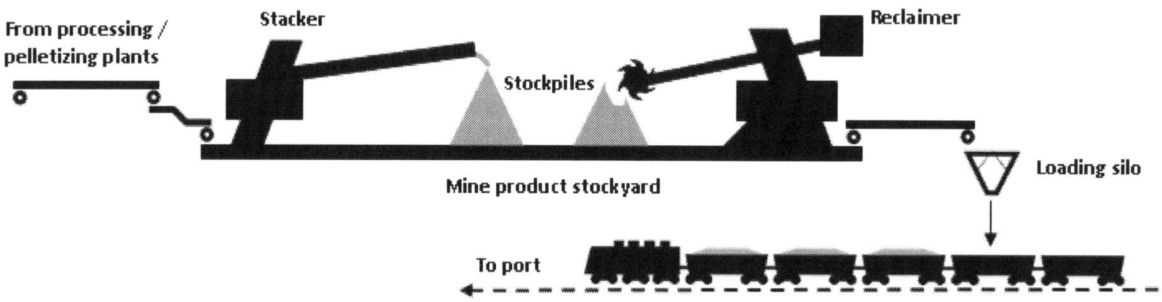

Figure 4. The port subsystem, with railroad and ocean interfaces

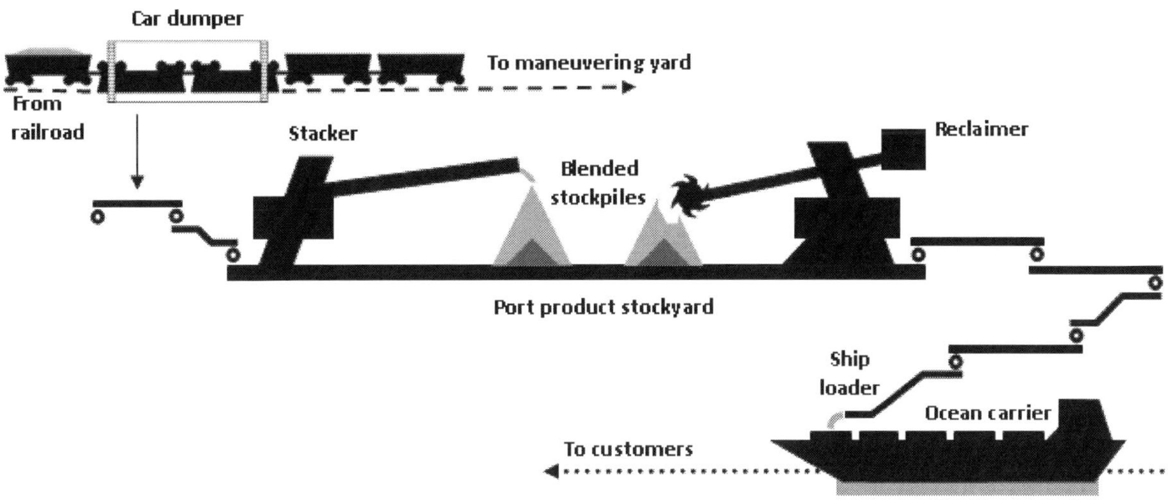

place, or to domestic supply stations, where additional processing – usually screening of lumpy ore products degraded during transport – may occur before serving local demand. Trains loaded with ore products may have to share the railway with trains used to convey different bulk cargoes, or even with passenger trains. Railroad capacity may thus be significantly constrained and impose important restrictions to the overall system operations.

The port receives ore products from a set of mine complexes and, according to a short-term program – which includes train and ship arrival, as well as stock levels and quality –, transfers those ore products to the product stockyard, pelletizing plants, screening plants, or even directly to awaiting ships. Additional blending may take place during stockpile formation or ship loading. Domestic demand may also be served by shipments originated at the port, often by rail or road.

Although some mining companies may also own and operate seaborne carriers, the majority of commercial contracts rely on third-party freight operators to transport ore products to offshore customers. Ports can ship ore products directly to customers located at, or close to, other offshore ports; alternatively, shipments may be directed to offshore supply stations, where additional blending and fine separation may take place, or where other pelletizing plants may be located. From offshore ports and supply stations, international demand can be served by train or road.

Some characteristics of commodity metal markets are essential to understanding decision-making in mining operations, especially when integrated approaches are considered. In the base metals market, such as iron ore, transportation costs amount to a significant fraction of the commodity's value. Usually, steel companies establish contracts with iron ore producers, in which prices are settled in accordance to chemical and physical characteristics of the products. Negotiated prices generally hold for approximately a year, and iron ore producers are obliged to supply a certain amount of iron ore to each contracted buyer (Newman et al., 2006).

Basic Problems in Mining Operations

An analysis of the literature devoted to optimization of mining operations is performed according to decision-level and functional dimensions which will further contribute to the global mining supply chain concept. At the decision-level dimension,

Management Science categorizes the decision-making process according to the following structure (Bradley, Hax & Magnanti, 1977):

1. **Strategic planning** deals with long-term decisions involving managerial policies and resource development. Decision analysis is large in scope and requires very aggregate information. In the mining context, strategic decisions include, for instance, the establishment of new mining ventures and capacity expansion of existing facilities.
2. **Tactical planning** seeks effective resource allocation to satisfy demand requirements and operation constraints on a medium-range time horizon. Decision analysis still requires significant information aggregation. In the mining context, tactical decisions include, for instance, developing mine or port production plans, or railroad transportation plans.
3. **Operations control** is concerned with short-term decisions, usually involving low-level programming and scheduling. Information generated at higher decision levels must be disaggregated to a useful degree. In the mining context, operational decisions involve, for instance, allocating and dispatching mine haulage equipment, and scheduling train traffic in constrained railroads.

At the functional dimension, mining operations typically involve the following:

1. **Open-pit and underground mine scheduling**, which involves determining an optimal production schedule over the life of the mineral deposit, from the feasibility study to the termination phase;
2. **Mine load and haulage equipment allocation and dispatching**, which involves effectively deploying trucks and shovels according to some performance criteria, while maintaining a steady, reliable ore feed to the processing plants;
3. **Processing**, which employs physical and chemical processes devoted to enhance ore quality by performing classification, concentration and agglomeration operations;
4. **Blending**, which involves mixing ore by appropriate stockpiling and reclaiming, both at mine and port stockyards, in order to deliver ore at the required quality specifications;
5. **Railway scheduling and dispatch**, which involves effectively routing and controlling train movement over a line, as well as planning meeting and passing of trains on single-line sections, while also attending to operational constraints;
6. **Port planning and scheduling**, which involves determining an optimal plan to satisfy shipment demand, while minimizing delays in serving the ship arrival queue, and determining effective routes to convey ore products from the stockyard to the ship holds.

It is important to notice that the review presented in the sections that follow does not intend to be thorough. Instead, the idea is to provide a broad notion of the problems, modeling approaches and solution procedures that pave the way to establishing the global mining supply chain concept.

Strategic Mine Planning

The initial decisions in a mining venture are the riskiest as they involve large quantities of financial resources and longer maturation time periods, and ultimately determine the success or failure of the organization. The main question posed here is to determine whether or not a proposed mining venture is feasible and, more importantly, if it is lucrative within shareholder expectations. Although the location of a mine complex is usually constrained to the vicinities of the ore reserve, many important aspects regarding environmental issues, logistics services and prices and even political uncertainties must also be considered.

Managerial decision-making is as important as market pricing and orebody grade when determining profitability, since strategic decisions must be weighed according to their economic consequences. Many different valuation techniques can be applied to decide upon project feasibility, all techniques being closely dependent on the proposed mine development plan. Discounted cash flow and risk/return analysis are often applied to evaluate return on investment according to orebody characteristics, capital and operating costs of equipment, and market trends on ore prices, transportation costs and expected demand levels (Runge, 1998). However, assigning resources to resolve the underlying uncertainties in project assumptions can make it economically infeasible. Both market and process uncertainty must be taken into account (Kamrad & Ernst, 2001): the first associated to output price and the second associated to the random variability inherent to the output's yield. An overview of strategic mine planning under risk assumptions is given by (Steffen, 1997).

A strategic mine plan can be composed of five stages (Kuchta et al., 2004). The first stage involves determining orebody grade extension and distribution by drilling and sampling the ore reserve. Using geostatistical methods (David, 1988) and quantitative resource assessment (Scott & Dimitrakopoulos, 2001), a three-dimensional discrete orebody block model can be developed to represent geological – grade, mineralogy, density and tonnage of component minerals – and economical attributes – operational costs and commodity prices (Topal, 2008). It is important to notice that quantifying the benefits from improved orebody data can be rather challenging since costs associated to data collection and analysis are significant. The second stage of the mine strategic plan involves deciding the most appropriate mining method: open pit, underground or a combination of both. The decision depends on geometry and depth attributes of the mineral deposit, as well as on the structural characteristics of the terrain. Open pit mining is thus more appropriate when the mineral deposit is relatively close to the surface. However, as the pit deepens, mining costs rise due to the need to establish gradually sloping pit walls to prevent waste material from sliding down into active regions. Also, haulage routes become more complex and the probability of reaching underground water increases. Hence, at some point the operation must either cease or proceed underground (Newman et al., 2006). The third stage involves designing the mine infrastructure, layout and necessary equipment in order to perform efficient operations. The fourth stage involves determining the mine production schedule itself. The fifth and last stage involves decommissioning the mine operations and restoring the site to an environmentally acceptable state (Caccetta & Kelsey, 2001).

Open Pit and Underground Mine Planning

A mine development plan encompasses long-term and short-term decisions regarding the sequence in which the mineral reserves should be exploited in order to maximize performance. This is probably one of the most capital-intensive phases of a mining venture. It is no surprise that the majority of the literature is devoted to address this problem efficiently. In open pit mines, the orebody is accessed from a large stretch which exposes the ore. Mining begins with an initial pit and proceeds with increasingly larger pits enclosing the previous ones until the orebody limits are reached. In underground operations, shafts and ramps provide access to the orebody and to the production area, thus constraining the mine layout. In both open pit and underground mines, however, the long-term development plan is used as a basis to determine shorter-term – quarterly, monthly or weekly – production plans. Recent reviews of models and algorithms for the open-pit mine production schedule problem is provided by (Caccetta, 2007) and (Osanloo et al., 2007). However, a brief analysis of some important aspects is presented in the following sections.

The Ultimate Pit Limit Problem

Determining the ultimate pit limit is a fundamental problem in mine planning. The optimal ultimate pit of a mine is the result of extracting the volume of material which provides the total maximum profit while satisfying integrity restrictions (e.g. wall slope) on the mine's physical structure. Throughout the life of a mining operation, the optimum pit must be constantly revised since changes in metal prices and operational costs may affect the long and short-term plans.

To design an optimal pit, the orebody block model can be represented by a graph $G=(N,E)$, where each node represents a block and each arc (i, j) represents the excavation dependency between nodes i and j. The economic value v_i associated to block i is usually assumed to be independent of the mining sequence. Lerchs and Grossmann (1965) have shown that determining the ultimate pit limit is equivalent to determine the maximal closure of graph G, which in turn is equivalent to solving a maximum flow problem in a network conveniently built from G (Picard, 1976), for which efficient polynomial-time algorithms exist. Hochbaum and Chen (2000) present a detailed study of the Lerchs-Grossmann algorithm, evaluating the effect of mine properties and ore distribution over the algorithm's running time. Underwood and Tolwinski (1998) develop a network flow algorithm based on the dual ultimate pit problem and compare it to the Lerchs-Grossmann algorithm, providing further insights to improve its efficiency. Those authors also point out that the economic value v_i should reflect the net present value of money generated by the undertaking (Tolwinski & Underwood, 1996). That means that the sequence in which the blocks are removed is rather relevant in determining the optimal plan, as it will be discussed next.

Mine Production Scheduling

The open pit mine production scheduling problem involves determining the optimal sequence in which the blocks within the orebody model should be removed in order to maximize net present value and satisfy a variety of physical and economical constraints. Physical constraints include block precedence, slope limitations, blending requirements, hauling, milling and processing capacities and their expected rate. Economical constraints relate to operation costs and (time variant) ore prices and demand. A common approach is to first determine the ultimate pit limit and then use it as a starting condition for the scheduling problem. However, Wang and Sevim (1995) point out that the overall mine production planning problem is actually comprised of a set of interdependent subproblems: a cutoff grade must be defined to determine the orebody extension, and hence, the ultimate pit limit; the ultimate pit limit is then used to develop a production schedule to determine revenues and costs, which can again be used to review the most profitable cutoff grades. Sattarvand and Niemann-Delius (2008) further argument that the ultimate pit limit and the mine scheduling problem should be addressed simultaneously, since the final shape of the pit should be a natural outcome of the optimal mining sequence.

Due to the large number of blocks in any practical-sized orebody model, mixed integer programming (MIP) formulations are usually too large to solve by conventional commercial integer programming software. Alternative approaches have thus been proposed, beginning with the nested Lerchs-Grossmann algorithm (Lerchs & Grossmann, 1965) – which generates a series of nested pits that are used to devise a production schedule – which is the basis of many industry mine optimization applications. Since then, many researchers have proposed different approaches to solve the problem. Tolwinski and Underwood (1996), for instance, model the evolution of the mine as a sequence of pits which is searched by a stochastic algorithm to find reasonably good approximations of the optimal schedule. Caccetta and Hill (2003) discuss recent solution approaches to the mine scheduling problem and propose a branch-and-cut algorithm that produces good solutions for realistic-sized problems. Ramazan and Dimitrakopoulos (2004) present an alternative MIP formulation, where ore blocks are modeled as binary variables and waste blocks are modeled as continuous variables, the latter representing the percentage of blocks to be mined in each period. That formulation reduces the number of integer variables but allows waste blocks to be partially mined, which significantly compromises optimality. Later, Ramazan (2007) proposes a variable aggregation approach, where blocks are combined into fundamental trees (any combination of blocks that can be profitably mined according to the problem restrictions) in such a way that optimality remains uncompromised. A linear programming model generates the fundamental trees for any given orebody model, allowing a modified MIP formulation to solve large scale scheduling problems, as illustrated in a case study on a large copper deposit. Boland et al. (2009) also address the problem by means of a variable aggregation approach, in which individual blocks are used only for processing decisions and disaggregation is performed using a linear programming relaxation of the original formulation. Metaheuristic approaches (Sattarvand & Niemann-Delius, 2008) have also been proposed.

Stochastic Approaches

Uncertainty and risk are present in many aspects of the mine production plan and directly affect cash flow management on mining operations. In particular, uncertain information regarding the block model grades can amount to significant deviations from optimal production plans obtained with deterministic schedulers. As *in situ* orebody variability and uncertainty are major reasons for those deviations, uncertainty and risk information should be incorporated in the orebody block model (Ramazan & Dimitrakopoulos, 2003). Decision-making in a risk-based model should then involve the following: integration of grade uncertainty into pit optimization and recoverable reserves; techniques for geological risk modeling in pit optimization and production scheduling; quantification of the risk inherent to mineral projects and ultimate pit limits; impact of high-risk grade zones in optimal pit limits; and risk analysis for production scheduling. Dimitrakopoulos and Ramazan (2004) propose a linear programming approach where decision variables represent the probability that a block will be mined in a given time period and the objective function seeks minimum deviation from expected grade and tonnage, which indirectly leads to a maximum net present value of the operation. The constraints include probability targets, equipment accessibility and mobility, grade blending and mill requirements, operational capacity, upper and lower bounds for ore quality, stripping ratio and wall slopes. Menabde et al. (2004) present a method for simultaneous optimization of the mining sequence and cutoff grade strategy for a set of conditionally simulated orebody models. Godoy and Dimitrakopoulos (2004) propose a simulation annealing algorithm that determines an optimal production schedule from a set of stochastically simulated orebody representations. Gholamnejad and Osanloo (2007) propose scheduling risky blocks later in the mining sequence, when more reliable information about the orebody may be available. Froyland et al. (2008) pose that, since the underlying stochastic process depends on the actual optimization decisions, stochastic programming models should be developed to explicitly incorporate information regarding different orebody geology scenarios.

It is important to notice that, although there has been significant effort to incorporate uncertainty into orebody representations, there is yet an increasing need for further understanding of the limitations of traditional, non-risk-based, mine scheduling approaches (Dimitrakopoulos, Farrelly & Godoy, 2002).

Considerations on Underground Mining Operations

Underground operations are significantly more challenging than open pit operations due to the reduced flexibility imposed by geotechnical and space constraints. Usually, a greater number of constraints and binary variables are required to develop a production schedule with mathematical programming techniques (Topal, 2008).

Smith et al. (2003) address the application of an ad hoc MIP approach based on variable aggregation and compare it to other approaches, such as spreadsheets and popular commercial mine schedulers used in a lead-zinc underground mine. Newman et al. (2006) provide a combined formulation for dealing with long- and short-term production scheduling in an underground iron ore mine. Long term decisions involve when to start mining each machine placement, while short term decisions involve the amount to be mined from each production block at each period, as well as finish dates for each line. To deal with the large size of the combined model, the authors develop different solution strategies, both exact and heuristics, such as variable reduction, variable aggregation and decomposition. Sarin and West-Hansen (2005) propose a Bender's decomposition scheme to determine the production schedule for an underground coal mine.

Mine Load and Haulage Equipment Allocation and Dispatching

Mine equipment operation is a major contributor to the overall operational cost of a mining project. Haulage cost is so expressive that simple spreadsheet applications are able to promote significant gains (Dessureault et al., 1999). The challenge is to effectively deploy trucks and shovels to work benches, in order to achieve productivity increase with a given fleet of equipment or, alternately, to obtain a significant reduction in the fleet size required to meeting production targets. Tangible benefits include minimizing re-handle, meeting blending constraints, and ensuring uniform processing plant feed rate. Intangible benefits include forecasting production, comparing truck and shovel performance, and generating on-line report information (White & Olson, 1986). The nature of haulage operations involves a fairly complicated queuing problem since shovels and trucks may have different production capabilities and queues may be formed both at servers and at dumping points (Kolonja et al., 1993).

A mine truck management system usually implements two functions: allocating a (limited) truck resource pool to loading points according to some production requirements; and dispatching and controlling those trucks in real-time, possibly reviewing allocation decisions with or without the intervention of a human operator. The most common approach is to solve the allocation problem via mathematical programming techniques and the dispatching problem via heuristics. Practical limitations often restrict the success of traditional allocation approaches, for instance: allocation is based on historical averaged data rather than on-line mine information, thus requesting additional resources to compensate uncertainties; human intervention is often required in dispatching decisions; and the dispatching algorithm should manage only minor perturbations to nominal operation, leaving major perturbations to new allocation runs.

Ta et al. (2005) propose a chance-constrained stochastic programming approach to truck allocation and dispatch, considering truck cycle time and truckload as the uncertain parameters. Truck allocation is decided using probability distribution information obtained from recent operational data. The dispatching algorithm monitors mine conditions and decides whether or not to reallocate trucks according to unpredicted perturbations. Çetin (2004) points out that due to the stochastic features and the interaction with the dispatch system, determining a global optimal solution is extremely complex. The author proposes a simulation approach to the truck dispatching problem where a set of heuristic rules, such as shovel waiting time, truck cycle time, earliest loading shovel, among others, is evaluated. Alarie and Gamache (2002) analyze the advantages and disadvantages of different solution strategies for the dispatching problem, highlighting important requirements of automated dispatching systems.

Railway Scheduling and Dispatch

Mining companies often use railroads to convey ore products from mines and processing plants to local customers or ports. In an integrated mining supply chain, routing and scheduling of trains is of major importance to the overall performance of the system, whether the railway operator is part of the mining company, or third-party service provider.

There has been extensive research in the realm of railway optimization, not necessarily exclusive to mining operations and ore transportation. Cordeau et al. (1998) present a literature review on train routing and scheduling problems. Routing involves determining operating policies for freight transportation and railcar fleet management, while scheduling addresses train dispatching and locomotive and car assignment. In the first category, Sayarshad and Ghoseiri (2009) propose a simulated annealing approach to solve the deterministic fleet size and freight car allocation problem according to yard capacity, expected demand, and number of loaded and empty cars at any given time and location. When different goods (e.g. ore, grains) are transported by the same railroad operator, various car types must be employed. Bojovic and Milenkovic (2008) propose an Analytic Hierarchy Process (Saaty, 1977) to determine an optimal rail fleet mix. Such mix must satisfy a known transportation demand, which involves not only selecting freight car types, but also determining the number of freight cars.

Train scheduling and dispatching is clearly one of the most complex problems in railway optimization. Salim and Cai (1995) consider a case of ore transportation by a single track railway system in Western Australia. Trains use passing loops either to overtake other trains, or to allow trains coming in the opposite direction to pass and avoid head-on collisions. The authors propose a scheduling model to generate optimal timetables for the arrival and departure at passing loops for every train in the system. The objective function considers the cost of stopping and waiting of each train, according to its characteristics (trains with high load, for instance, have higher stopping and restarting costs). Constraints ensure that no collisions occur; that no more than one train occupies a passing loop; and that a train must stop when two trains are at a loop. Some genetic algorithm formulations are proposed, all approximating a presumed global minimum. Vieira et al. (1999) address the train dispatching problem in a major Brazilian iron ore railroad by means of a fuzzy logic expert system which provides decision support to human dispatchers. The system is based on a set of rules which include refueling requirements, crew shifts, mandatory maneuvers, train priorities (passenger trains have precedence over cargo trains), track conditions and the monthly ore transportation demand. The approach naturally incorporates inherent uncertainties but may produce suboptimal results. In such cases, a limited search algorithm allows dispatchers to locally optimize train movements over a set of feasible movements.

Assigning and scheduling individual rolling stock elements is also a complex problem with significant impact on railroad operations. Ahuja et al. (2002) address the locomotive scheduling problem, in which a set of locomotives must be assigned to each train in a pre-planned schedule so as to provide enough power for service trips.

A mixed integer programming formulation seeks to determine: the set of active and deadheaded locomotives for each train; the set of light traveling locomotives from power sources to power sinks; and also direct train-to-train connections. A decomposition-based technique, coupled with a large-scale neighborhood search, allows solving the model in reasonable time. Significant economical savings were attained for a large railroad operator. A similar problem known as train coupling and sharing has found in multi-agent systems a novel approach to obtain optimal solutions for complex problem instances (Böcker et al. 2001). Fukusawa et al. (2002) address the freight car flow problem, which consists in determining, according to given train schedules and capacities, the full route of each car in the railroad network, as well as its load and unload sequence within a given time horizon. The problem is formulated as an integer multi-commodity flow program which is solved by a linear relaxation and a preprocessing variable aggregation phase. The authors report savings results for the largest Latin American railroad freight company.

Trains arriving at a railroad yard must be disassembled into individual cars to undergo unloading, cleaning and assembly on a new train. Those operations are performed by a set of switch engines, which are dedicated maneuvering locomotives. Scheduling switch engine operations aims at determining assignment sequencing in such a way that train assembly time and operational costs are minimized. The problem is strongly connected to the multiple pickup and delivery problem with time windows, which is used to model passenger and goods transportation. Sabino et al. (2006) propose an ant colony optimization algorithm to tackle the inherent complexity of the switch engine scheduling problem.

Port Planning and Scheduling

Port operations comprise a very important stage in a global mining supply chain. In the port scenario, traditional approaches of production planning and scheduling must deal with high infrastructure complexity, large decision flexibility and prohibitive penalties to unmet demand (especially in the case of ship docking and loading delays, when demurrage applies). This motivates research on methods to improve – and optimize, if possible – overall performance. The importance of the port stage is further aggravated due to the fact that it is the stage where ore sales are ultimately performed. Due to the high complexity and the difficulty on achieving efficient formulations with traditional mathematical programming efforts, current literature on port optimization is more devoted to simulation approaches.

A port for handling and shipping bulk products such as iron ore is a complex system. In a typical iron ore port, trains coming from the source mines arrive at the port railroad yard, where they are dismembered into car groups according to the type of ore product they carry. Switch engines drive the car groups to car dumpers, which tilt the cars and drop the ore in a feeder. From this feeder, the ore products are transferred to the stockyards through a network of conveyor belts. According to a ship docking schedule (ship queue) and a ship loading plan, specific ore products must be reclaimed from stockpiles and conveyed into each of the ship's holds, again through a network of conveyor belts. If the production schedule requires it, ore products unloaded from a train may bypass the stockyards and be directly transferred to the ship loaders. This decreases the operational cost, but limits the capacity of performing quality corrections by blending. Depending on the handled products, ports can use processing facilities to increase product quality, such as screening plants for separating ore fines from lumpy ore products. Large-scale iron ore ports, however, may also count on pelletizing plants aggregate pellet-feed ore, which would be otherwise unusable in the steelmaking industry. Thus, on train unloading, the pellet-feed ore may be transferred either to the stockyards for subsequent export, or to the pelletizing plants.

Managing ship arrival and departure is a very sensitive process due to the high penalties applied to operational delays. Arrival-time and service-time depend on ship size, cargo type and available equipment, thus imposing constraints on ship-to-berth allocation. Lee and Chen (2009) address the berth usage scheduling problem, in which the berthing time and position of every calling ship must be determined for a given time period. The approach considers the entire pier as a continuous space, in which ships must be berthed within one of several preferred sections, in a first-come-first-served basis. The authors propose an efficient neighborhood search heuristic to search within a berth allocation candidate space, which is later improved by a greedy approach. Van Asperen et al. (2003) present a simulation study to investigate the impact of different ship arrival processes on the efficiency of the loading and unloading operations. The arrival processes include: stock-controlled arrivals, which aim at maintaining target stock levels of materials in the storage facilities; equidistant arrivals, when ships arrive at regular intervals; and Poisson process arrivals, when ships arrive with exponentially distributed inter-arrival times.

Port simulation studies can also be performed when it is necessary to evaluate infrastructure development options to meet increased demands in tonnage and quality, as well as increased cost restrictions. El Sheikh et al. (1987) present a simulation study with the objective of estimating the number of berths required in the short- and medium-term planning problem for a third world port and also examining the impact of proposed handling improvements in overall performance. The number of berths depends on the expected demand, on the handling rates for different commodities, and on the expected level of service. Howard et al. (2005) discuss a simulation model to evaluate infrastructure development options and its effects on shipped ore quality for a mine-railroad-port system. In a case study, the simulation allowed a mining company to make informed investment decisions in choosing from reducing ore short-term variability, improving blending procedures in port stockpiling, or increasing accuracy of train composition estimates from the mine.

Port production planning is also a major decision problem in port operations. A high-level approach is presented by Everett (2001), who proposes a quadratic programming model to minimize a stress function that represents deviations from desired quality during ship loading. The model decides the amount of ore to be reclaimed and blended from each available stockpile in order to best approximate quality specifications. The general port production planning problem, on the other hand, involves determining the amount of different ore products to be received, stored, blended and loaded into ships, according to known supply from mines and railroad and known shipping demand. Optimal routes within the conveyor belt network must be chosen from a set of feasible operating routes according to the overall handling and conveying capacities. However, in order to ensure that ship demands will be satisfied within the required time restrictions, scheduling constraints must be integrated into the planning model (Dauzére-Péres & Lasserre, 1994), thus enabling consistency between decisions on different hierarchical levels. Drumond and Mateus (2000) present such an approach for a large Brazilian iron ore port. In the proposed model, each route selected at the production planning level corresponds to a task in the scheduling model; each non-concentrating peace of equipment in a route corresponds to an operation; and for each concentrating peace of equipment, the number of generated operations balances the number of routes sharing that equipment. Due to the large complexity of the model, only results for small problem instances are presented.

Blending

In the general sense, blending is defined as the "adding together continuously, in controlled proportions, of raw materials that are assumed to have completely constant proportions, i.e., time invariant" (Schofield, 1980, p. 8). In the mining industry, blending operations are found in different stages of the production chain and, though not always performed in a continuous manner and not necessarily associated with homogenization, they are fundamental in achieving ore product quality specifications. Those stages include the following (Everett, 2001):

1. Selecting orebody blocks with expected composition in one or more mines;
2. Selecting train and car loading sequences when transferring ore products from source mines to the port;
3. Selecting specific stockpiles where to unload an incoming train;
4. Selecting specific stockpiles to reclaim and load a particular ship.

In large scale mining operations, where several mines produce ore with different grades, there may be several opportunities for blending products to specific quality requirements and customer demand. It is important to notice, however, that ore blending and re-handling can be a major contributor to operating cost. Hence, there must be a tradeoff between cost and quality when blending is carried out for product quality alone (Howard et al., 2005).

Chanda and Dagdelen (1995) propose a linear goal programming approach to a short-term coal mine planning problem. The model minimizes deviations of individual quality parameters and tonnage from expected specification and demand, respectively, by determining the amount of ore to be mined from each block. Goal programming approaches, however, do not always guarantee pareto-optimal solutions (Melo et al., 2009). Everett (1997) presents a simulation model to support decision making in ore handling and stockpile formation in a port. The idea is to decide specific stockpiles which incoming trains should unload their contents into. The scenario is composed of three Australian mines which transfer ore products to a port through a single-line railroad. Ship loading is performed by reclaiming ore products from one or more stockpiles, according to blending requirements. In a later work, the author discusses algorithms and simulation models to support blending decisions on the four stages listed above (Everett, 2001).

Bazaraa et al. (2005) present a multi-period coal blending and distribution problem. Several mines produce coal at various rates with different quality specifications that can vary over time. Coal must be shipped to silo facilities for further processing and blending to predefined quality specifications before being shipped to customers. A linear network flow model is developed to determine optimal schedules for shipping coal from mines to silos, blending coal at the silos, and distributing coal to customers, subject to production capacity, storage and material flow balance, and quality requirement constraints, so as to satisfy customer demand at a minimum cost.

In a global mining supply chain, such as the one envisioned in this Chapter, more complex blending requirements are bound to be present since a number of mines, supply stations and ports all allow high flexibility in ore product specifications. This will be further discussed later.

Other Applications of Artificial Intelligence in the Mining Industry

Although not related to the main thread of this Chapter, there are many applications of artificial intelligence techniques to mining operational problems. Fuzzy logic expert systems, for instance, are a common choice in process control applications with a significant degree of uncertainty. Hall and Harris (1993) present a fuzzy logic system to improve productivity in autogenous mill and spiral iron recovery circuits. Santos et al. (1995) develop a fuzzy logic controller to assist slime thickener operations in a large Brazilian mining company. Both works derive rule-based expert knowledge from experienced people to implement the desired behavior into the control systems. Neural network techniques, on the other hand, are often chosen in pattern analysis applications. Singh and Mohan (2005) present a neural network expert system for ore sorting and classification, which assists operators in separating gangue material from large ore blocks. Classification is done by visual textural analysis of imagery taken at a processing plant feed.

Although the above examples are mostly concerned with operational control of industrial processes, higher-level decision making can also be supported by expert systems. Karadogan et al. (2008), for instance, present a fuzzy set application to support the selection of the most adequate mining method for an underground lignite mine. The method must be chosen from a set of self-supported, supported, and caving mining methods using a set of expert decision rules.

INTEGRATED APPROACHES AND THE GLOBAL MINING SUPPLY CHAIN

The supply chain concept is of fundamental interest in this text. There are several similar definitions and different levels of generality (Beamon, 1998; Pochet & Wolsey, 2006; Tsiakis & Papageorgiou, 2008). However, a definition which encompasses most aspects visited here could be stated as the following: a supply chain is an integrated process where a set of existing or potential organizations (or geographically dispersed sites of a single large company) work together to generate value by acquiring raw materials, converting these raw materials into specified end products, and delivering these end products to customers. Organizations are linked together by materials, information and financial flows, which can traverse the supply chain in both forward and backward directions.

The manufacturing industry has been the leading field for supply chain planning and optimization applications (Cohen & Lee, 1988). On the other hand, process-based industries, such as the mining industry, present specific characteristics that bring additional challenges to the development of intelligent decision-support systems (Crama et al., 2001). For instance, capital costs are far more significant than direct costs and thus efficient usage of equipment and facilities must be ensured at all times. Products are typically defined by the sequence of operations, what may lead to tightly intertwined planning and scheduling decision making. Also, the quality of raw materials and end products may be variable and nondeterministic, meaning that demand satisfaction may no longer be enforced, as it is often the case in discrete manufacturing. Process-based industries tend to move away from make-to-stock and towards make-to-order strategies, while offering a more diversified, customized line of products.

A supply chain is often viewed as the integration of two basic processes: production planning and inventory control, and distribution and logistics (Beamon, 1998). This integrated problem must be tackled by integrated analysis procedures in order to yield coherent and valuable solutions (Thomas & Griffin, 1996). This is of special interest when the ratio of distribution costs to production costs increases (Chandra & Fisher, 1994). In process-based supply chains, it is common to let the production schedule at the finishing plant drive the production requirements of intermediate supplying plants. However, as individual processes can be located at different geographic sites, the need for coordination is of utmost importance (de Matta & Miller, 2004). Freight transportation services must be closely coordinated with procurement, production and delivery systems in order to reduce raw materials, in-progress materials and finished products inventories (Kwon et al., 1998). Melo et al. (2009) present a recent and thorough review on supply chain management problems, with emphasis on facility location and related decisions.

The efficiency of supply chain systems mostly depends on management decisions, often based on intuition and experience. Due to changes in customer preferences, economy globalization and increasing competition, these decisions can often be far from optimal. Also, different stages in a supply chain are often supervised by different groups of people with different managing philosophies (Sarimveis et al., 2008). This is especially the case of the integrated mining industry. Advanced Planning Systems are a common approach to allow coordination of the planning effort over the whole supply chain. Optimization is one of the major characteristics of that approach, as well as the definition of alternatives, objectives and constraints for all planning tasks, the decomposition into planning modules, and their vertical and horizontal coordination by information flows. Traditional production planning applications, however, do not implement actual bottom-up coordination with respect to planning horizons, and no left-to-right coordination with respect to planning functions (Pochet & Wolsey, 2006). For the process-based mining industry, this is also especially the case.

Supply chain planning involves three integration dimensions: functional integration of primary activities – purchasing, manufacturing, warehousing, transportation – and related support activities; inter-temporal, or hierarchical, integration of primary activities on strategic, tactical and operational planning scopes; and spatial integration of primary activities (Pochet & Wolsey, 2006). Hierarchical integration is commonly considered over restricted functional domains, as it is the case when a tactical (medium-term) production plan provides goals and constraints for a series of operational (short-term) production schedules (Hax & Meal, 1975), or when scheduling constraints are integrated into the planning problem (Dauzére-Péres & Lasserre, 1994). Functional and spatial integration, on the other hand, consider coordinated decisions over the whole supply chain at the same decision level (Sarker & Diponegoro, 2009).

Some Related Work

Integrated approaches for dealing with mining supply chain scenarios are seldom found in the literature. Some related work, however, may present important and useful concepts to the ideas developed in this Chapter. For instance, Mateus et al. (1994) present a mathematical model applied to multi-period iron ore transportation planning at a Brazilian mining company, in a scenario which includes mine and port stockyards, loading facilities and railroads. The model does not address production issues, but focuses on transferring iron ore products from a set of mine stockyards to port stockyards in a way to minimize transportation costs and satisfy blending requirements, storage and transportation capacities, mine supply and port demand constraints. The problem is formulated as a linear network flow problem for which a Lagrangian decomposition solution approach is proposed. Pendharkar and Rodger (2000) broaden the scope by addressing the problem of simultaneously determining the production schedule for coal mines, the coal transportation schedule from mines to blending facilities, the blending schedule for blending facilities, and the allocation of blending facilities to markets. The problem is modeled as a multi-period nonlinear program, which aims at maximizing the total discounted net profit before taxes, while also satisfying production capacity, quality requirements, market demand and coal balance at blending facilities. A genetic algorithm is proposed and a hypothetical mining scenario is used to validate the model. This problem is similar to the linear coal blending and distribution problem discussed before (Bazaraa et al., 2005).

A previous section has discussed a work by Everett (2001), who presented a scenario composed of three Australian iron ore mines which transfer ore products to a port through a single-line railroad. Although decisions were not integrated in all stages, the idea of a mining supply chain – integrated mines, railroad and port – could already be envisioned. A management informing system is also proposed (Everett et al., 2001) to provide support in planning decisions regarding mining, transporting, stockpiling and shipping operations.

One may also see the basic elements of a mining supply in the reverse order: when a shipment of iron ore arrives at the destination port, it must be unloaded and transported to the final customer (the steelmaker), which may be located at the same port – at conveyor belt route distance – or farther – at railroad distance. Lu et al. (2005) present such an approach for the China's iron ore import logistics. The scenario shows the dry bulk logistics market dominated by steel-production companies, which led to inefficient allocation of transportation capacity and diseconomies of scale. Iron ore logistics centering ports are used to overcome such difficulties. The scenario is modeled as a mixed integer programming problem with the objective of minimizing order costs, shipping costs, and centralized inventory and port dispatching costs within a given planning horizon, while satisfying demand, capacity and inventory safety level restrictions. Results for a small size problem instance are provided.

The Global Mining Supply Chain

Large mining operations are often composed of several mines and processing plants, as well as stockyards, railroads, ports and supply stations, each with different capacities. According to the demand imposed by domestic and international customers, and depending on the characteristics of the ore products supplied by each mine, decisions such as mining, processing, handling, blending and shipping may involve different facilities at different stages of the supply chain. A global mining supply chain can thus be defined as **network of integrated facilities designed to process, using a variety of production techniques, and distribute, using a variety of transportation modals, bulk ore products from mines to customers, which can be (and usually are) at significant geographic distances**.

Below the main elements of a global mining supply chain are described in a network structure. The iron ore industry is taken as example, but similar approaches to other mineral industries could also be developed. In Figure 5, nodes represent pits, stockyards, processing plants and other facilities depicted in previous sections, organized into one or more mine complexes and ports, while arcs represent the links between them; road, rail, river and ocean transportation modals can be considered, as well as belt conveyors and ore ducts:

- **P** stands for the existing open or underground **pits** available for mining, which correspond to the primary supply nodes of the network;
- **CS** represents the milling facilities, where ROM is reduced to smaller size particles in **crushing and screening** processes;
- **RS** represents the **ROM stockpile**, which is often used as a buffer for downstream operations;
- **HS** represents the **homogenizing stockyard**, where stockpiles of homogenized ROM are formed to provide an uniform feed to the processing plant;
- **PR** represents the **processing plant**, where ore products are generated by working the ROM's physical and chemical processes;
- **PE** represents the **pelletizing plant**, where iron pellets are generated by working the pellet-feed ore in physical and chemical processes;
- **PS** represents the **product stockyard**, where ore products are stored prior to shipping;

Figure 5. A network model for the global mining supply chain

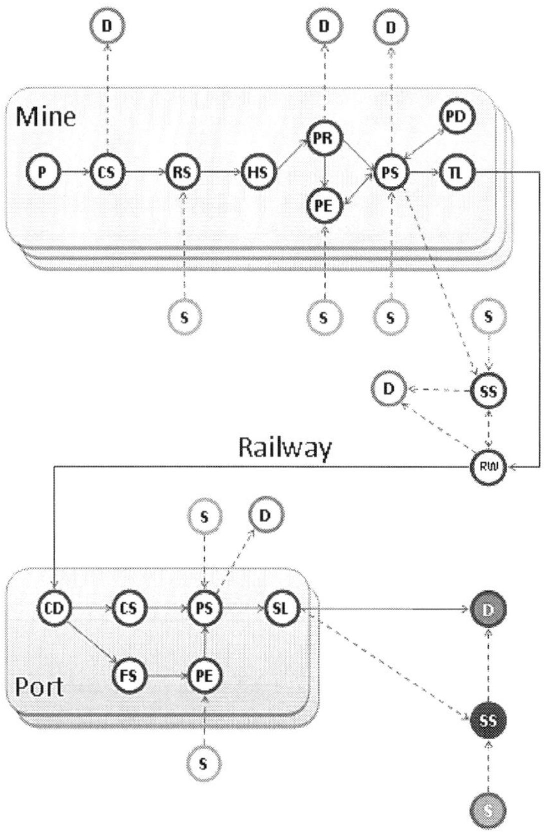

- **PD** represents the **product non-operational deposits**, which can be used to store ore products if the product stockyards' capacity is exceeded (stockpiles must be moved back to the product stockyards prior to shipping);
- **TL** represents the **train loading** system, which loads ore products into train cars according to specific loading plans;
- **RW** represents the **railway**, through which trains transport ore products from mines to ports, domestic customers or supply stations;
- **SS** represents the **supply stations**, which act as warehouses serving local customers and also as third-party sourcing stations;
- **CD** represents the **car dumpers** located at the port entrance, which unload the trains and transfer the ore products to the conveyor belt system;
- **FS** represents the **feed stockyards**, which store pellet-feed ore and other input materials in order to feed the pelletizing plants;
- **SL** represents the **ship loading** system, which loads ore products into ships' holds according to specific loading plans;
- **S** represents **sourcing nodes**, where additional ore products may be procured from third-party mining companies in order to satisfy quantity and/or quality requirements;
- **D** represents **demand nodes**, both domestic and offshore, which can be satisfied by means of many different transportation modals.

It is important to notice that end ore products in each mine may be considered intermediate products to the supply chain, as additional blending may occur either at the ports or at the supply stations. Supply stations may also operate as transshipment hubs, receiving ore products from the railway and dispatching them by road, or the other way around. Planning decisions are usually guided by expected revenue and variable and fixed costs modeled as linear or nonlinear functions on material flows and market conditions. In such a complex scenario, planning must be as detailed as lower is the decision level. However, the main opportunity, and also the main challenge in integrated planning of the mining supply chain is to explore the flexibility of the system to yield lower costs and increased product quality, especially in times of low demand.

Regarding strategic decisions, mining companies usually do not have a choice of location for their mines or markets (Pendharkar & Rodger, 2000), which means that facility location and network design decisions would be mainly focused on establishing new (or enhancing the capacity of existing) processing facilities, supply stations and transportation lines in order to satisfy increasing demand. Conversely, if market conditions deteriorate and demand decreases, it is important to decide whether or not to shutdown specific facilities in a temporary or permanent basis (Dias et al., 2006). Those decisions may have to be made according to budget limitations for each time period, which can also affect other strategic supply chain decisions (Dogan & Goetschalckx, 1999). Fixed costs, expected revenues and return on investment are thus the main drivers to strategic decisions in the mining supply chain.

Tactical decisions generally involve determining product portfolio for each mine, production amounts, utilization levels, and material flows according to production, storage and distribution capacity constraints. Ore product portfolio for mines and processing plants is somewhat restricted, but may show some variation in the medium-term, since the characteristics of the ROM feed may change according to the mine development plan. Ideally, however, that plan should consider the integrated mining supply chain plan, including the overall demand, stock levels, and blending opportunities, as well as individual costs and capacities. Different ROM characteristics may drive different production campaigns and yield different costs and productivity, thus impacting on the ore product mix. It is important to notice that ore processing commonly involves the generation of coproducts, thus imposing a production split constraint on the processing plants. At the design phase, the processing plant features are defined according to the characteristics of the expected ROM. This fact limits the production split and makes setup and changeover times and costs relatively irrelevant on the short-term plan. When more than one production circuit is available – such as the sinter-feed and pellet-feed circuits in some iron ore processing plants – there may be some split flexibility. For instance, if the mine supplies a thicker ROM, the processing plant can still employ additional grinding to generate finer ore products. However, if the mine supplies a finer ROM, the processing plant will certainly yield a higher proportion of fine products, since there is no way of compensating for that input. This asymmetry between supply and demand can lead to difficulties in satisfying quality range specifications. In such cases, the only way to satisfy demand in both quality and quantity is to explore the flexibility of the supply chain by blending ore products at mines and ports. Also, although production facilities may be relatively multi-purpose, it can be beneficial for the whole supply chain to operate specific plants in a dedicated mode, one that matches a typical ROM supply. Regarding utilization levels, processing plants usually operate at a specific nominal capacity level, where productivity is highest. Hence, accommodating production to demand level is often achieved by determining discrete idle time, when production stops and management seizes the opportunity to perform maintenance and refurbishment activities. Discrete production levels, which provide an acceptable relation between cost and productivity, can also be considered. Additionally, material flows can be viewed as discrete quantities when transportation is made by trains, cars or trucks. Fixed and variable costs and expected revenues are the main drivers to those tactical decisions.

At the operational level, the concept of mining supply chain planning and scheduling can be overwhelmingly complex. Short-term planning and scheduling in mines, railroads and ports are complex problems themselves, as it has been shown in previous sections. Detailed plans must be accommodated to specific equipment capacities and time constraints. For instance, the managing of the ship queue depends not only on the availability of berths, but also on the availability of the required ore products, which in turn depends on the delivery schedule of the railroad and, recursively, on the train loading at each mine. Also, stockyard allocation and stockpile formation – the equivalent of warehouse management in manufacturing supply chains – at mines and ports must be performed at a detailed level. End products tonnages must be allocated to specific stockyard areas in order to satisfy loading plans with optimal productivity. Space restrictions and product compatibility must be observed to avoid contamination. Setup costs and time would accrue when a given stockyard area receives a product that is incompatible with the one previously stored. Minimum stock levels must also be determined in order to accommodate uncertain events and, more importantly, to ensure rational utilization of handling equipment. For instance, if many small stockpiles exist, a greater number of different products could be stored, but the reclaimer would have to be repositioned several times in order to fulfill the loading plan. However, the alternative of having large, single-product stockpiles could reduce stockyard flexibility when different ore products are handled. At the ports, stockpiles may be formed according to specific shipments, which may also drive stockpile formation at the mines. At the operational level, blending can be used either to generate composite products or to make last-minute quality corrections, right before loading products into trains or ships. Also, since quality specifications are defined within a certain tolerance range, different ore products may be used to satisfy the same demand. Routing and scheduling decisions, at the mine, at the railway and at the port handling system, must be coordinated with stockyard management in order to satisfy the loading of trucks, trains and ships accordingly. Setup, changeover and variable costs, as well as time and productivity restrictions are the main drivers to operational decisions.

When inter-temporal integration comes into play, it is common to have strategic decisions defining high-level goals and constraints to tactical decisions, and the equivalent happens between tactical and operational decisions. If considered in isolation, decisions may turn out to be infeasible at lower levels, as higher-level decisions may consider averaged or aggregated data. For instance, a monthly production plan may generate infeasible daily production schedules if detailed operational time restrictions are larger than the averaged daily capacity. It is important to point out that, since considering simultaneous functional and inter-temporal integration can be computationally intractable, specific approaches must be developed. That will be the subject of the following Section. Figure 6 organizes the individual problems discussed in the text according to decision-level and functional integration dimensions.

A final word must again be said about uncertainty issues. All operations depicted above possess some degree of uncertainty and can significantly influence other downstream or upstream operations in the supply chain. For instance, unexpected downtime can decrease the nominal capacity and availability of all production and

Figure 6. Global mining supply chain problems organized according to functional and inter-temporal dimensions

	Mine	Railway	Port
Strategic	Strategic mine planning		
Tactical	Ultimate pit limit Production scheduling Storage & blending	Transportation planning	Production planning Storage & Blending
Operational	Mine equipment allocation Storage & blending Train loading	Train routing & scheduling Yard maneuvering	Production scheduling Storage & blending Ship queue management

distribution elements (Lin, 2009). Uncertainties inherent to the orebody model can influence the processing plant throughput, both in quantity and quality, which in turn affects the expected production portfolio and makes additional blending required at stockyards and loading points. In the more general context of the supply chain, demand, market prices, operational costs and financial risks – varying taxes, duties, exchange rates and transfer prices – may vary unpredictably and with different global effects (Germain et al., 2008). That would significantly affect the optimal production and distribution policies determined by solely deterministic approaches. Accounting for such uncertainties clearly brings additional complexity to an already difficult problem, since it would be necessary to model probability distributions or to consider a set of discrete outcome scenarios each with a specific probability of occurrence.

SOLUTION APPROACHES TO THE INTEGRATED MINING SUPPLY CHAIN

The problems addressed in this Chapter can be described in the general mathematical form as follows:

$$Z(x, y) = \min \{c(x) + d(y): (x, y) \in U\},$$

where $U = \{(x, y) \in R_+^n \times Z_+^p: u(x, y)\}$ is the set of feasible solutions. The x variables represent production and storage levels and material flows, while the y variables represent allocation, scheduling, capacity expansion and location decisions. Different configurations of the functions above describe specific problem categories, which vary in both structure and complexity. Solving the general optimization problem with nonlinear $c(.)$, $d(.)$ and $u(.)$ is a hard problem for which ad hoc solution methods are certainly required; also, proving a solution to be globally optimal or even close to optimal may be rather difficult. However, when those functions are linear and all decision variables are continuous, the general mathematical model is reduced to a simple linear programming model:

$$Z(x) = \min cx, x \in U,$$

with $U = \{x \in R_+^n: Ax \geq b\}$. As an example, a linear network model inspired on Figure 5 can be devised to determine minimum cost material flows and production and storage levels. Such a model can be textually stated as follows:

Minimize production, transportation and storage costs, subject to:

- *Supply and demand restrictions;*
- *Material balance restrictions in transshipment nodes;*
- *Blending restrictions;*
- *Inventory balance restrictions in stock nodes;*
- *Capacity restrictions in production plants, stockyards and transportation links;*
- *Nonnegativity restrictions.*

The number of linear variables depends on the number of commodities, on the number of layers in the network, on the arc density, and on the number of planning periods, but it is likely that any practical sized problem instances[1] would be solvable to optimality by general purpose optimization software. However, although such linear program could provide an efficient first approximation to the tactical mining supply chain planning problem, many important realistic features such as allocation decisions and fixed costs could only be addressed by the use of integer and binary variables. Hence, modeling lot sizing or production planning and scheduling problems would require the model above to be set into a mixed integer programming framework:

$Z(x, y) = \min cx + dy, (x, y) \in U,$

with $U = \{(x, y) \in R_+^n \times Z_+^p : Ax + By \geq b\}$. The above textual linear model should then be rewritten by modifying the objective function and adding the following constraints:

Minimize total (variable and fixed) costs, subject to:

- *Discrete production level restrictions;*
- *Production and distribution allocation restrictions;*
- *Integrality restrictions.*

Problems involving integer variables are much harder to solve, even for linear *c*(.) and *d*(.) functions. General purpose optimization software may or may not be able to solve reasonable sized problem instances to optimality, but realistically sized ones are bound to require alternative solution procedures to find good feasible solutions. A common approach is thus to try solving a problem with general purpose software either to optimality or until finding a solution within a predefined gap that satisfies the cost-benefit ratio required by the decision-maker (Melo et al., 2009). Routing and production scheduling problems also fit into this category, but are especially harder to solve since the number of integer and binary variables can be quite large. Specially tailored algorithms and heuristic procedures must be developed to attain efficient, approximate solutions. Typical scheduling problems could be described as follows:

Minimize costs and/or program time, subject to:

- *Decision precedence restrictions;*
- *Production, transportation and storage restrictions;*
- *Available production and transportation hours;*
- *Nonnegativity and integrality restrictions.*

Mathematical models with linear functions and all integer (binary) decision variables are used to model pure scheduling problems, such as the open pit mine planning problem, shown earlier in the text. Such models would have the form:

$Z(y) = \min dy, y \in U,$

where $U = \{y \in (0,1)^p: Ay \geq b\}$. Solving those problems can also be challenging due to the large number of binary variables, even for small mines.

It is important to notice that, although the mathematical framework for solving a mining supply chain planning problem would be the core of a decision-support system, all other informational aspects should be carefully taken into account. That means that not only solving the integrated model would be computationally challenging, but the sheer task of collecting and organizing all necessary information to define a feasible problem instance would demand considerable effort. A robust computational system should thus be envisioned to allow collecting, processing and aggregating operational data into useful input information to the mathematical models. Also, after obtaining a useful solution, the computational system should be able to provided useful output reports and support decision implementation and management, possibly at real-time. Without that software platform, an isolated mathematical approach would be futile.

Current literature shows the application of several alternative solution approaches to the analysis of individual supply chain and production-distribution problems, including Bender's decomposition (Dogan & Goetschalckx, 1999), Lagrangian decomposition (Tang & Liu, 2007), evolutionary metaheuristics (Nachiappan & Jawahar, 2007), control-theoretic (Sarimveis et al., 2008), game-theoretic (Esmaeili et al., 2009), among others. As discussed above, when some level of integration – functional or inter-temporal – is considered, the enhanced model complexity will certainly require alternative solution methods. One can thus develop the following approach:

1. **Establish the desired integration level, determining which functional and inter-temporal aspects to be considered.** Depending on the required analysis, not all aspects must necessarily be cast as decision variables, but can instead be modeled as known or fixed parameters. It is common, for instance, to perform supply chain planning at the same decision level dimension (e.g. tactical), where functional integration between mine, railway and port is considered. However, it may also be important to establish integrated plans at the same functional dimension, as is the case of integrated production planning and scheduling problems at mines or ports.
2. **Determine the links and interfaces between individual problems, according to the predefined integration approach.** Having defined the integration approach, all individual problems must be coupled together in the mathematical model, in a functional, hierarchical or inter-temporal manner. Proper relations between decision variables and constraints must be established to allow consistent decision flow throughout functional and decision level dimensions. Depending on the integration strategy, specific mathematical models may need to be developed, which are bound to be hard to solve to optimality due to the increased complexity.
3. **Solve the problem, or (more likely) develop specific solution procedures exploring problem structure and in accordance to available computational resources.** Previous sections have shown that solving some of the individual problems can be rather difficult. When such difficult problems are integrated into a single problem, traditional solution procedures may be futile. Branch-and-bound solution schemes, for instance, eventually become inefficient as the number of integer variables increases. In practice, however, even finding feasible solutions for such complex problems can be a difficult task. Also, it is important to take into

account error margin estimates of input data, since they may justify pursuing solutions within a reasonable gap from optimality, thus avoiding unnecessary computational effort (Cordeau et al., 2006).

The large number of integer variables is commonly the main issue. It has been shown that there is a clear separation on the variable set of mixed-integer programming models: continuous x variables represent production and storage levels and material flows, while integer or binary y variables represent allocation, scheduling and capacity expansion decisions. However, traditional decomposition based solution approaches may be difficult to develop due to the multilayer structure of the supply chain network and the interaction of decisions across several layers, which prevents decomposing the problem into smaller meaningful subproblems (Melo et al., 2009).

A solution technique which has been extensively used in integrated production lot sizing and scheduling problems and that could be applied to the proposed integrated problems in the global mining supply chain is the **relax-and-fix** strategy (Dillenberger et al., 1994). Relax-and-fix is a general-purpose methodology that solves the original mixed integer programming problem in a number of steps, each of which involving a subproblem of smaller complexity than the original problem. That is accomplished first by partitioning the integer variable set into a number of disjunctive subsets. The procedure then tries to solve a sequence of mixed integer subproblems in which the variables of only a single subset remain integer and all others are linearly relaxed. Since the number of integer variables is reduced at each stage, the subproblems can be solved with relative efficiency. As the series advances, each integer variable subset is permanently fixed to the optimal values obtained at the previous stage and the integrality constraint is forced upon the integer variables of the current stage. The variable set partitioning and the variable fixing criteria have a strong connection with the complexity of each subproblem. The traditional relax-and-fix strategy partitions the variable set according to time periods, but several other partitioning schemes have been studied, such as prioritizing the most valuable variables or randomly selecting them (Escudero & Salmeron, 2005), and defining partitions according to the scenario tree of a stochastic multistage problem (Beraldi et al, 2006). The relax-and-fix heuristic has been largely used as a method to obtain good primal bounds (feasible solutions) for hard mixed integer programs either on its own or as part of hybrid algorithms (Ferreira et al., 2009).

It is noteworthy that the integrated problems involved in optimizing the global mining supply chain can be rather complex if risk is incorporated in a stochastic, multistage version, even for a single decision level instance. It is very likely that employing efficient heuristic strategies would be the only alternative to solving real-world applications in reasonable time (Beraldi et al., 2006). Hence, the development of supply chain network models must be based on a sensible trade-off between realism, scope, complexity, and solvability (Melo et al., 2009).

FUTURE RESEARCH DIRECTIONS

This Chapter presented a general approach to the mining supply chain problem and its individual subproblems. Further research in solution methods to the integrated problems is clearly a priority, since realistically sized problem instances will certainly require special-purpose algorithms. Of particular interest is the solution of stochastic programming problems, which seem better suited to tackle all inherent uncertainties of mining operations.

As discussed previously, an integrated approach to the global mining supply chain would require a robust information management system, possibly integrated into an advanced planning and scheduling framework. Research in developing such computational systems would involve software engineering, telecommunication and information technology challenges, in an integrated effort to establish an intelligent global mining supply chain decision-support system.

CONCLUSION

Base metal mining is a highly capital intensive industry and, as such, effective usage of its resources and infrastructure is of utmost importance, especially in times of extreme economy activity. An analysis of the mining industry as a global supply chain – considering mines, railroads, ports, supply stations and customers – in an integrated perspective is thus imperative in promoting rational decision making for large-scale mining companies.

As the Background section has shown, most optimization approaches to mining problems address individual rather than integrated decisions. This is also true regarding decision-support systems, which are far more common in strategic mine planning software (Caccetta & Hill, 2003) than in the integrated mine-railroad-port chain. However, practical applications should consider not only the optimization techniques to planning and scheduling decisions, but also a real-time information technology platform to support activity execution and coordination in a global context (Soroor et al., 2009).

A great deal of study still remains to be done on solution procedures to deal with complex models, especially when the number of integer variables is large and there is no clear decomposition strategy to be applied over the problem structure. Approaches such as the relax-and-fix strategy are interesting candidates to such problems.

It is our intention that this Chapter may serve both as a review of the main problems involving operations research in mining, and as an introduction to a supply chain analysis of the global mining industry. It is fair to assume that similar approaches could be applied to analogous processes involving bulk commodities production and distribution systems, such as grain agribusiness. Significant economical gains would certainly accrue from such endeavors.

REFERENCES

Ahuja, R. K., Liu, J., Orlin, J. B., Sharma, D., & Shughart, L. (2002). *Solving real-life locomotive scheduling problems* (Tech. Rep. 4389-02). Cambridge, MA: Massachusetts Institute of Technology, Sloan School of Management.

Alarie, S., & Gamache, M. (2002). Overview of Solution Strategies Used in Truck Dispatching Systems for Open Pit Mines. *International Journal of Mining . Reclamation and Environment*, *16*(1), 59–76. doi:10.1076/ijsm.16.1.59.3408

Bazaraa, M. S., Jarvis, J. J., & Sherali, H. D. (2005). Linear programming and network flows. Hoboken, NJ: John Wiley & Sons, Inc.

Beamon, B. (1998). Supply chain design and analysis: Models and methods. *International Journal of Production Economics*, *55*, 281–294. doi:10.1016/S0925-5273(98)00079-6

Beraldi, P., Ghiani, G., Greco, A., & Guerriero, E. (2006). Fix and Relax Heuristic for a Stochastic Lot-Sizing Problem. *Computational Optimization and Applications*, *33*, 303–318. doi:10.1007/s10589-005-3055-2

Böcker, J., Lind, J., & Zirkler, B. (2001). Using a multi-agent approach to optimise the train coupling and sharing system. *European Journal of Operational Research*, *131*, 242–252. doi:10.1016/S0377-2217(00)00124-7

Bojovic, N., & Milenkovic, M. (2008). The best rail fleet mix problem. *Operations Research*, *8*(1), 77–87.

Boland, N., Dumitrescu, I., Froyland, G., & Gleixner, A. M. (2009). LP-based disaggregation approaches to solving the open pit mining production scheduling problem with block processing selectivity. *Computers & Operations Research, 36*(4), 1064–1089. doi:10.1016/j.cor.2007.12.006

Bradley, S. P., Hax, A. C., & Magnanti, T. L. (1977). Applied Mathematical Programming. Reading, MA: Addison-Wesley Publishing Company.

Caccetta, L. (2007). Application of Optimisation Techniques in Open Pit Mining. In A. Weintraub, C. Romero, T. Bjørndal, R. Epstein & J. Miranda (Eds.), Handbook Of Operations Research In Natural Resources, (pp. 547-559). New York: Springer Science+Business Media, Inc.

Caccetta, L., & Hill, S. P. (2003). An Application of Branch and Cut to Open Pit Mine Scheduling. *Journal of Global Optimization, 27*, 349–365. doi:10.1023/A:1024835022186

Caccetta, L., & Kelsey, P. J. (2001). Models for Mine Site Rehabilitation. In A. Jakeman (Ed.), *Proceedings of the Australian National University International Congress on Modelling and Simulation,* (pp. 2155-2160). Canberra, Australia: Modelling and Simulation Society of Australia and New Zealand Inc.

Çetin, N. (2004). *Open-pit truck-shovel haulage system simulation*. Unpublished Ph.D. Dissertation, School of Natural and Applied Sciences, Ataturk University, Turkey.

Chanda, E. K., & Dagdelen, K. (1995). Optimal blending of mine production using goal programming and interactive graphics systems. *International Journal of Surface Mining . Reclamation and Environment, 9*, 203–208. doi:10.1080/09208119508964748

Chandra, P., & Fischer, M. L. (1994). Coordination of production and distribution planning. *European Journal of Operational Research, 72*, 503–5717. doi:10.1016/0377-2217(94)90419-7

Cohen, M. A., & Lee, H. L. (1988). Strategic analysis of integrated production-distribution systems: Models and methods. *Operations Research, 36*(2), 216–228. doi:10.1287/opre.36.2.216

Cordeau, J.-F., Pasin, F., & Solomon, M. M. (2006). An integrated model for logistics network design. *Annals of Operations Research, 144*(1), 59–82. doi:10.1007/s10479-006-0001-3

Cordeau, J.-F., Toth, P., & Vigo, D. (1998). A Survey of Optimization Models for Train Routing and Scheduling. *Transportation Science, 32*(4), 380–404. doi:10.1287/trsc.32.4.380

Crama, Y., Pochet, Y., & Wera, Y. (2001). *A discussion of production planning approaches in the process industry,* (Tech. Rep. GEMME-0102). Liège, Belgique: Université de Liège.

Dauzére-Péres, S., & Lasserre, J.-B. (1994). An integrated approach in production planning and scheduling. *Lecture Notes in Economics and Mathematical Systems, 411.*

David, M. (1988). Handbook of Applied Advanced Geostatistical Ore Reserve Estimation. Amsterdam: Elsevier.

de Matta, R., & Miller, T. (2004). Production and inter-facility transportation scheduling for a process industry. *European Journal of Operational Research, 158*, 72–88. doi:10.1016/S0377-2217(03)00358-8

Dessureault, S., Scoble, M., & Dunbar, S. (1999). Intelligent Production Management in Mining Systems. In J. A. Meech, M. M. Veiga, M. H. Smith, & S. R. LeClair (Eds.), *Proceedings of The Second International Conference on Intelligent Processing and Manufacturing of Materials* (pp. 145-150), British Columbia University.

Dias, J., Captivo, M. E., & Clímaco, J. (2006). Capacitated dynamic location problems with opening, closure and reopening of facilities. *IMA Journal of Management Mathematics, 17*(4), 317–348. doi:10.1093/imaman/dpl003

Dillenberger, C., Escudero, L. F., Wollensak, A., & Zhang, W. (1994). On practical resource allocation for production planning and scheduling with period overlapping setups. *European Journal of Operational Research*, *75*, 275–286. doi:10.1016/0377-2217(94)90074-4

Dimitrakopoulos, R., Farrelly, C. T., & Godoy, M. (2002). Moving forward from traditional optimization: grade uncertainty and risk effects in open-pit design. *Mining Technology: Institute of Mining and Metallurgy Transactions section A, 1*(111), 82-88.

Dimitrakopoulos, R., & Ramazan, S. (2004). Uncertainty-based production scheduling in open pit mining. *Transactions of the Society for Mining, Metallurgy, and Exploration*, *316*, 106–112.

Dogan, K., & Goetschalckx, M. (1999). A primal decomposition method for the integrated design of multi-period production-distribution systems. *IIE Transactions*, *31*, 1027–1036.

Drumond, F. P., & Mateus, G. R. (2000). Efficient Production Planning and Scheduling Integrated Application. In F. Carter (Ed.) *Proceedings of the World Conference on Production and Operations Management* (pp. 165-174), Sevilla, Spain.

El Sheikh, A. A. R., Paul, R. J., Harding, A. S., & Balmer, D. W. (1987). A Microcomputer-Based Simulation Study of a Port. *The Journal of the Operational Research Society*, *38*, 673–681.

Escudero, L. F., & Salmeron, J. (2005). On a Fix-and-Relax Framework for a Class of Project Scheduling Problems. *Annals of Operations Research*, *140*, 163–188. doi:10.1007/s10479-005-3970-8

Esmaeili, M., Aryanezhad, M., & Zeephongsekul, P. (2009). A game theory approach in seller-buyer supply chain. *European Journal of Operational Research*, *195*, 442–448. doi:10.1016/j.ejor.2008.02.026

Everett, J. E. (1997). Simulation to reduce variability in iron ore stockpiles. *Mathematics and Computers in Simulation*, *43*, 563–568. doi:10.1016/S0378-4754(97)00045-1

Everett, J. E. (2001). Iron ore production scheduling to improve product quality. *European Journal of Operational Research*, *129*(2), 355–361. doi:10.1016/S0377-2217(00)00233-2

Everett, J. E., Howard, T., & Kamperman, M. (2001). The Development of a MIS for Iron Ore Mining Operations. In A. Harriger, (Ed.), *Proceedings of the Informing Science Conference*, Krakow, Poland.

Ferreira, D., Morabito, R., & Rangel, S. (2009). Solution approaches for the soft drink integrated production lot sizing and scheduling problem. *European Journal of Operational Research*, *196*, 697–706. doi:10.1016/j.ejor.2008.03.035

Froyland, G., Boland, N., & Dumitrescu, I. (2008). A multistage stochastic programming approach to open pit mine production scheduling with uncertain geology. In R. Beatson & R. Murray (Eds.), VII Joint Australia-New Zealand Mathematics Convention. Christchurch, New Zealand: University of Canterbury.

Fukusawa, R., Aragão, M. V. P., Porto, O., & Uchoa, E. (2002). Solving the freight car flow problem to optimality. *Electronic Notes in Theoretical Computer Science*, *66*(6), 1–11. doi:10.1016/S1571-0661(04)80524-3

Germain, R., Claycomb, C., & Dröge, C. (2008). Supply chain variability, organizational structure, and performance: The moderating effect of demand unpredictability. *Journal of Operations Management*, *26*, 557–570. doi:10.1016/j.jom.2007.10.002

Gholamnejad, J., & Osanloo, M. (2007). Incorporation of ore grade uncertainty into the push back design process. *Journal of the South African Institute of Mining and Metallurgy*, *107*, 177–185.

Godoy, M., & Dimitrakopoulos, R. (2004). Managing risk and waste mining in long-term production scheduling. *Transactions of the Society for Mining, Metallurgy, and Exploration*, *316*, 43–50.

Hall, M. B., & Harris, C. A. (1993). Fuzzy logic expert system for iron ore processing. In *Conference Record of the IEEE Industry Applications Society Annual Meeting* (vol. 3, pp. 2190-2199), Toronto, Canada.

Hax, A. C., & Meal, H. C. (1975). Hierarchical integration of production planning and scheduling. In M. A. Geisler (Ed.), Studies in Management Science (pp. 53-69). Amsterdam, The Netherlands.

Hochbaum, D. S., & Chen, A. (2000). Performance analysis and best implementations of old and new algorithms for the open-pit mining problem. *Operations Research*, *48*(6), 894–914. doi:10.1287/opre.48.6.894.12392

Howard, T. J., Carson, M. W., & Everett, J. E. (2005). Simulation modeling of iron ore product quality for process and infrastructure development. In A. Zerger & R. M. Argent (Eds.), *Proceedings of the International Congress on Modeling and Simulation,* (pp. 1251-1257), Melbourne, Australia.

Kamrad, B., & Ernst, R. (2001). An economic model for evaluating mining and manufacturing ventures with output yield uncertainty. *Operations Research*, *49*(5), 690–699. doi:10.1287/opre.49.5.690.10610

Karadogan, A., Kahriman, A., & Ozer, U. (2008). Application of fuzzy set theory in the selection of underground mining method. *Journal of the South African Institute of Mining and Metallurgy*, *108*, 73–79.

Kolonja, B., Kalasky, D. R., & Mutmansky, J. M. (1993). Optimization and dispatching criteria for open-pit truck haulage system design using multiple comparisons with the best and common random numbers. In G. W. Evans, M. Mollaghasemi, E. C. Russell, & W. E. Biles (Eds.), *Proceedings of the Winter Simulation Conference* (pp. 393-401), Los Angeles, CA.

Kuchta, M., Newman, A., & Topal, E. (2004). Implementing a Production Schedule at LKAB's Kiruna Mine. *Interfaces*, *34*(2), 124–134. doi:10.1287/inte.1030.0059

Kwon, O. K., Martland, C. D., & Sussman, J. M. (1998). Routing and scheduling temporal and heterogeneous freight car traffic on rail networks. *Transportation Research Part E (Logistics and Transpn. Rev.), 34*(2), 101-115.

Lee, Y., & Chen, C.-Y. (2009). An optimization heuristic for the berth scheduling problem. *European Journal of Operational Research*, *196*, 500–508. doi:10.1016/j.ejor.2008.03.021

Lerchs, H., & Grossmann, I. (1965). Optimum design of open-pit mines. *Canadian Institute of Mining Bulletin*, *68*, 17–24.

Lin, Y. (2009). Two-commodity reliability evaluation of a stochastic-flow network with varying capacity weight in terms of minimal paths. *Computers & Operations Research*, *36*, 1050–1063. doi:10.1016/j.cor.2007.12.005

Lu, J., Fu, M., & Sha, J. (2005). Research of import iron ore logistics system based on the minimum cost theory. In *Proceedings of the Second IEEE International Conference on Services Systems and Services Management* (pp. 391-396), Chongqing, China.

Mateus, G. R., Luna, H. P. L., Junior, J. F., Vieira, R. L. F., Costa, R. A. V., & Pratocinio, Z. K. G. (1994). *Iron ore transportation at MBR: characterization and modeling* (Tech. Rep. RT-009/94). Belo Horizonte, Brazil: Computer Science Department, Federal University of Minas Gerais.

Melo, M. T., Nickel, S., & Saldanha-da-Gama, F. (2009). Facility location and supply chain management–A review. *European Journal of Operational Research*, *196*, 401–412. doi:10.1016/j.ejor.2008.05.007

Menabde, M., Froyland, G., Stone, P., & Yeates, G. (2004). Mining Schedule Optimisation for Conditionally Simulated Orebodies. In R. Dimitrakopoulos & S. Ramazan (Eds.), Proceedings of Orebody Modeling and Strategic Mine Planning: Uncertainty and Risk Management, (pp. 357-342), Perth, Australia.

Nachiappan, S. P., & Jawahar, N. (2007). A genetic algorithm for optimal operating parameters of VMI system in a two-echelon supply chain. *European Journal of Operational Research*, *182*, 1433–1452. doi:10.1016/j.ejor.2006.09.040

Newman, A., Martinez, M., & Kuchta, M. (2006). A review of long- and short-term production scheduling at LKAB's Kiruna Mine. In J. W. Herrmann (Ed.), Handbook of Production Scheduling, (pp. 265-286). New York: Springer Science+Business Media, Inc.

Osanloo, M., Gholamnejad, J., & Karimi, B. (2007). Long-term open pit mine production planning: a review of models and algorithms. *International Journal of Mining. Reclamation and Environment*, *22*(1), 3–35. doi:10.1080/17480930601118947

Pendharkar, P. C., & Rodger, J. A. (2000). Nonlinear programming and genetic search application for production scheduling in coal mines. *Annals of Operations Research*, *95*, 251–267. doi:10.1023/A:1018958209290

Picard, J. C. (1976). Maximum closure of a graph and its applications to combinatorial problems. *Management Science*, *22*, 1268–1272. doi:10.1287/mnsc.22.11.1268

Pochet, Y., & Wolsey, L. A. (2006). Production planning by mixed integer programming. New York: Springer Science+Business Media, Inc.

Ramazan, S. (2007). The new Fundamental Tree Algorithm for production scheduling of open pit mines. *European Journal of Operational Research*, *177*, 1153–1166. doi:10.1016/j.ejor.2005.12.035

Ramazan, S., & Dimitrakopoulos, R. (2003). Production Scheduling Optimisation in a Nickel Laterite Deposit: MIP and LP Applications and Infeasibility in the Presence of Orebody Variability. In Carlton, (Ed.), *Proceedings of the Australasian Institute of Mining and Metallurgy International Symposium on Mine Planning and Equipment Selection* (pp. 1-5). Kalgoorlie, Wichita.

Ramazan, S., & Dimitrakopoulos, R. (2004). Recent applications of operations research and efficient MIP formulations in open pit mining. *Transactions of the Society for Mining, Metallurgy, and Exploration*, *316*, 73–78.

Runge, I. (1998). Mining Economics and Strategy. Littleton, CO: Society for Mining, Metallurgy and Exploration, Inc.

Saaty, T. (1977). Scenarios and priorities in transport planning: application to the Sudan. *Transportation Research*, *11*, 346–351. doi:10.1016/0041-1647(77)90044-2

Sabino, J. A., Stützle, T., Birattari, M., & Leal, J. E. (2006). ACO Applied to Switch Engine Scheduling in a Railroad Yard (extended abastract). In *Proceedings of the V International Symposium on Ant Colony Optimization and Swarm Intelligence* (pp. 502-503). Heidelberg: Springer Berlin.

Saker, B. R., & Diponegoro, A. (2009). Optimal production plans and shipment schedules in a supply-chain system with multiple suppliers and multiple buyers. *European Journal of Operational Research*, *194*, 753–773. doi:10.1016/j.ejor.2008.01.025

Salim, V., & Cai, X. (1995). Scheduling cargo trains using genetic algorithms. In *Proceedings of the IEEE International Conference on Evolutionary Computation* (pp. 224-227). Perth, WA.

Santos, R. R., Meech, J. A., & Ramos, L. T. S. (1995). Thickener operations at Carajas using a fuzzy logic controller Systems. *IEEE International Conference on Man and Cybernetics: Intelligent Systems for the 21st Century,* (vol. 2., pp. 1636-1639).

Sarimveis, H., Patrinos, P., Tarantilis, C. D., & Kiranoudis, C. T. (2008). Dynamic modeling and control of supply chain systems: a review. *Computers & Operations Research*, *35*, 3530–3561. doi:10.1016/j.cor.2007.01.017

Sarin, S. C., & West-Hansen, J. (2005). The long-term mine production scheduling problem. *Institute of Industrial Engineers Transactions*, *37*, 109–121.

Sattarvand, J., & Niemann-Delius, C. (2008). Perspective of metaheuristic optimization methods in open pit production planning. In J. Dubiński (Ed.), *XXI World Mining Congress* (pp. 143-155). London: Taylor & Francis Group.

Sayarshad, H. R., & Ghoseiri, K. (2009). A simulated annealing approach for the multi-periodic rail-car fleet sizing problem. *Computers & Operations Research*, *36*(6), 1789–1799. doi:10.1016/j.cor.2008.05.004

Schofield, C. G. (1980). Homogenization and Blending Systems. Clausthal, Germany: TransTech.

Scott, M., & Dimitrakopoulos, R. (2001). Quantitative analysis of mineral resources for strategic planning: implications for Australian geological surveys. *Natural Resources Research*, *10*(3), 159–177. doi:10.1023/A:1012536823294

Singh, V., & Mohan, R. S. (2005). Application of image processing and radial basis neural network techniques for ore sorting and ore classification. *Minerals Engineering*, *18*(15), 1412–1420. doi:10.1016/j.mineng.2005.03.003

Smith, M. L., Sheppard, I., & Karunatillake, G. (2003). Using MIP for strategic life-of-mine planning of the lead/zinc stream at Mount Isa Mines. In Application of Computers and Operations Research in the Minerals Industries, (pp. 465-473).

Soroor, J., Tarokh, M. J., & Shemshadi, A. (2009). Initiating a state of the art system for real-time supply chain coordination. *European Journal of Operational Research*, *196*, 635–650. doi:10.1016/j.ejor.2008.03.008

Steffen, O. K. H. (1997). Planning of open pit mines on a risk basis. Journal of the Southern African Institute of Mining and Metallurgy, March/April, 47-56.

Ta, C. H., Kresta, J. V., Forbes, J. F., & Marquez, H. J. (2005). A Stochastic Optimization Approach to Mine Truck Allocation. *International Journal of Surface Mining . Reclamation and Environment*, *19*(3), 162–175. doi:10.1080/13895260500128914

Tang, L., & Liu, G. (2007). A mathematical programming model and solution for scheduling production orders in Shanghai Baoshan Iron and Steel Complex. *European Journal of Operational Research*, *182*, 1453–1468. doi:10.1016/j.ejor.2006.09.090

Thomas, D. J., & Griffin, P. M. (1996). Coordinated supply chain management. *European Journal of Operational Research*, *94*, 1–15. doi:10.1016/0377-2217(96)00098-7

Tolwinski, B., & Underwood, R. (1996). A scheduling algorithm for open pit mines. *IMA Journal of Mathematics Applied in Business & Industry*, *7*, 247–270.

Topal, E. (2008). Early start and late start algorithms to improve the solution time for long-term underground mine production scheduling. *Journal of the South African Institute of Mining and Metallurgy*, *108*, 99–107.

Tsiakis, P., & Papageorgiou, L. G. (2008). Optimal production allocation and distribution supply chain networks. *International Journal of Production Economics*, *111*, 468–483. doi:10.1016/j.ijpe.2007.02.035

Underwood, R., & Tolwinski, B. (1998). A mathematical programming viewpoint for solving the ultimate pit problem. *European Journal of Operational Research*, *107*, 96–107. doi:10.1016/S0377-2217(97)00141-0

Van Asperen, E., Dekker, R., Polman, M., & Arons, H. S. (2003). Modeling ship arrivals in ports. In *Proceedings of the 2003 IEEE Winter Simulation Conference,* (pp. 1737-1744).

Vieira, P., Neto, E., Bessa, E., & Gomide, F. (1999). Railway dispatch planning and control. In *Proceedings of the 18th International Conference of the North American Fuzzy Information Processing Society,* (pp. 134-138), New York.

Wang, Q., & Sevim, H. (1995). Alternative to parameterization in finding a series of maximum-metal pits for production planning. *SME Transactions*, *298*, 178–182.

White, J. W., & Olson, J. P. (1986). Computer-based dispatching in mines with concurrent operating objectives. *AIME Transactions, 280.*

KEY TERMS AND DEFINITIONS

Grade: Refers to the physical and chemical quality characteristics, of the ROM or final ore product. Physical characteristics relate to particle size and structure, while chemical characteristics relate to the composition of the main desired and undesired elements.

Mass Efficiency: The ratio of the mass of final ore products generated by a processing plant to the mass of ROM fed into that processing plant.

Mine Complex: A mine complex is defined as the set of pits, which form the mine itself, and all other facilities used to process, store, and ship ore products.

ROM: Run-of-mine, which means all ore mined and not yet fully processed into end ore products.

ENDNOTE

1 For a problem instance containing 28 mines, 3 ports, 40 products and a planning horizon of 120 periods, which amounts to two million variables, an Intel Quad Core machine with 2.5GHz and 4GB RAM, determined the optimal solution in less than one minute. That problem size approximates a known Brazilian iron ore supply chain.

Chapter 9
Negotiation Policies for E-Procurement by Multi Agent Systems

Paolo Renna
University of Basilicata, Italy

ABSTRACT

The development of Information and Communication Technologies (ICT) allowed the enterprises to adopt an e-marketplace approach to Business to Business (B2B) applications. In particular, these kind of applications are demonstrating their capacity to provide real added value to manufacturing industries by allowing their global performance to increase. The implementation of the e-marketplace by firms is not considered an easy job because of the lack of automation: the human participation is still in all stages of the B2B process. The chapter proposes a three value added services: workflow design, Multi Agent System and negotiation approach. In particular, two negotiation, an auction and single round approaches with three customer behaviors are proposed. A simulation environment is developed in order to test the proposed approaches. The simulations have been conducted in several scenarios in order to highlight what is the best approach to perform.

INTRODUCTION

The development of Information and Communication Technologies (ICT) is deeply changing the way to do business for several manufacturing companies. Especially, Business-to-Business (B2B) applications are demonstrating the capacity to provide real value to manufacturing industries by allowing industry global performance to increase. According to Rangone et al. (2004) a basic classification of B2B solutions can concern their business model focus; indeed is possible to classify B2B solutions in those that are procurement oriented, and those that are supply chain oriented. This chapter focuses on procurement oriented solutions, that support the company in procurement transactions. These solutions are essentially referred as e-marketplace.

Several definitions of e-marketplace (EM) have been provided in the literature from 1998. However, a basic definition has been proposed by Grieger

DOI: 10.4018/978-1-61520-605-6.ch009

Table 1. E-procurement performance

Performance area	2001	2004	2006
Total Suppliers enabled	30	253	361
Total End-Users	1,000	2,309	1,381
Users: Current vs. Planned (%)	12%	43.5%	68%
Transactions per Month	1,340	5,244	2,977
Percentage of Indirect Spend managed by system	18%	37.6%	55%
Percentage of Services Spend managed by system	--	32.7%	29.3%

(2003) that is: "*an EM brings multiple buyers and sellers together (in a "virtual" sense) in one central market space. If it also enables them to buy and sell from each other at a dynamic price which is determinate in accordance with rules of the Exchange, it is called an electronic exchange; otherwise it is called a portal*".

In practice an e-marketplace is not an easy job, mainly for the lack of automation: the human participation is still present in all stages of the B2B process. Recently, to ride out this limitation, the diffusion of software agents is increased and, in few years, has lead to the automation of several activities in order to reduce time consuming and diminishing transaction costs. The used software agents differ form "traditional" software because they are autonomous or semi-autonomous, customizable and can be coordinated with other agents.

The e-procurement use by firms, as a recent research of the Aberdeen group (2006) testify, leads to the following improvements performance:

- Increased their spend under management by 36% ;
- Reduced their requisition-to-order cycles by 75% ;
- Reduced their requisition-to-order costs by 48% ;
- Reduced their maverick spend by 36%.

Table 1 reports the trend of the e-procurement applications upon 2001 and 2006.

In particular, the research concerns private neutral linear e-marketplaces owned by a third part where a set of registered buyers, customers, and a set of registered sellers, suppliers, are allowed to play procurement actions. Example of such e-marketplace are CPGmarket, Tribon Marketplace, ChemConnect, etc. According to Barrat et al. (2002) reduces waste and inefficiency in highly fragmented industries, by increasing visibility and a neutral knowledge base for both buyers and sellers. This business model is the most suitable for Small and Medium Enterprises (SMEs). In this context, the e-marketplaces can create value mainly through the matching mechanism among buyers and sellers where transaction attributes are determined at the moment of purchase.

Among the value added services (VASs) the following have a strongest importance:

- Production planning tools;
- Negotiation tools.

Production planning tools allow creating a link between commercialization and production activities providing a better service for the customer, that can gain reliable information about order availability and timing, and for the supplier, that can correctly plan resource use in order to achieve lower costs. On the other hand, negotiation tools allow making transactions able to take into account buyer's and seller's identities and goals, providing a better global satisfaction.

Although the importance of such tools in e-business applications is quite acknowledged, some problems may arise when it comes to design, implementation and cost evaluation.

The design and implementation of the above VASs can be developed by the following methodologies:

- multi agent system seems to be the most promising technology to be used in building these systems (Turowski, 2002);
- workflow analysis tools seems to be the most promising methodology to engineering e-business VAS design (Presley et al., 2001);
- open environment such as the Internet and the WWW, and open IT tools, such as Java and XML, are considered the reference IT technologies for developing these systems.

The chapter is structured as it follows: section 2 presents a literature overview on automated negotiation in B2B e-procurement; section 3 provides a general description of the proposed models through workflow management tools; the negotiation and planning models are described in sections 4 and 5; in section 6 the production planning algorithm is presented. The simulation environment that has been developed and the simulation results have been presented in sections 7 and 8. Finally, conclusions and further research paths are withdrawn in section 9.

LITERATURE REVIEW

Many researches investigated electronic services, like e-procurement, combined with intelligent decision support systems by the creation of intelligent distributed systems like Multi Agent Systems (MAS) [Maes et al., 1999, Kim and Lee, 2002]. Analyzing the literature about the intelligence agent in automated negotiation approach Bartolini et al. (2002) developed a negotiation protocol adopting FIPA from a software engineering prospective in a MAS environment. The generalized protocol developed has been implemented in JADE and give examples of rules for an English auction.

Benyoucef et al. (2002) proposed a rule-driven approach to represent, manage and explore negotiation strategies and coordination information. For that, they divide the behavior of negotiating agents into protocols, strategies and coordination. They develop simulation examples in the English and Dutch auctions. They also implemented simple coordination schemes across several auctions.

Oliver (1997) and Choin et al. (2001) developed genetic algorithms to find offers for agents to negotiate with other agents. The approaches based on genetic algorithms have a disadvantage because they require many trials to achieve good strategies in each round of negotiation.

Wurrman et al. (2002) examined the design space of auction mechanism and identify the core activities. Formal parameters qualifying the performance core activities enable precise specification of auction rules. This specification constitutes an auction description language that can be employed in the implementation of configurable marketplaces.

Dumas et al. (2005) proposed an approach to develop bidding agents that participate in multiple alternative auctions, with the goal of obtaining an item with a given probability. The approach consists of a prediction method and a planning algorithm. The prediction method exploits the history of past auctions in order to build probability functions capturing the belief that a bid of a given price may win a given auction. The planning algorithm computes a price, such that by sequentially bidding in a subset of the relevant auctions, the agent can obtain the item at that price with the desired probability. The approach addresses the case where the auctions are for substitutive items with different values. Experimental results showed that the approach increases the payoff of their users and the welfare of the market.

Chen et al. (2005) discussed the e-marketplace eAgora. It allows buyers and sellers to engage in multi-issue negotiations. eAgora implements several protocols based on a negotiation phase model constructed for this system. Its services include a software agent that generates and critiques offers.

The main restrictions are the following: no real-time information search on demand by an agent are incorporated, adaptive tracking of user profile based on the offers made; and profiling the opponent in order to infer the most likely directions for compromise and mutual benefits.

Kurbel and Loutchkor (2005) presented a model for multi-lateral negotiations of agents with fuzzy constraints on an electronic marketplace for personnel acquisition. The approaches developed involving negotiation protocol and negotiation strategies for bilateral negotiations. Using the bilateral model and the partial order of the set of all employees' agents created by the pre-selection procedure, the case of multi-lateral negotiations with many negotiation issues was considered.

Although the paper deals with the specific situation of an electronic marketplace for personnel acquisition, the proposed negotiation model can be used with appropriate modifications in the framework of other agent-based e-marketplaces where agents have contradictory aims and negotiation is required.

Cheng et al. (2006) presented a formal heuristic model for making trade-offs in automated negotiations in a third-party-driven e-marketplace. The tactics that the agents are to employ when making trade-offs are explicitly formulated as fuzzy inference systems, which are used to infer new offers at each round of negotiation. The experimental results demonstrate that the proposed automated negotiation algorithm is efficient in terms of the number of offers exchanged, the joint utility obtained, and the Pareto-efficiency of the negotiated contracts.

The automated negotiations formulated in this study do not consider the quantities that a buyer demands or the quantities that a supplier can provide

Saha (2006) presented an extended protocol for bilateral multi-issue negotiation. The paper showed that with this protocol self-interested agents are able to explore and reach win-win agreements without revealing its complete preference. The scenarios investigated regard two agents and two and four attributes in negotiation.

Rahwan et al. (2007) introduced a methodology for designing strategies for negotiating agents. The methodology provides a disciplined approach to analyzing the negotiation environment and designing strategies in light of agent capabilities, and acts as a bridge between theoretical studies of automated negotiation and the software engineering of negotiation applications. The methodology is defined STRATUM (for STRATegy via Unorthodox Methodology); the main function of this methodology is to guide the designer of negotiation strategies from domain and requirements analysis to producing modular high-level specifications of strategies.

Louta et al. (2008) proposed a dynamic multilateral negotiation model and a negotiation strategy based on a ranking mechanism. The contract generation algorithm of the Seller is coupled with a Buyer ranking mechanism that entails identification of the most suitable contract among the contracts proposed. The framework developed is limited to price and due date and is tested between one buyer and one seller.

Bandyopadhyay et al. (2008) proposed a reverse auction mechanism by Multi Agent System to automate transactions in e-marketplaces. The research developed concern the mixed-strategy equilibrium in capacity-constrained reverse auctions. The approach proposed involve only the price attribute and for a limited number of player.

From the analysis of the literature emerging the following limits:

- most of the research concerning only price or only one/two items in negotiation;
- few research concern a link between the negotiation process and the production planning activity;
- in most research to test the approaches has been developed a numerical example, in few papers a distributed agent architecture has been developed in order to test the approaches in a dynamic environment;
- most of the researches, deeply, investigated the agreement in a negotiation (often, two agents are involved), but the real value added and how the utility is distributed among suppliers and customers.

These lacks are the starting of the research presented in this chapter: an innovative Multi Agent Architecture in a private neutral e-marketplace is proposed. In particular, the research aims to understanding what kind of advantage customers and suppliers can gain by the implementation of e-marketplaces based on the proposed tools. The aims of this chapter are mainly three:

- the development of a Multi Agent Architecture by using UML and IDEF diagram to formalize the e-marketplace structure. The workflow is described form static and dynamic point of view.
- the development of a matching protocol among the involved agents; in particular, three mechanisms are proposed: negotiation, auction and single round. Moreover, three customer behaviors are implemented: time, imitation and hybrid.
- the development of a discrete event simulation environment, based on agents' formalism, to test the proposed B2B application. The simulation environment allows to test the architecture in very dynamic conditions.

The main trusts of the research are the following:

- the negotiation concerns multi-attributes and involving many agents;
- the production planning algorithm of the suppliers is linked to the negotiation mechanism in order to provide accurate information and, therefore, improve the agreement.
- The Multi Agent Architecture is implemented by open source tools (JAVA); this environment allows to evaluate the real advantage of the customers and suppliers involved in the network.

WORKFLOW DESIGN

The conceptual design of the e-marketplace is performed by the integration of IDEF0 and UML diagram in order to define the workflow and to simplify the engineering phase of the e-marketplace. Figure 1 shows the e-marketplace structure through the interaction with external actors. As the reader can notice, three kinds of actors have been located (Bruccoleri et al., 2005):

- the System Manager is the independent party that sets and manages the e-marketplace. Specifically, it sets e-marketplace tools and negotiation constraints that bounds the negotiation behavior of both sellers and buyers;
- the Customer is the generic registered e-marketplace client (hereafter also called customer) who is allowed to provide its negotiation strategies and its catalogue based order;
- the Supplier, is the generic registered e-marketplace merchant (hereafter also called seller), who provides catalogue information, negotiation strategies and planning information.

Figure 1. E-marketplace use cases

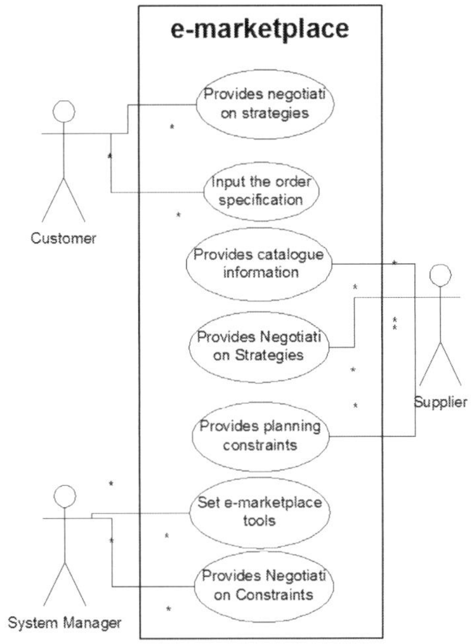

Once the generic customer inputs its order, the procurement transaction proceeds automatically being handled respectively by the customer and supplier agents. Therefore, a Multi Agent Architecture has been developed in order to implement the e-marketplace.

Figure 2 shows the class diagram of the Multi Agent Architecture. As therein showed, seven classes can be detected:

- *planner*; this class, operating in the simulation environment, has in charge data setting

Figure 2. Class diagram

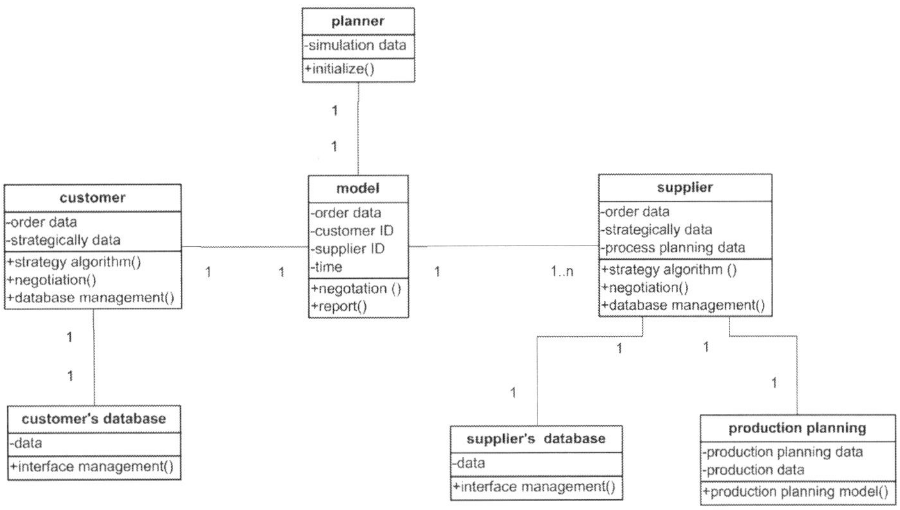

Figure 3. Sequence diagram

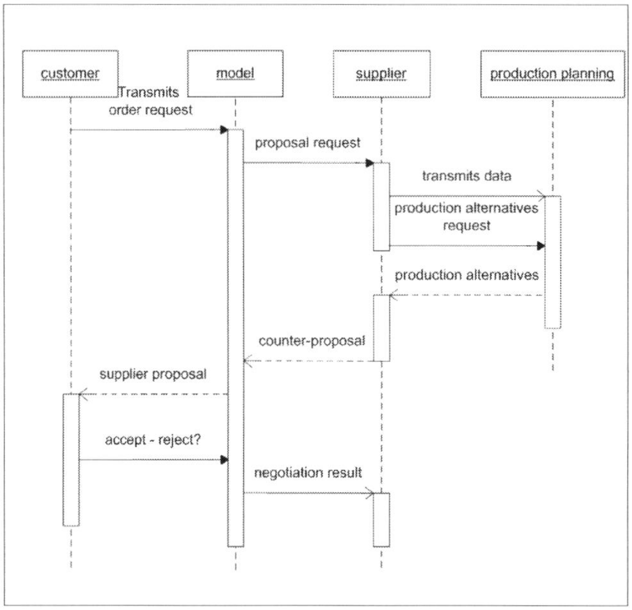

and the initialization of all the involved parameters;

- *model*; this class controls the events occurring among customers and suppliers. In particular, it supervises the orders data requested by the customers, the identification of the registered participants and the events timetable. Initially it operates by monitoring the negotiation, then collecting the final reports;
- *customer*; it is in charge of manage data regarding orders and to define the optimal strategy to adopt during the negotiation stage. It operates by using the negotiation algorithm to negotiate and manage the knowledge during the phases of interaction;
- *customers'database*; this class manages the customers' knowledge stored in a specific database by using an opportune interface linked to the previously described class;
- *supplier*; this class receives the orders data and knows the strategies regarding the suppliers as well as the process planning data. It operates by using the negotiation algorithm to negotiate and communicate with the production planning class in order to obtain the process planning alternatives;
- *production planning*; it manages both production planning data and orders features requested by the customers. It operates by using the production planning algorithm in order to provide the production planning alternatives to the suppliers' class.
- *suppliers' database*; this class manages the knowledge regarding the suppliers: it is stored in a specific database by an appropriate interface linked to the suppliers' class.

Figure 3 reports the sequence of all the actions occurring among the above described classes (the planner class is omitted to obtain visual simplicity). It is a preliminary investigation of a dynamic point of view; these actions are deeply investigated in the activity diagram (Figure 5).

Figure 4 shows the basic representation of the agents' interaction during the procurement

Figure 4. E-marketplace basic structure – level 0

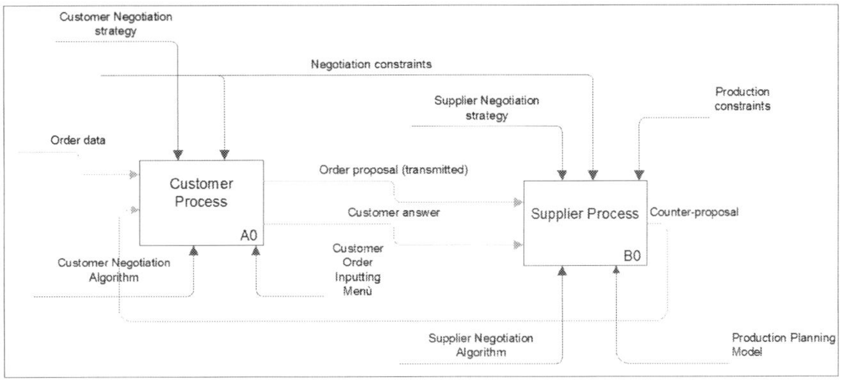

mechanism. As the reader can notice, in the IDEF0 diagram (Figure 4), it is possible to identify the customer and the supplier processes. The last one is triggered by the order data inputted by the customer through a specific Order Inputting Menu and it is constrained by the Customer Negotiation Strategy and by the Negotiation constraints. The output of this process can alternatively be the processed order proposal for electronic transmission to the supplier agent, or, during the negotiation,

Figure 5. Negotiation process activity diagram

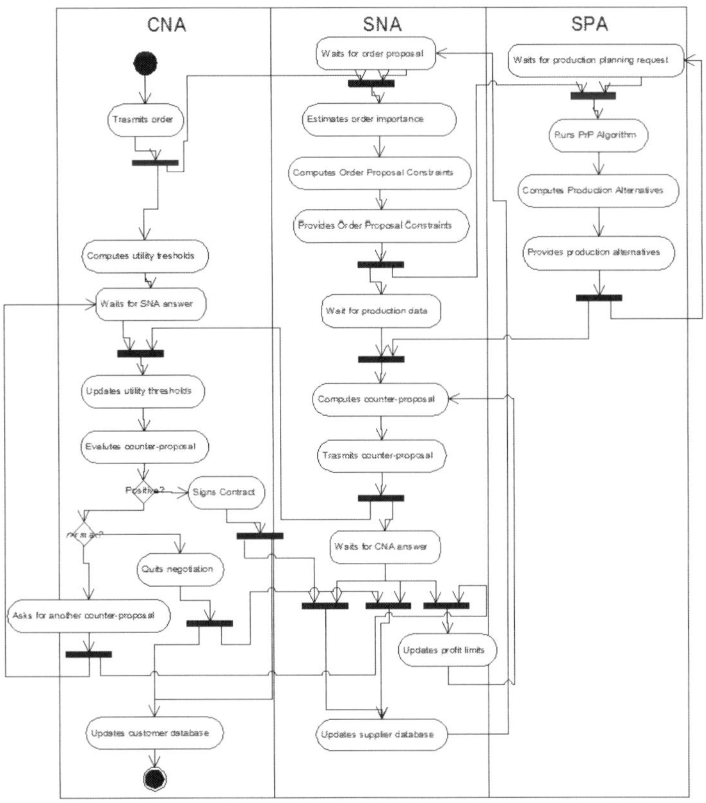

the customer answer to a supplier counter-proposal (obtained by using a proper Customer Negotiation Algorithm).

Differently, the supplier process is triggered by the customer order proposal and it is constrained by supplier Negotiation Strategy, Production Constraints and Negotiation Constraints. The supplier process elaborates a counter-proposal to submit to the customer process. Such process is undertaken by using proper supplier Negotiation Algorithms and Production Planning Models.

E-COMMERCE OPERATIONAL APPROACHES

The negotiation approach proposed in Bruccoleri et al. (2005) is fully described, because is the basic structure used as a start point to develop the proposed approaches: qualified, auction and single round.

Moreover, the original issue is the investigation of the customers' behavior integrated with the negotiation approach and the other proposed approaches.

Negotiation Process

Figure 5 shows, through an UML activity diagram, the detail of the negotiation process involving the agents Customer Negotiation Agent (CNA), Supplier Negotiation Agent (SNA) and Supplier Planning Agent (SPA). The negotiation process is characterized by the following constraints (*Negotiation constraints*):

- the negotiation is a multi-lateral one and it involves one customer and many suppliers (one to many approach);
- the negotiation is an iterative process with a maximum number of rounds, r_{max}; after that an agreement is reached or the negotiation fails;
- during each round the supplier can submit a new counter-proposal (N) to the customer while, at $r= r_{max}$, it can only accept (A), reject (R) or ask for last counter proposal. Obviously, the customers answer at generic round r, can be referred as $(A \vee R \vee N)_r$;
- the agreement is reached only if the customer accepts the supplier counter proposal at round $r< r_{max}$; in this case customer and supplier sign an electronic contract;
- supplier and customer behavior is assumed to be rationale according to their utility functions;
- the customer does not know suppliers' utility functions and vice versa; however suppliers and customers can only argue, by applying proper learning algorithms, the behavior of their counterparts.

The negotiation process starts with the order submitted by customer. The order is processed through the *Customer Order Inputting Menu* and it is delivered to the CNA. The order is represented by the array $(i, V_i, dd_i, p_i)_0$ being $i \in \{1,...,n\}$ the selected product from the supplier's catalogue, V_i the required quantity, dd_i the requested delivery date and p_i the asked price.

By the activity diagram of Figure 5 the following actions are carried out:

- *Transmits order*; the CNA transmits the order array $(i, V_i, dd_i, p_i)_0$ to the SNA;
- *Computes utility threshold;* The CNA computes its utility function and the lower threshold level according to the following expression:

$$Thu(r) = Thu_{\max} \cdot \left(1 - \frac{r-1}{r_{\max}-1}\right)^2 + F \cdot \left(\frac{r-1}{r_{\max}-1}\right) \cdot \left(1 - \frac{r-1}{r_{\max}-1}\right) + Thu_{\min} \cdot \left(\frac{r-1}{r_{\max}-1}\right)^2 \quad (1)$$

where:

- Thu_{max} is given by the sum of the maximum values reached by the considered utility functions (obtained when all the requested parameters are satisfied);
  - F is the utility function slope.
    - *Computes Order Proposal Constraints*; the SNA computes a feasible range of variation of both the required price (Δp_i) and due date (Δdd_i). In this chapter the values of these range are the following:

      $(\Delta\, dd_i) = \pm\ 5$;

      $(\Delta p_i) = +60\%$
    - *Provides Order Proposal Constraints*; the above values are transmitted to the SPA and they will be considered as bounds by the agents;
    - *Runs PrP;* the SPA runs the production planning (PrP) algorithm (described in section *Production Planning model*);
    - *Computes Production Alternatives;* as output of the PrP algorithm, the SPA computes an array of production planning alternatives PA_j (*j=1...m*) that associate a supplier profit (Pr_j) and an offered volume (V_j) to each combination of offered due date (dd_j) and price (p_j), that is $PA_j = (\mathrm{Pr}_j, V_j, dd_j, p_j)\ \forall j$, where $V_j \leq V_i$;
    - *Provides production alternatives; PA_j* is transmitted to the SNA;
    - *Computes counter-proposal*; if r=1, the SNA builds the set of alternatives $K_0 = \{1,2,..k,\ldots,n^*\}$ such as:

$$\mathrm{Pr}_k = \mathrm{Pr}_{\max} = \max_{j=1,\ldots,n}\left\{\mathrm{Pr}_j\right\} \forall k \in K_0 \quad (2)$$

and it searches within K_0 for the alternative j^* such as:

$$j^* \mid \min_{j\in K_0}\left(\frac{\left|dd_j - dd_i\right| + \left|p_j - p_i\right| + \left|V_j - V_i\right|}{3}\right) \quad (3)$$

On the other hand, if $r > 1$, the SNA applies a profit reduction strategy according to the negotiation round: it computes the new acceptable profit at the round r as reported in (4):

$$\mathrm{Pr}_r = \mathrm{Pr}_{\max} - \frac{PR_{\max} - PR_{\min}}{r\max} \cdot r \quad (4)$$

Afterwards SNA builds the set of production alternatives $K_r = \{1,2,..k,\ldots,m^*\}$ such that:

$$\mathrm{Pr}_k \geq \mathrm{Pr}_r\ \forall k \in K_r \quad (5)$$

and it finds the alternative j^* that minimizes the relation (3) with $j \in K_r$. The array (dd_{j^*}, p_{j^*}, V_{j^*}), both in cases $r = 1$ and $r > 1$, represents the supplier counter-proposal;

- *Transmits counter proposal*; the array (dd_{j^*}, p_{j^*}, V_{j^*}) is transmitted to the CNA. The SNA remains waiting for a CNA request;
- *Updates utility thresholds;* the CNA updates the utility function thresholds at the round r according to the expression (1).
- *Evaluates counter-proposal*; the CNA evaluates the utility related to the counter-proposal:

$$U^{c-p}_{r} = U_v + U_{dd} + U_p \quad (6)$$

where U_v, U_{dd}, U_p are respectively the utilities of the volume, the due date and the price, computed as:

$$U_v = \max\left(\left(\frac{V_{j*} - V_{min}}{V_i\text{-}V_{min},}\right);0\right), \tag{7}$$

where V_{min}=0.3*V_i.

$$U_{dd} = Max\left(Min\left(\frac{dd_{j*} - dd_{min}}{dd_i - dd_{min}};\frac{dd_{max} - dd_{j*}}{dd_{max} - dd_i}\right);0\right) \tag{8}$$

where dd_{max} and dd_{min}, are respectively dd_i +5 and dd_i -5,

$$U_p = Min\left(\left(\frac{p_i}{p_{j*}}\right);1\right) \text{ if } p_{j*}< p_{max} \text{ otherwise}$$
$$U_n = 0 \tag{9}$$

where p_{max}=1.6*p_i. If $U^{c\text{-}p}_r \geq Thu(r)$ the CNA accepts (A) the counter-proposal and it signs the agreement with SNA; afterwards they update their database with agreement data. Conversely, if $U^{c\text{-}p}_r < Thu(r)$ and $r<r_{max}$, CNA asks for a new counter-proposal (N) otherwise, if $r>r_{max}$, CNA rejects the proposal and quits the negotiation.

Qualified Contemporary Multi Negotiation (QCMN)

The negotiation approach has some drawbacks: the customer have to negotiate contemporary with all the suppliers, therefore the complexity to manage the process increase with the number of suppliers involved; the communication overhead caused by the number of messages that customer and suppliers exchange during the negotiation; finally, the time to reach an agreement is correlated to the number of the suppliers. A negotiation one to one can reduce the above drawbacks; in this case the customer negotiates with one supplier at time. If the negotiation fails the customer negotiates with another supplier. The negotiation performance depend by the sequence order of the suppliers, therefore an opportune strategy have to be implement.

The customer qualifies the supplier by the information of the past negotiation. The qualification process is performed by the customer that knows only two information: price and utility reached by the negotiation. The qualification criterion proposed is the computation of an equivalent price:

$$price\,eq = \sum_{order} price_{order} \cdot U^{c-p}{}_{order} \tag{10}$$

The expression is the total sum (only for the order that reach an agreement) of the price multiplied the customer utility. The *priceeq* is computed for each supplier.

The reasons of the above expression are the following:

- The price is the relevance of the transactions that reach an agreement between the customer and supplier. A high price can be due to a high volume requested or a rush due date.
- The utility denotes how the supplier satisfied the customer attributes.

Each time the customer needs to submit an order, it qualifies the supplier based on the expression (10). Therefore, the order will be submitted first to supplier in the high ranking position. If the negotiation with this supplier fails then the order will be submitted to the supplier in subsequent order. This approach leads to reduce the number of negotiation round by a sequential negotiation with the suppliers.

Figure 6. a) Neutral; b) Conceder; c) Waiting

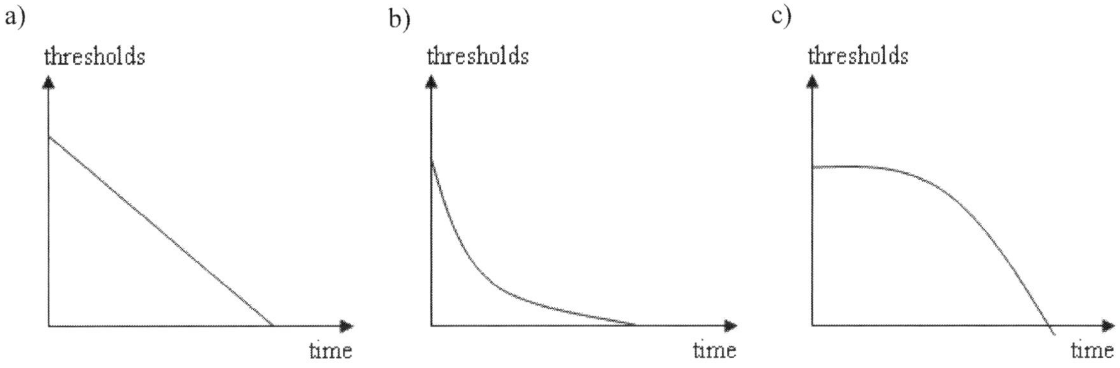

Auction

In this case, each supplier formulates the counter-proposal at the first round as reported in section *negotiation process*. For the subsequent round, the customer communicates the best counter-proposal; therefore, if the generic supplier doesn't formulate the best counter-proposal tries to improve its counter-proposal. The auction mechanism ends at the maximum number of round, and the customer assigns the order to the best counter-proposal if the utility of it is major of the minimum thresholds requested by the customer. Otherwise the negotiation ends without agreement. The approach leads to increment the number of round negotiation, but with a higher customer utility. The suppliers compete to acquire the order of the customer, but a supplier that submit the best counter-proposal doesn't reduce its profit to improve the counter-proposal (otherwise in negotiation approach the suppliers improve the counter-.proposal at each round).

Single Round

The approach is performed by a single round of the negotiation reported in section *negotiation process*, therefore the suppliers submit the best counter-proposal. The customer accept the counter-proposal if the utility customer is major of the minimum thresholds requested by the customer. Otherwise the negotiation ends without agreement. This approach leads to minimum round of negotiation (one), but the effects on the utility of customer and suppliers have to be investigated. This approach is used as a benchmark to evaluate the performance of the above approaches.

Customer Behavior

The customer in order to evaluate the suppliers' counter-proposal can implement a set of tactics that generate a new value of the utility threshold. The following tactics are proposed:

- *Time dependent;* the customer has a time deadline by which an agreement must be in place. This tactics model depending on the shape of the curve concession. Figure 6 shows the three models of this tactics.

The function of the tactic is the expression (1) that depends on the rounds of negotiation. The parameter F (function slope) sets the shape of the curve concession (figure 6). In this chapter, the neutral tactics is implemented (figure 6a).

- *Imitative*; in this case, the customer is not under pressure to reach an agreement. It may choose to use imitative tactics in order to reduce its utility satisfaction based on the concession of the supplier. In this case the utility thresholds is computed by the following expressions:

$$Thu(1) = Thu_{\max}, \text{ at first round of negotiation.} \quad (11)$$

$$Thu(r) = Max(Thu(r-1) - (U^{c\text{-}p}_{r} - U^{c\text{-}p}_{r-1})\cdot, Thu_{\min}) \text{ for } r>1. \quad (12)$$

The value $(U^{c\text{-}p}_{r} - U^{c\text{-}p}_{r-1})$ is the concession of the supplier between two successive rounds.

- *Hybrid*; the customer uses a mixed model between time and imitative. The threshold value is computed by the following expression:

$$Thu(r) == Min\begin{bmatrix} Max(Thu(r-1) - (U^{c\text{-}p}_{r} - U^{c\text{-}p}_{r-1})\cdot, Thu_{\min}), \\ Thu_{\max}\cdot\left(1-\frac{r-1}{r\max-1}\right)^2 + F\cdot\left(\frac{r-1}{r\max-1}\right)\cdot\left(1-\frac{r-1}{r\max-1}\right) + Thu_{\min}\cdot\left(\frac{r-1}{r\max-1}\right)^2 \end{bmatrix} \quad (13)$$

The three tactics model has been tested in all the negotiation mechanism proposed in order to evaluate the performance for both customers and suppliers.

PRODUCTION PLANNING ALGORITHM

The production planning proposed in Perrone et al. (2003) is adapted and implemented for each supplier. It has been assumed that the production plan activity proceeds by planning period of fixed length T, i.e. all the orders received within the period T can be re-planned in the period; once the period is over, re-planning is not allowed for orders of the previous period. The algorithm is depicted in the activity diagram of Figure 7.

- *Initializes algorithm parameter;* the SPA set the orders counter i = 1. Since the incoming order is surely a new one it needs to be planned for the first time (*In Negotiation*).
- *Set N.O. parameters*; the SPA sets the following Negotiation Order (*N.O.*) parameters:
  - $i^* = i$, index of the N.O.;
  - $Fmin_i^* = 0$, minimal fraction of production volume;
  - $Fmaxi^* = 1$, maximal fraction of production volume.

Considering the period $t < T$, if i is lower than the number of the arrived orders $N\_ord(t)$, there must exists other orders already planned. This means that the algorithm have to loads the existing information about this orders (*Load Planned Order*) and sets for each one ($j = 1,\ldots,N\_ord(t)$) the following *Firm Orders (F.O.)* parameters (*Sets F.O. Parameters*):

- $Fmin_j = Fmax_j = F^p_j$, i.e. the production volume fraction is fixed to the amount already planned (F^p_j);
- $dd^{int}_j = dd^{end}_j = d^p_j$, i.e. the due date lower (dd^{int}) and upper bounds (dd^{end}) are fixed to the agreed value (d^p_j);
- $p^L_j = p^H_j = p^p_j$, i.e. the lowest (p^L) and the highest (p^H) price are fixed to the amount agreed one (p^p_j);

Set initial N.O. constraints; the SPA sets the planning constraints according to Table 4, that is:

- $p_{i^*} = p^H_{i^*} = p_i^{(max)}$, $p^L_{i^*} = p_i^{(min)}$
- $dd_{i^*} = dd^{int}_{i^*} = dd_i^{(min)}$, $dd^{end}_{i^*} = dd_i^{(max)}$

Figure 7. Production planning process activity diagram

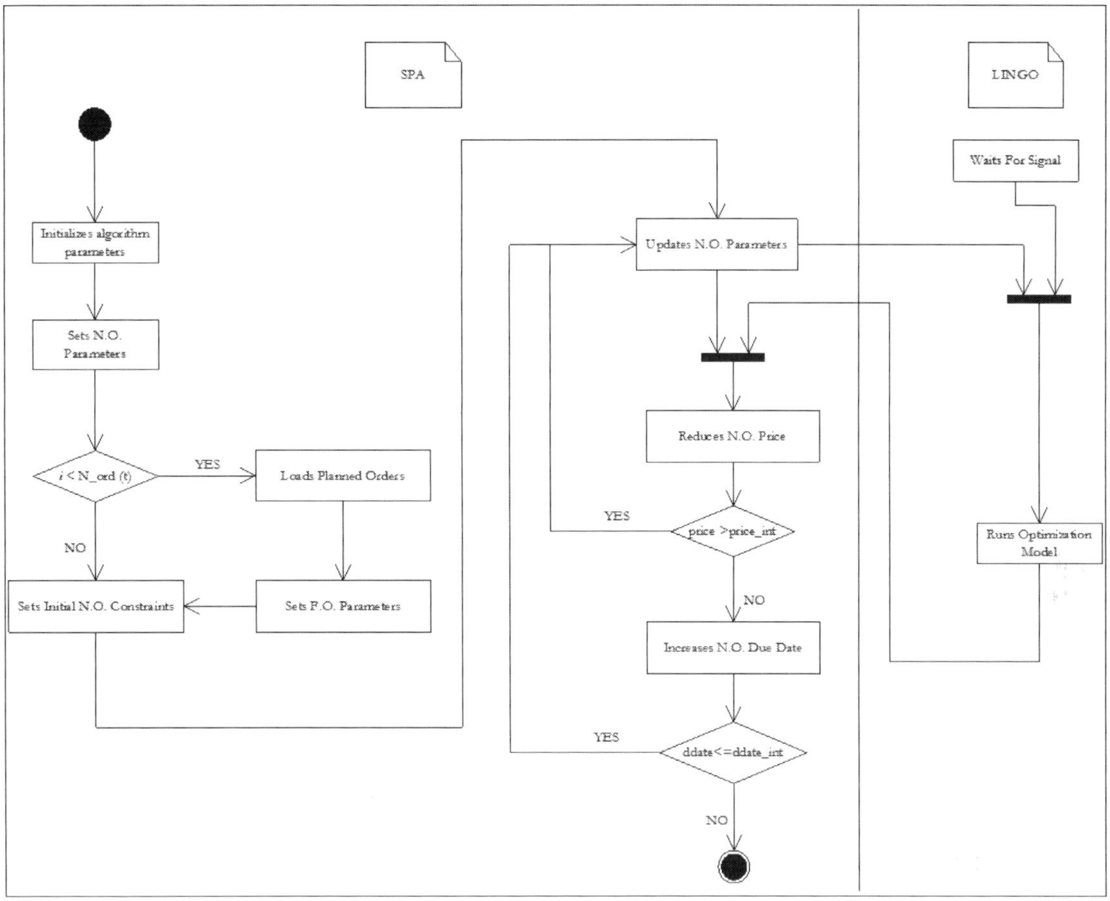

where $p_i^{(max)}$, $p_i^{(min)}$ and $dd_i^{(min)}$, $dd_i^{(max)}$ are respectively the bounds of p_i and dd_i intervals in Table 5.

- *Updates N.O. parameters;* the SPA passes p_i* and dd_i* to the LINGO Solver (*Runs optimization model*). Once the Lingo optimization ends, the LINGO Solver passes the volume and profit level associated to p_{i*} and $dd_{i*}$to the SPA and, at this point, it builds the production alternative array (*PA*).
- *Reduces N.O. price*; the SPA reduces the price p_{i*}' (previously calculated) according to the following expression:

$$p_{i*} = p_{i*}' - \alpha \cdot (p^H_{i*} - p^L_{i*}),\ \alpha \in [0,1] \qquad (14)$$

- *Increases N.O. due date*; the SPA increases the due date dd_i*' (previously calculated) as showed in:

$$dd_{i*} = dd_{i*}' - \alpha \cdot (dd^{end}_{i*} - dd^{int}_{i*}),\ \alpha \in [0,1]. \qquad (15)$$

THE PRODUCTION PLANNING MODEL

The production planning model is in the following deeply explained.

Indexes:

$i = 1, ..., m$ products (job);
$j = 1,...,n$ resources;
$t = 1,...,T$ time buckets;
$l = 1,...,L$ process plans;
Decision variables:
v_{il} 1 if job i is chosen by plan l, 0 otherwise
x_{il} fraction of the job i to be processed through plan l
y_{ijt} amount of resources j allocated to job i at time t
r_{jt} amount of ordinary manpower work to allocate to the resource j at time t
o_{jt} amount of overtime work to allocate to the resource j at time t
s_{jt} amount of resource j to outsource at time t
Model parameters:
p_i product price
d_i product due date
FC_{il} process plan l fixed cost when selected for job i
rs_{ilj} amount of type j resource needed for processing job i with work plan l
$nplan_i$ number of process plans for job i
$CRG_j j$ time unit cost when used during ordinary time
$COV_j j$ time unit cost when used during over time
$CAPR_{jt}$ resource j ordinary time capacity at time t
$CAPO_{jt}$ resource j overtime capacity at time t
$CAPS_{jt}$ resource j supplier capacity at time t
Objective function: Supplier Profit maximization

$$\max\left\{\sum_i \sum_l x_{il} \cdot p_i - C1 - C2 - C3\right\} \quad (16)$$

Constraints:

$$\sum_{t \le d_i} y_{ijt} \ge \sum_l x_{il} \cdot rs_{ilj} \;\; \forall\, i,j \quad (17)$$

$$\sum_l v_{il} \le 1 \;\; \forall\, i \quad (18)$$

$$x_{il} \le v_{il} \;\; \forall\, i,l \quad (19)$$

$$\sum_{l > nplan_i} v_{il} \le 0 \;\; \forall\, i \quad (20)$$

$$\sum_i y_{ijt} \le CAPR_{jt} + o_{jt} + s_{jt} \;\; \forall\, i,t \quad (21)$$

$$o_{jt} \le CAPO_{jt} \;\; \forall\, j,t \quad (22)$$

$$s_{jt} \le CAPS_{jt} \;\; \forall\, j,t \quad (23)$$

$$r_{ij} = \sum_i y_{ijt} - s_{jt} - o_{jt} \;\; \forall\, j,t \quad (24)$$

$$C1 = \sum_i \sum_l FC_{il} \cdot v_{il} \quad (25)$$

$$C2 = \sum_j \sum_t (CRG_j \cdot r_{jt}) \quad (26)$$

$$C3 = \sum_j \sum_y (COV_j \cdot o_{jt} + CSB_j \cdot s_{jt}) \quad (27)$$

$$y_{ijt} \ge 0 \;\; \forall\, i,j,t \quad (28)$$

$$r_{jt} \ge 0 \;\; \forall\, j,t \quad (29)$$

$$o_{jt} \ge 0 \;\; \forall\, j,t \quad (30)$$

$$x_{il} \ge 0 \;\; \forall\, i,l \quad (31)$$

$$s_{jt} \ge 0 \;\; \forall\, j,t \quad (32)$$

$$v_{il} \in \{0,1\} \ \forall \ i,l \qquad (33)$$

Constraint (17) forces the total amount of working time units of each resource to be at the least equal to the amount of working time units need, for the chosen process plan, to complete the job. It is to be highlighted that the job must be completed within the specified due date. Constraints (18) and (19) force each job to be assigned only to one work plan. Constraint (20) assures that plans not declared for job *i* cannot be chosen. (21 - 24) are production capacity constraints. Constraint (25) computes the cost of using a specific process plan, constraint (26) computes costs for regular time resource use, while constraint (27) calculates costs of using resources in overtime and outsourcing. Constraints (28 - 33) defines decision variables domain.

SIMULATION ENVIRONMENT

It has been developed a distributed simulation environment based on the proposed Multi Agent Architecture. It consists of a simulation environment, developed by using Java development kit package, able to test the functionality of the proposed approaches and to understand the related advantages and/or limits. The modeling formalism here adopted is a collection of independent objects interacting via messages. This formalism is quite suitable for Multi Agent Systems development. In particular, each object represents an agent and the system evolves through a message sending engine managed by a discrete event scheduler. Specifically, the following agents have been developed: the CNA, the SNA, the SPA, the *Scheduler Agent*, the *Model Agent* and *statistical package*. The former three agents have been deeply described in the previous sections. The *Scheduler Agent* is in charge with the system evolving by managing the discrete events of the simulation engine. Differently, *Model Agent* is in charge with the suppliers' and customers' interaction.

The statistical package, additionally, collects the generated data at the end of each simulation run and generates both reports and statistical analyses. Moreover, the customer and supplier agents have been provided with a local database to manage their own data. A proper interface has been developed to connect the Java code to this database. Finally, the SPA has been linked with a Lingo Solver by a proper interface. The simulator can be customized by the parameters reported in the class diagram in Figure 3. It generates a report of the simulations, in particular, the agreements reached for each requested order, the profit of each supplier and the rounds to reach an agreement.

The customer's order is defined by the following data:

- *order*, is the ordinal number referred to the orders;
- t_a is the time in which of the request of the customer has done. For each time bucket (T=30 periods) four orders are generated; t_a is obtained by a uniform distribution that takes into account the whole set of periods of the considered time bucket.
- dd_i is the requested due date. It is obtained by a uniform distribution among that takes into account both t_a and the ending period of the considered time bucket.
- V_i is the requested volume. It is obtained by a uniform distribution that takes into account the minimum and maximum values of the capacity for the considered period.
- p_i is the requested price. It is obtained by uniform distribution that takes into account the mark-up value, the average cost for each unit of product and the requested volume;
- *C_num* is a custom identification, it is obtained by uniform distribution that takes into account the customers type;

Table 2. Experimental class

	Low	High
Mark-up extracted	Uniform [0.8 – 1.2]	Uniform [1-1.4]
Volume extracted	Uniform [65 – 105]	Uniform [115-165]
Average Capacity utilization	30%	50%

- *Type* is the product identification, it is obtained by uniform distribution that takes into account the products type.

With the described simulation environment a test case consisting of one customer and five suppliers has been performed. Table 4 reports the suppliers' data (see paragraph 4 for the parameters reported in table).

Orders enter the supplier system during a time horizon of 120 periods (days) divided in buckets of 30 periods where is possible to make re-planning. For each time bucket, the customers insert three orders.

A several classes experimental has been considered by the following parameters:

- congestion of the suppliers by the volume that the customer can be requested;
- the price of the order submitted by the customer;
- the overlapping among the orders.

Table 5 reports the total hours for product unit in order to evaluate the average unit cost. Table 2 reports the value high and low for volume and mark-up.

The orders generation is obtained by the following steps:

1. Date of input for each order is random generated. (three orders in a time bucket);
2. Due date is random generated subject to the overlap constraint among the orders.

The due date of *i-th* order is random between t^i_a (*i-th* order) and t^{i+1}_a + overlap (*i-th* +*1* order).

3. It is generated the volume (low or high) and the volume of the order is obtained by the following expression:

$$Volume = (t_{arr} - dd) \cdot Volume\ extracted \quad (34)$$

4. It is generated the price of the order by the following expression:

$$price = average\ \cos t\ unit \cdot volume \cdot markup\ extracted \quad (35)$$

At the end of this process, it is generated the list of orders in terms of arrival time, due date, volume and price. Table 2 reports the value high and low for *"volume extracted"* and *"mark up extracted"*.

The two levels of mark-up mean: low level leads to price under and over the production costs of products; high level leads, always, to price over (equal for the lower bound of the distribution) the production costs of products.

The two levels of volume mean: low level leads to 30% of the average utilization of the network; high level leads to 50% of the average utilization of the network. The utilization of the network seems to be low, but the orders can be assign to one supplier, while the utilization regards the capacity of the network.

Each experiment class has been replicated in order to achieve a rate between the average performance and the corresponding 95% confi-

Table 3. Experiment classes

No.	Mark up	Volume	overlap
1	H	H	0
2	L	H	0
3	H	L	0
4	L	L	0
5	H	H	5
6	L	H	5
7	H	L	5
8	L	L	5
9	H	H	10
10	L	H	10
11	H	L	10
12	L	L	10

dence degree interval lower than 10%, for each performance index considered.

Table 3 shows the experimental plan, in which each parameter is extracted by a uniform distribution and the product type is selected by random distribution for each order.

For the overlap, three values have been considered: "*0*" means that among the orders no overlap periods are allowed, this state simplifies the production planning for the suppliers. "*5*" it is a medium level of overlapping among the orders. "*10*" it is a high level of overlapping among the orders.

Table 4 reports the suppliers' data. The data that characterize a supplier are the following:

- $CAPR$; it is the capacity in ordinary time;
- $CAPO$; it is the capacity in overtime;
- $CAPS$; it is the capacity in outsourcing;
- FC; it is the fixed cost of the process plan;
- CRG; it is time unit cost in regular time;
- COV; it is time unit cost in overtime;
- CSB; it is time unit cost in outsourcing.

The costs and the capacity are constant for each period t and resource j.

Moreover, one process plan for each product type is hypothesized, then $L=1$ (Table 5). These values simplify the production planning algorithm by reduce the computational time and allow to conduct a complete statistical analysis.

SIMULATION RESULTS

The performance measures considered are the following:

- average customer utility; it is the average of the equation (6).
- average total suppliers utility. It is the sum of the profit for all the suppliers.
- suppliers unbalanced index; it is the index of unbalanced profit among the suppliers computed by the following expression:

$$unbalanced = \sum_{i=1}^{N} \left| \frac{1}{N} - \frac{utility\ \sup plier_i}{total\ utility\ \sup plier} \right| \quad (36)$$

Table 4. Suppliers' data

	Supplier 1	Supplier 2	Supplier 3	Supplier 4	Supplier 5
CAPR	65	110	120	95	130
CAPO	60	115	135	85	130
CAPS	70	125	50	50	60
FC	267	214	90	136	154
CRG	42	46	50	50	49
COV	73	90	69	86	90
CSB	62	66	66	62	59

Table 5. Total hours for product unit

Product type	Total hours	Product type	Total hours
1	10	6	20
2	10	7	20
3	10	8	40
4	10	9	40
5	20	10	40

If the profit is distributed among suppliers equally, the expression (36) is equal to 0, otherwise the expression (36) is greater than 0. This performance is an important index that describes how the profit is shared among the suppliers. In fact, if the profit is unbalanced that means, some suppliers gain low profit from the e-marketplace.

- Average time to reach an agreement in terms of rounds of negotiation.

Tables 6,7,8 and 9 report the performance measures for each negotiation and customer's behavior approaches.

From the analysis of the table the following issues can be drawn:

- *customer utility;* the auction approach that stimulates the suppliers to improve the counter-proposal until the maximum number of rounds leads to obtain the best performance of customers' utility. The single round leads to the worst customers' utility because the suppliers submit only one proposal. The mixed customer's tactic leads to better performance compared with others two tactics, and in particular, the best value is obtained by the negotiation protocol.
- *Suppliers' utility*; for this performance, the mixed tactic and negotiation protocol leads to the best performance. As the reader can notice, the single round approach doesn't lead to better suppliers' utility, even if the supplier submits the proposal with maximum profit. This is because, the orders that reach an agreement are characterized by high profit for the suppliers, but many orders don't reach an agreement. Therefore, for the suppliers is opportune negotiating with the customers.

Table 6. Customers' utility

	Negotiation	Qualified	Auction	Single round
Imitative	24.18	21.40	**26.74 (+3.56%)**	16.35 (-38.86%)
Temporal	25.48	23.96	-	-
Mixed	25.82	23.93	-	-
Percentage difference	5.38%	11.96%		
Percentage difference	6.78%	11.82%		

Table 7. Suppliers' utility

	Negotiation	Qualified	Auction	Single round
Imitative	6.792.964	4.994.201	5.726.018 (-18.20%)	6.136.527 (-12.33%)
Temporal	6.636.737	5.572.544	-	-
Mixed	**6.999.910**	5.535.495	-	-
Percentage difference	-2.30%	11.58%		
Percentage difference	3.05%	10.84%		

- *Unbalanced index*; this performance is important, because if the unbalanced index is high means that some suppliers gain low profit by the use of the e-marketplace tool. The single round leads to better profit distribution among suppliers. The negotiation protocol and temporal tactics is the combination with a distribution of the profit nearer to the best value of single round approach, but this combination leads to better suppliers' profit.
- *Negotiation rounds;* The worst performance is, obviously, obtained by the auction approach that forces the suppliers to submit a new counterproposal until the timeline allowed. The value is minor of 300 (the maximum number of rounds possible) because in some cases the suppliers submit a counter proposal that satisfies all the attributes requested by the customer before the last round. The relevant result is the qualified approach; it reaches the agreement earlier than the single round. Therefore, this approach is suitable when the agreement must to be reach in a short time.

Table 8. Unbalanced suppliers' index

	Negotiation	Qualified	auction	Single round
Imitative	0.67	0.75	0.67 (19.64%)	**0.56 (-9.68%)**
Temporal	0.62	0.84	-	-
Mixed	0.66	0.94	-	-
Percentage difference	-7.46%	12.00%		
Percentage difference	-1.49%	25.33%		

Table 9. Average number of rounds

	Negotiation	Qualified	auction	Single round
Imitative	151.06	**32.05**	281.83 (779.34%)	60 (87.21%)
Temporal	157.06	50.07	-	-
Mixed	140.55	47.73	-	-
Percentage difference	3.97%	56.22%		
Percentage difference	-6.96%	48.92%		

Finally, the percentage differences reported in each table show that the qualified approach is more affected by the customer tactics than the negotiation approach.

The performance measures are combined in order to obtain a single value of performance of the proposed approaches. Therefore, the performance values are normalized and table 10 reports the normalized value of the performance measures. The normalization is computed by the following expressions:

$$\frac{\mathit{value} - \min(\mathit{value})}{\max(\mathit{value}) - \min(\mathit{value})} \tag{37}$$

Table 10. Normalized performance index

	Negotiation	Qualified	auction	Single round
	Normalized – customers' utility			
Imitative	0.75	0.49	**1.00**	0.00
temporal	0.88	0.73		
mixed	0.91	0.73		
	Normalized – suppliers' utility			
Imitative	0.90	0.00	0.36	0.57
temporal	0.82	0.29		
mixed	**1.00**	0.27		
	Normalized – unbalanced suppliers' index			
Imitative	0.71	0.50	0.71	**1.00**
temporal	0.84	0.26		
mixed	0.74	0.00		
	Normalized – number of rounds			
Imitative	0.52	**1.00**	0.00	0.89
temporal	0.50	0.93		
mixed	0.56	0.94		
	Normalized – total performance			
Imitative	0.72	0.50	0.52	0.61
temporal	0.76	0.55		
mixed	**0.80**	0.48		

Table 11. Overlap analysis

	Overlap 0	Overlap 5	Overlap 10	%	%
Customer utility	24.85	22.79	22.01	-8.29%	-11.43%
Suppliers' utility	14.511.174	19.413.770	29.286.580	33.78%	101.82%
Unbalanced index	0.62	0.64	0.61	3.22%	-1.61%
Negotiation rounds	138.16	144.91	152.14	4.89%	10.12%

It is used, if the high value of the performance leads to better performance.

$$\frac{\max(value) - value}{\max(value) - \min(value)} \tag{38}$$

It is used, if the high value of the performance leads to worst performance.

Therefore, for each performance class the value 1 identifies the best performance measure, instead the value 0 identifies the worst performance measure. Assume that, each performance measure has the same importance, the average of the performance measures ([0,1] interval) assigns the goodness of the approach. The normalized approach assumes that each performance have the same importance.

The best compromise among the performance measures is reached by the mixed customer's tactic and the negotiation protocol. Table 10, also, shows the value added services that each participant (supplier or customer) can gain by e-marketplace.

Table 11,12 and 13 report the performance measures in order to analyze how the parameters overlap, volume and mark-up affect the performance. The percentage difference is computed compared to "0" overlap.

- The overlap affects the suppliers' utility, in fact the total profit increases with the overlap degree. This is due to the opportunity of the customer to input order with high volume because the overlap among the orders increase the temporal horizon of each order. The same reason leads to reduce the customer utility and increment the time to reach an agreement with a minor strength. This is caused by the difficulties to planning orders overlapped. Finally, the unbalanced index doesn't affect by the overlap degree.
- The volume affects, obviously, the suppliers' utility. If the volume decreases, the total profit decreases too. The customer utility decreases with volume "high" because the increasing of suppliers' congestion leads to submit proposals with minor satisfaction for the customers. The low volume requested by the customers leads to increase the unbalanced index, while the time to reach an agreement increases with volume high.

Table 12. Volume analysis

	Volume high	Volume low	%
Customer utility	20.48	25.01	22.12%
Suppliers' utility	27.828.305	14.067.880	-49.45%
Unbalanced index	0.60	0.64	+6.67%
Negotiation rounds	150.75	137.88	-8.54%

Table 13. Mark-up analysis

	Mark - up high	Mark - up low	%
Customer utility	23.30	23.13	-0.73%
Suppliers' utility	22.992.394	19.201.275	-16.49%
Unbalanced index	0.62	0.62	0.00%
Negotiation rounds	142.22	146.41	2.95%

- The mark-up is a parameter with low influence on the performance measures. In case of mark-up low there is a reduction of the suppliers' utility. A low variance of the other performance measures is observed.

In briefly, from the point of view of the performance, the following issues can be drawn:

- The level of volume is the main factor for the customers' utility and unbalanced profit share;
- The level of overlap is the main factor for the suppliers' utility and negotiation rounds.
- The level of mark-up is the less important parameter.

These information can be used to investigate and implement the knowledge base for supplier and customers in order to select the opportune negotiation strategy.

CONCLUSIONS AND FUTURE DEVELOPMENT

This paper deals with real added value services in e-business applications. In particular, the value added services investigated regard: the protocol mechanism that suppliers and customers used to reach an agreement and the customer's tactics. The protocol mechanisms tested are the following: negotiation, qualified negotiation, auction and single round. Three customer's tactics has been proposed: temporal, imitative and mixed approaches. Moreover, a production planning algorithm was developed in order to provide the information necessary to improve the agreement process.

The e-marketplace environment is a "neutral e-marketplace" that is relevant for the Small and Medium Enterprises.

The implementation of the proposed approaches is based on three kinds of technologies that, in the author believes, are essential for added value services developing: agents, workflow management techniques and open source IT tools. Therefore, a Multi Agent Architecture was developed by UML and IDEF diagrams. The Simulation environment was developed by open source as JAVA package.

The experimental plan was conducted on three parameters: volume, overlap among the orders inputted by the customers and the mark-up level requested by the suppliers. A proper statistical engine extract the orders, therefore, a statistical analysis is necessary; Each experiment class has been replicated in order to achieve a rate between the average performance and the corresponding 95% confidence degree interval lower than 10%, for each performance index considered.

The performance measures evaluated are: customers' utility, suppliers' profit, unbalanced of the suppliers' profit and the time to reach the agreements.

The research suggests how, through simulation, to evaluate the real value of planning and negotiation tools in e-business environment and who, among customers and suppliers, get the main advantages from them, and, therefore, should pay for them.

Results of this research can be located at two levels. Concerning the specific dispute among customers and suppliers problem here addressed, the following conclusions can be drawn:

- Assumed the performance with the same importance, the negotiation protocol and mixed customer's tactic leads to the better results, and, therefore with the high valued added.
- A relevant result concern the time to reach the agreements. The qualified approach allows to reach an agreement with the lower time, even under the value of single round approach.
- The customers' tactics involved more performance measures variability in qualified case than the negotiation approach. Therefore, the negotiation approach is characterized by robustness to the customers' tactic.

Regards to the parameters volume, overlap and mark-up, the following conclusions can be drawn:

- The overlap among the orders influences suppliers' utility and negotiation rounds. The increasing of the overlap allows to input orders with high volume and therefore the suppliers can get more profit. Moreover, the overlap among the orders implies that the agreement is reached with more difficult.
- The volume influences, mainly the customers' utility. When the volume increasing the suppliers have to use all the available capacity, consequently the customer satisfaction is reduced.
- The mark-up is the parameter with low influence on the performance measures. The customers' utility and the unbalanced index are the same when the mark-up level changing. The suppliers' utility decreases when the mark-up level decreases.

At more strategic level, the research shows that:

- Agent technology is an enabling technology to approach and solve complex e-marketplaces environment.
- Discrete event simulation is a powerful tool for designing and testing distributed environment based on agent technology.
- Using open source technologies at the simulation package implementation phase allows code reusability for real platform development. In fact, the actual platform package can be built by utilizing the same code and architecture used for the simulation environment allowing time and investment cost saves. This reduces the risk related with investment in ICT and agent based technology in Business To Business applications.

Further researches will investigate the following aspects in depth:

- A proper local knowledge (for each agent) and an inferential engine that will enable the agents to select their best strategy in order to maximise their utility will be developed. The development of the inferential engine will be done by using a fuzzy logic approach, while the genetic algorithm methodology will be used to better fit the goal of the agents' utility optimisation.
- How a learning algorithm, based on past performances, can improve the agents' utility. The agent will be able to elaborate the past information coming from the negotiation output, and will try to forecast the opposite agents' counterproposals. The considered method will be obtained by using

a specific approach that uses the information obtained by the past negotiations to build new and more competitive strategies. Another possible method will be reinforcement learning (a common technique used by adaptive agents in MASs): its basic idea is to revise beliefs and strategies based on the success or failure of the observed performance. The reinforcement learning approach must be adapted to the production planning algorithms of the suppliers explained in this research.

- Furthermore, opportune approaches can be developed to enable both customers and suppliers to constitute a coalition using a Game Theory approach. The coalitional mechanism will be based on cooperative Game Theory and the simulation environment will be properly developed to investigate the added value given by this innovative approach.

REFERENCES

Aberdeen Group. (August 2006). *The E-Procurement Benchmark report.*

Bandyopadhyay, S., Rees, J., & Barron, J. M. (2008). Reverse Auctions with Multiple Reinforcement Learning Agents. *Decision Sciences*, *39*(1), 33–63. doi:10.1111/j.1540-5915.2008.00181.x

Barratt, M., & Rosdahl, K. (2002). Exploring business-to-business marketsites. *European Journal of Purchasing & Supply Management*, *8*, 111–122. doi:10.1016/S0969-7012(01)00010-7

Bartolini, C., Preist, C., & Jennings, N. R. (2002). A Software Framework for Automated Negotiation. In *Proceedings of AOSE 2002, International Workshop on Agent-Oriented Software Engineering,* Bologna, Italy, (LNCS 2585, pp. 88-100). Berlin: Springer Verlag.

Benyoucef, M. Alj, H., Levy, K., & Keller, R. K., (2002) A Rule-Driven Approach for Defining the Behaviour of Negotiation Software Agents. In J. Plaice, et al, (Eds.), *Proceedings of DCW 2002*, (LNCS 2468, pp. 165-181). Berlin: Springer Verlag.

Bruccoleri, M., Perrone, G., & Renna, P. (2005). Design and Evaluating Value Added Services in Manufacturing E-Marketplace. Amsterdam: Springer.

Chen, E., Kersten, G. E., & Vahidov, R. (2005). Agent-supported Negotiations in the E-marketplace. *International Journal of Electronic Business*, *3*(1), 28–49. doi:10.1504/IJEB.2005.006387

Cheng, C., Chan, C. H., & Lin, K. (2006). Intelligent agents for e-marketplace: Negotiation with issue trade-offs by fuzzy inference systems. *Decision Support Systems*, *42*, 626–638. doi:10.1016/j.dss.2005.02.009

Choi, S. P. M., Liu, J., & Chan, S. P. (2001). A genetic agent based negotiation system. *Computer Networks*, *37*, 195–204. doi:10.1016/S1389-1286(01)00215-8

Dumas, M., Aldred, L. J., Governatori, G., & ter Hofstede, A. H. M. (2005). Probabilistic Automated Bidding in Multiple Auctions. *Electronic Commerce Research*, *5*(1), 25–49. doi:10.1023/B:ELEC.0000045972.75533.90

Grieger, M. (2003). Electronic marketplaces: A literature review and a call for supply chain management research. *European Journal of Operational Research*, *144*, 280–294. doi:10.1016/S0377-2217(02)00394-6

Kim, W., & Lee, J. K. (2002). Agent and e-business models. *Decision Support Systems*, (Special Issue), 34.

Kurbel, K., & Loutchkor, I. (2005). A model for multi-lateral negotiations on an agent-based job marketplace. *Electronic Commerce Research and Applications*, *4*(3), 187–203. doi:10.1016/j.elerap.2005.01.002

Louta, M., Roussaki, I., & Pechlivanos, L. (2008). An intelligent agent negotiation strategy in the electronic marketplace environment. *European Journal of Operational Research*, *187*(3), 1327–1345. doi:10.1016/j.ejor.2006.09.016

Maes, P., Guttman, R. H., & Moukas, A. G. (1999). Agent that buy and sell: Transforming Commerce as we know it. *Communications of the ACM*, *42*(3), 81–91. doi:10.1145/295685.295716

Oliver, J. R. (1997). A machine learning approach to automated negotiation and prospects for electronic commerce. *Journal of Management Information Systems*, *13*(3), 83–112.

Perrone, G., Renna, P., Cantamessa, M., Gualano, M., Bruccoleri, M., & Lo Nigro, G. (2003). An Agent Based Architecture for production planning and negotiation in catalogue based e-marketplace. In *Proceedings of the 36th CIRP-International Seminar on Manufacturing Systems*, 03-05 June 2003, Saarbruecken, Germany

Presley, A., Sarkis, J., Barnett, W., & Liles, D. (2001). Engineering the Virtual Enterprise: An Architecture-Driven Modeling Approach. *International Journal of Flexible Manufacturing Systems*, *13*, 145–162. doi:10.1023/A:1011131417956

Rahwan, I., Sonenberg, L., Jennings, N. R., & McBurney, P. (2007). STRATUM: A methodology for designing heuristic agent negotiation strategies. *Applied Artificial Intelligence*, *21*(6), 489–527. doi:10.1080/08839510701408971

Rangone, A., & Bertelè, U. (2004). *Il B2B in Italia: Finalmente parlano i dati*. Rapporto dell'osservatorio B2B, Marzo 2004.

Saha, S. (2006). Improving Agreements In Multi-Issue Negotiation. *Journal of Electronic Commerce Research*, *7*(1), 41–49.

Turowski, K. (2002). Agent-based e-commerce in case of mass customization. *International Journal of Production Economics*, *75*, 69–81. doi:10.1016/S0925-5273(01)00182-7

Wurman, P. R., Wellman, M. P., & Walsh, W. E. (2002). Specifyng Rules for Electronic Aucions. *AI Magazine*, *23*(3), 15–23.

Chapter 10
A 3D Vision-Based Solution for Product Picking In Industrial Applications

Mirko Sgarbi
Scuola Superiore Sant'Anna, Italy

Valentina Colla
Scuola Superiore Sant'Anna, Italy

Gianluca Bioli
Scuola Superiore Sant'Anna, Italy

ABSTRACT

Computer vision is nowadays a key factor in many manufacturing processes. Among all possible applications like quality control, assembly verification and component tracking, the robot guidance for pick and place operations can assume an important role in increasing the automation level of production lines. While 3D vision systems are now emerging as valid solutions in bin-picking applications, where objects are randomly placed inside a box, 2D vision systems are widely and successfully adopted when objects are placed on a conveyor belt and the robot manipulator can grasp the object by exploiting only the 2D information. On the other hand, there are many real-world applications where the 3rd dimension is required by the picking system. For example, the objects can differ in their height or they can be manually placed in front of the camera without any constraint on the distance between the object and the camera itself. Although a 3D vision system could represent a possible solution, 3D systems are more complex, more expensive and less compact than 2D vision systems. This chapter describes a monocular system useful for picking applications. It can estimate the 3D position of a single marker attached to the target object assuming that the orientation of the object is approximately known.

INTRODUCTION

Vision systems are widely used in industrial applications for different tasks (Malamas, Petrakis, Zervakis, Petit, & Legat, 2003; Chin & Harlow, 1982; Nelson, Papanikolopoulos, & Khosla, 1996). For example, visual inspection systems allow 24-hour inline quality control of the parts during manufacturing processes. If some parts are out of tolerance

DOI: 10.4018/978-1-61520-605-6.ch010

or defective, they can be discarded at a very early stage or reallocated. Vision systems are used to verify the correctness of assembly operations by verifying the position and the orientation of each component. Code readers allow identification and full tracking of each product. Vision systems are used to create fully automated solutions for handling dangerous or toxic materials.

One of the most interesting fields where vision systems are successfully applied is the control of robots for pick and place operations. Thanks to increasing computational power and after years of extensive research, vision systems are now emerging and proving their validity in very complex applications like grasping parts that are randomly placed inside a bin. In this case, a 3D vision system has to recognize and localize the target object among many other ones, under any perspective transformation and even if it is partially visible. Typical 3D vision systems use either a single camera combined with a laser projector (Alshawish & Allen, 1995) or two cameras for stereo vision (Brown, Burschka, & Hager, 2003). On the other hand, less complex one-camera 2D vision systems (Cheng & Denman, 2005) are widely used in more structured pick and place applications, where the target objects lie, for instance, on a conveyor belt. These types of systems have reached a high level of reliability. However, most of them require the images to be taken with the target object located at a predefined distance from the camera. Perspective deformation errors of the image are compensated through the particular shape of the end-effectors. Actually the end-effectors are designed to allow a firm grasp despite of an imprecise localisation (obviously within reasonable limits).

Unfortunately, there are many real-world applications that cannot be represented by the above-described situation. The objects can differ in height or be manually placed in front of the camera without any constraints on their distance from the camera itself. Some production lines can adopt a mix of manual and automatic operations on the parts conveyed from one working unit to another one. For instance, automatic operations could be directly performed by a robot. It would be convenient to exploit the flexibility of the robot system in order to pick the object from a line, machine it and to place it back on the line. Unfortunately, picking objects on a conveyor that is used both by human operators and robot manipulators can be a very difficult task. For example, parts can be randomly placed over a conveyor belt or hung up on a conveyor chain without precise restrictions on orientation and position.

The detection of the picking point without a well-structured conveying system could, in practice, prevent efficient use of the automatic station, as human intervention is always required in order to pick the objects, place them on the machine and perform the reverse operation. An alternative approach consists in changing the entire production line by adding custom fixtures and specific constraints. This solution could result in high costs and poor flexibility. The optimal solution would be to adopt automatic units capable of substituting human intervention in pick and place operations. This goal can be achieved through the adoption of a 3D vision system to estimate the position and orientation of the part to be picked. As standard 3D systems are complex, expensive and cumbersome, other solutions have also been explored (Oh& Lee, 2007).

This chapter offers a solution that requires some minor constraints on the orientation and position of the objects that are placed on the conveyor by a human operator, i.e. the objects must be placed on the conveyor belt with approximately the same position and orientation and with a predefined side facing the camera. The proposed system estimates the 3D position of the object by using only one camera and only one marker on the target object. The object orientation has to remain almost the same for each picking operation. System accuracy must be sufficient for a correct grasp, but an exact object or marker location is not required.

The chapter is organised as follows: the following section provides the background on vision systems used to estimate the position and orientation of an object and it gives a brief overview of the cameras mathematical model and the essential pre-processing of the captured image. In the third section, typical applications are examined with a particular emphasis on production lines where manual and automatic operations are required on the same object at different manufacturing stages. Thereafter, the proposed system is described. The fourth section is focused on system calibration, while the fifth section describes the experimental setup that has been used to test system performance. The sixth section discusses the numerical results, while the last two sections are devoted to future work and concluding remarks.

BACKGROUND

As this chapter is focused on the application of machine vision for automatic target pose estimation, this section is intended to provide an overview of some basic concepts.

The Pose Estimation Problem

The combination of position and orientation with respect to a particular reference frame is often referred to as the *pose* of an object (Shapiro & Stockman, 2001). The pose estimation problem (Xu & Zhang, 1993) is a six degree-of-freedom problem in which the transformation from the object frame to the camera frame has to be found. The pose estimation problem is a fundamental goal in a wide variety of industrial applications. The increasing availability of fast and cheap software and hardware solutions (Delp, 1997) in image processing satisfies quite stringent performance criteria despite severe budget constraints.

In position-based visual servo systems (Hutchinson, Hager, & Corke, 1996), the pose data provided by one or more cameras are used for the positioning or tracking task of a robotic manipulator. The camera can be either mounted on the last link of the robot in a so-called eye-on-hand configuration, or fixed somewhere, in a so-called stand-alone configuration. Since the pose of the object is estimated relatively to the camera, the transformation between the robot and the camera must be known in order to control the robot's movements. Visual inspection systems use pose information to check if a part is in the right position or to allow further comparisons between measurements and the CAD model (Fernandez, Benchetrit, & Gachet, 2001). Pose data are also used in automatic assembly to mount the parts and to verify their final assembly (Nelson et al., 1996).

The pose estimation is sometimes linked to the recognition of the object itself. If the model of the object is known and different object types are present, the solution of the correspondence problem also includes the recognition of the object type (Richetin, Lapreste, & Rives, 1989). Other typical applications are passive navigation (Xiang & Navab, 2000), where visual landmarks allow the self-localization of the camera, and camera calibration (Tsai, 1987) in which the pose of a precise and well-known pattern is used to define a world reference frame with respect to the camera reference frame.

With a monocular vision system, as shown in Figure 1, the pose of an object can be determined from a minimum number of 3D points in the object model and their correspondences in the 2D image, given the camera projection model and the camera calibration parameters. The points can be object features like corners or some artificial markers. This problem has been approached by using both closed-form solutions (Ansar & Daniilidis, 2003) and numerical solutions (Diaz & Abderrahim, 2007). The closed-form solutions, also called *direct* or *exact* solutions, assume that a finite set of correspondences exists to solve a set of projection equations. Closed-form solutions are available for three (Haralick, Lee, Ottenburg, & Nolle, 1995) and four points (Horaud, Conio, Leboulleux, & Lacolle, 1989). Numerical solutions can address

Figure 1. Monocular vision system

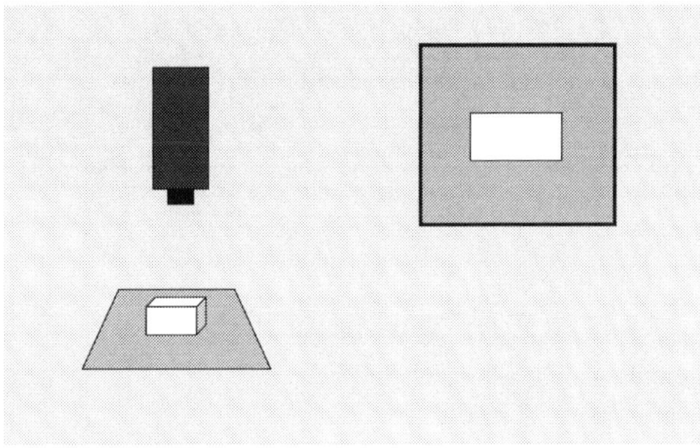

problems where several correspondences are available (Dementhon & Davis, 1995).

Finding the correspondences between object model points and image points can be another difficult problem when the pose is unknown. The simultaneous pose and correspondence determination has been addressed, for instance, in (David, DeMenthon, Duraiswami, & Samet, 2003).

Binocular stereo vision systems (Trucco & Verri, 1998) use images captured from different points of view to get depth information. As shown in Figure 2, the left and right images are slightly different and displacement between two corresponding points from one image to another, also known as disparity, is related to the depth of the point. A stereo system can either extract the 3D position of a single object point through triangulation or recover the surface of the object visible in both of the images. A stereo system is intrinsically more complicated than a monocular vision system. Moreover, it requires more space for the setup, more complex calibration procedures and more expensive hardware.

Other vision techniques, like depth from focus or defocus (Xiong & Shafer, 1993) or depth from shadows (Klette, Schluns, & Koschan, 1998) can be used to get the depth. These systems are rarely used in industrial applications because of their considerable complexity and poor reliability under generic and potentially difficult environmental conditions. For example, motorized lens systems could slow down the acquisition process and they could have lifetime problems in typical industrial

Figure 2. Binocular vision system

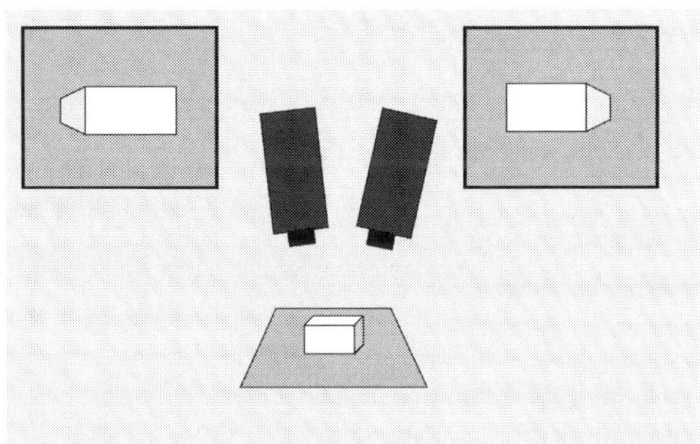

Figure 3. Pin hole camera

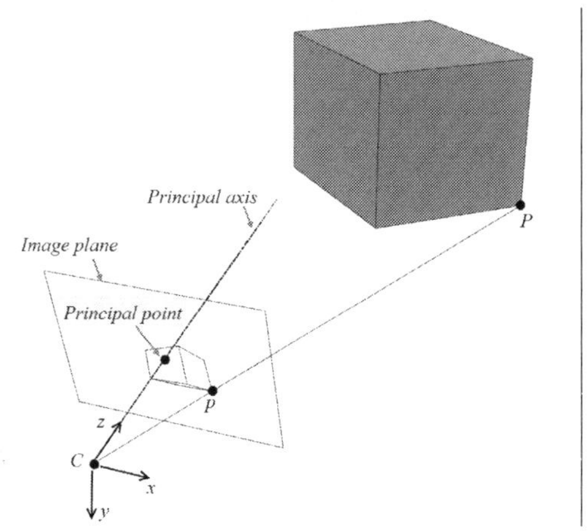

environments where dust, high humidity and wide temperature ranges are commonly found.

Camera Model

In first approximation, real camera devices act like a pin-hole camera where the image is obtained from projecting scene points through a single point onto an image plane. Although in the real device the image plane is behind the projection point, it is more intuitive to reason with the geometry depicted in Figure 3, where the point of projection corresponds to a viewpoint behind the image plane. The mathematical formulation remains unchanged.

The camera transforms a 3D scene point into an image point. By using homogeneous coordinates (Kline, 1972) this mapping can be described as:

$$\lambda\vec{p} = T \cdot \vec{P}$$

where $\vec{p} = (x, y, 1)$ is the 2D homogeneous coordinates for the image point, $\vec{P} = (X, Y, Z, 1)$ is the 3D homogeneous coordinates for the object point, T is a 3x4 matrix and λ is an arbitrary scale factor.

The distance *f* between the viewpoint and the image plane along the *principal axis* (or *optical axis*) is called *focal length*. The optical axis is conventionally aligned with the *z* axis of the camera coordinate system. The point of intersection between the principal axis and the image plane is called *principal point* (Clarke, Fryer, & Wang, 1998). In this context, if an object moves toward the viewpoint, its projection onto the image plane becomes bigger.

For a pin-hole camera the projective transformation becomes:

$$T = \begin{bmatrix} 1 & 0 & 0 & 0 \\ 0 & 1 & 0 & 0 \\ 0 & 0 & 1/f & 0 \end{bmatrix}$$

Each point its scaled by its individual depth:

$$(x, y) = \frac{f}{Z}(X, Y)$$

In the so-called *weak perspective projection* (Horaud, Dornaika, Lamiroy, & Christy, 1997), the assumption is made that the depth of the object is negligible and all the points get the same depth Z_0:

$$(x,y) = \frac{f}{Z_0}(X,Y)$$

Unfortunately, the focal distance alone is not enough to properly describe a real camera. Other parameters, like *pixel skew angle*, *pixel aspect ratio* and *principal point offset*, should be taken into account in order to obtain reliable measurements (Faugeras, 1993).

The parameters involved in the camera model are also known as *intrinsic parameters* while the parameters describing the pose of the camera with respect to a world coordinate frame are known as *extrinsic parameters*.

Taking into account both intrinsic and extrinsic parameters, the pinhole camera model can be described with the following equation:

$$T = \begin{bmatrix} f \cdot s_x & k & u \\ 0 & f \cdot s_y & v \\ 0 & 0 & 1 \end{bmatrix} \cdot \left[R \mid t \right]$$

where s_x / s_y is the aspect ratio, $k / f \cdot s_y$ is the skew and *(u,v)* are the coordinates of the principal point. This transformation cannot be perfectly described by a perspective transformation because of distortions which occur between the points on the object and the location of their images. The most important one is the *radial distortion* (Weng, Cohen, & Herniou, 1992), which becomes non-negligible for short focal lengths. These distortions can be modeled and various algorithms have been studied to reduce their effects (Heikkila & Silven, 1997). However, the model can only be an approximation of the real relationship.

Feature Extraction

Another fundamental task in computer vision is the extraction of features from the image. There are features which are invariant under perspective transformation. The *center of gravity* and *vertices* are examples of invariant features while the length of a line changes with the point of view. In order to extract interesting features, the image is first filtered in order to enhance the appearance of the object and to compensate for spatially-varying illumination. Usually this process makes the discontinuities more prominent. For example, the boundaries of an object appear as intensity discontinuities in the image.

Image filtering is a preliminary stage of the so-called *image segmentation* (Fu & Mui, 1981), that is, an operation aiming at creating a simplified version of an image which retains the significant information and is easier to analyze. During the segmentation process the image is partitioned into multiple pixel regions. A basic segmentation can be performed by using *edge operators* (Canny, 1986; Ziou & Tabbone, 1998) that detect and measure the magnitude and the orientation of these discontinuities. Other operators extract corners in the image (Ruzon & Tomasi, 2001) or other types of features (Wen & Oakley, 1995). Often, an image composed of disconnected edge elements is not interpretable from further processing algorithms. Additional processing is therefore required to get a meaningful segmentation of the image. The goal in this case is to make coherent features from many individual local edge elements. For example, the boundaries of the object can appear with holes due to noise or illumination problems. In this case, the segmentation algorithm has to join adjacent edges to create a meaningful boundary.

After a proper segmentation has been applied, the resulting boundaries can be analyzed in order to filter out what is not meaningful for the application. For instance, if the target is to find circles in the image, only closed regions must be considered. Other important features of the regions can be the area, the center, the length of the line or the perimeter and the diameter of the inscribed circle or the diameter of the outer circle.

A MONOCULAR 3D VISION SYSTEM FOR PICKING APPLICATIONS

This chapter is focused on the pose estimation problem in object picking where the pose of the object is required to align the end-effector of a robotic manipulator and move it in the picking position. Many production processes involve this task in the manufacturing chain. For instance, objects or components typically need to be picked from a container box, pallet or conveyor belt, to be packed, machined or delivered to another sector of the manufacturing chain.

The picking system could be placed at the end of a palletization line, at the entry of a particular production line in order to feed a CNC machine tool or along an inspection line for quality control. The production line can be completely automatic or semi-automatic. In the latter case, human operators execute manual operations like visual inspection, assembly, polishing, deburring, gluing, etc.

Picking Application

In a bin-picking application where the aim is to pick up randomly placed objects from a container box or pallet, 2D vision is insufficient because the objects will float together in the 2D image, and it is impossible to distinguish them. 3D vision systems are applied in this case (Rahardja & Kosaka, 1996). Picking from a pallet is another typical application. A pallet is a portable platform used in the transportation of goods. It can be structured or non-structured. In the first case, pieces are stacked in a regular manner in order to simplify the unloading phase. In the latter case, objects are randomly placed. Moreover, pieces can be loaded in a single or multiple layer configuration. Layers can be separated through paper or plastic inter-layers. For single layer pallets with regular parts on it, a 2D vision system can be adopted, otherwise some kind of 3D vision system has to be used to detect the object and to find its 3D pose (Allotta et al., 2003).

Vision systems are widely adopted in applications where the objects to be picked lie on a conveyor belt. Typically, a 2D vision system with a camera mounted above or beside the conveyor is used (Borangiu, 2002). The objects can be randomly placed on the belt or aligned and well-separated. The conveyor can either stop waiting for the end of the picking action or move at a constant speed. In the latter case, the robotic manipulator has to track the object in a different position from the position extracted from its image. Furthermore, objects can have the same shape and dimensions or differ in shape and/or dimensions. After the boundaries of the object have been calculated, the vision system can use a template matching algorithm to recognize the object type and estimate its 2D pose related to the picking point. Any commercial vision system for industrial purposes provides a toolkit for template matching. Typically, an image of a template object is captured and the operator has to give rules for the feature extraction and specify where the picking point is. The speed of the conveyors can range from very low to very high values. The higher the speed of the conveyor, the higher the performance required for the vision system and the robotic manipulator. Very slow conveyors are used in a wide range of applications when, for instance, the production cycle includes both manual operation and automatic operations on the transported objects (also in order to accomplish safety constraints).

Production Lines with Manual and Automatic Operations

A production line is a set of sequential operations that transforms input materials or semi-finished products into final products. Although a fully automated process (i.e. a process where no opera-

Figure 4. Production line with both human and robot operations

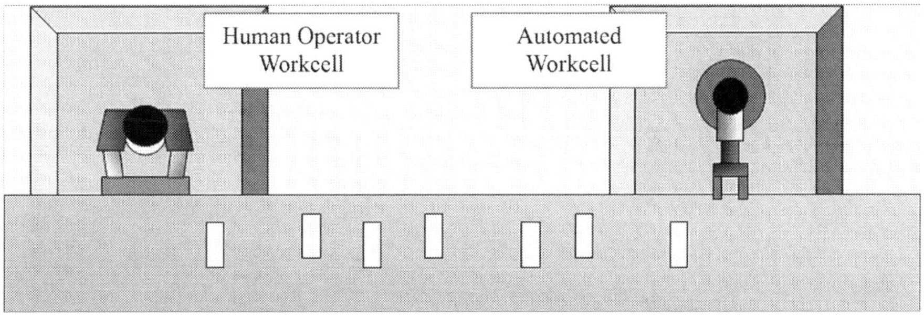

tion requires manual actions) is often the desired target solution, in many industrial contexts many operations are still executed by human operators. The reasons could be the difficulty to replicate some operations with an automatic system in a robust manner or the cost of a fully-automated production system is too high. In the leather goods industry, for example, a slow conveyor chain is the norm. Special pallets are attached to a chain and the different parts of the product are placed over a plate or hung up on forks. For each step along the line, the parts are picked, machined and placed back on the chain.

In similar contexts, three different approaches could be taken. The first one consists in changing the whole line in order to adopt, for instance, custom fixtures or special mechanisms. This approach could be very expensive and time-consuming. The second approach exploits separated workcells served by human operators that pick the components from the main line, load them in the workcell and place again the object on the conveyor after assembly. A third possible approach is to create robotic units that act like human operators. This latter solution gives greater flexibility and allows to save money by eliminating the manual picking and placing tasks and not having to replace the old line arrangements with more sophisticated structures.

The production line could be seen as a conveyor and a number of working units distributed along the line. Each unit can be replaced by a human operator or by a robotic system without affecting the rest of the line and without stopping the conveyor (see an example in Figure 4). For this solution, a vision system can be used to detect pieces on the conveyor and to recover their pose. This requires a slight retrofitting of the line. With this approach, another important issue must be considered. If only human operators are working on the line, the placement of the objects on the conveyed pallets are irrelevant and the objects can be randomly placed on the conveyor.

Under the hypothesis that both robotic manipulators and human operators work inline, as shown in Figure 4, the placing of the object assumes a considerable importance. However, in many cases the introduction of some constraints on the position and orientation of the object does not represent an obstacle. For example, it could be reasonable to ask the operator to place the object with a selected side facing the camera.

Requirements for the Vision System

The vision system should be:

- Compact. The system should be as compact as possible for flexible installation along the production line.
- Low cost. One of the objectives is to create a relatively inexpensive system.
- Simple. Setup and calibration should be easy to do.

- Fast. The pose estimation time should be negligible and should not affect the picking time.
- Robust. The system should remain reliable for a long time under typical industrial environmental conditions.

In order to satisfy the first requirement (i.e. the compactness), a monocular system is preferred. In order to reduce the cost of the system, hardware and software requirements should be minimized. Moreover, commercially available components should be preferred.

Setup and maintenance are important factors to take into account. In general, the higher the system complexity, the higher the setup time and, in particular, the calibration time. If the calibration is simple, an eventual relocation of the working unit does not constitute a problem for the maintenance service. Performance has to be taken into account when the picking time must be very short. The implementation of complex algorithms could result in a non-negligible latency.

Robustness is another key feature. Factors characterising normal operating conditions such as vibrations, dust and temperature changes must be tolerated by the system. The performance of the pose estimation algorithm should remain stable under reasonable limits if the illumination or the object colour change and if the manufactured object presents defects or a shape that differs from the standard.

The Vision System

The solution discussed in this chapter has been elaborated for production lines with both manual and automatic working units where the manufactured objects are carried by a conveyor belt or a conveyor chain at low speed.

In agreement with the above listed requirements, a monocular system is proposed and the following hypotheses are considered as valid:

- a selected side of the object is always orientated toward the camera;
- a circular marker is fixed in some manner on the visible side;
- the camera detects the marker always within a small area;
- object orientation does not substantially vary;
- the picking point is related to the centre of the circular marker.

This solution gives the possibility to use only one camera and only one marker to recover the object position. While using several markers for the pose estimation could be a problem, a single marker can be easily attached to the object. It can be a stick, the head of a screw, the head of a nail or any other part of the object.

The circular marker and the camera have to be positioned in order to satisfy the following conditions:

- the marker passes in front of the camera during the conveyor movement;
- when the marker is in front of the camera, the optical axis is normal or almost normal to the plane where the marker circle lies;
- when the marker is in front of the camera, its centre is close to the optical axis;
- when the object is in front of the camera, the projection of the marker in the image has to be visible inside the image;
- only one marker at a time can appear in the image.

The basic schema is depicted in Figure 5. The whole system includes the following components:

- a monochrome camera;
- one lens;
- an illumination system;
- a frame grabber;
- a PC.

Figure 5. Calibration of the monocular vision system

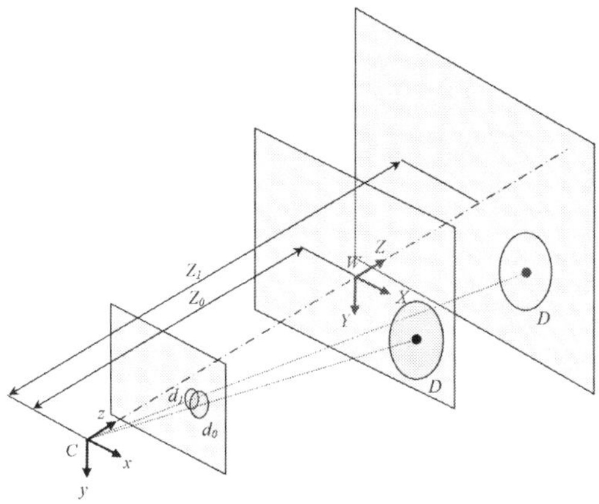

Only the vision system is described here, the control of the robotic manipulator is out of the scope of this chapter. Any industrial anthropomorphic robot with a proper end-effector mounted on its last link can be used to pick an object from a line. All the above-listed components are commonly available in an industrial context.

The software that runs on a PC has to elaborate the images coming from the camera and send the coordinates of the picking point to the robot controller. When the pose estimation is required, the software gets the image from the camera and extracts the required features. In this case, the software has to detect the boundary of the marker and to filter out disturbances as well as other unnecessary information. On that boundary, the software calculates the diameter of the circle and its centre. In general, the marker's projection assumes an elliptical shape. While the centre remains the same because it is an invariant feature under perspective transformation, the diameter is no longer calculable. To overcome this problem, the diameter of the outer circle is used. Filters can be applied in order to eliminate non closed boundaries, too small or too big ellipses, ellipses whose *circularity* (defined as the ratio of the area of the shape to the area of a circle having the same perimeter) is below a given threshold. The closer the marker to the camera, the bigger the circle in the image. Thus, the basic idea is to estimate the Z position of the marker centre through its diameter and to use the position of the marker centre in the image in order to estimate the X, Y coordinates, by supposing that a world reference system W is conveniently defined with the same orientation of the camera frame and is located at a distance Z_0 from the camera origin.

For the sake of simplicity, in the following the basic pin-hole camera model is exploited. The complexity of more realistic models is managed by well-known algorithms (Clarke & Fryer, 1998) that are implemented in almost any commercial system. When the marker, having a diameter D, lies in the XY plane of the world coordinate system, its projection is given by:

$$d_0 = D \cdot \frac{f}{Z_0}$$

If the distance of the marker plane becomes Z_1, the projected diameter is:

$$d_1 = D \cdot \frac{f}{Z_1}$$

Z_0 is an extrinsic parameter that is calculated during the calibration procedure. By combining the previous equations, the projected diameter d_1 can be calculated as:

$$d_1 = d_0 \cdot \frac{Z_0}{Z_1}$$

The desired output is the Z coordinate in the world reference system, which is given by:

$$Z = Z_1 - Z_0$$

Z can be calculated from the measure of d_1 by inverting the perspective projection:

$$Z = Z_0 \cdot \frac{d_0 - d_1}{d_1}$$

If $d_1 > d_0$, Z is negative and the object is closer to the camera.

For the other coordinates, the following reasoning will be used. If (x_0,y_0) is the position of the centre in the image plane when the marker lies on the XY plane of the world coordinate system, and (x_1,y_1) is the position of the centre in the image plane when the marker lies on a plane parallel to XY at a distance Z_1 from the origin of the camera reference frame, the relation between (x_1,y_1) and (x_0,y_0) is given by the following equations:

$$x_0 = x_1 \cdot \frac{d_0}{d_1}$$

$$y0 = y1 \cdot \frac{d0}{d1}$$

After the system has been calibrated, given a measure on the calibration plane, the corresponding measure in the world coordinates is known, thus the X and Y coordinates can be obtained from x_0 and y_0, respectively.

By referring to the basic pin-hole camera model, where only the focal distance is considered among the intrinsic parameters, the inverse projective transformation becomes:

$$X = x_1 \cdot \frac{d_0}{d_1} \cdot \frac{Z_0}{f}$$

$$Y = y_1 \cdot \frac{d_0}{d_1} \cdot \frac{Z_0}{f}$$

In the previous equations, the assumption is made that the marker always appears as a circle on the image plane, by neglecting the deformations due to the perspective model when the marker moves away from the optical axis or when it lies on a plane which is not normal to the optical axis. The error due to those assumptions will be evaluated afterwards. However, the aim of the proposed system is to provide an estimate that is useful for the picking system and not a reliable measure to be exploited for inspections and comparisons.

CALIBRATION

Camera calibration is required to extract metric information from 2D images. Calibration consists in finding the intrinsic and extrinsic parameters. By referring to the simplest camera model, the calibration procedure has to determine the focal distance, the position and the orientation of the camera with respect to a world coordinate system. The calibration procedure is performed by observing a calibration object whose geometry in 3D space is known with very good precision (Zhang, 1999). Usually, the calibration object is a simple grid composed of circular markers or a chessboard. In this context, the calibration of the extrinsic parameters concerns a world reference frame having its Z axis aligned with the optical axis (with reasonable approximation) and the X and Y axes aligned with the correspondent camera axes. From a theoretical point of view, only the distance Z_0 along the Z axis has to be found. In

Figure 6. Calibration of the monocular vision system

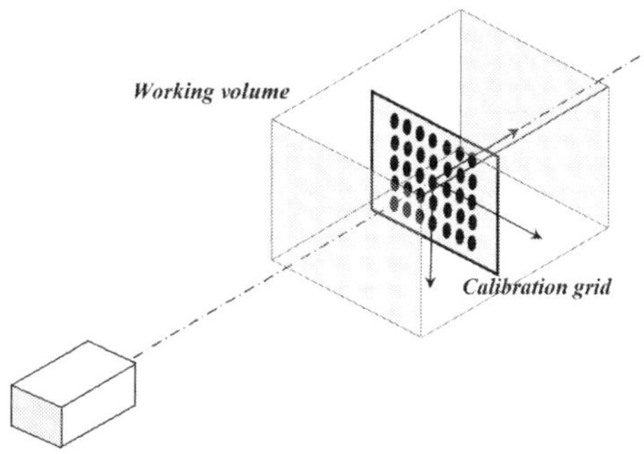

practice, a small rotation and a slight translation always exist.

After the camera has been calibrated, the system can give reliable measurements on the *XY* plane passing through Z_0. The error resulting from the measures at different *Z* positions mainly depends on the focal distance. The higher the focal distance, the lower the error.

Most of the available commercial systems implement sophisticated camera models and provide their own calibration software procedures for the identification of the intrinsic and extrinsic parameters. Those procedures are normally very easy to follow, thus custom calibration procedures will be avoided. The only requirement is to position the calibration grid as depicted in Figure 6 so that the origin of the world coordinate system is centred in the area where the marker passes through during a normal production cycle.

EXPERIMENTAL SETUP

An experimental setup has been created by using off-the-shelf components. This setup has been designed in order to test if the proposed method can accomplish the tasks in a typical and generic industrial context where highly tolerant end-effectors are used for standard pick and place operations. Therefore, low precision (up to some millimetres) is required in the estimation of the position of the object.

The whole setup is depicted in Figure 7 and it includes:

- A desktop PC with Windows©;
- A frame grabber integrated in the Matrox© acquisition board;
- A standard industrial analog CCD camera (produced by TELI©) with a resolution of 768x576 pixels;
- Lens with a fixed 12 mm focal distance;
- A calibration grid of 7x7 circular markers;
- A digitizer arm (MicroScribe3D, produced by Immersion Corporation);
- Some sample objects with a white circular marker;
- LED lights located beside the camera.

The Matrox© MIL library has been installed on the PC. The MIL library offers all the functionalities required for:

- retrieving the image data from the frame-grabber;
- extracting the boundaries;
- selecting and extracting the desired features (centre and outer diameter);
- calibrating the system by calculating the intrinsic and extrinsic parameters.

Figure 7. Picture of the experimental setup

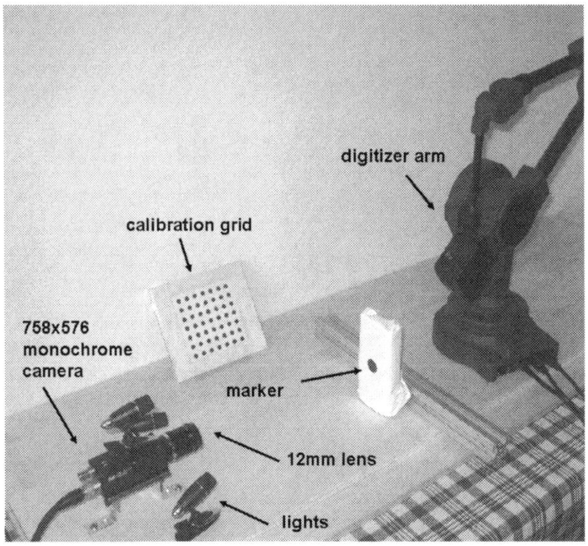

The calibration algorithm can take into account the complete set of intrinsic parameters that were described in the background section and even the distortion effects. Illumination has been conditioned through a simple set of LED positioned near the camera lens in order to reduce the problems caused by varying environmental conditions. The frame-grabber is attached to one of the internal PCI slots of the PC. A single cable that links the camera to the frame grabber transmits the video signal and the power.

Several tests have been performed by using small wooden objects where a white disc has been attached. The diameter of the marker is 17 mm. The output of the system is compared to the correspondent position measured through the digitizer arm. The camera has been horizontally mounted at a distance of about 270 mm from the origin of the adopted reference frame. The same 7x7 grid is used to calibrate both the camera and the arm digitizer. The calibration of the arm is made using three points: one on the origin, one on the X direction and one on the XY plane where Y is positive. As it is foreseen that system behaviour depends only on the Z position and the distance from the optical axis, for the sake of simplicity, the tests have been conducted on the XZ plane for positive X values so that the distance from the optical axis matches the X coordinate.

The experiments have been executed as described in the following. Measures were taken by varying the Z position in the range [-50,+50] mm and the X position in the range [0,+40] mm with a null Y component. The marker surface is maintained normal to the optical axis. Each measure consists in three steps: firstly the object is moved and blocked somewhere in the working zone. Second, the centre position $\vec{P}_d$ of the marker is measured through the digitizer arm. Third, the position $\vec{P}_{est}$ of the marker is estimated by means of the vision system.

NUMERICAL RESULTS

In order to evaluate the goodness of the proposed approach, the error ε, which is defined as the distance between the estimated position $\vec{P}_{est}$ and the position $\vec{P}_d$ measured by means of the digitizer arm, has been determined for each test.

Figure 8. Error function

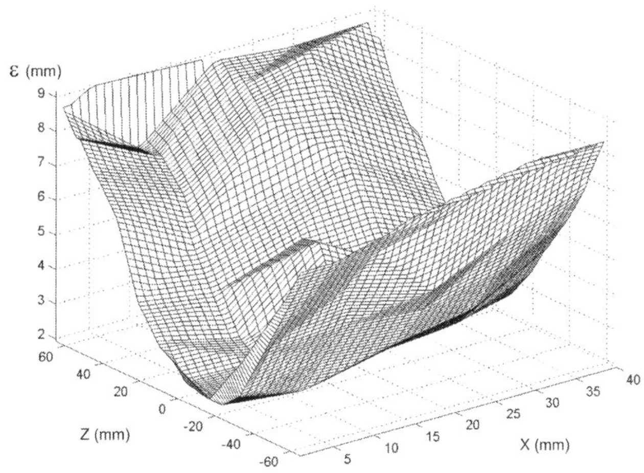

$$\varepsilon = \left\| \vec{P}_d - \vec{P}_{est} \right\|$$

Figure 8 shows a 3D graph that illustrates the error trend as a function of the Z position and the X position, which represents the distance from the optical axis.

The error increases with increasing values of the distance of the point from the origin of the work reference frame, while it remains almost constant along the X axis, thanks to the adoption of the outer circle diameter. The maximum value ε_{max} of the error is about 9.2 mm. By assuming that Z lies in the range [-40,+40] mm, the system gives the position with a precision of +/- 5 mm. Figure 8 also shows that the error function can be formulated as a function only of Z and that, around the origin of the world reference frame, this function can be approximated through a second order polynomial.

As shown in Figure 9, knowing the error function shape allows a simple *a-posteriori* correction that provides good results. It can be noticed that with the error compensation, the maximum residual error is ε_{max} = 1.3 mm, with a considerable performance improvement.

FUTURE RESEARCH DIRECTIONS

The system for picking applications described in the previous sections estimates the position of the object and not its orientation. There are cases where the orientation of the part is fundamental because the gripper does not work if the object is not aligned properly. In these cases, the lack of information concerning the object's orientation makes this system useless. Future work aims at estimating the orientation of the objects while preserving system robustness and compactness with minimal constraints. For instance, the use of three markers could solve the problem in some cases but, in general, this represents a non-negligible modification of the object which is not always possible. Other pose estimation methods based on a single camera, as the one described in (David et al., 2003), require both to have the knowledge of the object model and to solve the correspondence problem between features on the 3D model and the features on the image.

Another area where enhancements can be studied investigated concerns the precision of the system and its operative limits. Figure 9 shows that, by adopting the polynomial error correction,

Figure 9. Error function after using an a posteriori correction

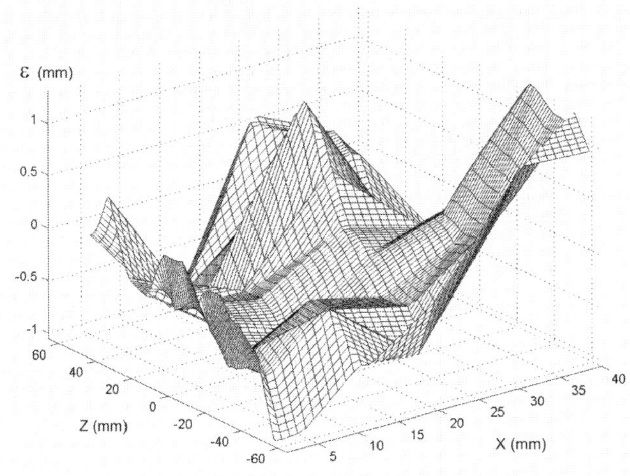

the residual error is a non-linear function which depends on the distance between the point and the optical axis and the distance between the point and the camera. The error function could be modeled by using conventional or advanced methods (such as soft-computing algorithms, e.g. based on neural networks (Haykin, 1999)) and thus a more refined method could improve error compensation.

Finally, specific calibration algorithms can be developed in order to accomplish both the calibration of the camera and the characterization of the shape of the error function.

Beside the pose estimation problem, another important issue in a typical production line is the object identification. Objects differing in size, colour or other characteristics, can be machined and conveyed along the line. The control system of the robot manipulator may need to know what kind of object is facing the camera in order to decide the picking point, the target placing position or the configuration of the machining process. Although identification can be done by using radio frequency identification (Landt, 2005) or bar codes (Palmer, 1991), the same marker used to estimate the position can be exploited to this purpose. For instance, coded markers (Naimark & Foxlin, 2002) can be adopted.

CONCLUSION

A monocular 3D vision system capable of finding the position of an object to be picked from a production line is introduced. The system can be adopted in production lines with both robotic manipulators and human operators. In similar contexts, a human operator cannot guarantee to place an object on a conveyor with precise orientation without using special fixtures.

The addition of special hardware on the line, such as any modification in general, can be very expensive and can reduce the flexibility of a production process. On the other hand, an automated picking system that works after the human operator needs to know the object position and orientation of the object. If this information is unavailable, automatic picking cannot be performed or, at least, additional human resources are required to serve the unit. In this case, the solution of the pose estimation problem through a vision system becomes a crucial point.

The proposed systems achieves the above discussed task and works under the hypotheses that the objects are located in front of the camera approximately with the same orientation and that slight errors in the position estimation are compensated by proper grippers. Compared to stereo systems, a monocular system which uses standard components and a simple marker is compact, inexpensive and easy to install system. Calibration is quite easy and system accuracy proves to be acceptable for picking operations even when using basic hardware.

REFERENCES

Allotta, B., Colombo, C., Bertocci, F., Scarselli, L., Sgarbi, M., Bioli, G., et al. (2003). A system based on stereo vision for unpacking pieces of wood with an industrial robot. *Proceedings of the 1st International Workshop on Advances in Service Robotics*, Italy, *3*, 1448-1457.

Alshawish, H. M. M., & Allen, C. R. (1995). 3D object recognition using coded light projection for robot assembly applications. In *Proceedings of the 21st International Conference on Industrial Electronics, Control and Instrumentation*, Orlando, FL, *2*(6), 1240-1247.

Ansar, A., & Daniilidis, K. (2003). Linear pose estimation from points or lines. *IEEE Transactions on Pattern Analysis and Machine Intelligence*, *25*(5), 578–589. doi:10.1109/TPAMI.2003.1195992

Borangiu, T. (2002). Visual conveyor tracking for "pick-on-the-fly" robot motion control. In *Proceedings of the 7th IEEE International Workshop on Advanced Motion Control*, Maribor, Slovenia, (pp. 353-358).

Brown, H. M. M., Burschka, D., & Hager, G. D. (2003). Advances in computational stereo. *IEEE Transactions on Pattern Analysis and Machine Intelligence*, *25*(8), 993–1008. doi:10.1109/TPAMI.2003.1217603

Canny, J. (1986). A computational approach to edge detection. *IEEE Transactions on Pattern Analysis and Machine Intelligence*, *8*(6), 679–714. doi:10.1109/TPAMI.1986.4767851

Cheng, F. S., & Denman, A. (2005). A study of using 2D vision system for enhanced industrial robot intelligence. *Proceedings of the IEEE International Conference on Mechatronics and Automation*, Ontario, Canada, *3*, 1185-1189.

Chin, R. T., & Harlow, C. A. (1982). Automated visual inspection: A survey. *IEEE Transactions on Pattern Analysis and Machine Intelligence*, *4*(6), 557–573. doi:10.1109/TPAMI.1982.4767309

Clarke, T. A., Fryer, J. F., & Wang, X. (1998). The principal point and CCD cameras. *The Photogrammetric Record*, *16*(92), 293–312. doi:10.1111/0031-868X.00127

Clarke, T. A., & Fryer, J. G. (1998). The development of camera calibration methods and models. *The Photogrammetric Record*, *16*(91), 51–66. doi:10.1111/0031-868X.00113

David, P., DeMenthon, D., Duraiswami, R., & Samet, H. (2003). Simultaneous pose and correspondence determination using line features. *Proceedings on the IEEE Computer Society Conference on Computer Vision and Pattern Recognition*, Madison, WI, *2*, 18-20.

Delp, E. J. (1997). Image processing software and hardware. *IEEE Signal Processing Magazine*, *15*, 8–11.

Dementhon, D. F., & Davis, L. S. (1995). Model-based object pose in 25 lines of code. *International Journal of Computer Vision*, *15*(1-2), 123–141. doi:10.1007/BF01450852

Dhome, M., Richetin, M., Lapreste, J. T., & Rives, G. (1989). Determination of the attitude of 3D objects from a singleperspective view. *IEEE Transactions on Pattern Analysis and Machine Intelligence*, *11*(12), 1265–1278. doi:10.1109/34.41365

Diaz, J. C., & Abderrahim, M. (2007). Modified SoftPOSIT algorithm for 3D visual tracking. In *Proceedings of the IEEE International Symposium on Intelligent Signal Processing*, Madrid, Spain, (pp. 1-6).

Faugeras, O. (1993). Three-dimensional computer vision: a geometric viewpoint. Cambridge, MA: MIT Press.

Fernandez, J., Benchetrit, D., & Gachet, D. (2001). Automated visual inspection to assembly of frontal airbag sensors of automobiles. In *Proceedings on the 8th IEEE International Conference on Emerging Technologies and Factory Automation*, Antibes Juan-Les-Pins, France, *2*, 631-634.

Fu, K. S., & Mui, J. K. (1981). A survey on image segmentation. *Pattern Recognition, 13*(1), 3–16. doi:10.1016/0031-3203(81)90028-5

Haralick, R. M., Lee, D., Ottenburg, K., & Nolle, M. (1995). Analysis and solutions of the three point perspective pose estimation problem. In *Proceedings of the IEEE Computer Society Conference on Computer Vision and Pattern Recognition*, Maui, Hawaii, (pp. 592-598).

Haykin, S. (1999). Neural networks: A comprehensive foundation (2nd ed.). Upper Saddle River, NJ: Prentice Hall.

Heikkila, J., & Silven, O. (1997). A four-step camera calibration procedure with implicit image correction. In *Proceedings on the IEEE Computer Society Conference on Computer Vision and Pattern Recognition*, San Juan, Puerto Rico, (pp. 1106-1112).

Horaud, R., Conio, B., Leboulleux, O., & Lacolle, B. (1989). An analytic solution for the perspective 4-point problem. *Computer Vision Graphics and Image Processing*, *47*, 33–44. doi:10.1016/0734-189X(89)90052-2

Horaud, R., Dornaika, F., Lamiroy, B., & Christy, S. (1997). Object pose: The link between weak perspective, paraperspective, and full perspective. *International Journal of Computer Vision*, *22*(2), 173–189. doi:10.1023/A:1007940112931

Hutchinson, S., Hager, G. D., & Corke, P. I. (1996). A tutorial on visual servo control. *IEEE Transactions on Robotics and Automation*, *12*(5), 651–670. doi:10.1109/70.538972

Klette, R., Schluns, K., & Koschan, A. (1998). Computer vision. Singapore: Springer.

Kline, M. (1972). Mathematical thought from ancient to modern times. New York: Oxford University Press.

Landt, J. (2005). The history of RFID. *IEEE Potentials*, *24*(4), 8–11. doi:10.1109/MP.2005.1549751

Malamas, E. N., Petrakis, E. G. M., Zervakis, M. E., Petit, L., & Legat, J. D. (2003). A survey on industrial vision systems, applications and tools. *Image and Vision Computing*, *21*(2), 171–188. doi:10.1016/S0262-8856(02)00152-X

Naimark, L., & Foxlin, E. (2002). Circular data matrix fiducial system and robust image processing for a wearable vision inertial self-tracker. In *Proceedings of ACM/IEEE International Symposium on Mixed and Augmented Reality*, Darmstadt, Germany, (pp. 27-36).

Nelson, B. J., Papanikolopoulos, N. P., & Khosla, P. K. (1996). Robotic visual servoing and robotic assembly tasks. *IEEE Robotics & Automation Magazine*, *3*(2), 23–31. doi:10.1109/100.511777

Oh, J. K., & Lee, C. H. (2007). Development of a stereo vision system for industrial robots. In *Proceedings of the International Conference on Control, Automation and Systems*, Seoul, Korea, (pp. 659-663).

Palmer, R. C. (1991). The bar code book. Peterborough, NH: Helmers Publishing.

Rahardja, K., & Kosaka, A. (1996). Vision-based bin-picking: Recognition and localization of multiplecomplex objects using simple visual cues. In *Proceedings of the IEEE/RSJ International Conference on Intelligent Robots and Systems*, Osaka, Japan, *3*, 1448-1457.

Ruzon, M., & Tomasi, C. (2001). Edge, junction, and corner detection using color distributions. *IEEE Transactions on Pattern Analysis and Machine Intelligence*, *23*(11), 1281–1295. doi:10.1109/34.969118

Shapiro, L. G., & Stockman, G. C. (2001). Computer vision. London: Prentice Hall.

Trucco, E., & Verri, A. (1998). Introductory techniques for 3D computer vision. Upper Saddle River, NJ: Prentice Hall.

Tsai, R. Y. (1987). A versatile camera calibration technique for high-accuracy 3D machine vision metrology using off-the-shelf TV cameras and lenses. *IEEE Journal on Robotics and Automation*, *3*(4), 323–343. doi:10.1109/JRA.1987.1087109

Wen, Y., & Oakley, J. P. (1995). Implementation of oriented filters for arc detection. In *Proceedings of the 6th British conference on Machine vision*, Birmingham, UK, *2*, 483-492.

Weng, J., Cohen, P., & Herniou, M. (1992). Camera calibration with distortion models and accuracy evaluation. *IEEE Transactions on Pattern Analysis and Machine Intelligence*, *14*(10), 965–980. doi:10.1109/34.159901

Xiang, Z., & Navab, N. (2000). Tracking and pose estimation for computer assisted localization in industrial environments. In *Proceedings on the 5th IEEE Workshop on Applications of Computer Vision*, Palm Springs, CA, (pp. 214-221).

Xiong, Y., & Shafer, S. A. (1993). Depth from focusing and defocusing. In Computer Vision and Pattern Recognition, (pp. 68-73).

Xu, W., & Zhang, L. (1993). Pose estimation problem in computer vision. In *Proceedings of the IEEE Region 10 Conference on International Computer, Communication, Control and Power Engineering*, Beijing, China, *2*, 1138-1141.

Zhang, Z. (1999). Flexible camera calibration by viewing a plane from unknown orientations. In *Proceedings of the 7th IEEE International Conference on Computer Vision*, Kerkyra, Greece, *1*, 666-673.

Ziou, D., & Tabbone, S. (1998). Edge detection techniques: An overview. *International Journal of Pattern Recognition and Image Analysis*, *8*(4), 537–559.

KEY TERMS AND DEFINITIONS

Pose: This term is commonly used in computer vision to refer to the combination of the position and orientation of an object with respect to a particular reference frame.

Pose Estimation: A six degree-of-freedom problem where the transformation from the object frame and the camera frame has to be found.

Bin Picking: Picking up objects randomly arranged in a bin.

Pin-Hole Camera: A very simple camera where light rays from an object pass through a small hole to form an image.

Intrinsic Parameters: All the parameters that characterize the camera model. Typical intrinsic parameters are the focal length, the pixel skew angle, the pixel aspect ratio and the principal point offset.

Extrinsic Parameters: Parameters which describe the position and orientation of the camera with respect to a world coordinate frame.

Image Segmentation: The process of partitioning an image into multiple regions, i.e. a set of pixels. The aim of segmentation is to have a simplified version of an image which retains significant information and is easier to analyze.

End-Effector: The device mounted at the moving end of a robotic arm. The design of the end-effector depends on the application. For example, a gripper for pick and place operations or a spindle for machining operations.

Camera Calibration: The process where the intrinsic and extrinsic parameters of a camera are found using proper calibration procedures.

Chapter 11
Developing a Collaborative Supply Chain Management Platform:
A Service-Oriented Approach

Yves Wautelet
Hogeschool-Universiteit Brussel, Belgium

Youssef Achbany
Université catholique de Louvain, Belgium

Jean-Charles Lange
Université catholique de Louvain, Belgium

Vi Tran
Université catholique de Louvain, Belgium

ABSTRACT

Service-oriented computing is becoming increasingly popular. It allows designing flexible and adaptable software systems that can be easily adopted on demand by software customers. Those benefits are from primary importance in the context of supply chain management; that is why this paper proposes to apply ProDAOSS, a process for developing adaptable and open service systems to an industrial case study in outbound logistics. ProDAOSS is conceived as a plug-in for I-Tropos - a broader development methodology - so that it covers the whole software development life cycle. At analysis level, flexible business processes are generically modelled with different complementary views. First of all, an aggregate services view of the whole applicative package is offered; then services are split using an agent ontology - through the i framework - to represent it as an organization of agents. A dynamic view completes the documentation by offering the service realization paths. At design stage, the service center architecture proposes a reference architectural pattern for services realization in an adaptable and open manner. The paper finally presents the implemented platform for a particular service – manage transport – so that the reader can realize how the developments have been achieved.*

DOI: 10.4018/978-1-61520-605-6.ch011

1. INTRODUCTION

Today's enterprise information systems have to match with their operational and organizational environment. Unfortunately, software project management methodologies are traditionally inspired by programming concepts rather than by organizational and enterprise ones. In order to reduce as much this distance, Agent-Orientation is increasingly emerging and has been the object of more and more research over the last 10 years. Its success comes from the fact that it better meets the increasing complexity and flexibility required to develop software applications built in open-networked environments and deeply embedded into human activities.

The gap between the traditional software engineering approaches and multi-agent software systems using artificial intelligence concepts remains nevertheless important. A series of development methodologies for designing multi-agent systems (MAS) have appeared over the years; those methods have their own characteristics and use various models to analyse the environment, to design the system and also various languages to implement the proposed solution. This paper is part of the effort to dispose of methodologies covering the whole development life cycle: from the analysis of industrial (multiactor) cases to the development of MAS software applications by extending the I-Tropos software process (Wautelet, 2008) notably through service orientation. This plug-in is the Process for Developing Adaptable and Open Service Systems (ProDAOSS, see (Achbany, 2009; Achbany et al., 2008b)). Contributions also includes the development of the supply chain platform itself.

A supply chain is the set of all actors and relations between them which participate in the process of delivering value to a customer as a product or service. It includes all the processes from raw materials to delivery and is viewed as a network of information, materials and financial flows. In this paper we particularly focus on the application of our software development methodology on outbound logistics - i.e. the process related to the movement and storage of products and goods from the supplier to the end user - with a strong focus on transportation. The idea underlying the developed collaborative application package is to offer the most advanced services to optimize the transportation chain on a global basis. The paper will present the analysis and design stages of the application package development.

The paper is structured as follows. Section 2 discusses the state of the art including actor collaboration and the services approach in supply chain management as well as the related work. Section 3 briefly presents the development process – ProDAOSS – and its application domain in the context of this book chapter – outbound logistics. Section 4 overviews the application development: the analysis and design stages are depicted in detail and the implementation is presented through screenshots. Section 5 presents the future research directions. Finally, section 6 concludes the paper.

2. STATE OF THE ART

This section envisages actor collaboration and the services approach in supply chain management, it also browses the work related to the one developed in this chapter.

2.1 Actor Collaboration in Supply Chain Management

By nature, supply chain management is an interesting area for the development of industrial multi-actor software systems since it involves a series of collaborating or competing companies with tens of roles played by hundreds of individuals. The benefits that can be taken from such systems can lead to avoid waste of resources for all the involved actors. Collaborative decision will tend to avoid local equilibriums (at actor level) and wastes in the global supply chain optimisation, giving opportunities to achieve the greatest value that the chain can deliver at lowest cost (see Pache, & Spalanzani, 2007, Samii, 2004). Such a result can only be achieved through actors' collaboration. As noticed by (Moyaux T., Chaib-draa B., & D'Amours S., 2006), the term collaboration is confusing since various interpretations take place in the context of supply chains, they distinguish 3 overlapping levels of collaboration in real supply chains:

- Information centralization is the most basic technique of information sharing. Applied to outbound logistics it can be the shipper announcing its transportation needs or the carrier sharing all its planned transportations. Moyaux et al. distinguish information sharing from centralisation by the fact that the latter is "the multi-casting in real-time and instantaneously of the market consumption information" (Moyaux T., Chaib-draa B., & D'Amours S., 2006) while the former is the sharing of the demand and supply information between companies;
- Vendor Managed Inventory (VMI) and Continuous Replenishment Program (CRP) are collaboration techniques in which retailers do not place orders because wholesalers use information centralization to decide when to replenish them;
- Collaborative Planning, Forecasting and Replenishment (CPFR) enhance VMI and CRP by incorporating joint forecasting. CPFR includes only two levels of supply chain, retailers and wholesalers, it allows companies to electronically exchange a series of written comments and data which include past sales trends, scheduled promotions and forecasts. It shares more information than only demand information. This allows the participants to coordinate joint forecasts by focussing on the differences in forecasts.

Our interpretation of the term collaboration in the context of this paper is at first information centralisation, indeed by announcing the demand and supply concerning transportation (in terms of transport services, logistic requests, transportations, etc.), the actors collaborate for answering to the requirements of the other actors at best conditions. Indeed, collaboration on the basis of information sharing is not incompatible with competition. Since an actor role can be played by different - probably competing - companies, the demander can use the best possible offer or the supplier can make interesting offers to dispose of optimally filled transportations. At second by sharing this information benefits can be issued in terms of global optimization. The whole outbound logistic chain can be optimized in terms of load balancing, cross dockings, tours, time, etc. Moreover, real-time information about the development of the services allows internal follow up and (re) optimization.

2.2 The Services Approach in Supply Chain Management

In the context of the development of a multi-actor collaborative application package we used a service-oriented approach for developing the MAS. This has been done following the ProDAOSS methodology overviewed later in this paper, for a full description using the Software Process Engineering Meta Model (OMG, 2008) see (Achbany, 2009). The services approach in the context supply chains is of particular interest because:

- Supply chain software and data collection will, in the future, take the form of a utility accessible on demand;
- Service-oriented architecture (SOA) could have a profound effect on the way supply chain software is designed, sold and

implemented in the future. Common workflow based developments progressively leave place to generic customizable systems. In that perspective, SOA is designed to support rather than dictate business processes. This trend is for example also present in ERP systems development;

- Software developments' integration is from strategic importance in nowadays business context, SOA with its inherent flexibility ease these procedures.

2.3 Related Work

The application of MAS technology onto supply chain management has already driven plenty of literature. Indeed, the subject covers several aspects that is why we will focus here on the application of MAS development processes onto supply chain cases.

(Penserini & al., 2004, Penserini, Kolp, & Spalazzi, 2007) focuses on the socially-driven approach of the Tropos software development methodology in order to build up an agent-based information system prototype in supply chain management. Their approach however remains limited since their supply chain model is (too) simplified and mostly focuses on the application of architectural styles and design patterns at architectural and detailed design levels rather than on the whole development life cycle.

(Govindu, & Chinnam, 2007) proposes a generic process-centered methodological framework called Multi-Agent Supply Chain Framework (MASCF), to simplify MAS development for supply chains applications. Their idea is to map elements from the Supply Chain Operations Reference (SCOR, see (SCC, 2005)) model to the Gaia development methodology (Wooldridge at al., 2000) for multi-agent supply chain system development. Mapping the SCOR and the Gaia concepts is interesting for analysts but it supposes modelling supply chain processes using this formalism while it would be strain forward to dispose of a fully integrated generic and flexible methodology for MAS development in supply chain management. That is why we choose to extend a current methodology with elements and axioms adapted for such flexible industrial developments rather than to map existing supply chain modelling elements to software development elements.

In (Ghenniwa & al, 2006), the idea that of an electronic marketplace (eMarketplace) as an architectural model to develop collaborative supply chain management and integration platform is defended. In their architecture, the eMarketplace exists as a collection of economically motivated software agents of service-oriented cooperative distributed systems. The authors discuss how coordination approaches asauctions and and multi-issues negotiation can be developed in the context of this eMarketplace. The objective here is to enable business entities to obtain efficient resource allocation while preserving long-term relationships. This idea is also present in our software system but our platform is developed using the services approach and focuses on outbound logistics.

3. A SERVICE-ORIENTED DEVELOPMENT PROCESS APPLIED TO OUTBOUND LOGISTICS

This section presents the service-oriented software process used to develop our platform and describes it in details. The typical outbound logistics workflow is also presented. Our developments are based on the presented conceptual model.

3.1 The Process

Service-oriented computing is becoming increasingly popular when developing large and flexible software solutions. Since our goal is to dispose of a methodology for developing large industrial software application packages this was adopted in our process called ProDAOSS (i.e., Process for Developing Adaptable and Open Service Systems). Indeed, services are used as fundamentals to drive the software process. They are distinguished and documented at analysis stages (through the FaMOS framework, see (Wautelet, Kolp, & Achbany, 2008)) then converted into service centers at design stage for an agent oriented implementation. More precisely:

- The organisation is firstly described as a set of organisational services in the Strategic Services Diagram;
- Organizational services are split in a series of goals, tasks and resources depicted in the Strategic Dependency and Rationale Diagrams;
- Organisational services realization paths are documented by the Dynamic Service Hypergraph;
- Organisational services are designed as service centers in the architectural design discipline;
- Services' realization environment is open and adaptable through the use of a reinforcement learning alogrithm (Sutton, & Barto, 1998) and a probabilistic reputation model (Fouss, Achbany, & Saerens, 2008).

The process description that follows uses the *SPEM*(*Software Process Engineering Metamodel*) notation. Each discipline (architectural design, detailed design and implementation) is modeled as a workflow in which each *ProcessRole* performs a series of *Activities*. *WorkProducts* (Documents or Models) are inputs or outputs of those activities. A complete specification of the SPEM notation can be found in (OMG, 2008) while the SPEM description of the disciplines is described in (Wautelet, 2008; Wautelet, Kolp & Achbany, 2006; Achbany, Wautelet & Kolp, 2008).

3.2. Generic Process Description

As exposed previously, ProDAOSS is a process plug-in for the development of adaptable and open service systems. Thanks to the flexibility it offers it could be adopted/adapted in any MAS development methodology, however for the reasons exposed in (Achbany, Wautelet, & Kolp, 2008) it will be exposed here as a plug-in to the I-Tropos methodology.

ProDAOSS defines a series of custom models and other WorkProducts in various disciplines – i.e. development stages – for service-oriented development. The I-Tropos disciplines in which the ProDAOSS' models are included are depicted in Figure 1 as a SPEM disciplines process package. Refinements brought to these disciplines are studied into the next section, we summarize the contributions of the plug-in as:

- At Organisational Modelling and Requirements Engineering stages, the inclusion of the FaMOS framework (Wautelet, Achbany, & Kolp, 2008). This framework allows service-oriented analysis by refining the traditional i* approach;
- At Architectural Design stage, the adjunction of the service center architecture defines a custom architectural pattern for the development of the analyzed services as service centers;
- At Detailed Design, Task Allocation and Reputation algorithms subscribes the adaptable and open characters of the process' designed solution.

These disciplines addendums are studied in further details into the rest of this section.

3.2.1. Analysis

I-Tropos' analysis disciplines are Organisational Modelling and Requirements Engineering. During the first one, the emphasis is put on understand-

Figure 1. The ProDAOSS process: Main WorkProducts contributions

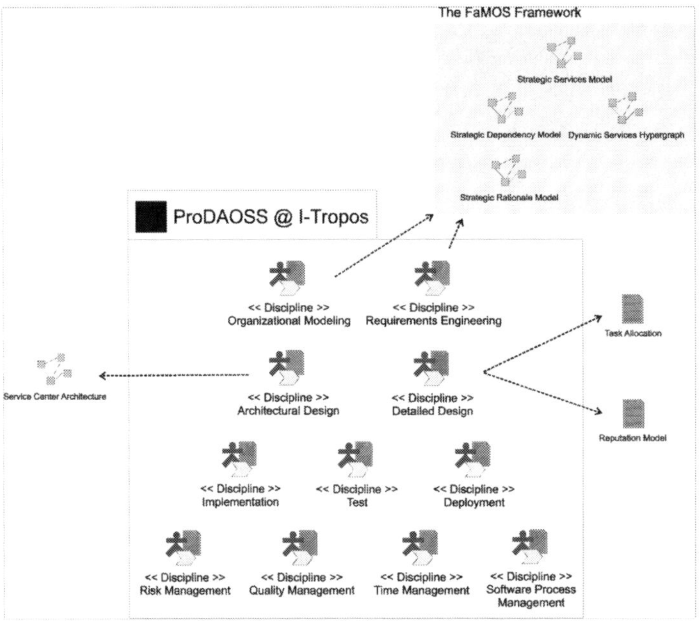

ing the motivation and rationale that underlie system requirements. The second extends models created previously by including the system tobe, modelized as one or more actors. The system dependencies contribute in meeting goals of the stakeholders identified in Organizational Modeling discipline.

Those stages are related so that the models developed during them are the same but with a different focus so that they are similarly structured in terms of process realisation. Figures 2 and 3 depict the WorkDefinion of Organisational Modelling and Requirements Engineering respectively, after the Stakeholder Analysis, the Strategic Analysis formalizes their behaviour, the models introduced

Figure 2. Organizational modeling WorkDefinitions

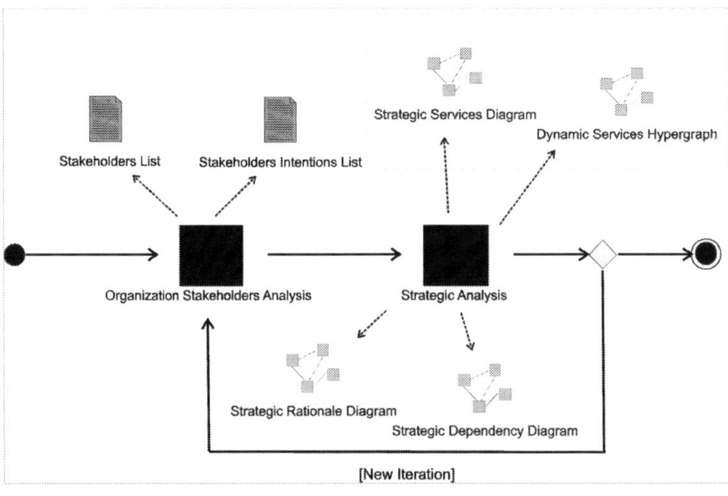

Figure 3. Requirements engineering WorkDefinitions

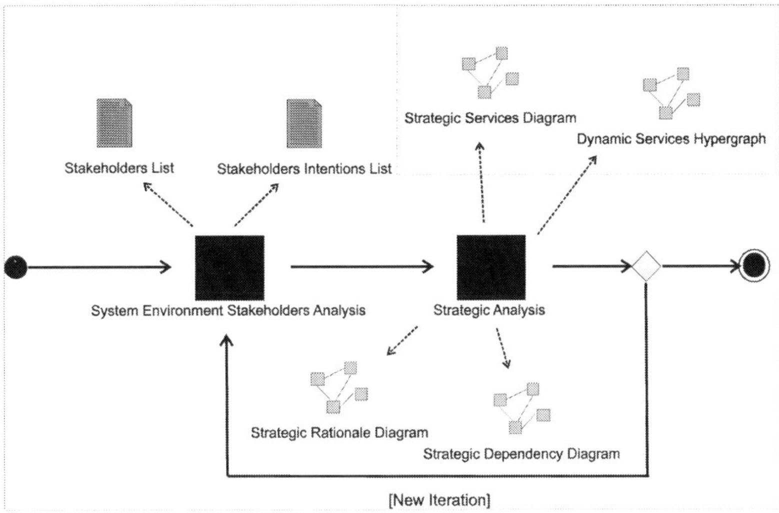

by the ProDAOSS process are included in the red rectangle.

Refinements of ProDAOSS compared to the original process mostly concentrate on the Strategic Analysis. The later is, indeed, concerned with the adoption of the FAMOS framework (Wautelet, Achbany, & Kolp, 2008). FAMOS offers a multiple-view service oriented modelling framework, its adoption into the I-Tropos process constitutes the way to include service oriented computing into I-Tropos' MAS analysis. Models are developed in a more structured way than in the original process since the Strategic Services Diagram offers a more aggregate and complete view of the software system to develop. The Strategic Dependency and Strategic Rationale Diagrams represent the rationale of the agents in a static manner while the Dynamic Services Hypergraph represent the paths of atomic activities to fulfil a service.

3.2.2. Architectural Design

The Architectural Design discipline aims to build the system's architecture specification, by organizing the dependencies between the various sub-actors identified so far, in order to meet functional and non-functional requirements of the system.

The most important advancement in matter of architectural design in the ProDAOSS process is the inclusion of the Service Center Architecture (SCA). The SCA is a design architecture for implementing open and adaptable MAS. In a model-driven approach, the services distinguished at analysis stages (OM and RE) are mapped to the SCA as service centers.

As shown in Figure 4, the ProDAOSS process refines the I-Tropos architectural design disciplines with the SSD and DSH in the Strategic Analysis, but also with the SCA in the System Architecture Design WorkDefinition. The Strategic Analysis is similar to the one described in the previous section. This WorkDefinition has been included into this discipline because the (eventual) inclusion of an Architectural Style into the project refines the developed analysis diagrams.

3.2.3. Detailed Design

The Detailed Design discipline aims at defining the behavior of each architectural component in further detail. This discipline is concerned with the specification of the agent micro level taking

Figure 4. Architectural design WorkDefinitions

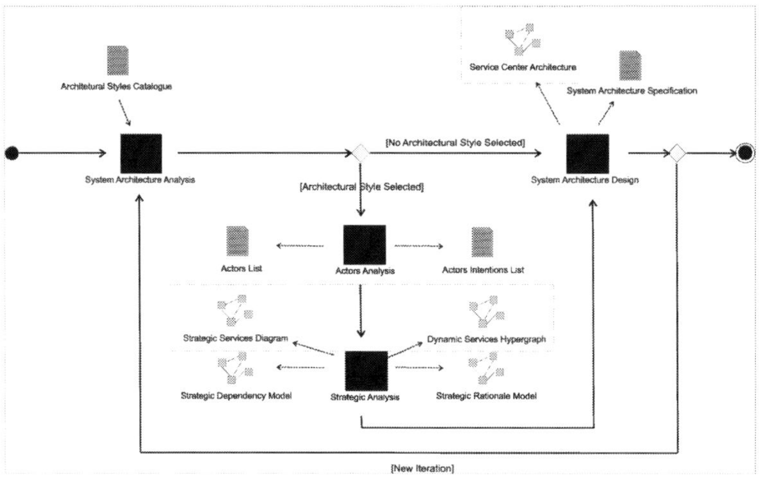

into account the implementation platforms. The objective is to perform a design that will map directly to the code.

In the ProDAOSS perspective, the detailed design discipline mainly describes the SCA into further details notably through the use of task allocation and reputation algorithms.

As shown in Figure 5, the ProDAOSS process refines the I-Tropos detailed design disciplines with the Task Allocation and Reputation Models. Those are taken as input into the Agent Behavior Description.

3.3. Outbound Logistics

The model presented hereafter is a process and interaction view of a supply chain with respect to outbound logistics. Outbound logistics is the process related to the movement and storage of products from suppliers to end users. The model incorporates the processes realized at the actor's level and the interaction between the different actors. The individual actors, however, can play different roles. For example, a producer will sometimes be a shipper, when he fulfills orders from its clients, and sometimes be a final client, when he has ordered shipments from its own suppliers. Therefore, the model is presented in terms of roles played by actors.

In the processes involved in outbound logistics, four major roles can be identified:

1. ***The shipper:*** This role is taken by an actor who needs transportation for a shipment. He has received an order from a client and has to deliver this order. The shipment has its own characteristic in terms of volume, transportation constraints, pickup and delivery dates, transportation mode and the shipper has its own expectation on the price he should be charged. The shipper will turn towards one or several carriers to realize that shipment.
2. ***The final client:*** This role is taken by an actor who has ordered a shipment from a supplier and awaits its delivery. The client is generally not involved in the choice of the carrier, but is impacted by the planned delivery date and the delays that could result in the realization of the transportation.
3. ***The carrier:*** This role is taken by an actor owning transportation capacity who receives transportation orders from shippers for the realization of shipments to final clients. The core business of the actor having this role is

Figure 5. Detailed design WorkDefinitions

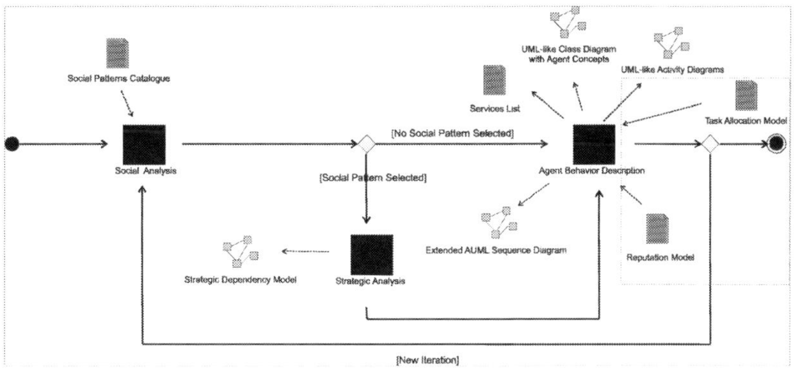

the transportation. As such, he must realize these orders while maximizing his profit.

4. ***The logistic infrastructure manager:*** This role is taken by an actor who owns the logistic infrastructure which is needed by carriers such as hubs, rail tracks, airports, docks,… The utilization of this infrastructure is rented to the carriers while other services could be offered (transshipments, cross-docking, storage,…).

Our generic process and interaction diagram focuses on collaboration. Collaboration consists of two mechanisms.

First, information sharing allows visibility between the actors in the supply chain. This information can generate value for each actor since it allows them to better plan their own operations and optimize their own decisions. Information sharing can appear at every stage. At strategic level, announcing its own planned requirements and developments will allow the other actors to position themselves better in their own strategic decisions. Tactically, it raises the opportunity to better fit the needs of its clients and suppliers. Operationally, it allows all actors to be kept informed of the realization of one's operations and adapt more rapidly to problems encountered by one actor that could affect the others.

A second and more subtle aspect of collaboration is coordination. If information sharing has been a current practice in a supply chain for a long period of time, and trust has emerged between the actors, they can consider moving further to coordinate decisions. In this case, decisions are not only taken based on internal and shared information in order to improve its own efficiency, but are discussed and taken in order to optimize some processes more globally. The goal is not only to take the decisions that are favorable for one actor but also decisions that favorable for its clients and suppliers. In this case, it is the efficiency of the chain itself that can improve.

For each role, the processes involved in outbound logistic can be categorized into different levels. At strategic level, the actors take long term decisions determining their relations to the other actors, fixing their long term objectives, aiming at matching their capacity to the market requirements and defining adequate corporate policies. At tactical level, the actors take medium term decisions which aim at achieving the strategic goals. Those decisions define the structure in which operations will take place. At operational level, day-to-day decisions are taken which direct the logistic flows. At each level, interactions occur between the actors playing the different roles.

On the generic process and interaction diagram, the processes and interactions are defined in a generic way, which could be extended further by defining for each process the tasks that are involved. This tasks definition gives rise to an extended model which is beyond the scope of this chapter. The processes are categorized by roles and levels.

At the strategic level, each player is involved with his own strategic analysis and defines its strategic goals. He determines his offer, his markets, his objectives and future investments, and his corporate strategies. He realizes global supply and demand planning and turns to long term business partners.

The shipper will evaluate his future transportation needs. He will determine the type of transportation he will require the characteristics in terms of volume, specific constraints, reliability, quality, schedules, frequency and expected fares. He will scan the market for appropriate carriers to take care of his transportation activities and may eventually enter into long term contracts with those carriers.

The carrier will decide on which market he will be present and with which offer. In this model, we use the term « presence » to refer to a strategic offer made by a carrier. A presence is the expression of this offer. It can define the destinations he serves, the capacity he makes available on specific destinations pairs or in specific regions, the transportation mode that is involved, the type of clients he is willing to work with and possibly the tariffs he offers. The presence allows his offer to be visible by the market. He can find shippers reacting to his presence definition with whom he may enter into long term contracts.

The infrastructure manager is involved in a similar strategic process. He evaluates the capacity he plans to offer to the carriers, and the policies for accepting carriers on its infrastructure. He determines how the carriers will use this capacity, under what conditions, and how this will be priced. He determines the development of his infrastructure.

The final client has generally little impact on strategic outbound logistics decisions. In our model, he does not choose the carriers that will be selected, or the paths that will be used by the shipments, although, in practice, he could be involved by his suppliers in those decisions.

At the tactical level, more concrete decisions are taken which aim at implementing the strategy defined at the corporate level. The business units adapt this strategy in terms of policies under which the operations will happen. For example, the shipper may determine rules for allocating concrete requests for shipment to specific carriers, or to realize possible trade-off between reliability, speed and price of the transportation for each kind of shipment. The carrier will define how he will implement his announced presence, by defining connections and paths. Connections are links between two locations on which he plans to have some transports. Defining connections structure his network and allows him to reserve the required capacity on those connections. He will also define paths, which are decisions about how to route the shipments from the origin to the destinations by using a series of connections. The paths can be dependant on the type of shipments.

The operational level deals with concrete shipments that have to be realized. The operational level itself can be decomposed into the scheduling phase and the monitoring phase.

In the scheduling phase, the players plan the short term realization of the outbound logistics. The shipper receiving orders from his clients will determine his day-to-day transportation needs. In our model, those needs are defined as « logistic requests ». Such requests identify the shipment, the origin and the destination, the requested pickup and delivery dates and other constraints specific to the shipment. Once the logistic requests have been identified for a scheduling horizon, the shipper tries to match those requests to a set of carriers, based on the tactical policies of the business unit. The shipper builds, for each request, one or several « transportation calls », which are calls that will be addressed to the carriers to realize a part of the journey of the shipment. The shipper then submits these calls to the relevant carrier.

The carrier receives the transportation calls, and will accept or refuse them. If he accepts them, he may want to reserve the required capacity on the path that will be used by the shipment. According to his tactical policies, he may already allocate the shipment to a series of connections. This allocation allows him to evaluate the delivery date which he can confirm to the shipper. As a new scheduling horizon begins, the carrier's scheduler lists the shipments allocated, or to be allocated, to each specific connection. He verifies the feasibility of the allocation, and modifies it if needed. He creates « transports » to realize the shipments. In our model, a transport is a specific movement of a logistic unit (truck, train, barge,...) on a connection. It has a capacity, a schedule, a list of stops and the logistic unit is identified. The shipments are then allocated to specific transports. The path of each shipment is verified and required terminal operations, such as transshipments, are identified and communicated to the involved infrastructure managers.

The infrastructure manager receives an infrastructure load from the carriers, and schedules his operations and teams to meet such load. In an efficient supply chain, infeasibilities of one player are communicated to the others for schedule adaptations. When communication is absent, infeasibilities generally result in delays.

The monitoring phase consists of checking if everything goes as planned, which is rarely the case. The shipper wants to check where his shipments are, if the delivery dates are still confirmed, if the shipments encounter problems on their route to be aware as soon as possible of any delay or loss of the shipments. The final client should have the same visibility on the shipments he awaits.

The logistic players, i.e. the carriers and the infrastructure managers, can react by taking correctives measures in case of problems in the realization of the schedules. However, this requires, first, a great visibility on the realization of the transports, second, a great flexibility in rescheduling, and, third, good communication with the other logistic partners (See Figure 6).

4. THE COLLABORIATIVE PLATFORM DEVELOPMENT

This section introduces the application of the ProDAOSS process onto the development of a collaborative platform for outbound logistics actors. The development of such a package is however too large to be fully developed here. That is why we will, in the Strategic Services Diagram, document the entire application package in terms of organizational services and focus on only one aspect, Manage Transports, in the rest of the presentation. So that, the Strategic Rationale Diagram, the Dynamic Service Hypergraph - at analysis stage - and the Service Center Architecture - at design stage - will only document this service.

4.1. Analysis

For the application analysis, the ProDAOSS process uses the models included in the FaMOS framework (Wautelet, Achbany, & Kolp, 2009). Firstly, we describe all the services included in the application package through the Strategic Services Diagram. Afterwards, the Strategic Rationale Diagram and the Dynamic Service Hypergraph focus on the service *Manage Transport*.

The Strategic Services Diagram (SSD) of Figure 7 introduces all the services of the application package. This view allows all project stakeholders to share a common aggregate view of the services including their dependency relationships. As evoked previously, due to a lack of space, we will, in the context of this paper, focus on the Manage Transports service. To fulfil this service, the Customer depends of the Scheduler.

The Strategic Rationale Diagram (SRD) of Figure 8 depicts the actors, goals, tasks and resources involved into the Manage Transport service realization. This diagram offers a com-

Figure 6. The outbound logistics workflow

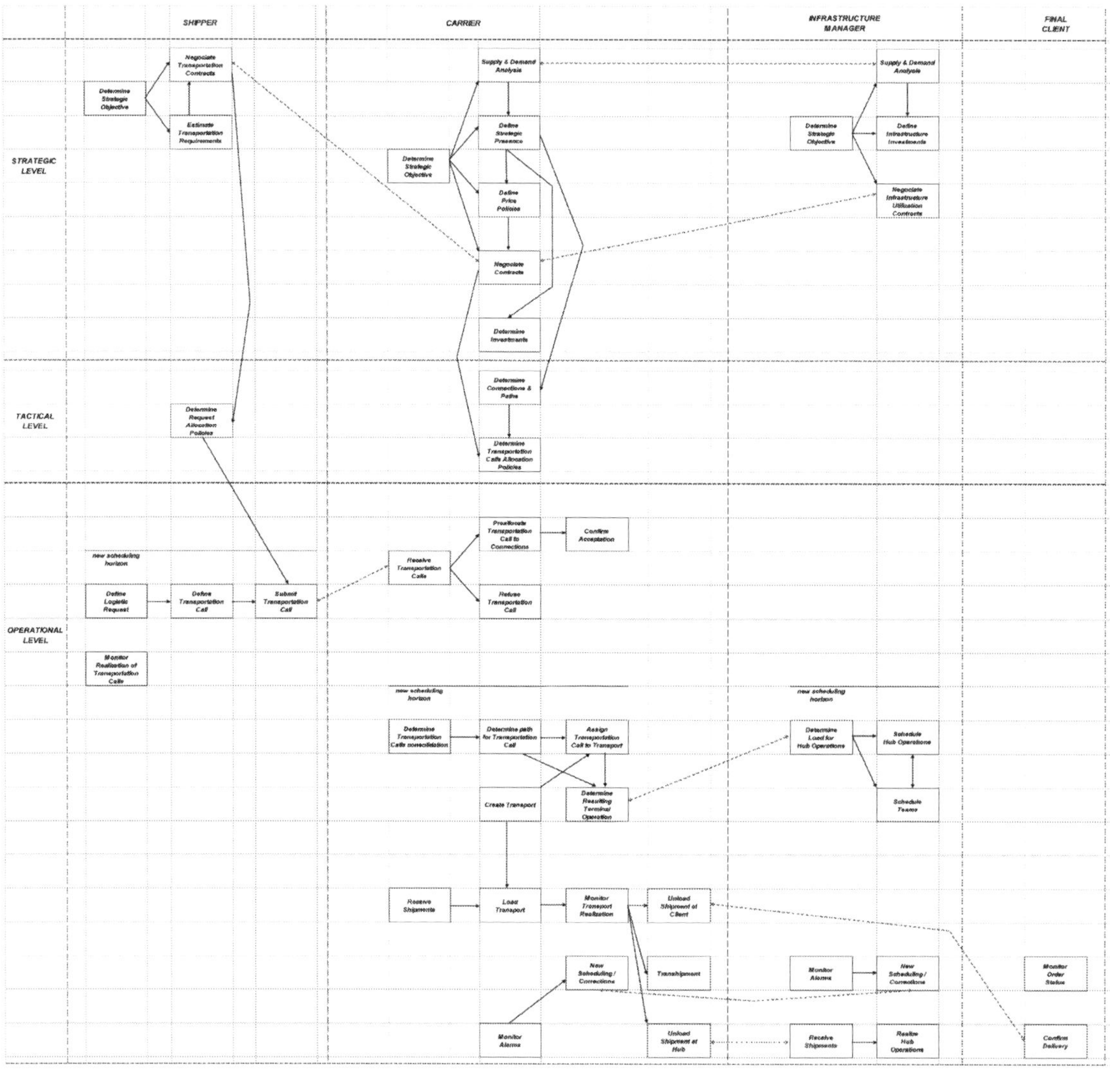

plete static view of the actors' dependencies for resources, goals and tasks. This view is from primary importance for designing the MAS since the agents that will instantiate the involved actors will be the ones that will be assigned the tasks realization into the Manage Service service center at design stage (more information is given into the next section). Seven actors are involved in the process of managing transports with the scheduler as the central one. The scheduler main task is to plan a new horizon: this is achieved by creating transports and assigning the logistic requests transmitted by the Order Representative to those transports (for more information see the realization paths in the paragraph below).

Finally, the Dynamic Service Hypergraph (DSH) offers a complementary view to the SRD presented above by showing the service realization paths. This diagram is also of primary importance in the context of the service center architecture at design stage (see next section).

Figure 9 represents fulfilment paths for the Manage Transport service. Each node is a step in service provision and each edge corresponds to

Figure 7. Application package strategic services diagram

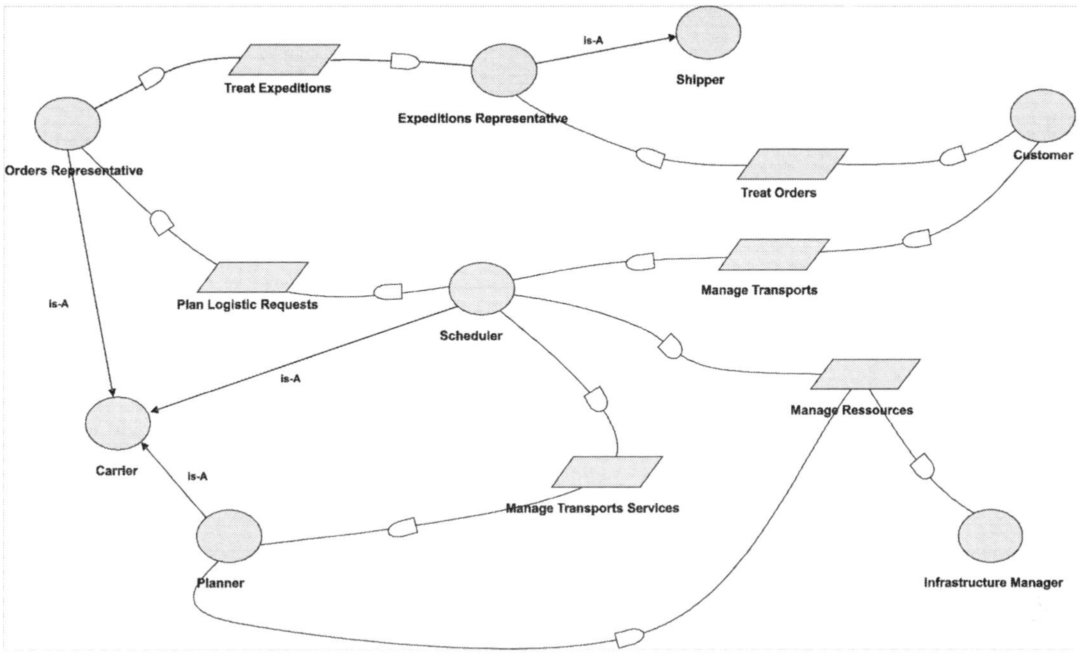

Figure 8. The Manage Transport's strategic rationale diagram

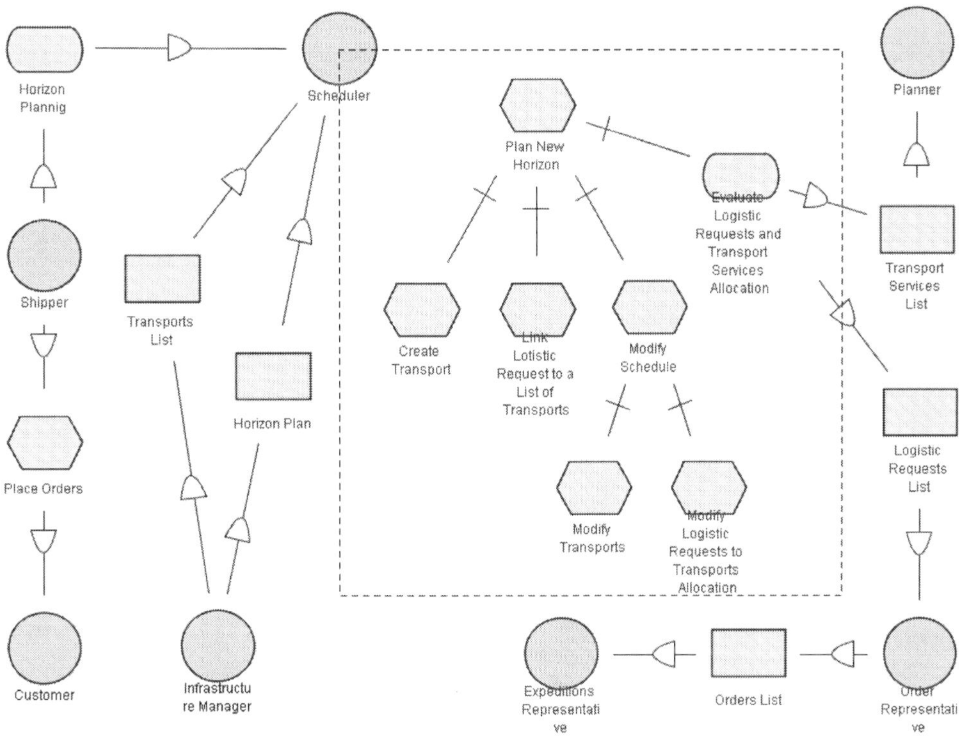

Figure 9. The Manage Transport's dynamic service Hypergraph

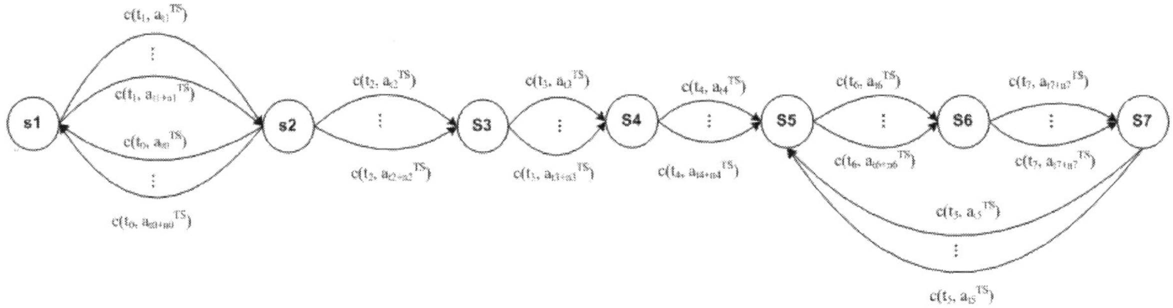

the execution of a task t_k by a specialist agent $a_{k,u}$, where u ranges over specialists that can execute t_k according to the criteria set in the service request. A function of the criteria set $c(t_k,a_{k,u})$, labels each edge and represents the QoS advertised by the specialist $a_{k,u}$ for performing the task t_k. Note that different paths offering different QoS are available. Indeed, as shown in Table 1 path $< t_0, t_1, t_2, t_3, t_4, t_6, t_7 >$ offers alternative ways of fulfilling the service. The following list gives a description of each task used in the Hypergraph; tasks are here limited to a scheduler agent:

4.2 Design

The ProDAOSS process proposes, at architectural design stage, an open, distributed, self-organizing and service-oriented MAS architecture called Service Center Architecture (SCA) first defined in (Achbany, 2009; Jureta et al., 2007b; Jureta et al., 2007a) and specialized/ instancied here for an application to outbound logistics. SCA allows unbiased service provision driven by multiple concerns and tailored to user expectations. Deploying such a system places very specific criteria on the properties of MAS architectures that can support it:

1. A simple architecture would minimize the variety of interactions between the participating agents. Ensuring interoperability would thus not involve high cost on the providers.
2. Internal functions as task allocation, reputation computation, etc, ought to remain outside the responsibility of the entering and leaving agents to avoid bias. The architecture must therefore integrate a special class of agents that coordinate service provision

Table 1. Description of the tasks

Start Node	End Node	Task	Description
S_1	S_2	t_0	It modify requests allocation.
S_1	S_2	t_1	It verifies the logistic requests or transportation services allocation.
S_2	S_3	t_2	It modifies requests allocation.
S_3	S_4	t_3	It creates transport.
S_4	S_5	t_4	It links the logistic requests to a series of transports.
S_5	S_6	t_5	It evaluates the schedule feasibility.
S_6	S_7	t_6	It modifies transports.
S_7	S_5	t_7	It modifies logistic requests attribution to transports.

and that are internal to (i.e., do not leave) the system.

3. Since there is no guarantee that agents will execute tasks at performance levels advertised by the providers, reputation calculation and task allocation should be grounded in empirically observed agent performance and such observation be executed by internal agents.
4. Varying quality of service (QoS) requests and change in the availability of agents require task allocation to be automated and driven by QoS, reputation scores and other relevant considerations (e.g., deadlines).
5. To ensure continuous optimization of system operation, task allocation within the architecture should involve continuous observation of agent performance, the use of available information to account for agent reputation or behavior, and exploration of new options to avoid excessive reliance on prior information.

A MAS architecture answering those 5 criteria is proposed. This organizes agents into groups, called service centers. Each service center specializes in the provision of a single service. Within each center, a task allocation and reputation algorithm are integrated in the proposed architecture.

Figure 10. Logical architecture

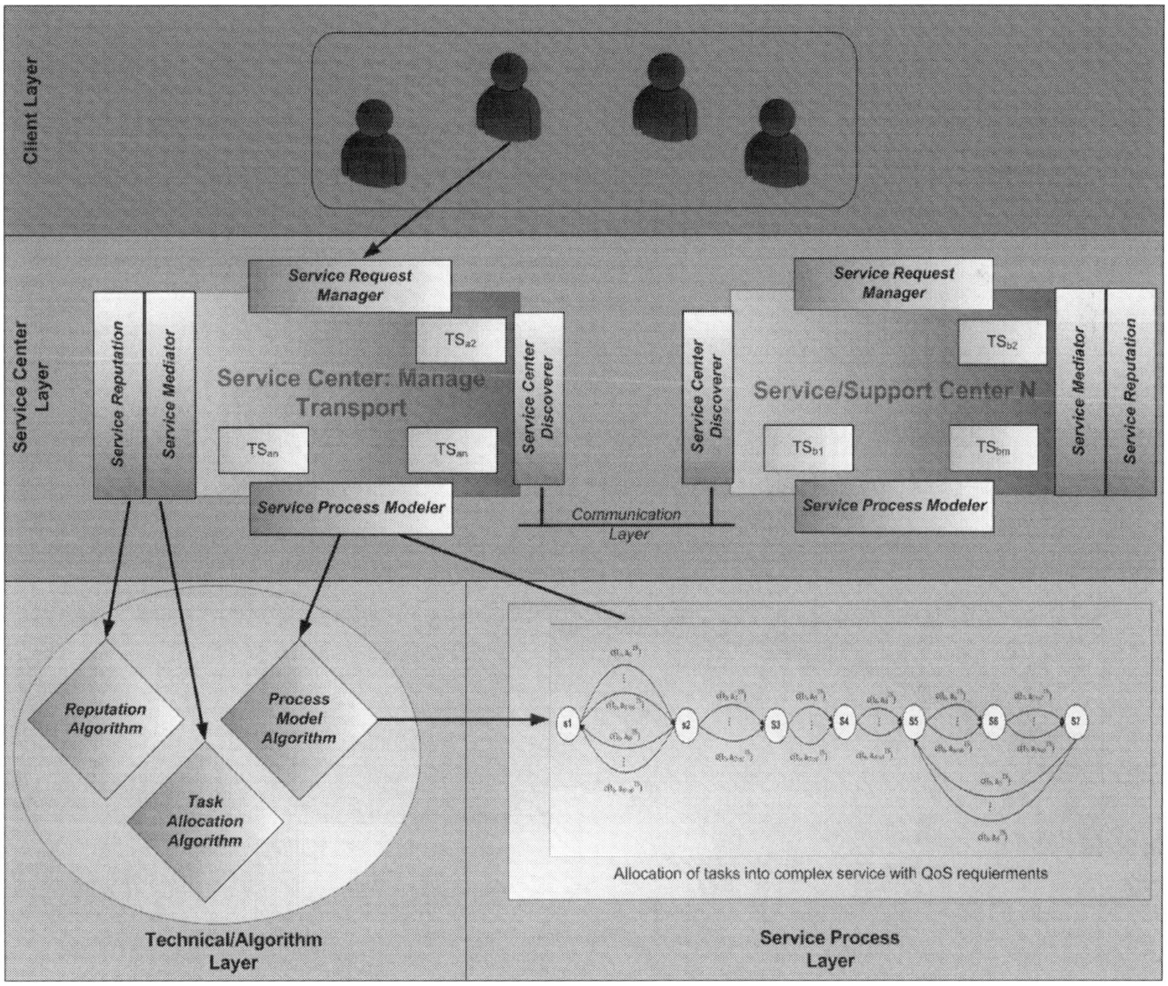

The suggested architecture, the task allocation algorithm and the reputation model allow the building of open, adaptable, distributed, service-oriented MAS, adaptable to changes in the environment by accounting on both experience acquired through system operation and optimal exploration of new task allocation options.

As presented in Figure 10, the logical architecture proposed in this section falls four layers.

4.2.1 Upper Layer: User Client

The upper layer represents user clients (humans or applications) that interact with the middle layer and transmit client requests. More information can be found in the fifth part of this chapter called Implementation.

4.2.2 Middle Layer: Service Center

The middle layer contains the various service centers based on the Service Center Architecture (see (Achbany, 2009; Jureta et al., 2007a; Jureta et al., 2007b) for more details) connected by a communication layer. The Service Center Architecture (SCA) groups services into service centers (SC). Each SC contains all distinct tasks, executed by Task-Specialist agents (TS), needed to provide a service corresponding to a service request originating from the user. As represented in Figure 10, five special agents are also present in each SC:

- The Service Request Manager is responsible to receive and manage a service request. It represents the public interface of a SC. A service request gives: (i) a formula which describes the goal state or a set of goal states (that is, the important thing to do is to achieve any state in which this formula is true), (ii) a set of hard constraints on quantitative QoS criteria and (iii) a QoS criterion to optimize. When a request is received by this agent, it dispatches the request to the service mediator which allocates the different task to the given TS by optimizing the QoS criterion and by respecting the set of hard constraints.
- The Service Mediator composes tasks by observing past performance of individual TS, then subsequently using (and updating) this information through the task Allocation Algorithm (TA)(see (Achbany et al., 2006; Achbany et al., 2008a; Jureta et al., 2007a)). Combining the SCA and the TA brings the following benefits: (a) Adaptability to changes in the availability and/or performance levels of TS is ensured, as the algorithm accounts for actual performance observed in the past and explores new compositions as new TS appear. (b) Continuous optimization over various criteria in the algorithm allows different criteria to guide service composition, while exploitation and exploration ensure Merdiator continually revise composition choices. (c) By localizing composition decisions at each Mediator, the architecture remains decentralized and permits distribution of resources. (d) The architecture and algorithm place no restrictions on the openness of or resource distribution in the system.
- The Service Process Modeler manages the hypergraph corresponding to the Dynamic Service Hypergraph that represents the execution of the service. Indeed, a service can be understood as a process, composed of a set of tasks ordered over the graph representing the service. Each SC has a Task Directory (TD) which is a repository referencing the tasks available in this SC. Each entry of TD is a Task Directory Entry describing a task and all information about it.
- The Service Reputation is responsible to model and compute the reputation scores of all TS. To compute these reputations, it

uses a reputation algorithms presented in (Achbany, 2009 ; Fouss et al., 2008). The computation of the reputation scores is done on the basis of the feedbacks given by the users and these scores represent the internal quality of the TS in executing the corresponding task.

- The Service Center Discovery has the role to explore the network in order to discover the other SC to compose services from different centers in more complex services. The SC representing the service Manage Transport and the corresponding Dynamic Service Hypergraph illustrating its process model are shown in Figure 10. For each task (i.e., each edge composing the Dynamic Service Hypergraph) of the service, we have a set of TS able to execute this task. For the Manage Transport service, we have seven tasks and thus seven set of TS. Each TS is characterized by a reputation score computed by the service reputation. The Mediator uses this reputation score and others quality constraints given by the users in their service requests in order to allocate tasks to fulfil the service.

4.2.3 Bottom Layer: Technical/ Algorithm and Service Process

The bottom layers are the technical/algorithm layer and the service process layer. Technical/algorithm layer contains all the algorithms and technical processes used and executed by the special agent of the service center, i.e., the service process modeler executes a process model algorithm to build the service process corresponding to the global goal of its service center, the service mediator uses a task allocation algorithm (Achbany et al., 2006; Achbany et al., 2008a; Jureta et al., 2007a) to allocate tasks to TS and the service reputation uses the reputation algorithm (Achbany, 2009; Fouss et al., 2008) to compute the reputation score of the TS. By keeping these algorithms and others technical processes in a lower level, we keep independent these technical features from the service center and these technical aspects can be easily changed and adapted without any changing to the higher level, i.e., the service center itself. The service process layer contains business processes, i.e., compositions of tasks in simple services or compositions of simple services in a more complex service. For the service Manage Transport, this layer contains its process model represented by the Dynamic Service Hypergraph as shown in Figure 10. These layers communicate between each other in order to answer clients' requirements.

4.3 Implementation

This section briefly presents a series of interfaces issued of the implementation of the system's Manage Transport service. The Manage Transport service itself is briefly described here with focus on the role of the agents and how they interact (more information about the business workflow is provided in section 3.3). The accent is put on the operational level, for full platform interfaces documentation – including strategic and tactical levels – reefer to (Wautelet, Lange & Tran, 2008). Implementation is based onto the analysis and design disciplines exposed previously.

When a user gets connected to Manage Transport service, the User is instantiated and displays the interface depicted on Figure 11. It allows the new coming user to register on the application. The information provided by the users is handled by the Database agent who checks the validity. Once this has been done, the carrier can perform different types of actions related to the transports. At any moment during the session the user can use a navigation-bar to switch from one to another section.

As overviewed in section 3.3, the shipper determines his transportation needs with respect to the orders he received from his clients so that the shipper agents transmit those needs to the carrier agents in the form of logistic requests. Figure 12

Figure 11. Manage Transport service, logon window

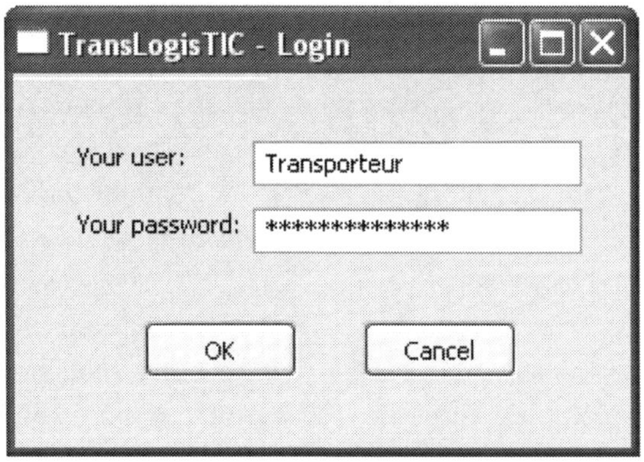

illustrates a shipper window for logistic requests management; note that this window is not part of the Manage Transport service. When identified for a scheduling horizon, those logistic requests are matched to a set of carriers on the basis of the tactical policies of the business unit (which is part of the agent's beliefs).

The shipper agent builds, for each request, one or several transportation calls which are submitted by shipper agents to relevant carrier agents. Figure 13 illustrates the carrier interface for the transportation calls management (they are, on the interface, refeered to as "AT" meaning "Appel à transport", French translation of "transportation call"). Indeed, when one of them receives some

Figure 12. Treat Orders service, logistic requests window

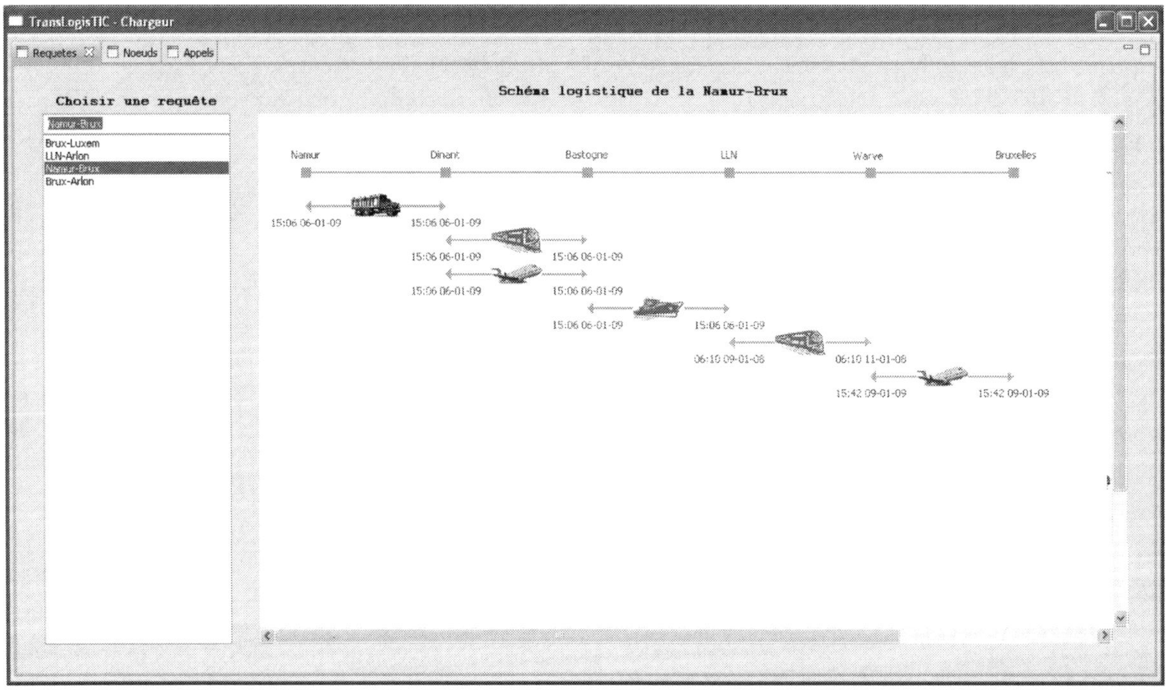

Figure 13. Manage Transport service, transportation calls window

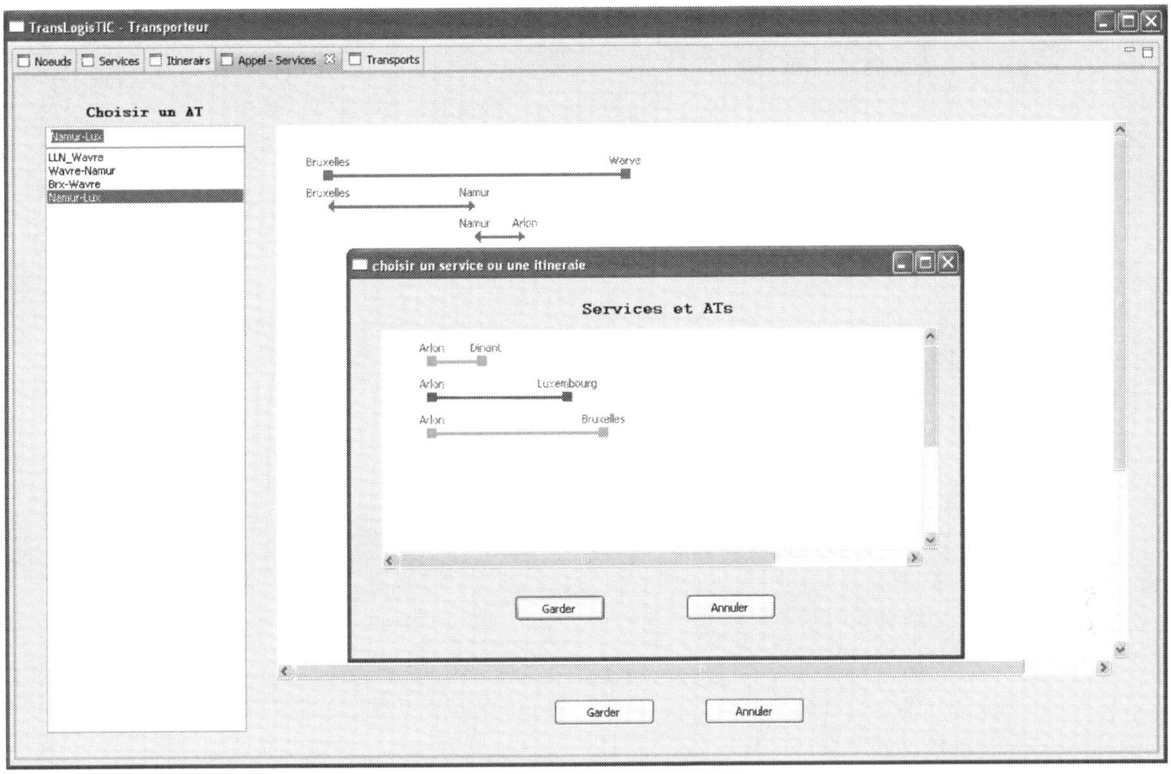

and accepts them he has to reserve the capacity on the path that will be used by the shipment. This window allows him to allocate the shipments to a series of connections.

Facing a new scheduling horizon, the carrier needs to verify feasibility of allocated shipments (by constraint checking in the agents' knowledge base); modifications are eventually performed. The agent then creates transports and places shipments onto to realize them. The interface presented in Figure 14 depicts the transports creation elements at a carrier's disposal. The path of each shipment is verified and required terminal operations such as for example transshipments are computed and transferred to infrastructure manager agents.

5 FUTURE RESEARCH DIRECTIONS

Future research directions include further developments of the service-oriented software development methodology as well as extensions of the supply chain platform itself.

First of all, limitations and further developments of the software process include:

- The process needs to gain experience with its use. In this chapter, we have applied it on a case study. By doing so, we have tailored the process to a particular project and shown how our framework can help the design of MAS. However, it should be tested on more case studies and knowledge base should evolve from project to project;
- Process description does not systematically include guidance for using software patterns, the reuse of the Commercial Off

Figure 14. Manage Transport service, transports window

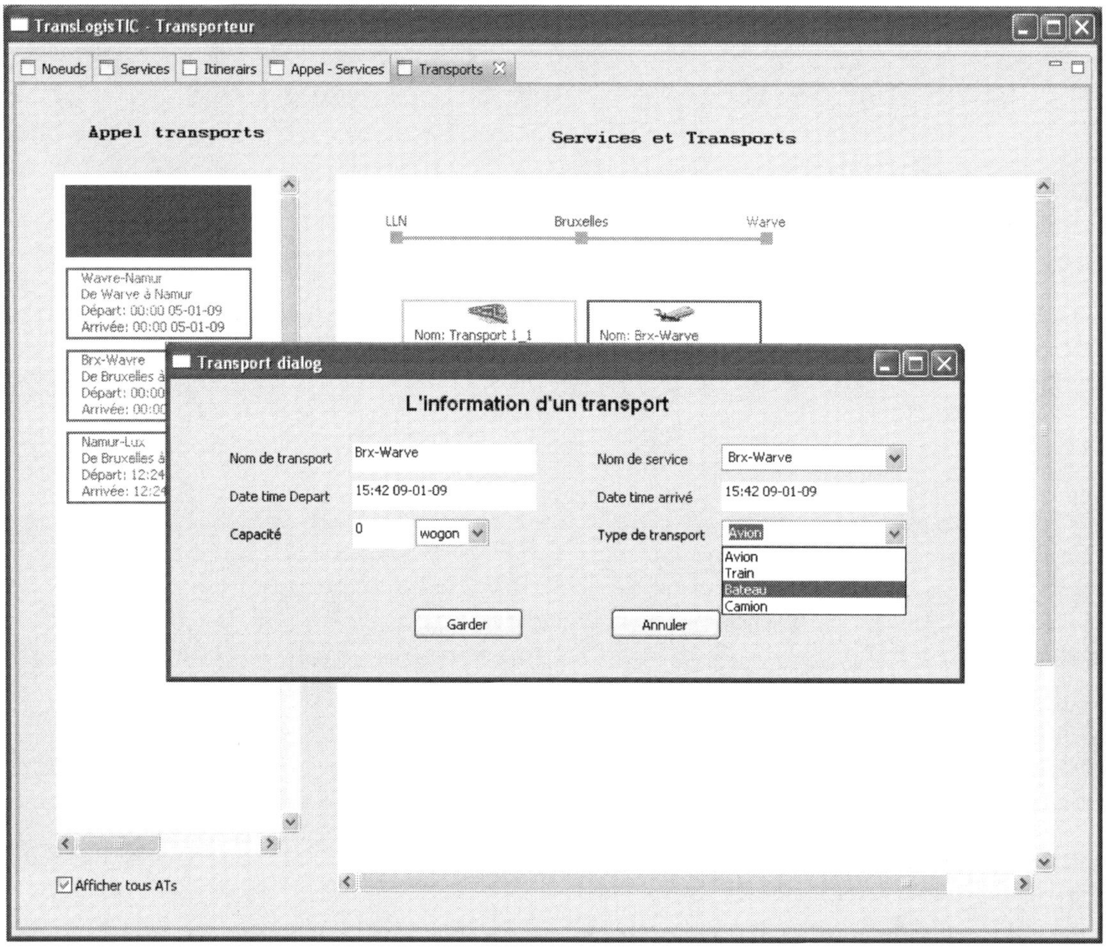

The Shelf (COTS) and open source components; an explicit and systematic approach for selecting and applying those features could be included into the process rather than using an ad hoc manner. Attempts to deal with such issues can be found in (Wautelet, Achbany, Kiv, & Kolp, 2009) and other extensions are currently under development;

- The methodology is mostly designed for large, user intensive, organizational projects. It is especially suited for companies IS development rather than for stand alone developments;
- The methodology could include a systematic guidance for organizational change management. Nowadays organizations are constantly evolving: reorganisations, merges, acquisitions, etc. lead to a need to constantly manage change. ProDAOSS in its actual version is better able to deal with such issues than traditional waterfall methodologies because it has been conceived as a plugin for I-Tropos which is iterative and possess organizational modelling capabilities. Nevertheless we suggest that to adequately manage change and organizational evolutions, one could include a dedicated

discipline for a systemic approach of this topic.

Furthermore, on the platform itself, a series of complementary developments can provide interesting results for global supply chain optimization, more precisely:

- Including connections to third party software and more particularly Enterprise Resource Planning (ERP) systems but also to Fleet Management Systems (FMS), Warehouse Management Systems (WMS) or Transportation Management Systems (TMS) could increase the level of available information (see (Wautelet, Achbany, Kiv, & Kolp, 2009)). The correct exploitation of this information could lead to better optimization of the supply chain;
- Development and inclusion of a series of optimization algorithms including fleet management, logistic containers optimization (see (Lange, & Semal, 2009)), infrastructure resources allocation optimization are currently under development and will be incorporated in the research platform;
- Extension of the platform to more actors and more particularly the inclusion of upstream and downstream to outbound logistics processes for more global optimization constitute promising areas for global supply chain optimization.

6 CONCLUSION

Service-oriented architecture (SOA) promises to change the way business and technology vendors buy, sell, deploy, and manage application portfolios. For the first time, business users will be able to summon applications to support a business process rather than launch a business process constrained by the application. That is why the application of such technology into Supply Chain Management is of primary interest. Indeed, since the environment tends to be more and more competitive and rapidly evolving, flexibility in information systems functionalities is a natural trend of the marked. That is why a supply chain actor requiring functionality to support any supply chain process that can be accessed directly from the vendor's library in the form of a service could represent the business model of the coming years.

More pragmatically, this paper has presented the application of the ProDAOSS process i.e. a process for developing adaptable and open service systems onto the development of an outbound logistics platform. The framework is conceived as a plug-in for the I-Tropos methodology so that it covers the whole development life cycle. In the context of this paper, only the analysis and design stages were presented. At analysis level, flexible business processes were modelled with different complementary views. First of all, an aggregate services view of the whole applicative package is offered; then services are split using an agent ontology – through the i* framework – to design a flexible multi-agent system. A dynamic view completes the documentation by offering the service realization paths. At design stage, the service-center architecture proposes a reference architectural pattern for services realization in an adaptable and open manner. The implementation discipline as presented in the chapter has given the reader a sample of the developed platform by presenting several interfaces of the Manage Transport service part of the Outbound Logistics workflow and the agent-based reasoning underneath.

The application package integration in industrial context is currently under study. It has been conceived to work with a Geographic Information System (GIS) developed by a third party for realtime positioning of transports and dynamic horizon (re)planning in function of collected data. New challenges and further developments will come from the deployment of the platform in such an environment.

ACKNOWLEDGMENT

Most of the research on outbound logistics including the collaborative platform development made at UCL/CESCM and the contents of this paper have been initiated by the Walloon region under the auspices of the TransLogisTIC project (www.translogistic.be). We gratefully acknowledge the region and the project industrial partners for their support.

REFERENCES

Achbany, Y. (2009). *A multi-agent framework for open and dynamic service-oriented systems:investigation and application to web services*. PhD thesis, Université catholique de Louvain, Louvain School of Management (LSM), Louvain-La-Neuve, Belgium.

Achbany, Y., Fouss, F., Yen, L., Pirotte, A., & Saerens, M. (2006). Optimal tuning of continual online exploration in reinforcement learning. In *proceedings of the International Conference on Artificial Neural Networks (ICANN)*.

Achbany, Y., Fouss, F., Yen, L., Pirotte, A., & Saerens, M. (2008a). Tuning continual exploration in reinforcement learning: An optimality property of the boltzmann strategy. *Neurocomputing*, *71*(13-15), 2507–2520. doi:10.1016/j.neucom.2007.11.040

Achbany, Y., Wautelet, Y., & Kolp., M. (2008b). *Process for developing adaptable and open service systems*. Technical Report.

Faulkner, S., Kolp, M., Wautelet, Y., & Achbany, Y. (2006). A formal description language for multi-agent architectures. In *proceedings of the 17th International Workshop on Agent-Oriented Information System*, Hakodate, Japan, 2006.

Fouss, F., Achbany, Y., & Saerens, M. (2008). A probabilistic reputation model. Technical Report.

Ghenniwa, H., Dang, J., Huhns, M., & Shen, W. (2006). emarketplace model: An architecture for collaborative supply chain management and integration. In B. Chaib-draa & J.P. Mller (Eds.) Multiagent-based Supply Chain Management, (pp. 29–62). Berlin: Springer.

Govindu, R., & Chinnam, R. (2007). Mascf: A generic process-centered methodological framework for analysis and design of multi-agent supply chain systems. *Computers & Industrial Engineering*, *53*, 584–609. doi:10.1016/j.cie.2007.06.003

Jureta, I. J., Faulkner, S., Achbany, Y., & Saerens, M. (2007a). Dynamic task allocation within an open service-oriented mas architecture. In *Proceedings of the 2007 International Conference on Autonomous Agents and Multiagent Systems (AAMAS'07)*.

Jureta, I. J., Faulkner, S., Achbany, Y., & Saerens, M. (2007b). Dynamic web service composition within a service-oriented architecture. In *Proceedings of the International Conference on Web Services (ICWS'07)*.

Kolp, M., & Wautelet, Y. (2007). *DesCARTES Architect: Design CASE Tool for Agent-Oriented Repositories, Techniques, Environments and Systems*. Louvain School of Management, Université catholique de Louvain, Louvain-la-Neuve, Belgium. Retrieved from http://www.isys.ucl.ac.be/descartes

Kruchten, P. (2003). The Rational Unified Process: An introduction. Boston: Addison Wesley.

Lange, J. C., & Semal, P. (2009). Return flows of empty logistic containers for a single factory. In *proceedings of the 8ième Congrès International de Génie Industriel*, accepted for publication.

Moyaux, T., Chaib-draa, B., & D'Amours, S. (2006). Supply chain management and multi-agent systems: An overview, in multiagent-based supply chain management, (pp 1–27). Chaib-draa B. & Mller J. P. (eds.). Berlin: Springer.

Object Management Group. (2007). *The Software Process Engineering Metamodel Specification. Version 1.1*. Retrieved from http://www.omg.org/technology/documents/formal/spem.htm

OMG, (2008). Software and systems process engineering meta-model specification. Version 2.0.

Pache, G., & Spalanzani, A. (2007). La gestion des chanes logistiques multi-acteurs: perspectives stratgiques. Grenoble, France: Presses Universitaires de Grenoble (PUG).

Penserini, L., Kolp, M., Spalazzi, L., & Panti, M. (2004). Socially-based design meets agent capabilities. In *Proceedings of the IEEE/WIC/ACM International Conference on Intelligent Agent Technology - IAT 2004*. Beijing, China: IEEE CS Press.

Samii, A. K. (2004). Stratgie logistique, supply chain management: Fondements – méthodes - applications. Dunod.

Supply Chain Council. (2005). *Supply-chain operations reference-model (scor) - version 7.0 [overview]*.

Sutton, R. S., & Barto, A. G. (1998). Reinforcement learning: An introduction. Cambridge: MIT Press.

Wautelet, Y. (2008). *A Goal-Driven Project Management Framework for I-Tropos Multi-Agent Iterative Software Development*. PhD thesis, Louvain School of Management, Université catholique de Louvain, Louvain-la-Neuve, Belgium.

Wautelet, Y., Achbany, Y., Kiv, S., & Kolp, M. (2009). A Service-Oriented Framework for Component-Based Software Development, An i* Driven Approach. In *proceedings of the 11th International Conference on Entreprise Information Systems (ICEIS)*, Milan.

Wautelet, Y., Achbany, Y., & Kolp, M. (2008). A service-oriented framework for mas modeling. In *proceedings of the 10th International Conference on Entreprise Information Systems (ICEIS)*, Bacelona.

Wautelet, Y., Kolp, M., & Achbany, A. (2006). *I-Tropos, An Iterative SPEM-Centric Software Project Management Process*. Working Paper IAG Series 13/06, Louvain School of Management, Université catholique de Louvain, Louvain-la-Neuve, Belgium.

Wautelet, Y., Lange, J.-C., & Tran, V. (2008). BW3 Plateforme de services, Groupe de travail WP3.4 Mise au point des services à la chaîne d'approvisionnement en amont et aval. TACHE 3.4.1 Etude des données collectées automatiquement, développement des applications. TransLogisTIC project, TLTIC-BW3-T341, Walloon Region.

Wooldridge, M., Jennings, N. R., & Kinny, D. (2000). The Gaia methodology for agent-oriented analysis and design. *Journal of Autonomous Agents and Multi Agent Systems*, *3*(3), 285–312. doi:10.1023/A:1010071910869

Yu, E. (1995). *Modeling Strategic Relationships for Process Reengineering*. PhD thesis, University of Toronto, Department of Computer Science, Toronto, Canada.

Selected Readings

Chapter 12
Applying Dynamic Causal Mining in Health Service Management

Yi Wang
Nottingham Trent University, UK

ABSTRACT

This article describes an application that illustrates the role of data mining technology in identifying hidden causal knoledge from health and medical data repositories. Across the health care and medical enterprises, a wide variety of data is being generated at a rapid rate. Current information technologies tends to focus on a more statical side of causal knowledge and do not address the dynamic causal knowledge. This article shows that the dynamic causal relation data can be captured for treatment, payment, operations purposes and administrative directed insights. Accessing this currently unrealized knowledge potential would enable the delivery of actionable knowledge to medical practitioners, healthcare system managers, policy planners and even patients to make a significant difference in overall healthcare.

LITERATURE REVIEW

Medical and Health Management

Patient record management systems is desired in clinical settings (Abidi, 2001; Heathfield & Louw, 1999; Jackson, 2000). The major reasons include physicians' significant information needs (Dawes & Sampson, 2003) and clinical information overload. Hersh (1996) classified textual health information into two main categories: patient-specific clinical information and knowledge-based information.

Knowledge management capabilities have been incorporated in many clinical systems since the 1980s in order to provide a better understanding and management basis. In the *HELP* system, decision logic was stored to allow it to respond to new data entered (Kuperman, Gardner, & Pryor, 1991). The *SAPHIRE* system performs automatic indexing of radiology reports by utilizing the UMLS Metathesaurus (Hersh, Mailhot, Arnott-Smith, & Lowe, 2002). The clinical data repository at Columbia-Presbyterian Medical Center (Friedman, Hripcsak, Johnson, Cimino, & Clayton, 1990) is a database used for decision support (Hripcsak, 1993). Another clinical data repository is the University of Virginia Health System (Schubart & Einbinder, 2000). Case based reasoning also has been proposed in Montani and Bellazzi (2002). Janetzki, Allen, and Cimino (2004) use a natural language processing approach to link electronic health records to online information resources.

APPLYING DATA MINING IN MEDICAL AND HEALTH MANAGEMENT

Data mining has been used to extract diagnostic rules from breast cancer data (Kovalerchuk, Vityaev, & Ruiz, 2001). Data mining has also been applied to clinical databases to identify new medical knowledge (Hripcsak, Austin, Alderson, & Friedman, 2002; Prather, Lobach, Goodwin, Hales, Hage, & Hammond, 1997).

Dreiseitl, Ohno-Machado, Kittler, Vinterbo, Billhardt, and Binder (2001) compare five classification algorithms for the diagnosis of pigmented skin lesions. This is similar to the measurement of algorithmic performances in other areas applications (Yang & Liu, 1999). For example, Acir and Guzelis (2004) apply support vector machines in automatic spike signal detection in ElectroEncephaloGrams (EEG). Kandaswamy, Kumar, Ramanathan, Jayaraman, and Malmurugan (2004) use artificial neural network to classify lung sound signals into different categories.

Chrisman et al. (2003) incorporated biological knowledge and expression data using a Bayesian framework. Imoto et al. (2003) incorporated protein-protein interaction, protein-DNA interactions, and transcriptional factor binding site information. Text mining is used to study of protein interaction network inference (Iossifov et al., 2004).

System Dynamics has a number of strengths that make it especially useful in health care settings (Roberts & Hirsch, 1976). Some of the System Dynamics work in health care has been done by Pugh-Roberts Associates (Hirsch & Miller, 1974). The other area in which early work took place was community mental health (Levin & Roberts, 1976).

The Soundview-Throgs Neck community examined the forces contributing to a rapid rise in heroin addiction (Levin, Kirsch, & Roberts, 1972). A generic model of ambulatory care was developed to analyse economic performance of organization (Hirsch & Bergan, 1973). Another model developed for the State of Minnesota examined the factors affecting the success of new organizations, called Human Service Boards (Hirsch, Bergan, & Frohman, 1974). Two comprehensive System Dynamics models were developed to aid manpower policy formulation in dentistry (Hirsch & Killingsworth, 1975) and nursing (Bergan & Hirsch, 1976).

One model reflected the effects of such factors as the "technology gap" between the hospital and nearby community hospitals (Hirsch, Forsyth, Bergan, & Goodman, 1976). A model helped a medical school examine problems in its relationships with affiliated hospitals and restructure those relationships accordingly (Stearns, Bergan, Roberts, & Cavazos, 1978). Models created for project changes in areas' populations, health care consequences of those changes, and shifts in utilization patterns as a result of changes in resources available, insurance coverage, and various health policies (Hirsch & Henderson, 1977). Several modelling efforts, in fact, were designed to stop at the point where a set of causal diagrams had been completed (Stearns, Bergan, Roberts, & Quigley, 1976).

Dynamic Causal Mining

The Dynamic Causal Mining (DCM) algorithm was discovered in 2005 (Pham, Wang, & Dimov, 2005) using only counting algorithm to integrate with *Game Theory*. It was extended in 2006 (Pham, Wang, & Dimov, 2006) with delay and feedback analysis, and was further improved for the analysis in *Game Theory* with *Formal Concept* analysis (Wang, 2007). *DCM* enables the generation of *dynamic causal rules* from data sets by integrating the concepts of *Systems Thinking* (Senge, Kleiner, Roberts, Ross, & Smith, 1994) and *System Dynamics* (Forrester, 1961) with *Association mining* (Agrewal et al., 1996). The algorithm can process data sets with both categorical and numerical attributes. Compared with other *Association Mining* algorithms, *DCM* rule sets are smaller and more dynamically focused. The pruning is carried out based on polarities. This reduces the size of the pruned data set and still maintains the accuracy of the generated rule sets. The rules extracted can be joined to create dynamic policy, which can be simulated through software for future decision making. The rest of this section gives a brief review of *Association Mining* and *System Dynamics*.

Association Mining and Process Mining

Association mining was discovered by Agrewal, Mannila, Srikant, Toivonen, and Inkeri (1996). It was further improved in various ways, such as in speed (Agrewal et al., 1996; Cheung, Han, Ng, & Fu, 1996) and with parallelism (Zaki, Parthasarathy, Ogihara, & Li, 1997) to find interesting associations and/or correlation relationships among large sets of data items. It shows attributes value conditions that occur frequently together in a given dataset. It generates the candidate *itemsets* by joining the large *itemsets* of the previous pass and deleting those subsets which are small in the previous pass without considering the transactions in the database. By only considering large *itemsets* of the previous pass, the number of candidate large *itemsets* is significantly reduced.

The idea of process mining was investigated in contexts of software processes and workflow (Agrawal, Gunopulos, & Leymann, 1998). Cook and Wolf (1998) propose methods for process discovery in case of software sequential processes. Herbst and Karagiannis used a hidden Markov model in the context of workflow management and other sequential processes (Herbst, 2000; Herbst & Karagiannis, 1999, 2000). Maruster, van der Aalst, Weijters, van den Bosch, and Daelemans (2001) suggest a technique for discovering the underlying process from hospital data assuming that the workflow log does not contain any noisy data. A heuristic method that can handle noise is presented in Weijters and van der Aalst (2001).

System Thinking and System Dynamics

System thinking is based on the belief that the component parts of a system will act differently when isolated from its environment or other parts of the system.

Systems thinking is about the interrelated actions which provide a conceptual framework or a body of knowledge that makes the pattern clearer Senge et al. (1994). It is a combination of many theories such as soft systems approach and system theory (Coyle, 1996). Systems thinking seeks to explore things as wholes, through patterns of interrelated actions.

System Dynamics can be defined as "a qualitative and quantitative approach to describe model and design structures for managed systems in order to understand how delays, feedback, and interrelationships among attributes influence the behaviour of the systems over time" (Coyle, 1996), or "the purpose of the model is to solve a problem,

Figure 1. DCM process

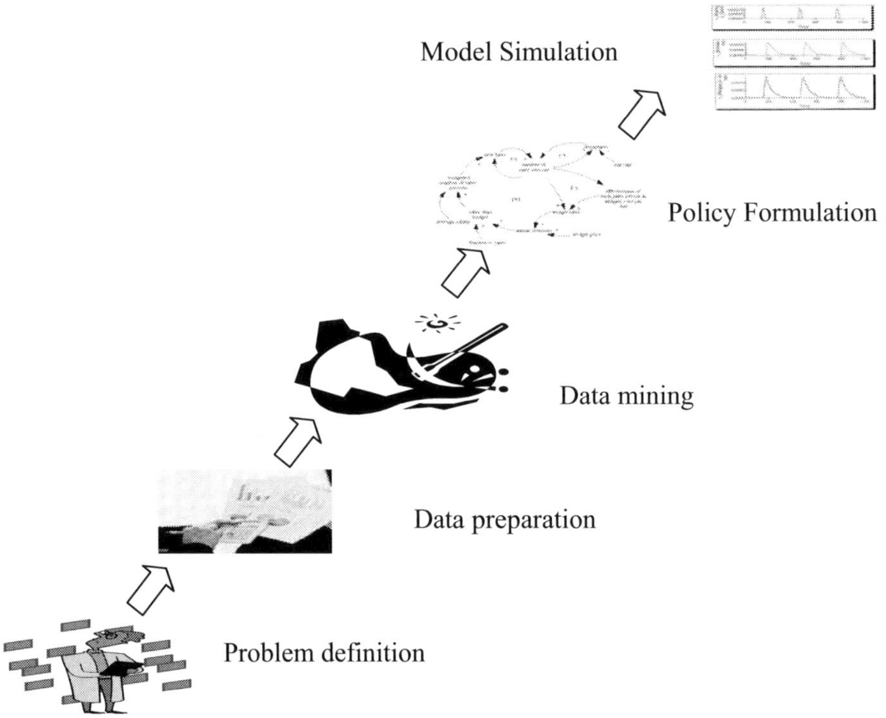

not simply to model a system. The model must simplify the system to a point where the model replicates a specific problem"(Sterman, 1994).

System dynamics is a tool to visualize and understand such patterns of dynamic complexity, which is built up from a set of system archetypes based on principles in *System thinking* (Sterman, 2000). *System dynamics* visualizes complex systems through causal loop diagrams. A causal loop diagram consists of a few basic shapes, that together describe the action modelled.

System dynamics addresses two types of behaviour, *sympathetic* and *antipathetic* (Pham et al., 2005). *Sympathetic behaviour* indicates an initial quantity of a target attributes starts to grow, and the rate of growth increases. *Antipathetic behaviour* indicates an initial quantity of a target attributes starts either above or below a goal level and over time moves toward the goal.

BASIC CONCEPTS OF DYNAMIC CAUSAL MINING

This section provides a general framework for *DCM.* It is an iterative and continual process of mining rules, formulating policies, testing, and revision of the models. The stages are depicted in Figure 1.

- **Stage 1: Problem definition.** In this phase, the problem is identified and the key variables are given. Also, the time horizon is defined so that the cause and effects can be identified.
- **Stage 2: Data preparation.** Data are collected from various sources and a homogeneous data source is created to eliminate the representation and encoding differences.
- **Stage 3: Data mining.** This stage involves transforming data into rules. This thesis suggests using *DCM* as a data mining tool.

The details of *DCM* are explained in the following sections and next article.

- **Stage 4: Policy formulation.** Policies are groups of the rules extracted by mining techniques. Policies improve the understanding of the system. The interactions of different policies must also be considered since the impact of combined policies is usually not the sum of their impacts alone. These interactions may reinforce each other or have an opposite effect. The policy can be used for behaviour simulation to predict the future outcome.
- **Stage 5: Model Simulation.** This stage tests the accuracy of the policies. The policies will predict results for new cases so the managers can alter the policy to improve future behaviour of the system. It is necessary to capture the appropriate data and generate a prediction in real time, so that a decision can be made directly, quickly, and accurately.

DATASET

To find dynamic causality among a set of attributes means to identify correlation and interdependencies between them. The *DCM* algorithm is a way of describing the state of a target system as it evolves in time. It discovers dynamic causality in a data set by matching the dynamic behaviour between separated attributes.

Time Stamp

Definition 1: *A dynamic time stamp is created from two time stamps. Consider two time stamps t_i and t_{i+1}. The dynamic time stamp Δt_i is equal to the difference between two consecutive time stamps.*

$$\Delta t_i = t_{i+1} - t_i \qquad (1)$$

Time stamps are used for identifying the range of variables. The size of each time stamp is selected by the specific need and may vary in different situations. For instance, a time stamp for an increase in production may be in the order of months, while for a change in a cell may be in the order of milliseconds. The time stamp also can help to determine how detailed the variables need to be. The attribute may increase or decrease dramatically if the time stamp is in the order of seconds, however it may be assumed to be constant if the time stamp is in the order of years. All the time stamps should be of uniform length. In order to carry out *DCM*, the time stamps are summarised or partitioned into equal-sized time stamps. A time stamp is useful for describing and prescribing changes to the systems and objects.

Data

Definition 2. *A dynamic attribute is the change or the difference between two attribute values with consecutive time stamps. The two types of value do not have the same nature. Let D denote a data set which contains a set of n records with attributes $\{A_1, A_2, A_3, \ldots A_m\}$, where each attribute is of a unique type (for example; cost of medicine, treatment volume, inventory volume, etc). Each attribute is associated with a time stamp t_i, where $i = \{1,2,3,\ldots n\}$. Let D_{new} be a new database constructed from D such that dynamic attribute $\Delta A_{m,\Delta t_i}$ in D_{new} is given by:*

$$\Delta A_{m,\Delta t_i} = A_{m,t_{i+1}} - A_{m,t_i} \qquad (2)$$

Table 1. Original database D

Time	A_1	A_2
1	9	2
2	17	3
3	10	12
4	4	16
5	7	24

Table 2. Derived database Dnew

Δt	ΔA_1	ΔA_2
Δt_1	+8	+1
Δt_2	-7	+9
Δt_3	-6	+4
Δt_4	+3	+8

where m identifies the attribute of interest.

The classical *Association Mining* algorithms can be applied only to data in the original form (attribute form), for example, in the market basket problem (Agrawal et al., 1993) the focus is on the items of each purchase. On the other hand, *DCM* is interested in the dynamic changes between data. To apply *DCM*, the records are arranged in a temporal sequence ($t = 1, 2, \ldots, n$). Definition 2 is only for numerical attributes and the causality between categorical attributes in *D* can be identified by examining the differences of corresponding changes in attribute values. An example of such a database is shown in Table 1. A_1 and A_2 represent attributes, such as from a tax database like income and tax.

Definition 3. *In the case of categorical attributes, the dynamic causal attributes can be identified by joining the polarities of corresponding changes in attribute values. Let* D_{new} *be a new data set constructed from D such that attribute* $\Delta A_{m,\Delta t_i}$ *in* D_{new} *is given by:*

$$\Delta A_{m,\Delta t_i} = \text{join}(A_{m,t_{i+1}}, A_{m,t_i}) \quad (3)$$

where join is a function combining $A_{m,t_{i+1}}$ and A_{m,t_i}. For example, (t=1, $A_{m,t_{i+1}}$ = Red) and (t=2, A_{m,t_i} = Blue) then (Δt =1, $\Delta A_{m,\Delta t_i}$ = RedBlue).

The *attribute* is gathered first and the *dynamic attribute* is derived from the *attribute*. The *dynamic attribute* identifies the significant relationship between the dynamics of the *attribute*.

MEASUREMENTS

Since the input of the *DCM* algorithm can be quite large, it is important to prune away the redundant attributes.

Definition 4. *A polarity indicates the direction of a change of an attribute. There are three types of polarity (+, -, 0); where + indicates an increase, - indicates a decrease, and 0 indicates neutrality, that is, no change at all.*

Definition 5. *A polarity combination is a joint set of two or more polarities. The simultaneous presence of combinations (+,+) and (-,-) indicates sympathetic changes and will produce a sympathetic rule. The simultaneous presence of combinations (+,-) and (-,+) indicates antipathetic changes and produces antipathetic rules. There are four different combinations of polarity (+,-) (antipathetic negative), (+,+) (sympathetic positive), (-,+) (antipathetic positive), and (-,-) (sympathetic positive) to indicate the degree of causality.*

This differs from the classical causal loops relation which has only + and -, due to a simultaneous increase of an attribute set not automatically leading to a simultaneous decrease of the same set.

Definition 6. *A support is the ratio of records of a certain polarity combination over the total number of records in the dynamic attributes. Three supports are applied in DCM; sympathetic support, antipathetic support, and single support. For data set* D_{new} *and any two attributes* $\Delta A_{1,\Delta t_i}$ *and* $\Delta A_{2,\Delta t_i}$ *the three kinds of supports are defined as follows:*

Sympathetic Support

$$(\Delta A_{1,\Delta t_i}, \Delta A_{2,\Delta t_i}) = \frac{freq\,(+,+)}{n} \quad (4a)$$

or

$$\frac{freq\,(-,-)}{n} \quad (4b)$$

Figure 2. Graph of two dynamic attributes

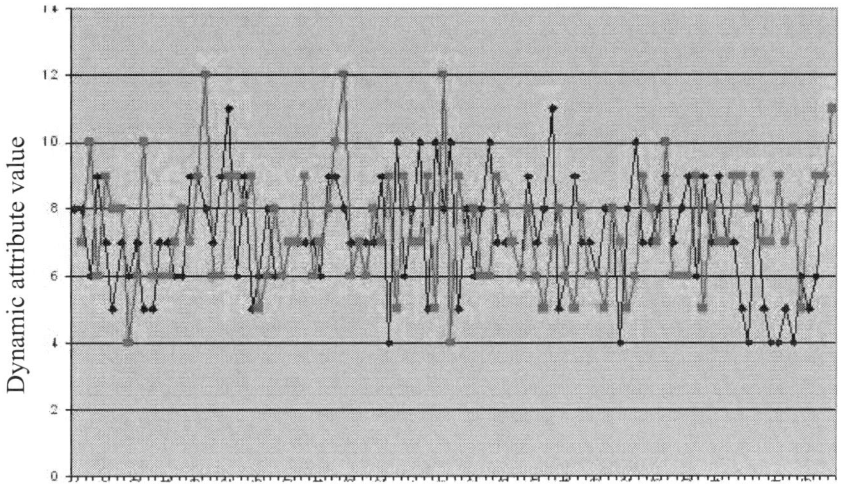

Antipathetic Support

$$(\Delta A_{1,\Delta t_i}, \Delta A_{2,\Delta t_i}) = \frac{freq\,(+,-)}{n} \qquad (5.a)$$

or

$$\frac{freq\,(-,+)}{n} \qquad (5.b)$$

Single Attribute Support

$$(\Delta A_{m,\Delta t_i}) = \frac{freq(+)}{n} \qquad (6.a)$$

or

$$\frac{freq(-)}{n} \qquad (6.b)$$

or

$$\frac{freq(0)}{n} \qquad (7)$$

Where freq (+,+) is a function counting the number of times where an increase in $\Delta A_{1,\Delta t_i}$ is associated with a simultaneous increase in $\Delta A_{2,\Delta t_i}$, freq (-,-) is a function counting the number of times where an decrease in $\Delta A_{1,\Delta t_i}$ is associated with a simultaneous decrease in $\Delta A_{2,\Delta t_i}$, freq (-,+) is a function counting the number of times where an decrease in $\Delta A_{1,\Delta t_i}$ is associated with a simultaneous increase in $\Delta A_{2,\Delta t_i}$, and freq (+,-) is a function counting the number of times where an decrease in $\Delta A_{1,\Delta t_i}$ is associated with a simultaneous decrease in $\Delta A_{2,\Delta t_i}$.

All *supports* relate to the frequencies of the occurring patterns. For a given user specified *support*, the problem of *DCM* is to find all rules where the support is greater than the user defined *support*. The *support* is the frequency of occurrences of attribute sets that support a rule.

Figure 2 shows the results of the comparison between two dynamic attributes. The *X*-axis represents the time stamp and the *Y*-axis represents the value of the dynamic attributes. Figure 3 shows the graphical representation of two dynamic attributes, where the polarity combination is indicated.

Definition 7. *A support level is the value threshold for each dynamic attribute. Every record in a dynamic attribute must have an absolute value larger or equal to the support level in order to be considered as candidate for the dynamic rule.*

Figure 3. Illustration of polarity combination

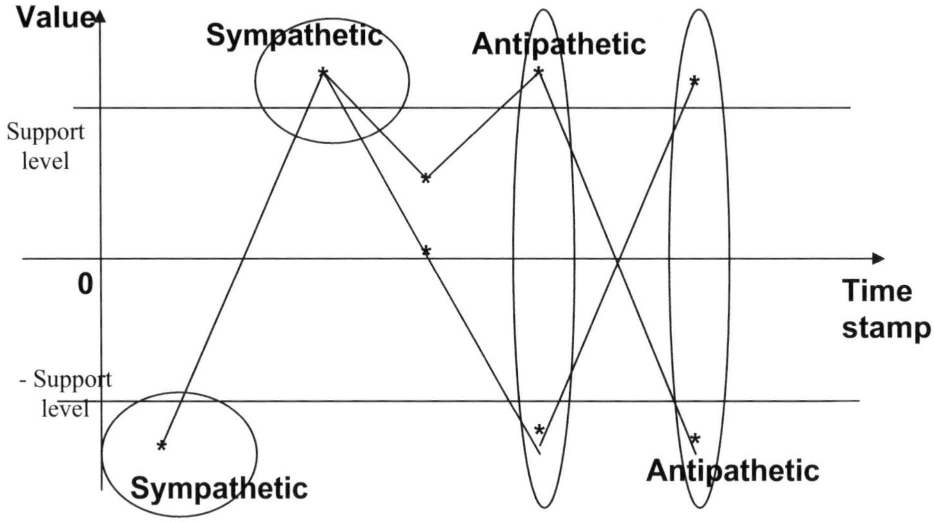

Table 3. Derived database Dnew with arrows indicating support counting direction

Δt	ΔA_1	ΔA_2
Δt_1	+8	+1
Δt_2	-7	+9
Δt_3	-6	+4
Δt_4	+3	+8

Table 4. Counting result

	(+,+)	(-,-)	(+,-,)	(-,+,)
Supports($\Delta A_{1,} \Delta A_{2)}$	2/4	0	0	2/4

For a given *support level*, a positive and negative value of the *support level* can be then drawn as shown in Figure 3. The support is the occurrence of the polarity combination above the value of the support level.

Definition 8. *A frequent dynamic set is a pair of dynamic attributes which contain a polarity combination with frequency occurrence above a user-defined support threshold.*

Theorem 1. If a pair of dynamic attributes ($\Delta A_{1,}$ ΔA_2) is infrequent, then either one individiual dynamic attribute is infrequent or both dynamic attributes are infrequent.

Proof: If a dynamic attribute set is frequent, then this indicates that both the dynami attributes are above the user-defined threshhold.

The above theorem is a consequence of Definition 7 for the frequent dynamic set. This observation forms the basis of the pruning strategy in the search procedure for frequent dynamic sets, which has been leveraged in many Association Mining algorithms (Zaki, 2000), that only the single dynamic attribute found to be frequent needs to be extended as candidate for the rule.

Theorem 2. Confidence measure is not useful in DCM analysis.

Proof: The confidence measure is not used here because the total numbers of records in an attribute is equal to the total number of time stamps. Thus it makes the confidence equal to the

Figure 4. Notation of a dynamic causal rule

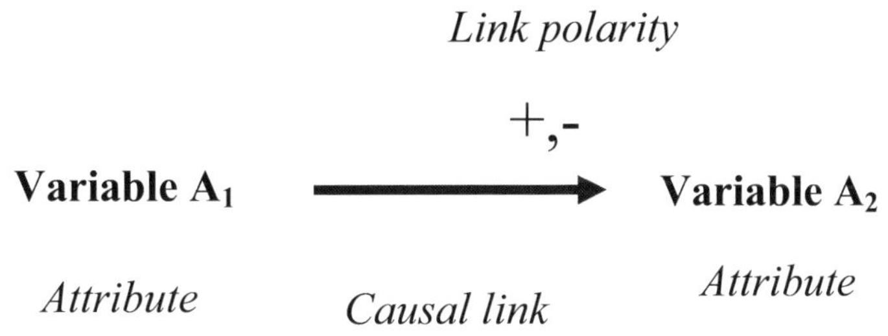

support. Instead, multiple supports are introduced to further reduce the running time and size of the relevant rules.

RULE REPRESENTATION

A dynamic causal rule consists of variables connected by arrows denoting the causal influences among the attributes. Figure 4 shows an example of the notation. Two attributes, A_1 and A_2, are linked by a causal arrow. Each causal link is assigned a polarity and the link indicates the direction of the change.

In *System Dynamics,* a symbol $x \rightarrow^{+} y$ can be interpreted as $\delta y/\delta x > 0$ and $x \rightarrow^{-} y$ can be interpreted as $\delta y/\delta x < 0$. This analogy is applied in *DCM* and the *dynamic causal rules* produced by *DCM* can form causal diagrams, which will be used to simulate future behaviour.

Definition 9. *A dynamic causal rule is derived from a frequent dynamic attribute set. A dynamic causal rule can be either strong or weak. A weak rule is a set of attributes with polarity that partially fulfils equation (4), (5), or (6). A strong rule is a set of attributes with polarity that completely fulfils equation (4), (5), or (6). There are two types of strong rule, sympathetic and antipathetic.*

Figure 2 shows that variables A_1 and A_2 are causally dependent. If *any* two variables A_1 and A_2 are truly causally related, then a change of A_1 causes a change of A_2. This article focuses on the discovery of the causality with no time delay, which means that attributes are in same time period or interval. This implies that attributes should occur within the same time stamp.

Theorem 3. The support for a strong rule is less than or equal to the half of the total time stamps.

Proof: Each dynamic attribute can have only one type of polarity at one time stamp. If the occurrence of one polarity is huge, the occurrence of the other two polarities will diminish. Following the definition of strong rules, the occurrence of the polarities + and – have both to be highly frequent. Since support indicates the times each polarity pair occurred over the total time stamp, the support cannot be more than ½ of the total time stamps.

A *sympathetic* rule causes an increase or decrease in the output of a target system. It reinforces a change with more change in the same direction. An *antipathetic* rule represents an adjustment to achieve a certain goal or objective. It indicates a system attempting to change from its current state to a goal state. This implies that if the current state is above the goal state, then the system forces it down. If the current state is below the goal state, the system pushes it up. An *antipathetic* rule provides useful stability but resists external changes.

Dynamic Policy

Definition 10. *A dynamic policy consists of one or more dynamic causal rules. In a dynamic policy with several dynamic causal rules, each rule should share at least one dynamic attribute with other rules.*

Definition 11. *A dynamic policy is single if it consists only of one rule and if that rule does not share any common attribute with other rules in a dynamic policy.*

Given a *single dynamic policy* where an attribute A_1 is dynamically causally related to another attribute A_2, then such a rule can be represented as the following function.

$$\Delta A_2 = k \Delta A_1 \quad (8)$$

where

$$k_t = \frac{\Delta A_{2,t}}{\Delta A_{1,t}} \quad (9)$$

Given A_1^{new} indicating new values added in A_1, a ΔA_1^{new} can be calculated based on A_1^{new}, and based on equation 8 a ΔA_2^{new} can be calculated. If given a value $A_{2,t=0}$ *a new set of* A_2 can be derived, where $A_{2,t=2} = A_{2,t=1} + \Delta A^{new}_{2,t=1}$, $A_{2,t=3} = A_{2,t=2} + \Delta A^{new}_{2,t=2}$ *, and so forth.*

Definition 12. *A dynamic policy is serial if each rule shares only one attribute with one other rule in a dynamic policy. A serial dynamic policy is open if each attribute in the policy has only one causal link, in other words, if there is a start and an ending attribute. A serial dynamic policy is closed if there is no start or ending attribute.*

Given an *open serial dynamic policy* where an attribute A_1 is dynamically causally related to another attribute A_2, and A_2 is dynamically causally related to another attribute A_3, then such a rule can be represented as the following function.

$$A_{2,T} = k_{1,T} A_{1,T} \text{ and } A_{3,T} = k_{2,T} A_{2,T} = k_{2,T}\, k_{1,T} A_{1,T} \quad (10)$$

where

$$T = (1, 2, 3, \ldots t),\ k_{1,t} = \frac{\Delta A_{2,t}}{\Delta A_{1,t}} \text{ and } k_{2,t} = \frac{A_{3,t}}{A_{2,t}} \quad (11)$$

Definition 15. *A dynamic policy is complex if the policy consists of a mixture of open and closed serial dynamic policies.*

For a complex or closed dynamic policy there are attributes which are connected with more than one attribute. For instance, A_1 is connected with A_2, A_3, and A_4. A_2, A_3 and A_4 are not connected. Then A_1 can be represented with:

$$A_{2,T} * A_{3,T} * A_{4,T} = C_T$$

where

$$T = (1, 2, 3, \ldots t) \text{ and } C_t = A_{2,t} * A_{3,t} * A_{4,t}$$

ILLUSTRATIVE EXAMPLE

- **Stage 1:** To establish a dynamic data set. The dynamic set is calculated based on the dataset illustrated in Table 5 and Equation 1. Table 6 shows the dynamic set after the subtraction. $A_1, A_2 \ldots A_n$ could represent treatment volume, doctor capacity, and so forth, which all varies through time.
- **Stage 2:** To prune the dynamic data set based on the specified *support*. Pruning is carried out to remove columns (attributes) where the level of support is below the minimum set. In this example, the *support* is set to 2, which means columns with two or more 0s are removed (value 0 indicates no dynamics

Table 5. Input dataset

Time	A_1	A_2	A_3	A_4	A_5	A_6	A_7
1	9	2	10	22	20	100	13
2	17	3	10	28	20	100	13
3	10	12	10	27	20	100	6
4	4	16	10	22	20	100	2
5	7	24	10	14	20	3	10
6	6	18	10	11	20	100	5
7	6	18	10	5	20	100	4
8	11	21	10	5	20	100	9
9	20	13	10	1	20	100	2
10	21	8	10	8	5	100	8

Table 6. Dynamic dataset

ΔA_1	ΔA_2	ΔA_3	ΔA_4	ΔA_5	ΔA_6	ΔA_7
+8	+1	0	+6	0	0	0
-7	+9	0	-1	0	0	-7
-6	+4	0	-5	0	0	-4
+3	+8	0	-8	0	-97	+8
-1	-6	0	-3	0	97	-5
0	0	0	-6	0	0	-1
+5	+3	0	0	0	0	+5
+9	-8	0	-4	0	0	-7
+1	-5	0	+7	-15	0	+6

Table 7. Pruned dataset

ΔA_1	ΔA_2	ΔA_4	ΔA_7
+8	+1	+6	0
-7	+9	-1	-7
-6	+4	-5	-4
+3	+8	-8	+8
-1	-6	-3	-5
0	0	-6	-1
+5	+3	0	+5
+9	-9	-4	-7
+1	-5	+7	+6

in the attribute at the corresponding time stamp and more occasionally the set value of 0 indicates the attribute is not dynamic). Table 7 illustrates the "pruned" dynamic data set.

- **Stage 3:** To create dynamic rules, the dynamic *supports* are counted and calculated according to Equations (4), (5), and (6). Table 8 shows the dynamic supports for the pairs of attributes in Table 7. In this example, the user-specified support is set to 0.3, which means any attribute pair with dynamic support with value larger than or equal to 0.3 is a dynamic rule. Table 7 shows how the *DCM* rules are generated. Note that (F1&F7) is the only strong sympathetic rule because when one of the attributes increases its value, the other will automatically increase its value and vice versa.

Table 8 shows the attribute pair A_1 and A_7 and dynamic attributes ΔA_1 and ΔA_7.The strokes indi-

Table 8. Dynamic support

	(+,+)	(-,-)	(-,+)	(+,-)
F1&F2	0.3	0.1	0.2	0.2
F1&F4	0.2	0.3	0	0.2
F1&F7	0.3	0.3	0	0.1
F2&F4	0.1	0.2	0.1	0.3
F2&F7	0.2	0.2	0.1	0.2
F4&F7	0.1	0.5	0.1	0

Table 9. The rules generated

(+,+)	(F1&F2), (F1&F7)
(-,-)	(F1&F4), (F1&F7), (F4&F7)
(+,-)	(F2&F4)

Figure 5. Plot for A1,new and A2,new

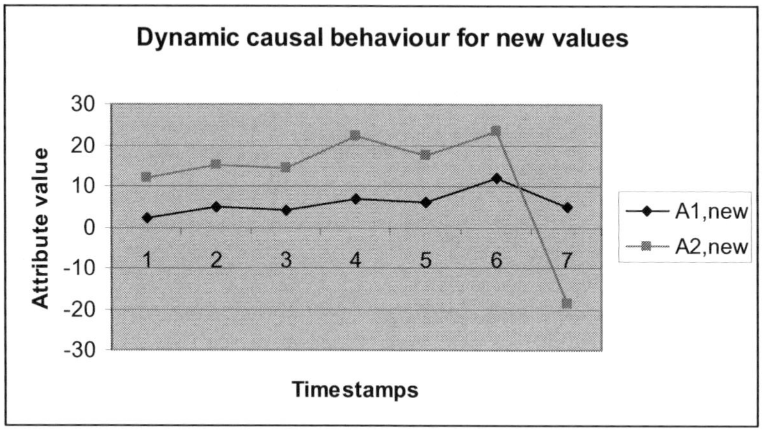

Figure 6. The steps of DCM

Part 1: – **Preprocessing**: Removal of the "least" causal data from database
Part 2: – **Mining**: Formation of a rule set that covers all training examples with minimum number of rules
Part 3: – **Checking:** Check if an attribute pair is self contradicting (sympathetic and antipathetic at the same time)

Input: The original database, the values of the pruning threshold for the neutral, sympathetic and antipathetic *supports*.
Output: Dynamic sets
Step 1: Check the nature of the attributes in the original database (numerical or categorical). Initialize a new database with dynamic attributes based on the attributes and time stamps from original database.

Step 2: Initialize a counter for each of the three polarities.

Step3: Prune away all the dynamic attributes with *supports* above the input thresholds.

cate redundant attributes and redundant *dynamic attributes*. Given a new attribute $A_{1,new}$ as shown in Table 9, from it $\Delta A_{1,new}$ and $\Delta A_{2,new}$ can be calculated as shown in Figure 6. Figure 4.3 shows the plots of $A_{1,new}$ and $A_{2,new}$ and it is clear that the two plots are causally related.

MINING ALGORITHM

Problem Formulation

Let D denote a database which contains a set of n records with attributes $\{A_1, A_2, A_3, \ldots A_n\}$, where each attribute is of a unique type (sale price, production quantity, inventory volume, etc). Each attribute is linked to a time stamp t. To ap-

Figure 7. The checking step of DCM

Input: The mined database, the values of the pruning threshold for the supports of the polarity combinations.
Output: Dynamic sets
Step 1 . Check w eather a r ule is s elf-contradictory (a rule i s both sympathetic and antipathetic).
Step 2. If step 1 returns true then
Retrieve the attribute pair form the preprocessed database
Step Initialize a counter that includes polarity combination
Step 4. For the pair of attributes
Count t he o ccurrence of polarity combination with t wo records each time.
Prune away the pairs if the counted support is below the input threshold.

ply *DCM*, the records are arranged in a temporal sequence ($t = 1, 2,..., n$). The causality between attributes in *D* can be identified by examining the polarities of corresponding changes in attribute values. Let D_{new} be a new data set constructed from *D*. A generalized dynamic association rule is an implication of the form $A_1 \rightarrow^p A_2$, where $A_1 \subset D$, $A_1 \subset D$, $A_1 \cap A_2 = \phi$ and *p* is the polarity.

The implementation of the *DCM* algorithm must support the following operations:

1. To add new attributes.
2. To maintain a counter for each polarity with respect to every dynamic value set. While making a pass, one dynamic set is read at a time and the polarity count of candidates supported by the dynamic sets is incremented. The counting process must be very fast as it is the bottleneck of the whole process.

Algorithm Description

DCM makes two passes over the data as shown in Figure 6 and Figure 7. In the first pass, the *support* of individual attributes is counted and the frequent attributes are determined. The dynamic values are used for generating new potentially frequent sets and the actual *support* of these sets is counted during the pass over the data. In subsequent passes, the algorithm initializes with dynamic value sets

Table 10. Pruned results

Data set	Single Support						
	0.05	0.10	0.15	0.20	0.25	0.30	0.35
Adult	5%	20%	27%	74%	100%	100%	100%
Bank	11%	20%	60%	94%	100%	100%	100%
Cystine	5%	33%	70%	100%	100%	100%	100%
Market basket	6%	10%	50%	86%	100%	100%	100%
Mclosom	1%	13%	38%	72%	90%	100%	100%
ASW	1%	8%	40%	68%	70%	97%	100%
Weka-base	6%	25%	52%	86%	100%	100%	100%

Table 11. Original data sets

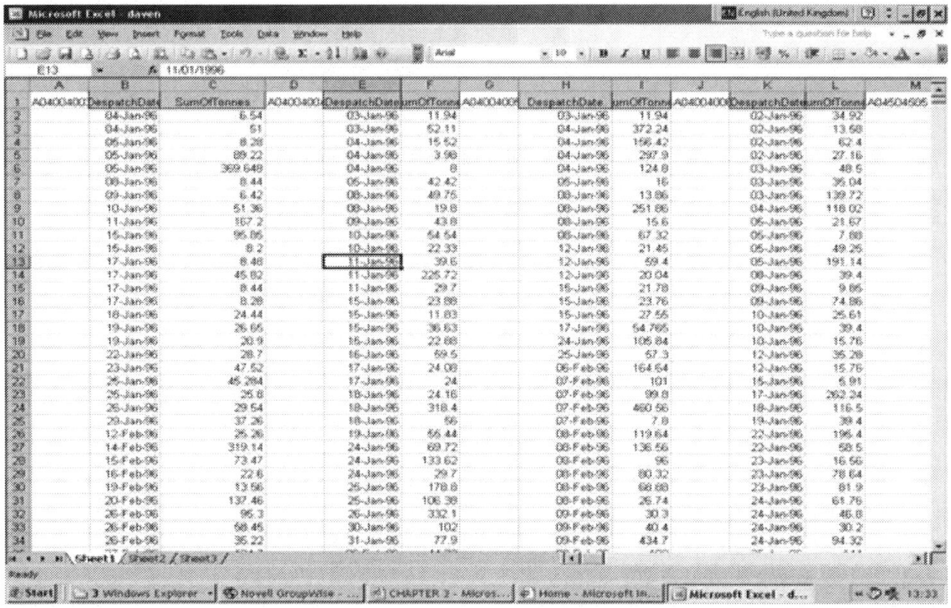

	A	B	C	D	E	F	G	H	I	J	K	L	M
1	A0400400	DespatchDate	SumOfTonnes	A0400400	DespatchDate	umOfTonne	A0400400	DespatchDate	umOfTonne	A0400400	DespatchDate	umOfTonne	A04504505
2		04-Jan-96	6.54		03-Jan-96	11.94		03-Jan-96	11.94		02-Jan-96	34.92	
3		04-Jan-96	51		03-Jan-96	52.11		04-Jan-96	372.24		02-Jan-96	13.58	
4		05-Jan-96	8.28		04-Jan-96	15.52		04-Jan-96	156.42		02-Jan-96	62.4	
5		05-Jan-96	89.22		04-Jan-96	3.98		04-Jan-96	297.9		02-Jan-96	27.16	
6		05-Jan-96	369.648		04-Jan-96	8		04-Jan-96	124.8		03-Jan-96	48.5	
7		08-Jan-96	8.44		05-Jan-96	42.42		05-Jan-96	16		03-Jan-96	35.04	
8		09-Jan-96	6.42		08-Jan-96	49.75		08-Jan-96	13.86		03-Jan-96	139.72	
9		10-Jan-96	51.36		08-Jan-96	19.8		08-Jan-96	251.86		04-Jan-96	118.02	
10		11-Jan-96	167.2		09-Jan-96	43.8		08-Jan-96	15.6		05-Jan-96	21.67	
11		15-Jan-96	95.86		10-Jan-96	54.54		08-Jan-96	67.32		05-Jan-96	7.88	
12		15-Jan-96	8.2		10-Jan-96	22.33		12-Jan-96	21.45		05-Jan-96	49.25	
13		17-Jan-96	8.48		11-Jan-96	39.6		12-Jan-96	59.4		05-Jan-96	191.14	
14		17-Jan-96	45.82		11-Jan-96	225.72		12-Jan-96	20.04		08-Jan-96	39.4	
15		17-Jan-96	8.44		11-Jan-96	29.7		15-Jan-96	21.78		09-Jan-96	9.86	
16		17-Jan-96	8.28		15-Jan-96	23.88		15-Jan-96	23.76		09-Jan-96	74.86	
17		18-Jan-96	24.44		15-Jan-96	11.83		15-Jan-96	27.56		10-Jan-96	25.61	
18		19-Jan-96	26.65		15-Jan-96	36.63		17-Jan-96	54.765		10-Jan-96	39.4	
19		19-Jan-96	20.9		16-Jan-96	22.88		24-Jan-96	105.84		10-Jan-96	15.76	
20		22-Jan-96	28.7		16-Jan-96	59.5		25-Jan-96	57.3		12-Jan-96	35.28	
21		23-Jan-96	47.52		17-Jan-96	24.08		06-Feb-96	164.64		12-Jan-96	15.76	
22		25-Jan-96	45.284		17-Jan-96	24		07-Feb-96	101		15-Jan-96	6.91	
23		25-Jan-96	25.8		18-Jan-96	24.16		07-Feb-96	99.8		17-Jan-96	262.24	
24		26-Jan-96	29.54		18-Jan-96	318.4		07-Feb-96	460.56		18-Jan-96	116.5	
25		29-Jan-96	37.26		18-Jan-96	56		07-Feb-96	7.8		19-Jan-96	39.4	
26		12-Feb-96	25.26		19-Jan-96	56.44		08-Feb-96	119.64		22-Jan-96	196.4	
27		14-Feb-96	319.14		24-Jan-96	69.72		08-Feb-96	136.56		22-Jan-96	58.5	
28		15-Feb-96	73.47		24-Jan-96	133.62		08-Feb-96	96		23-Jan-96	16.56	
29		16-Feb-96	22.6		24-Jan-96	29.7		08-Feb-96	80.32		23-Jan-96	78.64	
30		19-Feb-96	13.56		25-Jan-96	178.8		08-Feb-96	68.68		23-Jan-96	81.9	
31		20-Feb-96	137.46		25-Jan-96	106.38		08-Feb-96	26.74		24-Jan-96	61.76	
32		26-Feb-96	95.3		26-Jan-96	332.1		09-Feb-96	30.3		24-Jan-96	46.8	
33		26-Feb-96	58.45		30-Jan-96	102		09-Feb-96	40.4		24-Jan-96	30.2	
34		26-Feb-96	35.22		31-Jan-96	77.9		09-Feb-96	434.7		24-Jan-96	94.32	

based on dynamic values found to be frequent in the previous pass. After the second of the passes, the *causal rules* are determined and they become the candidates for the dynamic policy. In the *DCM* process, the main goal is to find the strong dynamic causal rule in order to form a policy.

Table 12. Dynamic attributes

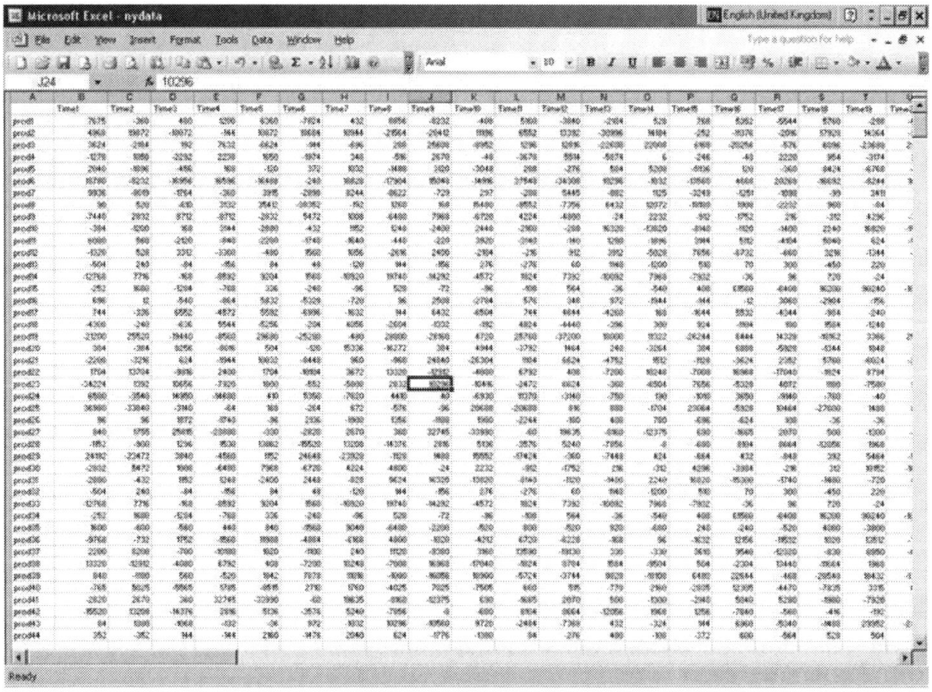

Figure 8. Rule plot with support level = 0

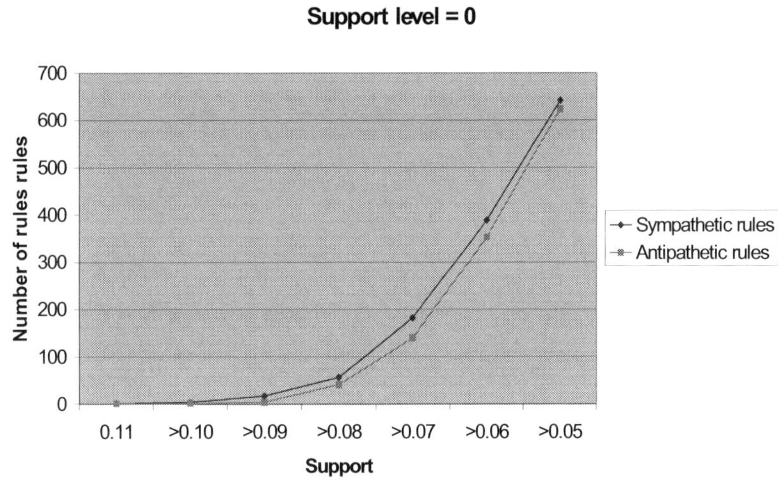

It also represents a filtering process that prunes away static attributes, which reduces the size of the data set for further mining.

The process time would be n^2-n, where n is the number of attributes. This becomes a huge problem if n becomes too large. It is obvious that the task becomes much simpler if the size of n could be reduced before the search. Table 10 shows the pruned percentage of the total database based on the support. The support level is set to 0.

EXPERIMENT

Data Preparation

The overall aim is to identify hidden dynamic changes. It was taken from a local hospital The original data was given as shown in Table 11. The only data of interest are the data with changes, for example sale amounts of a medicine, the time stamp, and so forth. The rest of the static data, such as the weight and the cost of the product,

Figure 9. Rule plot with support level = frequent

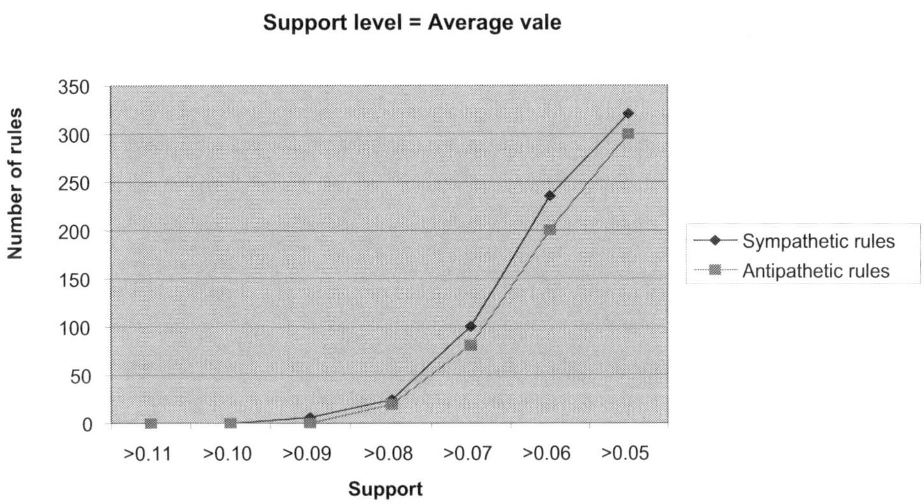

Table 13. Result generated by the algorithm

Strong rules	Support
Sympathetic	
{C15276179, F030008}	0,093
{J08008008, F060010}	0,089
{A04004004, A05005005}	0,086
{A05005006, C10251104}	0,084
{A04004004, F100020}	0,082
Antipathetic	
{A05005008, C15276179}	0.092
{C10251104, F070010}	0.083
{A05005008, F030008}	0.082

can be removed. This data consists of real life data. This dataset contains 65,536 attributes of metal manufacturing, with eight records in each attribute.

After cleaning the data, the *dynamic attributes* are found as shown in Table 12. The *dynamic attribute* is calculated by finding the difference between sales amounts in one month and sales amounts in the previous month.

In the next step, the neutral attributes are pruned. The idea of pruning is to remove redundant dynamic attributes; thus fewer sets of attributes are required when generating rules. The first pruning is based on the single attribute *support*. In this case, the single attribute *support* is defined to be 0.5, which means that if an attribute with polarity +, -, or 0 occurs in more than half of total time stamps, it will be pruned. In this case, 429 attributes remain for the rule generation.

In this experiment, dynamic sets are compared based on a simultaneous *time stamp*. Then the support of *sympathetic* and *antipathetic* rules for each dynamic set is calculated. The *support* is used as the threshold to eliminate unsatisfactory dynamic sets and to obtain the rules from the satisfactory sets. In Table 12 the left most colum indicates the codes for each attributes. The top row indicates the times stamps and the rest of table is dynamic attributes.

The algorithm was run based on the procedures described in previous sections. Figure 8 shows the plot of sympathetic and antipathetic support.

The x-axis represents the support and the y-axis represents the number of rules. This database shows that there are more sympathetic rules than antipathetic rules. The figure shows that increasing support will lead to exponential growth of the rules. As the support reaches 0.05 or 5, as it indicates on the figure, the number of rules is 630. Most of these rules are redundant and have no meaning due to the low support. Figure 9 shows the rule plot with support equal to average value, where the +support = the average of all positive records and –support = the average of all negative records. The number of rules has decreased by applying the support level.

Table 13 shows the extracted strong rules with support level equal to average value and support larger than 0.08. C15276179, A04004004... are the code for each attributes. There are only dynamic pairs so there is no need to do the simulation

DISCUSSION

Apriori provides some form of causal information, that is, suggesting a possible direction of causation between two attributes, but there is no basis to conclude that the arrow indicates direct or even indirect causation. The *DCM* algorithm, on the other hand, shows causality between attributes. Thus, where association rule generation techniques find surface associations, causal inference algorithms identify the structure underlying such associations.

Each type of relationship generated by the *DCM* algorithm provides additional information. The *DCM* algorithm finds four kinds of relationships, each of which deepens the user's understanding of their target system by constructing the possible models. For example, $A_1 \rightarrow^{+} A_2$ provides more information than $A_1 \rightarrow A_2$ because the latter indicates that A_1 coexists with A_2. The condition of the rule is not stated (whether sympathetic or antipathetic). A genuine causality such as $A_1 \rightarrow^{+} A_2$ provides useful information because it indicates that the relationship from A_1 to A_2 is strictly sympathetic causal.

The *rules* extracted by *DCM* can be simulated by using software to model the future behaviour. The *rules* extracted by association algorithm cannot be simulated.

CONCLUSION AND SUMMARY

This article provides insight into how to use dynamic causal mining approaches to medical and health management research can be addressed through management strategies specifically designed to respond.

The algorithm shows opportunities for more extensive examination of dynamics between attributes as well as produces a framework for such analysis. A mining framework will provide medical and health management researchers, as well as researchers interested in other complex social phenomena, with insights into determining the best research design given a specific context.

This article addresses important and challenging issues of causal analysis within e-government and implementation of causality extraction and construction algorithms for medical and health management data. This article also studies the computational costs of causal, and implements a multipruning approach for building models that incurs minimal computational cost while maintaining accuracy and unsupervised causality extraction algorithms to reduce the reliance on medical and health management data.

REFERENCES

Abidi, S.S.R. (2001). Knowledge management in healthcare: Towards knowledgedriven decision-support services. *International Journal of Medical Informatics, 63,* 5-18.

Acir, N., & Guzelis, C. (2004). Automatic spike detection in EEG by a two-stageprocedure based on support vector machines. *Computers in Biology and Medicine,34*(7), 561-575.

Agrawal, R., Gunopulos, D., & Leymann, F. (1998). Mining process models from workflow logs. In *Sixth International Conference on Extended Database Technology* (pp. 469-483).

Agrawal, R., Mannila, H., Srikant, R., Toivonen, H., & Inkeri, A. (1996). *Fast discovery of association rules, advances in knowledge discovery and data mining.* The Association for the Advancement of Artificial Intelligence and the MIT Press (pp. 307-328).

Bergan, T., & Hirsch, G.B. (1976). *Analysis and Planning for improved distribution of Nursing personnel and services. A national model of supply, demand, and distribution.* Cambridge, MA: Pugh-Roberts Associates, Inc.

Cheung, D., Han, J., Ng, V.T., & Fu, A.W. (1996). Fast distributed algorithm for mining association rules. In *International Conference on Parallel and Distributed Information Systems*, Tokyo, Japan (pp. 31-42).

Cook, J.E., & Wolf, A.L. (1998, November). Event-based detection of concurrency. In *Proceedings of the Sixth International Symposium on the Foundations of Software Engineering (FSE-6),* Orlando, FL (pp. 35-45).

Coyle, R.G. (1996). *System dynamics modelling: A practical approach.* London: Chapman and Hall.

Dawes, M., & Sampson, U. (2003). Knowledge management in clinical practice: Asystematic review of information seeking behavior in physicians. *International Journal of Medical Informatics, 71,* 9-15.

Dreiseitl, S., Ohno-Machado, L., Kittler, H., Vinterbo, S., Billhardt, H., & Binder, M. (2001). A comparison of machine learning methods for the diagnosis of pigmented skin lesions. *Journal of Biomedical Informatics, 34,* 28-36.

Flood, R. (1999). *Rethinking the fifth discipline. Learning within the unknowable.* London: Ruthledge.

Forrest, S., Hofmeyr, S.A., Somayaji, A., & Longstaff, T.A. (1996). A sense of self for unix processes. In *1996 IEEE Symposium on Security and Privacy* (pp. 120-128). IEEE Computer Society. *Building the virtual state. Information technology and institutional change.* Washington, DC: Brookings Institution Press.

Forrester, J.W. (1961, July-August). Industrial Dynamics: A Major Breakthrough for Decision Makers. *Harvard Business Review,* 37-66.

Fountain, J.E. (2001).

Friedman, C., Hripcsak, G., Johnson, S.B., Cimino, J.J., & Clayton, P.D. (1990). Ageneralized relational schema for an integrated clinical patient database. In *Proceedings of the 14th Annual Symposium on Computer Applications in Medical Care* (pp. 335-339).

Heathfield, H., & Louw, G. (1999). New challenges for clinical informatics: Knowledge management tools. *Health Informatics Journal, 5*(2), 67-73.

Herbst, J. (2000). A machine learning approach to workflow management. In *11th European Conference on Machine Learning.* Berlin, Germany: Springer (LNCS 1810, pp. 183-194).

Herbst, J., & Karagiannis, D. (1999, August), An inductive approach to the acquisition and adaptation of workflow models. In M. Ibrahim & B. Drabble (Eds.), *Proceedings of the IJCAI'99 Workshop on Intelligent Workflow and Process Management: The New Frontier for AI in Business* (pp. 52-57). Stockholm, Sweden.

Herbst, J., & Karagiannis, D. (2000). Integrating machine learning and workflow management to support acquisition and adaptation of workflow models. *International Journal of Intelligent Systems in Accounting, Finance and Management, 9,* 67-92.

Hersh, W., Mailhot, M., Arnott-Smith, C., & Lowe, H. (2002). Selective automated indexing of findings and diagnoses in radiology reports. *Journal of Biomedical Informatics, 34,* 262-273.

Hripcsak, G. (1993). Monitoring the monitor: Automated statistical tracking of a clinical event monitor. *Computers and Biomedical Research, 26*(5), 449-466.

Hripcsak, G., Austin, J.H., Alderson, P.O., & Friedman, C. (2002). Use of natural language processing to translate clinical information from a database of 889,921 chest radiographic reports. *Radiology, 224*(1), 157-163.

Hirsch, G.B., & Bergan, T.A. (1973). Simulating ambulatory care systems: The relationship between structure and behaviour. In *Proceedings of the Summer Computer Simulation Conference,* Montreal, Quebec, Canada (pp. 877-884).

Hirsch, G.B., Bergan, T.A., & Frohman, A.L. (1974). Organizational change in the provision of human services. In *Proceedings of the Summer Computer Simulation Conference,* Houston, Texas.

Hirsch, G.B., Forsyth, B.R., Bergan, T.A., & Goodman, M.R. (1976). Projecting a teaching hospital's future utilization: A dynamic simulation approach. *Journal of Medical Education, 51*(11), 937-939.

Hirsch, G.B., & Henderson, J.D. (1977). A simulation model for "what if?" questions in plan development. *American Journal of Health Planning, 2*(1), 40-46.

Hirsch, G.B., & Killingsworth, W.R. (1975). A new framework for projecting dental manpower requirements. *Inquiry, 12*(2), 126-142.

Hirsch, G.B., & Miller, S. (1974). Evaluating HMO policies with a computer simulation model. *Medical Care, 12*(8), 668-681.

Jackson, J.R. (2000). The urgent call for knowledge management in medicine. *The Physician Executive, 26*(1), 28-31.

Janetzki, V., Allen, M., & Cimino, J.J. (2004). Using natural language processing to link from medical text to online information resources. In *Proceedings of Medinfo* (pp. 1665).

Kandaswamy, A., Kumar, C.S., Ramanathan, R.P., Jayaraman, R., & Malmurugan, N. (2004). Neural classification of lung sounds using wavelet coefficients. *Computers in Biology and Medicine, 34,* 523-537.

Kovalerchuk, B., Vityaev, E., & Ruiz, J.F. (2001). Consistent and complete data and expert mining in medicine. In K.J. Cios (Ed.), *Medical data mining and knowledge discovery.* New York: Physica-Verlag.

Kuperman, G.J., Gardner, R.M., & Pryor, T.A. (1991). *The HELP system.* New York: Springer-Verlag.

Levin, G., Hirsch, G.B., & Roberts, E. (1972). Narcotics and the community: A system simulation. *American Journal of Public Health, 62*, 8-6.

Levin, G., & Roberts, E.B. (1976). *The dynamics of human service delivery.* Cambridge, MA: Ballinger Publishing Company.

Maruster, L., van der Aalst, W.M.P., Weijters, T., van den Bosch, A., & Daelemans, W. (2001, October 25-26). Automated discovery of workflow models from hospital data. In B. Kröse et al. (Eds.), *Proceedings 13th Belgium- Netherlands Conference on Artificial Intelligence (BNAIC'01),* Amsterdam, The Netherlands (pp. 183-190).

Montani, S., & Bellazzi, R. (2002). Supporting decisions in medical applications: The knowledge management perspective. *International Journal of Medical Informatics, 68,* 79-90.

Pham, D.T., Wang, Y., & Dimov, S. (2005). Intelligfent manufacturing strategy selection. In *Proc 1st International Virtual Conference on Intelligent Production Machines and Systems* (pp. 312-318). Oxford: Elsevier.

Pham, D.T., Wang, Y., & Dimov, S. (2006). Incorporating delay and feedback in intelligent manufacturing strategy selection. In *Proc 2nd International Virtual Conference on Intelligent Production Machines and Systems* (pp. 246-252). Oxford: Elsevier.

Prather, J.C., Lobach, D.F., Goodwin, L.K., Hales, J.W., Hage, M.L., & Hammond, W.E. (1997). Medical data mining: Knowledge discovery in a clinical data warehouse. In *Proceedings of the AMIA Annual Symposium Fall 1997* (pp. 101-105).

Richardson, G.P. (1996). System dynamics. In S.I. Gass & C.M. Harris (Eds.), *Encyclopedia of operations research and management science* (pp. 656-660). Boston, MA: Kluwer Academic Publishers.

Richardson, G.P. (1999). Citation for winner of the 1999 Jay W. Forrester Award: Jac Vennix. *System Dynamics Review, 15 4,* 375-377.

Richardson, G.P., & Andersen, D.F. (1995). Teamwork in group model building. *System Dynamics Review, 11*(2), 113-137.

Roberts, E.B., & Hirsch, G.B. (1976, Winter). Strategic modelling for health care managers. *Health Care Management Review,* 69-77.

Schubart, J.R., & Einbinder, J.S. (2000). Evaluation of a data warehouse in an academic health sciences center. *International Journal of Medical Informatics*, *60*, 319-333.

Senge, P., Kleiner, A., Roberts, C., Ross, R., & Smith, B. (1994). *The fifth discipline fieldbook.* New York: Doubleday

Stearns, N.S., Bergan, T.A., Roberts, E.B., & Cavazos, L.F. (1978). A systems intervention for improving medical school-hospital interrelationships. *Journal of Medical Education, 53,* 464-472.

Stearns, N.S, Bergan, T.A., Roberts E.B., & Quigley, L. (1976, Fall). Systems intervention: New help for hospitals. *Health Care Management Review, 1,* 9-18.

Sterman, J.D. (1994). Learning in and about complex systems. *System Dynamics Review, 10*(2-3), 291-330.

Sterman, J. D. (2000). *Business dynamics: Systems thinking and modelling for a complex world.* Boston, MA: Irwin McGraw-Hill.

Yang, Y., & Liu, X. (1999). A re-examination of text categorization methods. In *Proceedings of the 22nd Annual International ACM Conference on Research andDevelopment in Information Retrieval* (SIGIR'99) (pp. 42-49).

Wang, Y. (2007). Integration of data mining with game theory. *Journal of International Federation for Information Processing, 207/2006,* 275-280. Boston, MA: Springer

Weijters, T., & van der Aalst, W.M.P. (2001, October 25-26). Process mining: Discovering workflow models from event-based data. In B. Kröse et al. (Eds.), *Proceedings 13th Belgium-Netherlands Conference on Artificial Intelligence (BNAIC'01),* Amsterdam, The Netherlands (pp. 283-290).

Zaki, M.J., Parthasarathy, S., Ogihara, M., & Li, W. (1997). New parallel algorithms for fast discovery of association rules. *Data Mining and Knowledge Discovery, 1*(4), 343-37.

This work was previously published in International Journal of Healthcare Information Systems and Informatics, Vol. 3, Issue 4, edited by J. Tan, pp. 17-38, copyright 2008 by IGI Publishing (an imprint of IGI Global).

Chapter 13
Using Simulation Systems for Decision Support

Andreas Tolk
Old Dominion University, USA

ABSTRACT

This chapter describes the use of simulation systems for decision support in support of real operations, which is the most challenging application domain in the discipline of modeling and simulation. To this end, the systems must be integrated as services into the operational infrastructure. To support discovery, selection, and composition of services, they need to be annotated regarding technical, syntactic, semantic, pragmatic, dynamic, and conceptual categories. The systems themselves must be complete and validated. The data must be obtainable, preferably via common protocols shared with the operational infrastructure. Agents and automated forces must produce situation adequate behavior. If these requirements for simulation systems and their annotations are fulfilled, decision support simulation can contribute significantly to the situational awareness up to cognitive levels of the decision maker.

INTRODUCTION

Modeling and simulation (M&S) systems are applied in various domains, such as

- supporting the analysis of alternatives,
- supporting the procurement of new systems by simulating them long before first prototypes are available,
- supporting the testing and evaluation of new equipment by providing the necessary stimuli for the system being tested,
- training of new personnel working with the system, and many more.

The topic of this chapter is one of the most challenging applications for simulation systems, namely the use of simulation systems for decision support in general, and particularly in direct support of operational processes. In other words, the

DOI: 10.4018/978-1-60566-774-4.ch014

decision maker is directly supported by M&S applications, helping with

- "what-if" analysis for alternatives,
- plausibility evaluation for assumptions of other party activities,
- consistency checks of plans for future operations,
- simulation of expected behavior based on the plan and trigger the real world observations for continuous comparison (are we still on track),
- manage uncertainty by simulating several runs faster than real time and display variances and connected risks,
- trend simulation to identify potentially interesting developments in the future based on current operational developments, and additional applications that support the meaningful interpretation of current data.

While current decision support systems are focused on data mining and data presentation, which is the display of snap-shot information and historical developments are captured in most cases in the form of static trend analyses and display curves (creating a common operating picture), simulation systems display the behavior of the observed system (creating a common executable model). This model can be used by the decision maker to manipulate the observed system "on the fly" and use it not only for analysis, but also to communicate the results very effectively to and with partners, customers, and supporters of his efforts. As stated by van Dam (1999) during his lecture at Stanford: *"If a picture is worth a 1000 words, a moving picture is worth a 1000 static ones, and a truly interactive, user-controlled dynamic picture is worth 1000 ones that you watch passively."* That makes simulation very interesting for managers and decision makers, encouraging the use of decision support simulation systems. Another aspect is that of complex systems: nonlinearity and multiple connections. In order to understand and evaluate such system, traditional tools of operational research and mathematics have to be increasingly supported by the means of modeling and simulation. The same is true for decisions in complex environments, such as the battlefield of a military decision maker or the stock market for an international investment broker.

To this end, the simulation system must be integrated into operational systems as a decision support service. In order to be successful, not only the technical challenges of integration, discrete and other simulation technologies, into operational IT systems must be solved. It is also required that the simulation system fulfills additional operational and conceptual requirements as well. Simulation systems are more than software. Simulation systems are executable models, and models are purposeful abstractions of reality. In order to understand if a simulation system can be used for decision support, the concepts and assumptions derived to represent real world objects and effects in a simplified form must be understood. The conceptualization of the model's artifacts is as important as the implementation details of the simulation. As stated in Tolk (2006): *interoperability of systems requires composability of models!*

The author gained most of his experience in the military sector, integrating combat M&S into Command and Control (C2) systems. The development of the Levels of Conceptual Interoperability Model (LCIM) capturing the requirement for alignment on various levels to support decision support is a direct result of the experiences of integrating M&S services as web-services into service-oriented C2 systems (Tolk et al., 2006). It is directly related to the recommendations found in the North Atlantic Treaty Organization (NATO) Code of Best Practice for C2 Assessment (NATO, 2002) that was compiled by a group of international operational research experts in support of complex C2 analysis. It was also influenced by the recommendations of the National Research Council (2002, 2006), as using simulation for procurement decision or for analysis and using this analysis for decision support are closely related topics.

Furthermore, the growing discipline of agent-directed simulation (ADS) is very helpful in providing new insights and methods (Oren et al, 2000). ADS consists of three distinct yet related areas that can be grouped under two categories. First, *agent simulation* (or simulation for agents), that is simulation of systems that can be modeled by agents in engineering, human and social dynamics, military applications, and so on. Second, agents for simulation can be grouped under two sub-categories, namely *agent-based simulation*, which focuses on the use of agents for the generation of model behavior in a simulation study; and *agent-supported simulation*, which deals with the use of agents as a support facility to enable computer assistance by enhancing cognitive capabilities in problem specification and solving.

The vision of using simulation systems in general, and discrete event simulation systems in particular, for decision support is that a decision maker or manager can utilize an *orchestrated set of tools* to support his decision using reliable simulation systems implementing agreed concepts using the best currently available data. It does not matter if the decision support system is used in the finance market, where the stock market is simulated on a continuous basis, always being adjusted and calibrated by the real stock data, or if it used to support a traffic manager in guiding a convoy through a traffic jam during rush hour to the airport while constantly being updated by the recent traffic news. The technologies described here support the military commander in making decisions based on the best intelligence and surveillance data available by a sensor, as well as to the surgeon using a detailed model of the human body in preparation of a risky surgery. While the application fields are significantly different, the underlying engineering methods are not.

The section will start by presenting the relevant work, focusing on the special insights from the military domains before generalizing them for other applications. The main part is built by enumerating and motivating the requirements for simulation systems when being used for decision support, as identified by the National Science Foundation and related organizations. Finally, some examples are given and current developments are highlighted.

RELEVANT WORK

The area of related and relevant work regarding decision support systems in general and the use of simulation systems for decision support in general is huge. A book chapter can never suffice for a complete explanation. Therefore, the focus of this section is to highlight some of the most influencing works leading to formulation of the requirements for simulation systems. Additional information is contained in the section giving examples of decision support simulations in this chapter.

The need for using simulation systems in addition to traditional decision support systems is best derived from the work documented in the NATO Code of Best Practice for C2 Assessment (NATO, 2002). After having operated under more or less fixed strategic and doctrinal constraints for several decades, in which NATO and the Warsaw Pact faced each other in a perpetual lurking position, NATO suddenly faced a new operational environment for their decisions when the Warsaw Pact broke apart. While in the old order the enemy was well known – down to the equipment, strategy, and tactics – the new so-called "operations other than war" and "asymmetric operations" were characterized by uncertainty, incompleteness, and vagueness. At the same time, developments in information technology allowed the efficient distribution of computing power in the form of loosely coupled services. Consequently, the idea was to use an orchestrated set of operational tools – all implemented as services that can be loosely coupled in case of need – to support the decision

Figure 1. Command and Control Improvements

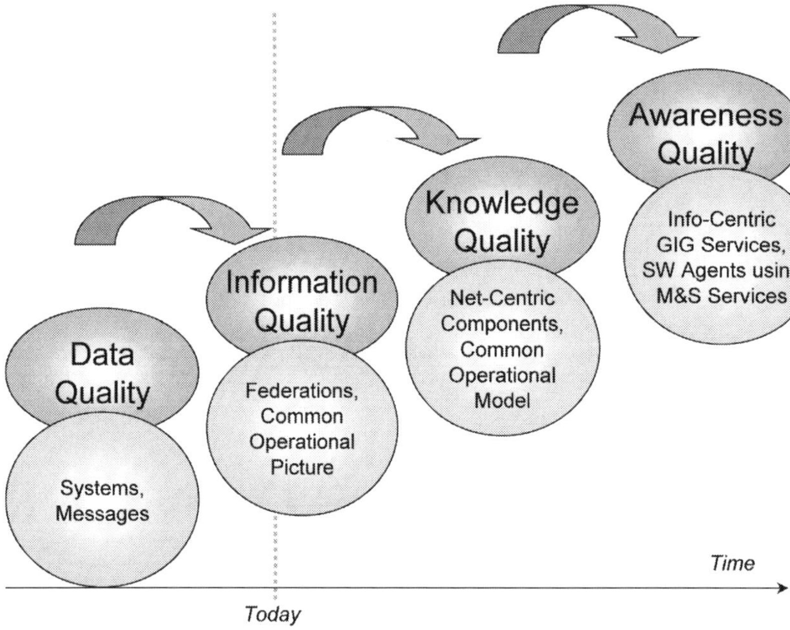

maker with analysis and evaluation means in an area defined by uncertainty, incompleteness, and vagueness regarding the available information. In order to measure improvement in this domain, the value chain of Net Centric Warfare was introduced; see among others (Alberts and Hayes, 2003):

- *Data* is factual information. The value chain starts with *Data* Quality describing the information within the underlying C2 systems.
- *Information* is data placed into context. *Information* Quality tracks the completeness, correctness, currency, consistency, and precision of the data items and information statements available.
- *Knowledge* is procedural application of information. *Knowledge* Quality deals with procedural knowledge and information embedded in the C2 system such as templates for adversary forces, assumptions about entities such as ranges and weapons, and doctrinal assumptions, often coded as rules.
- Finally, *Awareness* Quality measures the degree of using the information and knowledge embedded within the C2 system. Awareness is explicitly placed in the cognitive domain.

C2 quality is improved by an order of magnitude when a new level of quality is reached in this value chain. Figure 1 depicts this. C2 quality is improved by these developments as follows:

- Data quality is characterized by stand-alone developed systems exchanging data via text messages as used in most C2 systems. Having the same data available at the distributed locations was the first goal to reach.
- By the introduction of a common operational picture, data is put into context, which evolves the data into information. The collaborating systems using this common operational picture result in an order of magnitude of improvement of the Command and Control quality, as decision

makers share this common information. As stated before: a picture is worth a 1,000 words.

- The next step, which is enabled by service-oriented web-based infrastructures, is the use of simulation services for decision support. Simulation systems are the prototype for procedural knowledge, which is the basis for knowledge quality. Instead of just having a picture, an executable simulation system can be used.
- Finally, using intelligent software agents to continually observe the battle sphere, apply simulations to analyze what is going on, to monitor the execution of a plan, and to do all the tasks necessary to make the decision maker aware of what is going on, C2 systems can even support situational awareness, the level in the value chain traditionally limited to pure cognitive methods.

Traditional decision support systems enable information quality, but they need the agile component of simulation in order to support knowledge quality as well. In other words, numerical insight into the behavior of complex systems as provided by simulations is needed in order to understand them.

In order to support the integration of decision support simulations, it is necessary to provide them as services. However, this task is not limited to technical challenges of providing a web service or a grid service, but the documentation of the service and the provided functionality is essential to enable the *discovery*, *selection*, and *composition* of this service in support of an operational need. The papers (Tosic et al., 2001) and (Srivastava and Koehler, 2003) summarize the state of the art of service composition. Pullen et al. (2005) show the applicability for M&S services. Additionally, what is needed are annotations. Annotations give meaning to services by changing them into semantic web services. The reader is referred to (Agarwal et al., 2005) and (Alesso and Smith, 2005) for more information on this topic.

In order to identify what information is needed to annotate operational M&S services, the Levels of Conceptual Interoperability Model (LCIM) was developed. The closest application to the topic of this book chapter is documented by Tolk et al. (2006). The LCIM exposes layers of abstractions that are often hidden: the conceptualization layer leading to the model, the implementation layer leading to the simulation, and technical questions of the underlying network. Each layer is tightly connected with different aspects of interoperation. We are following the recommendation given by Page and colleagues (Page et al., 2004), who suggested defining composability as the realm of the model and interoperability as the realm of the software implementation of the model. Included in the technical challenge of integrating networks and protocols, the following three categories for annotations emerge:

- *Integratability* contends with the physical/ technical realms of connections between systems, which include hardware and firmware, protocols, networks, etc.
- *Interoperability* contends with the software and implementation details of interoperations; this includes exchange of data elements via interfaces, the use of middleware, mapping to common information exchange models, etc.
- *Composability* contends with the alignment of issues on the modeling level. The underlying models are purposeful abstractions of reality used for the conceptualization being implemented by the resulting systems.

The LCIM increases the resolution by adding additional sub-layers of interoperation. The layer of integratability is represented by the *technical layer*, which ensures that bits and bytes can be exchanged and correctly interpreted. The *syntactic layer* allows mapping all protocols to a common structure. The *semantic layer* defines the meaning of information exchange elements. Syntax and semantics belong to the interoperability realm. In the *pragmatic layer*, the information exchange elements are grouped into business objects with a common context. Annotations on the *dynamic layer* capture the processes invoked and the system state changes taking place when business objects are exchanged between systems. Finally, the relevant constraints and assumptions are captured in the *conceptual layer*, which completes the composability realm.

The LCIM supports a structured way to annotate M&S services. Dobrev et al. (2007) show how this model can be used to support interoperation in general applications. Zeigler and Hammonds (2007) use it to compare it with their ideas on using ontological means in support of interoperation. It was furthermore applied for the Department of Defense, the Department for Homeland Security, The Department of Energy, and NATO. These annotations are necessary requirements to allow *discovery*, *selection*, and *composition* of services.

These annotations should be interpreted as a machine understandable version of the underlying conceptual model of the M&S service. Robinson (2008) defines the conceptual model as *"a non-software specific description of the simulation model that is to be developed, describing the objectives, inputs, outputs, content, assumptions, and simplifications of the model."* He furthermore points out that there is a significant need to agree on how to do develop conceptual models and capture information formally. What is needed in support of composable services is therefore to capture *objectives, inputs, outputs, content, assumptions, and simplifications* of the model in the *technical, syntactical, semantic, pragmatic, dynamic, and conceptual* category. The discipline of model-based data engineering (Tolk and Diallo, 2008) is a first step into this direction.

To understand why these annotations are so important, it is necessary to understand how machines gain understanding. Zeigler (1986) introduced a model for understanding a system within another observing system. Figure 2 shows the three premises that need to be supported by the annotations describing the M&S services. The system – or the M&S service – is herein described by its *properties* that are grouped into propertied *concepts* (the basic simulated entities and attributes), the *processes* (the behavior of simulated entities and how their attributes change), and *constraints* (assumptions constraining the values of the attributes and the behavior of the system).

- The first premise is that the observing system has a *perception* of the system to be understood. This means that the properties and processes must be observable and perceivable by the observing system. The properties used for the perception should not significantly differ in scope and resolution from the properties exposed by the system under observation.
- The second premise is that the observing system needs to have a *meta-model* of the observed system. The meta-model is a description of properties, processes, and constraints of the expected behavior of the observed system. Without such a model of the system, understanding is not possible.
- The third premise is the *mapping* between observations resulting in the perception and meta-models explaining the observed properties, processes, and constraints.

In other words, machine understanding is the selection process of the appropriate meta-model to explain the observed properties, processes, and constraints. This corresponds to the selection of appropriate M&S services to support a decision. The properties and propertied concepts are described by syntax, semantic, and pragmatic annotations, processes by dynamic annotations, and

Figure 2. Premises for System's Understanding

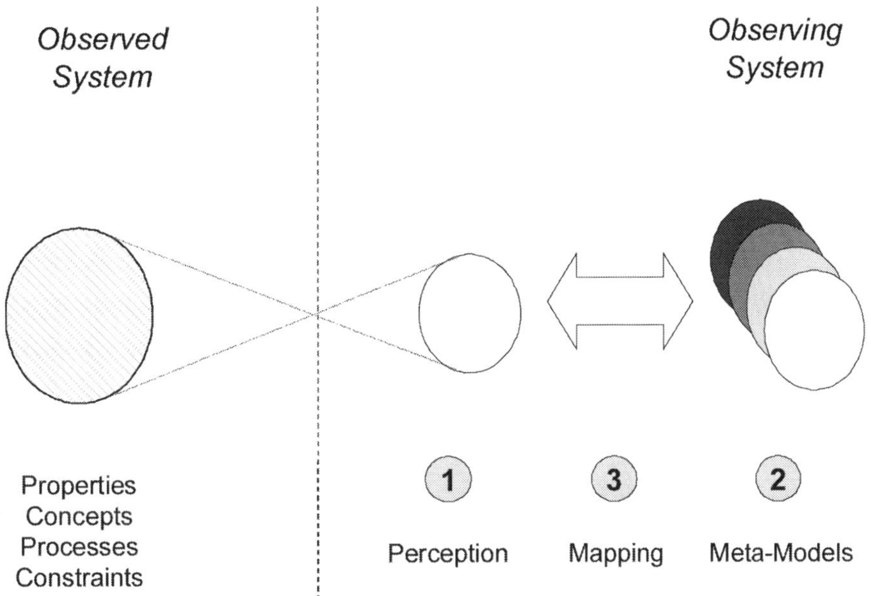

constraints by conceptual annotations capturing objectives, inputs, outputs, content, assumptions, and simplifications in addition to implementation details and technical specifications. No matter if these annotations are used to discover, select, and orchestrate M&S functionality as operational services or if they are used by intelligent agents to communicate their use, they are necessary for every application beyond the traditional system developments that are often intentionally not reused in their requirements. This section can therefore also serve as a guideline for what is needed to annotate legacy systems that shall be integrated into a net-centric and service-oriented environment to contribute to a system of systems.

Table 1 can be used as a checklist to ensure that all information is captured or obtainable for a candidate simulation system for decision support.

All this related work sets the frame for describing M&S services to support their discovery and orchestration for integration as an orchestrated set of tools into the operational infrastructure used by the decision maker. The following section will describe the requirements for the simulation systems themselves in more detail.

Table 1. Checklist points for decision support simulation annotations

Annotation Categories	Levels of Interoperation	System Characteristics	Conceptual Model Characteristics
• Integratability	• Technical	• Properties	• Objectives
• Interoperability	• Syntactic	• Concepts	• Inputs
• Composability	• Semantic	• Processes	• Outputs
	• Pragmatic	• Constraints	• Content
	• Dynamic		• Assumptions
	• Conceptual		• Simplifications

REQUIREMENTS FOR SIMULATION SYSTEMS

This section will explain the necessary requirements for simulation systems when they are to be used as decision support systems. These requirements may not be sufficient for all application domains, so additional application domain expertise is needed for informed selection. While the focus in the last section was annotation, the focus here will be the content and completeness of the simulation system.

This section will start with general requirements for all simulation systems to be applied to decision support and will finish with additional requirements in the case a federation of simulation systems is to be applied, which is the more likely scenario. As the NATO Code of Best Practice (NATO, 2002) points out: it is highly unlikely that one tool or simulation system will be able to deal with all questions describing the sponsor's problem; the use of an orchestrated set of tools should be the rule.

This section extends and generalizes the findings documented in Tolk (1999) and referenced in NRC (2002). While the principle results are still valid, the development in the recent years, in particular in the domain of agent-based models in support of behavior modeling and of computer generated forces contributed significantly to solutions in challenging areas that need to be incorporated. The section on current developments in this chapter will focus on these developments in more detail.

Modeling of Relevant System Characteristics

Models are purposeful abstractions from reality. This means that they simplify some things, leave others out, use assumptions, etc. When using a simulation system as a decision support simulation, it is crucial that all relevant system characteristics are captured. This includes all aspects of the system: modeled entities (properties and concepts), modeled behavior and interactions (processes), and modeled constraints. The reason is trivial: if something important is not part of the model, it cannot be considered for the analysis, nor can it be part of the recommended solution.

The artifacts used for documentation of the system (and annotation) during the conceptualization phase should capture the necessary information. As defined by Robinson (2008) in his overview work on conceptual modeling, the characteristics of a conceptual model are *objectives, inputs, outputs, content, assumptions, and simplifications*. A practical way to accomplish this task has been captured in the contributions of Brade (2000), which will be addressed in the section on verification and validation.

Example: A simulation system shall be used to support the decision of where to install additional gas stations in a town. It models the cars used in this town, the behavior of the car drivers, and the gas stations already in use within this town. The idea is to use simulation based optimization to find out how many new gas stations should be built and where.

In order to be able to use the simulation system, additional system characteristics may have to be captured, such as

- Under which circumstances are drivers willing to go to neighboring towns to buy gas to fill up their cars? (Assumption that drivers in the town will use gas stations in this town)
- How will the competition react? Will they build new stations? Will they close down stations? (Assumption that only the company conducting the study actively changes the gas supply infrastructure)
- Are there additional influences that are relevant, such as the overall driving behavior based on current average oil prices? (Assumption that decision rules used by simulated entities follow a closed world assumption)

Even if this is not implemented, the simulation can still be used in support of analysis, but the expert must be very well aware of what the simulation systems simulates and how. In other words, an awareness of the assumptions and constraints affecting the validity of the simulation results is necessary.

In summary, it is essential that the simulation system can support the decision to be made by ensuring that all concepts, properties, processes, and constraints identified to be relevant in the problem specification process are implemented. The NATO Code of Best Practice (NATO, 2002) gives guidance for the problem specification process. The conceptual models used for the simulation development document the respective characteristics of the simulation.

Ability to Obtain All Relevant Data

Closely related is the second premise that must be fulfilled: the relevant data needed for the simulation system initialization and execution must be obtainable. Even if a simulation system is complete in describing all concepts, properties, processes, and constraints, the model can be practically useless if the necessary data to drive these models cannot be provided. The quality of the solution is driven by the quality of the model and the quality of the data.

The NATO Code of Best Practice (NATO, 2002) gives guidance with respect to obtaining data and ensuring the necessary quality of data. Among the identified factors for good data are the reliability of sources and the accuracy of data. Additional factors are the costs to obtain data, how well the data is documented, if and how the data have been modified, etc.

Another aspect that increases in importance in the area of net centricity and service- oriented architectures is the alignment of protocols for data storage and exchange in operational systems and decision support simulation systems. The optimal case is that decision support simulation systems and the embedded operational system use the same data representation. If this is not the case, data mediation may be a possible solution to mapping the existent operationally available data to the required initialization and input data. However, it must be pointed out that data mediation requires the mapping of data is complete, unambiguous, and precise. To this extent, Model-based Data Engineering was developed and successfully applied (Tolk and Diallo, 2008).

An aspect unique to M&S services is the need that modeled data are conceptually connected to operationally available data. As models are abstractions of reality, some data may be "academic" abstractions that theoretically are constructible, but are difficult to observe or to obtain. In particular statistical measure of higher order, such as using a negative polynomial bivariate intensity probability distribution function to model the movement of entities as a fluid, often make perfectly sense when developing the model, but may be very hard to feed with real world data.

Example: A simulation system shall be used to support a decision maker with evacuation decisions during a catastrophic event (Muhdi, 2006). Most evaluation models currently used are flow-based models. The data available in a real emergency, however, is discrete, describing exit obstacles, individuals, and other data that need to be converted into this model (and potentially mapped back in support of creating elements of a plan that needs to be shared using the operational infrastructure).

In summary, it is essential that data needed by the model can be obtained and mediated. The data will be used to initialize the simulation systems and as input data during execution.

Validation and Verification of Model, Simulation, and Data

Validation and verification are processes to determine the simulation's credibility. They deal with answering questions such as "Does the simulation system satisfy its intended use? Can the simulation system be used to evaluate specific questions? How close does the simulation system come to reality?" In other words, validation and verification are the processes of determining if a simulation is correct and usable to solve a given problem.

The US Department of Defense defined validation and verification for military use in their M&S instruction (DoD, 1996). Validation is the process of determining the degree to which a model or simulation is an accurate representation of the real world from the perspective of the intended uses. Verification is the process of determining that a model or simulation implementation accurately represents the developer's conceptual description and specifications. In other words, validation determines if the right thing is coded while verification determines if the thing is coded right. Validation determines the behavioral and representational accuracy; verification determines the accuracy of transformation processes.

There are many papers available dealing with the necessity to validate and verify models and simulation before using them for decision making. The interested reader is pointed to the overview of methods and tools provided by Balci (1998) and several specific papers by Sargent (1999, 2000, 2007). The work of Brade (2000) making practical recommendations regarding artifacts was already mentioned in a previous section.

It seems to be obvious that simulation systems designed to be used as decision support simulation systems must be validated and verified. This is true for the models, the simulations, and the data. If this is not the case, the results will not be credible and reliable and as such not applicable to support decisions.

It is not trivial but is at least possible to accomplish verification and validation for physical processes and models. However, the simulated entities and processes are not limited to such physical processes. Cognitive processes and decision models need to be modeled as well. Moya and Weisel (2008) point out the resulting challenges.

Example: To show the necessity of verification and validation, two examples of simulation failures in operational environments are given that are directly applicable to decision support simulation systems as well.

Simulation in Testing: During Operation Iraqi Freedom, Patriot missiles shot down two allied aircraft & targeted another. On March 23, 2003, the pilot and co-pilot aboard a British Tornado GR4 aircraft that was shot down by a U.S. Patriot missile died. On April 2, 2003, another Patriot missile downed a U.S. Navy F/A-18C Hornet which was flying a mission over Central Iraq. The evaluation report identified one of the causes of these failures stemmed from using an invalid simulation to stimulate the Patriot's fire control system during its testing.

Simulation in Engineering: Another catastrophic event in spring 2003 was the Columbia disaster. The space shuttle had been damaged by foam debris during takeoff. NASA engineers decided, based on their professional judgment, that the damage would not endanger the shuttle when returning to earth. They were wrong and the shuttle broke apart when entering the atmosphere, killing the crew and throwing the shuttle program significantly back. What is of interest for the readers of this chapter is that the simulation available to the experts predicted the disaster, but the results were not deemed reliable and credible by the experts. Obviously they were mistaken.

In summary, it is necessary to only make use of validated and verified models and data for decision support simulation systems. It is essential that the decision maker is supported with reliable and credible information.

Creating Situation Adequate Behavior

One of the most challenging premises is to fulfill the requirement for situation adequate behavior.

This premise addresses the behavior of simulated entities, which is represented by the processes of the system characteristics. The premise has a very practical side and a resulting challenge. Many simulation systems used in other application domains, in particular for training and testing, also require that the simulated entities behave as they would in the real world. If this behavior is connected with human decision making, it is quite often humans in the loop making the decision.

A typical military computer assisted exercise comprises not only the training audience, but also soldiers representing the subordinates, partners, and superior commands, as well as the opposing forces. To ensure that soldiers "train as they fight," the units are commanded by military experts. The simulation computes the movement, the attrition, the reconnaissance, and other processes that are based on physical aspects. It is more the rule than the exception that more soldiers are needed to support the simulation system than are trained in an event. The use of agents to generate the orders is mandatory for decision support; otherwise the manpower would increase to the point of no longer being practical or feasible.

Example: If training on the brigade level is conducted, approximately 800 orders have to be created in order to drive a simulation model. Taking into account that not only the orders for the brigade are needed, but also for the neighbored units and – last but not least – the orders for the enemy increases this number by the factor of four to six resulting in the number of 3,000 to 5,000 orders to be created for just one alternative. This is accomplished by a group of 500 to 600 soldiers. As this many personnel can never be supported by a brigade headquarter that wants to use the simulation for decision support, the majority of these orders must be generated by means of behavioral representation in modeling and simulation.

In summary, intelligent software agents representing human behavior in simulation systems must ensure that the simulated entities behave correctly. Scripted and rule driven approaches are not sufficient. The conference on behavioral representation in modeling and simulation (BRIMS) is a good source of current research and proposed solutions. Yilmaz et al. (2006) are giving a good overview of such use of agents in serious games as well as in simulation systems.

Additional Issues When Using Federations of Simulation Systems

The first four premises must be fulfilled by every simulation system that will be used for decision support. However, as pointed out several times in this chapter, the application of an orchestrated set of tools in order to evaluate all relevant aspects of a model is the rule. If several simulation systems need to be used to provide the required functionality, some concerns need to be addressed that are unique to federations of simulation systems.

The main challenge is to orchestrate simulations not only regarding their execution, but also to conceptually align them to ensure that the federation delivers a consistent view to the decision maker fulfilling all requirements that have been captured. The LCIM can support this challenge. A simulation federation in itself is a complex system of systems. Current simulation interoperability standards are not sufficient to support the necessary consistency. Besides several publications by the author in this domain, this view is shared by many other experts in the field, such as Zeigler and Hammonds (2007) show in their survey. Yilmaz (2007) proposed the use of meta-level ontology relations to measure conceptual alignment.

The objective of these alignments is to harmonize the three elements essential for simulation result consistency, which are the concepts underlying the simulated entities (resolution and structure), the internal decision logic used to generate the behavior of the simulated entities, and the external measure of performance used to evaluate the accomplishment. If this is not the case, the results will be counter-intuitive at best, and inconsistent and wrong at worst. As shown in Muguira and Tolk (2006), even if all federates are validated and correct, the federation may still expose structural variances, making the result unusable for decision support.

Example: The triangle of concepts, internal decision logic, and external evaluation logic must be harmonized regarding all three aspects, or structural variance can result in non-credible results.

- *Concepts and decision logic:* Simulation *A* represents a fish swarm as a cubicle; simulation *B* uses a statistical distribution within a bowl. If the decision logic of simulation *A* is used to support a decision in simulation *B*, the decision is based on the wrong assumptions and is likely to be wrong.
- *Concepts and evaluation logic:* If the measure of merit requires inputs not exposed by the federation, or if the structure and resolution are significantly different in the federated simulation systems, the evaluation is wrong.
- *Decision and evaluation logic:* One of the most observed reasons for strange behavior in the results of federations is that the measure of merit used for the evaluation and the measure of merit used to optimize the decisions internally are not harmonized. If the decision logic targets to maximize the amount of fish captured in each event and the evaluation logic checks if the overall regeneration of fish is ensured as well, it is likely that structural variances will occur.

In summary, it must be ensured that the simulation systems are not only coupled and technically correct (based on currently available simulation interoperability standards), but that they are aligned regarding concepts, internal decision logic, and external evaluation logic as well.

Summarizing all five premises dealt with in this chapter, the following enumeration lists the questions that need to be answered to ensure that the requirements are fulfilled:

- Are all concepts having a role in solving the problem identified and simulated in the simulation?
- Are the properties used to model the propertied concepts in the necessary resolution and the necessary structure?
- Are all identified processes (entity behavior and overarching processes) modeled?
- Are the assumptions and constraints identified for the operational challenge to be decided upon reflected appropriately by the simulation system?
- Can operational data and author authoritative data sources provide all data needed for the initialization of the simulation system?
- Can operational data provide all data needed as input data during the execution of the simulation system?
- Do the operational infrastructure and the decision support simulation system share the same data model, or – if this is not the case – can model-based data engineering be applied to derive the necessary mediation functions? Are possible semantic losses resulting from the mapping acceptable?
- Is the data obtainable in the structure and resolution (and accuracy) needed, or – if this is not the case – can the data be transformed into the required format?
- Are all potential M&S services and simulation systems validated and verified?
- Are the data validated and verified?
- Is the behavior of all simulated entities situation adequate?
- In case of personnel intensive simulation systems, can the human component be replaced with intelligent software agents to produce the required decisions (or can it be ensured that always enough persons are available to support the application)?

- Are the represented concepts (simulated entities) sufficient to produce the properties needed for the measures of merit of the decision logic and the evaluation logic?
- Are the measures of merit used for the internal decision logic aligned with the external evaluation logic?

This list builds the core of questions the developer of decision support simulation systems must be able to answer positively. Additional application specific questions are likely and need to be captured for respective development or integration projects as requirements.

EXAMPLES OF DECISION SUPPORT SIMULATION APPLICATIONS

The previous sections dealt with the necessary annotation for M&S services and the requirements for simulation systems when being used for decision support. This section gives some selected references to examples of using simulation for decision support. While these examples are neither complete nor exclusive, they do show that decision support simulation is already applied in various fields.

Kvaale (1988) describes the use of simulation systems in support of design decisions for a new generation of fast patrol boats. This application is the traditional use of simulation in support of the procurement process: alternatives are simulated and compared using a set of agreed to measures of merit. Although this application is not driving the support using operational data directly obtained from operational systems, it is one of the first journal papers describing the use of simulation systems for decision support.

Everett (2002) describes the design of a simulation model to provide decision support for the scheduling of patients waiting for elective surgery in the public hospital system. The simulation model presented in this work can be used as an operational tool to match hospital availability with patient need. To this end, patients nominated for surgery by doctors are categorized by urgency and type of operation. The model is then used to simulate necessary procedures, available resources, resulting waiting time, and other decision parameters that are displayed for further evaluation. Therefore, the model can also be used to report upon the performance of the system and as a planning tool to compare the effectiveness of alternative policies in this multi-criteria decision health-care environment.

Truong et al. (2005) present another application domain for decision supporting use of simulation: fisheries policy and management decisions in support of optimizing a harvesting plan for the fishing industry. As in many application areas, the behavior of fish and the effects of harvesting are not fully understood, but can be captured to sufficient detail to build a simulation that reflects the known facts in sufficient detail. This enables simulation-based optimization using the simulation to obtain quasi-objective function values of possible alternatives, in the example particular fishing schedules. This idea is applicable in similar environments with uncertain and imprecise data that exposes some trends that can be captured in simulations.

Power and Sharda (2007) summarized related ideas recently in their work on model-driven decision support systems. Following their definition, model-driven decision support systems use algebraic, decision analytic, financial, simulation, and optimization models to provide decision support. Like this chapter, they use optimization models, decision theory, and other means of operational analysis and research as an orchestrated set of tools in which simulation is embedded in an aligned way.

Decision support systems, as well as the use of simulation systems, have a relatively long history in the military domain. An example is given by Pohl et al. (1999) who present the results of a project sponsored by the Defense Advance Project

Research Agency (DARPA). The Integrated Marine Multi-Agent Command and Control System (IMMACCS) is a multi-agent, distributed system. It is designed to provide a common tactical picture as discussed earlier in this chapter and an early adapter of the agent-based paradigm for decision support. Between 1999 and 2004, the Office for Naval Research (ONR) sponsored a series of workshops on decision support systems in the United States. Furthermore, Wilton (2001) presented an overview of decision support simulation ideas integrated with C2 devices for the training of soldiers.

Management related military applications are regularly discussed at the annual International Command and Control Research and Technology Symposia (ICCRTS), which features a special track on decision support. The work presented here is often focused on cognitive aspects of sense-making and aims more at increasing the shared situational awareness than on a common technical framework. Many principles are not limited to the military domain but are applicable to all forms of agile organizations without fixed external structures. An example is the analysis of requirements of cognitive analysis to support C2 decision support system design by Potter et al. (2006).

The books edited by Tonfoni and Jain (2002), Phillips-Wren and Jain (2005), and Phillips-Wren et al. (2008) are valuable references for examples of using means of artificial intelligence and intelligent software agents in support of decision making using simulation systems. The use of ontological means to ensure composability of models and interoperability of simulations is the topic of several additional publications.

CURRENT DEVELOPMENTS

As with the previous section of this chapter, it is extremely difficult to decide which of the current developments should be highlighted, as every development in the discipline of M&S improves the usability of resulting systems for decision support. The focus of contributions in this section is therefore relatively small. As before, the idea is not to be restrictive but to give examples.

The military community used the Simulation Interoperability Workshops to work on the development of a technical reference model (TRM) for coupling of command, control, communications, computers, intelligence, surveillance, and reconnaissance (C4ISR) with M&S systems (Griffin et al., 2002). Figure 3 shows a generalization of the model, as already recommended by Tolk et al. (2008).

The model focuses on data exchange requirements and categories. The data is categorized as:

- simulation specific management data unique to the decision support simulation system,
- operational initialization data describing the data needed for initialization of both systems describing concepts, properties, processes, and constraints,
- dynamic exchange of operational data describing information that captures the input and output data of both worlds during execution, and
- operational system specific management data unique to the IT infrastructure used by the decision maker.

Unfortunately, the standardization work on the TRM was never completed, so that besides the final report of the study group and several contributing workshop papers no standard in support of embedding decision support simulations into operational IT infrastructures exists. Work in this domain would be very helpful to the M&S community.

As pointed out before, the US Department of Defense is working on a series of strategies and standards to enable net-centric operations. Another

Figure 3. Generalization of the C4ISR Technical Reference Model

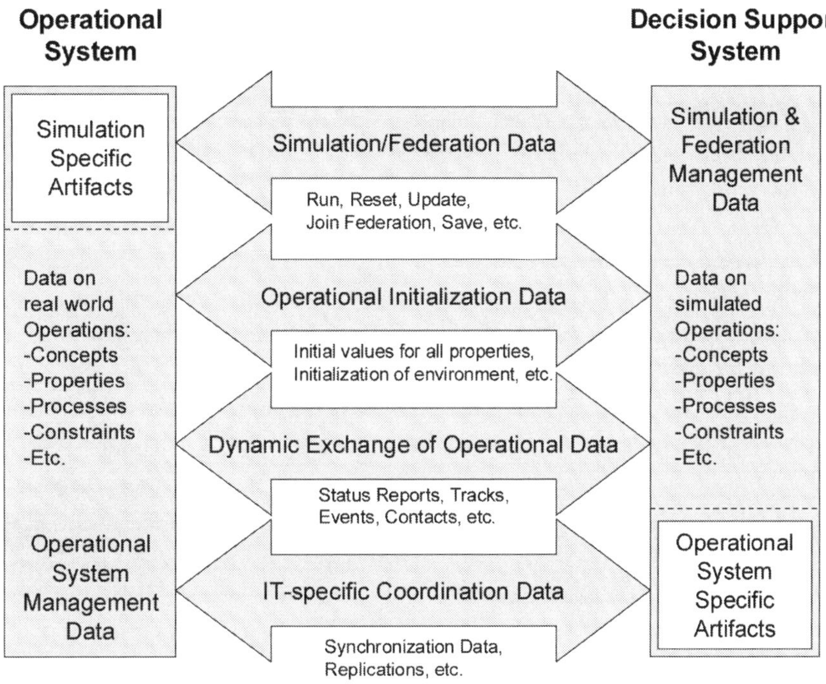

standard developed under the roof of the Simulation Interoperability Standardization Organization (SISO), the Base Object Model (BOM) Standard, is currently being evaluated to be used for the registration of M&S services. The standard is defined in two documents, the "Base Object Model (BOM) Template Standard" and the "Guide for Base Object Model (BOM) Use and Implementation" (SISO, 2006). The first document provides the essential details regarding the makeup and ontology of a BOM, the companion document gives examples and best practice guidelines for using and implementing the new standard. In summary, the BOM standard provides a standard to capture the artifacts of a conceptual model. Furthermore, it can be used to design new simulation systems as well as integrating legacy simulations. The conceptual model elements defined by the BOM standard contain descriptions of concepts, properties, and processes. The description is not only static, but the interplay is captured in the form of state machines as well. The BOM template is divided in five categories and reuses successful ideas of the current simulation interoperability standard "High Level Architecture" (IEEE 1516) and supports:

- Model Identification by associating important metadata with the BOM. Examples include the author of the BOM, the responsible organization, security constraints, etc.
- Conceptual Model Definition by describing patterns of interplay, state machines representing the aspects of the conceptual model, entity types, and event types.
- Modeling Mapping by defining what simulated entities and processes represent what elements of the conceptual model.
- Object Model Definition by recording the necessary implementation details (objects, interactions, attributes, parameter, and data types as defined by IEEE 1516)

- Additional Supporting Tables in the form of notes and lexicon definitions.

The BOM standard has successfully been applied in several US military research projects. Outside the US Department of Defense, its use has not yet been documented sufficiently to speak of a broadly accepted standard. The potentials, however, are impressive, as shown by Searle and Brennan (2006) in their educational notes for NATO.

As mentioned at the beginning of this section, many other developments are of high interest to decision support simulation developers. The increasing use of agent-directed simulation is one aspect. The human behavior representation in M&S is another. Complex systems in knowledge-based environments (Tolk and Jain, 2008) are another domain of interest, in particular how to cope with uncertainties or how to apply ontological means in support of complex system interoperation. Enumerating all interesting fields lies beyond the scope of this chapter.

In summary, the developer of decision support simulation systems or the engineer tasked with the integration of simulation systems for operational decision support must follow developments in all levels of interoperation: from technical innovations enabling better connectivity (such as optical memories or satellite based internet communications) via improvement in the interoperability domain (such as new developments in the domain of semantic web services) to conceptual questions (including standardizing artifacts in machine understandable form). As systems developed for this domain need to be highly reliable and credible, the engineer needs not only to be highly technically competent, but also needs to follow the code of ethics of the profession, as wealth – and sometimes even survival – will depend on the work and efforts produced.

SUMMARY

Decision support of operational processes is the most challenging application for simulation systems. In all other application domains, the necessity for credible and reliable results is lower than for real world operation decision support. While in all other domains there is always the chance to react and counteract to insufficient M&S functionality, a wrong recommendation in support of real world operations can lead to significant financial trouble or even the loss of lives.

This chapter summarized the requirements for simulation systems when being used for such applications. It showed the necessary annotation to allow the discovery, selection, and orchestration of M&S systems as services in service-oriented environments. It also listed the premises for simulation system functionality, focusing on completeness of concepts, properties, processes, and constraints, obtainability of data, validation and verification, and the use of means of knowledge management. The current developments continue to close gaps so that the use of simulation in the context of operational decision support will soon enable support to even the cognitive levels of group decision making and common situational awareness.

REFERENCES

Agarwal, S., Handschuh, S., & Staab, S. (2005). Annotation, Composition and Invocation of Semantic Web Services. *Journal of Web Semantics*, *2*(1), 1–24.

Alberts, D. S., & Hayes, R. E. (2003). Power to the Edge, Command and Control in the Information Age. *Department of Defense Command and Control Program; Information Age Transformation Series*. Washington, DC.

Alesso, H. P., & Smith, C. F. (2005). *Developing Semantic Web Services*. Wellesley, MA: A.K. Peters, Ltd.

Balci, O. (1998). Verification, Validation, and Testing. In J. Banks, (Ed) *Handbook of Simulation*. New York: John Wiley & Sons.

Brade, D. (2000). Enhancing M&S Accreditation by Structuring V&V Results. *Proceedings of the Winter Simulation Conference*, (pp. 840-848).

Department of Defense. (1996). *Modeling & Simulation Verification, Validation, and Accreditation* (US DoD Instruction 5000.61). Washington, DC: Author.

Dobrev, P., Kalaydjiev, O., & Angelova, G. (2007). *From Conceptual Structures to Semantic Interoperability of Content*. (LNCS Vol. 4604, pp. 192-205). Berlin: Springer-Verlag.

Everett, J. E. (2002). A decision support simulation model for the management of an elective surgery waiting system. *Journal for Health Care Management Science*, *5*(2), 89–95. doi:10.1023/A:1014468613635

Family of Standards for Modeling and Simulation (M&S) High Level Architecture (HLA) – (a) IEEE 1516-2000 Framework and Rules; (b) IEEE 1516.1-2000 Federate Interface Specification; (c) IEEE 1516.2-2000 Object Model Template (OMT) Specification IEEE 1516-2000, *IEEE Press.*

Griffin, A., Lacetera, J., Sharp, R., & Tolk, A. (Eds.). (2002). *C4ISR/Sim Technical Reference Model Study Group Final Report (C4ISR/Sim TRM),* (SISO-Reference Document 008-2002-R2). Simulation Interoperability Standards Organization, Orlando, FL.

Kvaale, H. (1988). A decision support simulation model for design of fast patrol boats. *European Journal of Operational Research*, *37*(1), 92–99. doi:10.1016/0377-2217(88)90283-4

Moya, L., & Weisel, E. (2008) The Difficulties with Validating Agent Based Simulations of Social Systems. *Proceedings Spring Multi-Simulation Conference, Agent-Directed Simulation*, Ottawa, Canada.

Muguira, J. A., & Tolk, A. (2006). Applying a Methodology to Identify Structural Variances in Interoperations. *Journal for Defense Modeling and Simulation*, *3*(2), 77–93. doi:10.1177/875647930600300203

Muhdi, R. A. (2006). Evaluation Modeling: Development, Characteristics and Limitations. *Proceedings of the National Occupational Research Agenda* (NORA), (pp. 87-92).

National Research Council. (2002). *Modeling and Simulation in Manufacturing and Defense Acquisition: Pathways to Success*. Washington, DC: Committee on Modeling and Simulation Enhancements for 21st Century Manufacturing and Defense Acquisition, National Academies Press.

National Research Council. (2006). *Defense Modeling, Simulation, and Analysis: Meeting the Challenge*. Washington, DC: Committee on Modeling and Simulation for Defense Transformation, National Academies Press.

North Atlantic Treaty Organization. (2002). *NATO Code of Best Practice for Command and Control Assessment*. Revised ed. Washington, DC: CCRP Press

Ören, T. I., & Numrich, S. K. S.K., Uhrmacher, A.M., Wilson, L. F., & Gelenbe, E. (2000). Agent-directed simulation: challenges to meet defense and civilian requirements. In *Proceedings of the 32nd Conference on Winter Simulation,* Orlando, Florida, December 10 – 13. San Diego, CA: Society for Computer Simulation International.

Page, E. H., Briggs, R., & Tufarolo, J. A. (2004). Toward a Family of Maturity Models for the Simulation Interconnection Problem. In *Proceedings of the Spring Simulation Interoperability Workshop*. New York: IEEE CS Press.

Phillips-Wren, G., Ichalkaranje, N., & Jain, L. C. (Eds.). (2008). *Intelligent Decision Making –An AI-Based Approach.* Berlin: Springer-Verlag.

Phillips-Wren, G., & Jain, L. C. (Eds.). (2005). *Intelligent Decision Support Systems in Agent-Mediated Environments.* The Netherlands: IOS Press.

Pohl, J. G., Wood, A. A., Pohl, K. J., & Chapman, A. J. (1999). IMMACCS: A Military Decision-Support System. *DARPA-JFACC 1999 Symposium on Advances in Enterprise Control*, San Diego, CA.

Potter, S. S., Elm, W. C., & Gualtieri, J. W. (2006). Making Sense of Sensemaking: Requirements of a Cognitive Analysis to Support C2 Decision Support System Design. *Proceedings of the Command and Control Research and Technology Symposium.* Washington, DC: CCRP Press.

Power, D. J., & Sharda, R. (2007). Model-driven decision support systems: Concepts and research directions. *Journal for Decision Support Systems*, *43*(3), 1044–1061. doi:10.1016/j.dss.2005.05.030

Pullen, J. M., Brunton, R., Brutzman, D., Drake, D., Hieb, M. R., Morse, K. L., & Tolk, A. (2005). Using Web Services to Integrate Heterogeneous Simulations in a Grid Environment. *Journal on Future Generation Computer Systems*, *21*, 97–106. doi:10.1016/j.future.2004.09.031

Robinson, S. (2008). Conceptual modelling for simulation Part I: definition and requirements. *The Journal of the Operational Research Society*, *59*, 278–290. doi:10.1057/palgrave.jors.2602368

Sargent, R. G. (1999). Validation and verification of simulation models. *Proceedings of the Winter Simulation Conference*, *1*, 39–48.

Sargent, R. G. (2000). Verification, Validation, and Accreditation of Simulation Models. *Proceedings of the Winter Simulation Conference*, (pp. 50-59).

Sargent, R. G. (2007). Verification and validation of simulation models. *Proceedings of the Winter Simulation Conference*, (pp. 124-137).

Searle, J., & Brennan, J. (2006). General Interoperability Concepts. In *Integration of Modelling and Simulation* (pp. 3-1 – 3-8), (Educational Notes RTO-EN-MSG-043). Neuilly-sur-Seine, France

Simulation Interoperability Standards Organization (SISO). (2006). *The Base Object Model Standard; SISO-STD-003-2006: Base Object Model (BOM) Template Standard; SISO-STD-003.1-2006: Guide for Base Object Model (BOM) Use and Implementation.* Orlando, FL: SISO Documents.

Srivastava, B., & Koehler, J. (2003). Web Service Composition - Current Solutions and Open Problems. *Proceedings ICAPS 2003 Workshop on Planning for Web Services.*

Tolk, A. (1999). Requirements for Simulation Systems when being used as Decision Support Systems. [), IEEE CS press]. *Proceedings Fall Simulation Interoperability Workshop*, *I*, 29–35.

Tolk, A., & Diallo, S. Y. (2008) Model-Based Data Engineering for Web Services. In Nayak R. et al. (eds) *Evolution of the Web in Artificial Intelligence Environments*, (pp. 137-161). Berlin: Springer.

Tolk, A., Diallo, S. Y., & Turnitsa, C. D. (2008). Implied Ontological Representation within the Levels of Conceptual Interoperability Model. [IDT]. *International Journal of Intelligent Decision Technologies*, *2*(1), 3–19.

Tolk, A., Diallo, S. Y., Turnitsa, C. D., & Winters, L. S. (2006). Composable M&S Web Services for Net-centric Applications. *Journal for Defense Modeling and Simulation*, *3*(1), 27–44. doi:10.1177/875647930600300104

Tolk, A., & Jain, L. C. (Eds.). (2008). *Complex Systems in the Knowledge-based Environment.* Berlin: Springer Verlag.

Tonfoni, G., & Jain, L. C. (Eds.). (2003). *Innovations in Decision Support Systems*. Australia: Advanced Knowledge International.

Tosic, V., Pagurek, B., Esfandiari, B., & Patel, K. (2001). On the Management of Compositions of Web Services. In *Proceedings Object-Oriented Web Services (OOPSLA)*.

Truong, T. H., Rothschild, B. J., & Azadivar, F. (2005) Decision support system for fisheries management. In *Proceedings of the 37th Conference on Winter Simulation*, (pp. 2107-2111).

van Dam, A. (1999). Education: the unfinished revolution. [CSUR]. *ACM Computing Surveys*, *31*(4). doi:10.1145/345966.346038

Wilton, D. (2001). The Application Of Simulation Technology To Military Command And Control Decision Support. In *Proceedings Simulation Technology and Training Conference* (SimTecT), Canberra, Australia.

Yilmaz, L. (2007). Using meta-level ontology relations to measure conceptual alignment and interoperability of simulation models. In *Proceedings of the Winter Simulation Conference*, (pp. 1090-1099).

Yilmaz, L., Ören, T., & Aghaee, N. (2006). Intelligent agents, simulation, and gaming. *Journal for Simulation and Gaming*, *37*(3), 339–349. doi:10.1177/1046878106289089

Zeigler, B. P. (1986) Toward a Simulation Methodology for Variable Structure Modeling. In Elzas, Oren, Zeigler (Eds.) *Modeling and Simulation Methodology in the Artificial Intelligence Era.*

Zeigler, B. P., & Hammonds, P. E. (2007) *Modeling & Simulation-Based Data Engineering: Introducing Pragmatics into Ontologies for Net-Centric Information Exchange*. New York: Academic Press.

KEY TERMS AND DEFINITIONS

Decision Support Systems: Are information systems supporting operational (business and organizational) decision-making activities of a human decision maker. The DSS shall help decision makers to compile useful information from raw data and documents that are distributed in a potentially heterogeneous IT infrastructure, personal or educational knowledge that can be static or procedural, and business models and strategies to identify and solve problems and make decisions.

Decision Support Simulation Systems: Are simulation systems supporting operational (business and organizational) decision-making activities of a human decision maker by means of modeling and simulation. They use decision support system means to obtain, display and evaluate operationally relevant data in agile contexts by executing models using operational data exploiting the full potential of M&S and producing numerical insight into the behavior of complex systems.

Integratability: Contends with the physical/technical realms of connections between systems, which include hardware and firmware, protocols, networks, etc. If two systems can exchange physical data with each other in a way that the target system receives and decoded the submitted data from the sending system the two systems are *integrated.*

Interoperability: Contends with the software and implementation details of interoperations; this includes exchange of data elements via interfaces, the use of middleware, and mapping to common information exchange models. If two systems are integrated and the receiving system can not only decode but understand the data in a way that is meaningful to the receiving system, the systems are *interoperable.*

Composability: Contends with the alignment of issues on the modeling level. The underlying models are purposeful abstractions of reality used for the conceptualization being implemented by the resulting systems. If two systems are interoperable and share assumptions and constraints in a way that the axioms of the receiving system are not violated by the sending system, the systems are *composable*.

Conceptual Modeling: Is the process of defining a non-software specific formal specification of a conceptualization building the basis for the implementation of a simulation system (or another model-based implementation) describing the objectives, inputs, outputs, content, assumptions, and simplifications of the model. The conceptual model conceptual model is a bridge between the real world observations and the high-level implementation artifacts.

Validation and Verification: Are processes to determine the simulation credibility. *Validation* is the process of determining the degree to which a model or simulation is an accurate representation of the real world from the perspective of the intended uses. Validation determines the behavioral and representational accuracy. *Verification* is the process of determining that a model or simulation implementation accurately represents the developer's conceptual description and specifications. Verification determines the accuracy of transformation processes.

Model-Based Data Engineering: Is the process of applying documented and repeatable engineering methods for *data administration* – i.e. managing the information exchange needs including source, format, context of validity, fidelity, and credibility –, *data management* – i.e. planning, organizing and managing of data, including defining and standardizing the meaning of data and of their relations -, *data alignment* – i.e. ensuring that data to be exchanged exist in all participating systems, focusing a data provider /data consumer relations -, and *data transformation* – i.e. the technical process of mapping different representations of the same data elements to each other – supported by a common reference model.

This work was previously published in Handbook of Research on Discrete Event Simulation Environments: Technologies and Applications edited by E. Abu-Taieh & A. El Sheikh, pp. 317-336, copyright 2010 by Information Science Reference (an imprint of IGI Global).

Chapter 14
Genetic Algorithm to Solve Multi-Period, Multi-Product, Bi-Echelon Supply Chain Network Design Problem

R. Dhanalakshmi
IBM India Pvt. Ltd, India

P. Parthiban
National Institute of Technology, India

K. Ganesh
Tata Consultancy Services Limited, India

T. Arunkumar
Vellore Institute of Technology, India

ABSTRACT

In many multi-stage manufacturing supply chains, transportation related costs are a significant portion of final product costs. It is often crucial for successful decision making approaches in multi-stage manufacturing supply chains to explicitly account for non-linear transportation costs. In this article, we have explored this problem by considering a Two-Stage Production-Transportation (TSPT). A two-stage supply chain that faces a deterministic stream of external demands for a single product is considered. A finite supply of raw materials, and finite production at stage one has been assumed. Items are manufactured at stage one and transported to stage two, where the storage capacity of the warehouses is limited. Packaging is completed at stage two (that is, value is added to each item, but no new items are created), and the finished goods inventories are stored which is used to meet the final demand of customers. During each period, the optimized production levels in stage one, as well as transportation

DOI: 10.4018/jisscm.2009062902

levels between stage one and stage two and routing structure from the production plant to warehouses and then to customers, must be determined. The authors consider "different cost structures," for both manufacturing and transportation. This TSPT model with capacity constraint at both stages is optimized using Genetic Algorithms (GA) and the results obtained are compared with the results of other optimization techniques of complete enumeration, LINDO, and CPLEX.

INTRODUCTION

To exploit economies of scale and order in large lots, the important issue in supply chain is to optimize the inventory level by considering various costs in maintaining a high service level towards the customer. Since, the cost of capital tied up in inventory is more, the inventory decision in the supply chain should be coordinated without disturbing the service level. The coordination of inventory decision within an entity is viable, but not between the entities. So the integration of the entities to centralize the inventory control is needed.

Several factors influence inventory policy and practice for the firm includes distribution savings demand seasonality in production and purchasing economies with the desired level of customer service. Transportation is the spatial linkage for the physical flows of a supply chain. Today supply chain has to examine the effective use of supply chain with value added processes in distribution center with minimum inventory levels for delivering the product in time. The gateway of the supply chain is to know our customer and serve the needs by considering the aspects of speed and reliable service with the value propositions. The effective supply chain management needs to improve customer service, reduction of costs across the supply chain, optimal management of existing inventory with optimized manufacturing schedules. The impact of distribution inventory will enhance the customer value in the form of lower costs across the supply chain.

This project presents the Genetic Algorithm Distribution Inventory Model (GADIM) approach for optimizing the inventory levels of supply chain entities with the consideration of two-echelon plant to warehouse and warehouse to customers. This model is especially suitable for inventory control and cost reduction by unlocking the hidden profits through the genetic algorithm optimization. The generalized GADIM evaluate the transportation link between the two entities and finding the best route for transporting the product from plant to warehouse and then to customer. The demand of the customer is known in advance, so that the production rate and the inventory levels can be adjusted in the plant and warehouse. Since the logistics plays an important role in the supply chain that the two important sectors transportation and distribution inventory has taken into consideration for redesigning the allocation and routing through optimization. This approach is based on genetic algorithm. It searches the population of solutions of an optimization problem towards the improvement by simulating the natural search and selection process associated with natural genetics.

The model comprises the genetic algorithm optimization where the inventory level at the plant and warehouse are optimized according to the production capacity of the same. The distribution cost plays an important role in optimizing the inventory level, so that the total cost for the entire supply chain is minimized. The GADIM approach assure in minimizing the total cost of the supply chain by optimizing the inventory levels in accordance with production capacity of plant and warehouse.

LITERATURE SURVEY

The literature review in aspect of distribution inventory management of supply chain, states clearly that no optimal inventory policy has been developed for a serial Supply Chain in view of the complexity of the problem. Hence the objective of the study is to optimize the inventory level for a two-echelon single product serial supply chain using Genetic Algorithm. So as to minimize the total supply chain cost comprising the distribution and production related cost.

Cakravastia (2002) developed a analytical model for the supplier selection process in designing a supply chain network. The constraints on the capacity of each potential supplier are considered in process. The assumed objective of the supply chain is to minimize the level of customer dissatisfaction, which is evaluated by two performance criteria. (1) Price and (2) delivery lead time. The overall model operates at two levels of decision making: the operational level and the chain level. An optimal solution in terms of the models for the two levels can be obtained by using a mixed integer programming technique.

Abdul-Jalbar et al. (2005) addresses a multi-echelon inventory system with one-warehouse and N-retailers. The demand at each retailer is assumed to be known and satisfied by the warehouse. Shortages are not allowed and lead times are negligible. Costs at each facility consist of a fixed charge per order and a holding cost.

The goal is to determine single-cycle policies which minimize the average cost per unit time, that is, the sum of the average holding and setup costs per unit time at the retailers and at the warehouse. They propose an O (N log N) heuristic procedure to compute efficient single-cycle policies. This heuristic is compared with other approaches proposed by Graves and Schwarz (1977) and Muckstadt and Roundy (1987). They carry out a computational study to test the effectiveness of the heuristic and to compare the performance of the different procedures. From the computational results, it is shown that the new heuristic provides, on average, better single-cycle policies than those given by the Muckstadt and Roundy (1993).

Gunasekaran et al. (2004) have studied that Supply chain management has been a major component of competitive strategy to enhance organizational productivity and profitability. In recent years, organizational performance measurement and metrics have received much attention from researchers and practitioners.. Performance measurement and metrics have an important role to play in setting objectives, evaluating performance, and determining future courses of actions. Performance measurement and metrics pertaining to SCM have not received adequate attention from researchers or practitioners. We developed a framework to promote a better understanding of the importance of SCM performance measurement and metrics.

Yokoyama and Lewis (2003) discusses a new model and its solution procedure for the commodity distribution system consisting of distribution centers and consumer points are discussed. Demand is assumed to be a random variable that obeys a known, stationery probability distribution. An integrated optimization model is built where both the order-up-to-R policy, which is one of the typical inventory policies of periodic review models, and the transportation problem are considered simultaneously. The assignment of consumer points to distribution centers is not fixed. The problem is to determine the target inventory and the transportation quantity in order to minimize the expectation of the sum of inventory related cost and transportation cost. Simulation and linear programming are used to calculate the expected cost, and a random local search method is developed in order to determine the optimum target inventory. He uses a genetic algorithm to test and compare with the proposed random local search method. The model and effectiveness of the proposed solution procedure are clarified by computational experiments.

A decision support system for supplier selection is developed by S. H. Ghodsypour and O' Brien (1998). In this article an integration of analytical hierarchy process and linear programming is proposed to consider both tangible and intangible factors in choosing the best suppliers and placing the optimum order quantities among them such that the value of purchasing becomes maximum. This model can be applied for supplier selection with or without capacity constraints.

Badri (1999) proposes the use of Analytic Hierarchy Process (AHP) and multi-objective goal programming methodology as aid in making location- allocation decisions. The methodology presented can help facility planning authorities to formulate viable location strategies in the volatile and complex global decision environment.

A brief review of literature for modeling of multistage supply chain can be found in the paper by Beamon (1999). The above mentioned works leave scope for development of an optimization technique based on a meta-heuristic approach. Thus we select the Genetic Algorithm to address the problem of "total cost optimization in a bi echelon multi-product and multi period supply chain environment". We thereby also aim to study the other optimization techniques so as to compare and provide with the best optimization technique.

RESEARCH GAP ANALYSIS

Aim

The aim of the article is to optimize the total cost of a two stage multi product multi period supply chain environment using Genetic Algorithm Distribution Inventory model by taking into consideration all the constraints and assumptions. It also aims to provide a comparative study among the existing optimization techniques like LINDO, CPLEX, Complete Enumeration and GADIM so as to provide the vendor with the best optimization technique depending on time and data constraints.

Objectives

The main objective of the present work is to demonstrate the importance of GADIM with relevance to supply chain management and logistics. There has been a growing interest in GA with respect to SCM and logistics. The main reason being that supply chain and logistics management is a collaborative function and a number of players are involved in the decision making process. So the approach used should support group tasking and GA is particularly suitable in that aspect. Moreover, in logistics decision making a number of non quantifiable criteria comes into picture and GA handles this very effectively.

Another objective of this work is to incorporate customer in the logistics decision making process. Costs are often used as the main factor in logistics, where as enough attention is not paid to the various quantitative and qualitative customer service elements. The objective here is to develop a customer centric approach.

The present work is limited to evaluation of alternative links and nodes in a supply chain. This project aims to optimize the total cost of a multi product two stage multi period supply chain using GA. This work shall improve the productivity and minimize the losses of the two-stage SCM thereby leading to optimized route based on the customer demand.

Scope

Supply chain is the emerging area with lot of problems in the production-distribution echelon and the problem has been less addressed and solved. Many Researchers have pursued their research in solving the production -distribution inventory management for a single echelon system (only supplier / manufacturer / distributor / retailer). But, (However) the Multi echelon distribution inventory management has received less attention from solution approaches. Some of the Researchers have tried this problem with the linear programming and Traditional Optimization technique. This problem hasn't approached with the Non Traditional Optimization Technique. Thus we aim to approach this problem by using the meta- heuristic approach Genetic Algorithm. This method is capable of handling large data and particularly in this case where we shall deal with multi-product multi-period and two stage namely production, warehousing and customer. Thereby on obtaining the result we shall compare it with the existing optimization techniques of complete enumeration, LINDO and CPLEX.

Problem Structure

In many multi-stage manufacturing supply chains, transportation related costs are a significant portion of final product costs. It is often crucial for successful decision making approaches in multi-stage manufacturing supply chains to explicitly account for non-linear transportation costs. A two-stage supply chain that faces a deterministic stream of external demands for a multi product is considered. A finite supply of raw materials, and finite production at stage one has been assumed. Items are manufactured at stage one and transported to stage two, where the storage capacity of the warehouses is limited. Packaging is completed at stage two (that is, value is added to each item, but no new items are created), and the finished goods inventories are stored which is used to meet the final demand of customers. During each period, the optimized production levels in stage one, as well as transportation levels between stage one and stage two and routing structure from the production plant to warehouses and then to customers, must be determined.

Problem Definition

The investigation of Supply Chain Distribution Inventory Management has done using a model GADIM with a focus on inventory level at plant and warehouse, the production rate and production cost at the plant, and the transportation costs between the entities at all cross-levels. The three entities plant, warehouse and customers are considered in this supply chain model. The model is generalized by consideration of 'Npl' number of plants,'Nw' number of warehouses and 'Nc' number of customers. The model GADIM suits for any kind of industry, which follows the three-entity supply chain model with finite number of plants, warehouses and customers. The Total demand of the customer is satisfied by the stock in the warehouse which inturn procure the product stock from the plant entity. The time factor and cost level in transporting the product between the three entities varies according to the logistics flow structure from the plant to the warehouse and then to customer of the supply chain model. So the delay factor may occur in accordance with the optimal transportation route. Since, there is a limited inventory level in the plant and warehouse it has a possibility of maintaining the Backlog Inventory to satisfy the demand of the customers in the forthcoming period of this GADIM supply chain. The purpose of GADIM of supply chain is to provide a nearer optimal production rate for the plant, inventory level for the plant and warehouse by minimizing the total logistics costs of the entire supply chain. In designing the model, three "subsystems" need to be analyzed: (1) the production rate at each plant for all periods (2) the inventory at each plant and warehouse for all periods. (3) route between entities where the product is transferred from plant to warehouse and then to customer for all periods. A synthesis of these subsystems is used to minimize the total costs of supply chain.

Methodology

The methodology of Genetic Algorithm begins with defining the objective function. In our problem the objective function is to minimize the total cost. To obtain the objective function the constraints namely production cost, transportation cost, storage cost and customer demand are considered. The first stage involves initialization. The initial chromosomes are randomly generated. Next, these chromosomes are evaluated and after evaluation the best chromosome are selected. Then the selected chromosomes undergo crossover and mutation operations. After the mutation process the entire cycle is iterated till the best result is obtained. The results obtained are compared with the results of other optimization techniques of complete enumeration, LINDO, CPLEX and the comparative result is tabulated.

Assumptions

The model assumes that a single product with a stochastic demand.

The Total demand for all periods should be less than the Total production capacity and warehouse capacity.

The Total production rate for all period is equal to total demand of all period.

Time taken for transporting the product between the entities is homogenous and not taken into consideration.

Constraints

- Manufacturing plant
  - Product cost of each plant
  - Inventory cost of each plant
  - Products (production) capacity of each plant
  - Inventory capacity of each plant
- Ware house
  - Total cost from each plant to each warehouse
  - Inventory cost of each warehouse
  - Inventory capacity of each warehouse
- Suppliers
  - Total cost from each warehouse to each customer
  - Demand by each customer for one period

Taking into account the above said constraints, the following outputs are obtained

1. Total cost for all periods in order to satisfy the demand constraint.
2. Optimal production rate in each period for all plant
3. Optimal warehouse stock in each period for all warehouses.
4. Optimal Routing (from plant to warehouse to customers) for all periods.

Notations

The definitions of notations are:

- Np: Number of period
- Npl: Number of plant
- Nw: Number of warehouse
- Nc: Number of customer
- Pci : Production cost at i th plant i= 1 …..Npl
- ICCPi: Inventory carrying cost at i th plant i= 1 …..Npl
- ICCWj : Inventory carrying cost at j th warehouse j =1 ….Nw
- CDEMpk: Demand at the K th customer at the period ‘p’ p=1… Np
- PCAPi: Production capacity at the i th plant
- WCAPj: Warehouse capacity at the j th warehouse
- BO: Backordering cost
- ICAPi: Inventory capacity at the i th plant
- TCPWpij: Unit Transportation cost from source ‘i’ to destination ‘j’ for period ‘p’.
- TCWCpjk: Unit Transportation cost from source ‘j’ to destination ‘k’ for a period ‘p’.
- Prpi: Production rate at i th plant for a period ‘p’.
- UTPWpij: Unit transferred from i th plant to j th warehouse for a period ‘p’.

- UTWCpjk: Unit transferred from j th warehouse to k th customer for a period ‘p’.
- WHSpj: Warehouse stock at j th warehouse for a period ‘p’.
- USCpk: Unsatisfied demand at k th customer for a period ‘p’.
- ICPp: Total inventory cost for all plant for a period ‘p’.
- ICWp: Total inventory cost for all warehouse for a period ‘p’.
- TPCp: Total production cost for a period ‘p’.
- PCPi: Production cost at i th plant
- PWTCp: Plant to warehouse total transportation cost for a period ‘p’.
- WCTCp: Warehouse to customer total transportation cost for a period ‘p’.
- SCp: Shortage cost for a period ‘p’.
- IP: Inventory in plant.
- IW: Inventory in Warehouse.
- TC: Total cost comprising production, inventory (plant and warehouse), transportation (plant to warehouse to customer) and shortage costs for all period.

Mathematical Model

The objective function of GADIM is to minimize the total cost of the supply chain. The model is interpreted in linear programming form as given below.

$$\text{Min}\, TC = \sum_{p=1}^{Np}\left(ICP(p) + ICW(p) + TPC(p) + PWTC(p) + WCTC(p) + SC(p)\right) \quad (1)$$

Subject to Constraints,

$$ICP = \sum_{i=1}^{Npl}\left(IP(i) * ICCP(i)\right)$$

$$ICW = \sum_{j=1}^{Nw}\left(IW(j) * ICCW(j)\right)$$

$$TPC = \sum_{i=1}^{Npl} PCP(i)$$

$$PCP(i) = \sum_{j=1}^{Nw}\left(UTPW(i,j) * PC(i)\right)$$

$$PWTC = \sum_{j=1}^{Nw}\sum_{i=1}^{Npl}\left(UTPW(i)(j) * TCPW(i)(j)\right)$$

$$WCTC = \sum_{k=1}^{Nc}\sum_{j=1}^{Nw}\left(UTWC(j)(k) * TCWC(j)(k)\right)$$

$$SC = \sum_{k=1}^{Nc}\left(USC(k) * BO\right)$$

The production rate should be less than or equal to the production capacity of each plant

$$\Pr(i) \leq PCAP(i)$$

The Total Production rate is equal to total customer demand

$$\sum_{p=1}^{Np}\sum_{i=1}^{Npl}\Pr(i)(p) = \sum_{p=1}^{Np}\sum_{k=1}^{Nc} CDEM(k)(p)$$

The sum of units transferred from each plant to all warehouse should be less than or equal to production rate of the particular plant

$$\sum_{j=1}^{Nw} UTPW(i)(j) \leq \Pr(i)$$

The sum of units received from all plants to the warehouse should be less than or equal to the warehouse capacity

$$\sum_{i=1}^{Npl} UTPW(i)(j) \leq WCAP(j)$$

Warehouse stock is equal to the sum of units received from all plant.

$$WHS(j) = \sum_{i=1}^{Npl} UTPW(i)(j)$$

The warehouse stock should not exceed the warehouse capacity

$$WHS(j) \leq WCAP(j)$$

The sum of units transferred from warehouse to all customers should be less than or equal to the warehouse stock

$$\sum_{k=1}^{Nc} UTWC(j)(k) \leq WHS(j)$$

The sum of units received to customer from all warehouse should be less than or equal to customer demand

$$\sum_{j=1}^{Nw} UTWC(j)(k) \leq CDEM(k)$$

Unsatisfied customer demand for the current period

$$USC(k) = CDEM(k) - \sum_{j=1}^{Nw} UTWC(j)(k)$$

Next period demand is

$$CDEM(k) = USC(k-1) + CDEM(k)$$

METHODOLOGY OF GADIM

Our approach is based on Genetic Algorithm. GA's are inspired by natural evolution. GA's search populations of solutions of an optimization problem towards improvement by simulating the natural search and selection process associated with natural genetics. GA's encodes an optimization problem into a set of genetic characteristics. Each solution is encoded in a string (called chromosome) of binary digits or integers. The method uses three natural operators selection, crossover and mutation to search populations of solutions towards improvement. GA's starts with a random population of solutions to explore the solution space of a problem. Then GA's attempt to improve the solutions through a number of iterations called generations. The performance of each solution to the problem is evaluated by fitness function, which corresponds to the objective function of the optimization problem. Each individual contributes to the next generation in proportion to its fitness. The selection of chromosomes for reproduction is biased so that the fitter one tends to reproduce more often than the less fit ones. Following parental selection, crossover and mutation are applied. Crossover combines materials from parents to produce their children. It also provides pressure for improvement or exploration. On the other hand, mutation makes small local changes of feasible solutions to provide diversity of population for wide exploration of feasible solutions. GA is a global search methodology whose convergence although possibly slow, has been proven (Holland, 1975). The mutation operator is usually defined to ensure that the generated solutions will not get trapped in some local minimum. Moreover, the final solution does not depend on the initial solutions. Hence, in most badly the initial population is randomly generated. The chance of reaching a very good solution, if not optimal one, is very high and increases, of course, with the number of generations.

The Procedure of the Classical GA is as follows

1. Initialize a set of feasible solutions (i.e., initialize a population of chromosomes) randomly.
2. Evaluate each chromosome in the population by a fitness function.
3. Select chromosomes for reproduction.
4. Apply crossover and mutation on the selected chromosomes to produce new chromosomes
5. Evaluate the new chromosomes.
6. If the stopping condition is reached, return the best solution; if not, go to 3.

Figure 1.

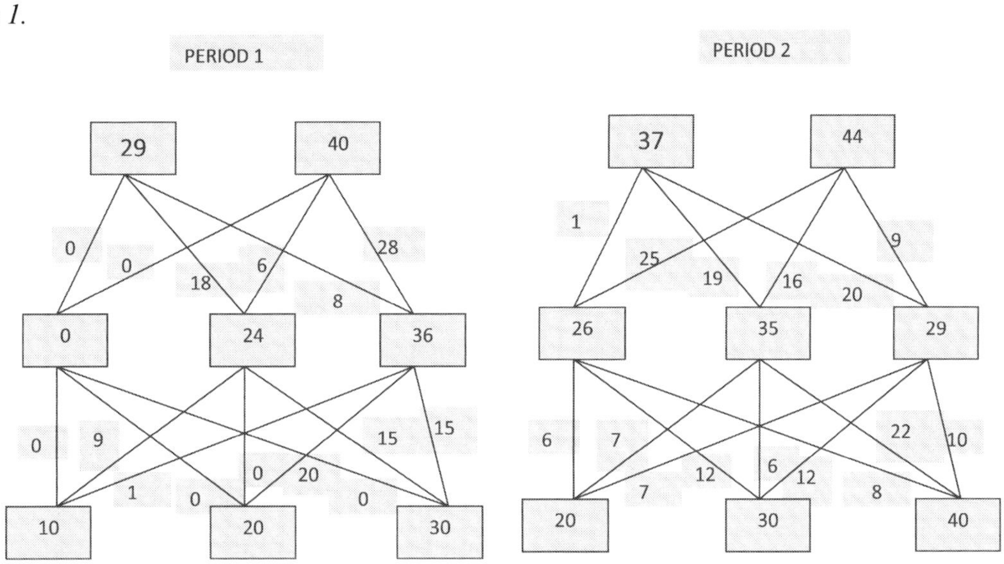

For the single product GADIM, the chromosomes considered includes the production rate of the plant for all period, the inventory level at the warehouse for all period, total demand of customer for all period with the amount of product transferred from plant to warehouse and then to customer for all period. The chromosome is randomly generated by satisfying the feasibility condition for the process of Genetic Algorithm.

The example of the randomly created chromosome for the two periods, with the assumption of 2 plants, 3 warehouses and 3 customers in GADIM supply chain is considered for illustration.

The 10 Population size chromosomes are considered. The pop size randomly generated chromosomes are evaluated using the specified objective function (Eqn 1) after evaluating all the chromosomes. Select the better parent chromosome through the selection process. The Roulette wheel selection approach has been used in the selection process, where the fitness function is calculated as the inverse of the objective function i.e., 1/ (1+TC). The Probability occurrence of each chromosome in the selection process is calculated and inturn the cumulative probability. The generated random numbers in the range 0-1 is compared with the cumulative probability to select the better parent chromosome for the next crossover process. The crossover operator suggested by Vignaux and Michalewicz (1991) for the transportation problem is considered with slightly modification for the crossover process for the present problem. For the crossover operation the production rate of the plant, inventory level of the warehouse, unit product transferred from plant to warehouse and then to customer. The crossover operator is suggested as a matrix for all the above measure where the transformation of the matrix with the viable feasibility condition in each chromosome is considered.

Steps in Crossover

Assume that two matices X1 and X2 are selected as parents for the crossover operation. The crossover is performed in three steps.

- Step1. Create two temporary matrices 'D' and 'R' as follows. Matrix 'D' keeps rounded average values from both parents, and matrix R keeps track of whether any rounded is necessary.

$$D = \left[\left(X1 + X2 \right) \div 2 \right]$$

$$R = \left(X1 + X2 \right) \bmod 2$$

- Step 2. Divide (Split) matrixes 'R' into two matrices R1 and R2. It has many possible ways to divide 'R' into R1 and R2 while satisfying the condition.

$$R = R1 + R2$$

- Step 3. Then we produce two offspring of Y1 and Y2 as follows.

$$Y1 = D + R1$$
$$Y2 = D + R2$$

The illustrative example of the crossover operation for the pair of chromosome as shown below. The chromosome for the crossover is divided into two sections, 1. plant to warehouse for all period 2. warehouse to customer for all period (See Tables 1 through 5).

1. Plant to warehouse for all period:

Similarly the chromosomes are crossover for the warehouse to customer.

The next process mutation is carried over for the crossover chromosome in using the mutation operator devised by Vignaux and Michalewicz (1991). So the procedure of mutation operation is performed in three steps.

- Step 1. Make a sub matrix from a parent matrix. Randomly select (i = 1 to Npl for all period) rows and (j = 1 to Nw for all period) columns to create a (p*q) sub matrix 'Y'
- Step 2. Reallocate a commodity for the sub matrix. The available amount of commodity and demands for the sub matrix are determined. We can use the initialization procedure to assign new values to the sub matrix such that all constraints are satisfied.
- Step 3. Replace the appropriate elements of the parent matrix by new elements from the reallocated sub matrix Y.

Table 1. First chromosome (X1): Second chromosome (X2)

1	0	0	7	0
0	4	0	0	0
2	1	4	0	5
0	0	6	0	0
0	0	5	0	3
0	4	0	0	0
0	0	5	7	0
3	1	0	0	2

Table 2. The temporary matrices 'D' and 'R' are

1	0	1	1	1
0	0	0	0	0
0	1	1	1	1
1	1	0	0	0

Table 3. Matrix D Matrix R

0	0	2	3	1
0	4	0	0	0
1	0	4	3	2
1	0	3	0	1

Table 4. Then divide R into R1 and R2 as follows: Matrix R1 Matrix R2

0	0	1	0	1
0	0	0	0	0
0	1	0	1	0
1	0	0	0	0
1	0	0	1	0
0	0	0	0	0
0	0	1	0	1
0	1	0	0	0

Table 5. Finally, two offspring Y1 and Y2 are: Offspring Y1 Offspring Y2

0	0	3	3	2
0	4	0	0	0
1	1	4	4	2
2	0	3	0	1
1	0	2	4	1
0	4	0	0	0
1	0	5	3	3
1	1	3	0	1

Select the chromosome randomly for the mutation process.

The illustrative example for the mutation from plant to warehouse for all period is (See table 6.).

Select two rows (2, 4) and three columns (2, 3, 5) randomly. The corresponding sub matrix and the reallocated sub matrix are (See Tables 7 and 8):

Similarly the chromosomes are mutated for warehouse to customer for all period.

The process is repeated with the mutated chromosome as an initial chromosome for the specified number of generations.

Table 6. Chromosome X

0	0	5	0	3
0	4	0	0	0
0	0	5	7	0
3	1	0	0	2

Table 7. Sub matrix Y; after reallocation

4	0	0	2	0	2
1	0	2	3	0	0

Table 8. Then the offspring after mutation is offspring

0	0	5	0	3
0	2	0	0	2
0	0	5	7	0
3	3	0	0	0

Other Methodologies

Complete Enumeration

Enumerative schemes have been considered in many shapes and sizes. The idea is fairly straight forward. Within a finite search space, or a discredited infinite search space, the search algorithm starts looking at objective function values at every point in the space, one at a time. Although the simplicity of this type of algorithms is attractive and enumeration is a very human kind of search (when the number of possibilities is small), such schemes must ultimately be discounted in the robustness race for one simple reason lack of efficiency. Many practical spaces are too large to search one at a time and still have a chance of using the information to some practical end. Even the highly touted enumerative scheme dynamic programming breaks down on problems of moderate size.

It is a method which is very time taking and elaborative. As this method involves one to go through all the various permutations and combinations it becomes virtually impossible to handle very large data. It is a very tedious as it is a manual process and the results obtained are near optimal and not necessarily the best.

LINDO

LINDO (Linear, INteractive, and Discrete Optimizer) is a software package tool and it is a convenient, but powerful tool for solving linear, integer, and quadratic programming problems. These problems occur in areas of business, industry, research and government. Specific applications areas where LINDO has proven to be of great use would include product distribution, ingredient blending, production and personnel scheduling, inventory management... The list could easily occupy the rest of this help file.

The guiding design philosophy for LINDO has been that, if a user wants to do something simple, then there should not be a large setup cost to learn the necessary features of LINDO. At the other extreme, LINDO has been used to solve real industrial linear, quadratic, and integer programs of respectable size. For commercial applications, LINDO is frequently used to solve problems with tens of thousands of constraints and hundreds of thousands of variables.

There are three basic styles of using the LINDO software. For small to medium sized problems, LINDO is simple to use interactively from the keyboard. Entering a model is quite easy to do. It's also possible to use LINDO with files created elsewhere, containing scripts of commands and input data, and producing files for reporting purposes. Finally, custom-created subroutines may be linked directly with LINDO to form solution containing both your code and the LINDO optimization libraries.

The LINDO software is designed to be simple to learn and to use. This is particularly true for small problems.

Table 9. The transportation cost from plant to warehouse is

1	2	3
4	5	6

Table 10. The transportation cost from warehouse to customer is

1	2	3
4	5	6
7	8	9

CPLEX

CPLEX Interactive Optimizer is a command-line interactive program, provided in executable, ready-to-use form. It packs all the power and speed of CPLEX into an easy-to-use, easy-to-learn format, featuring a simple user interface and an extensive help system. New users become proficient and productive immediately. Just read in a problem, issue the "optimize" command and review results. ILOG CPLEX is designed to help solve corporate business problems, using the most sophisticated analytical techniques. Major companies and software providers in supply-chain planning, network design, transportation logistics, utilities and a variety of other industries depend on CPLEX in mission-critical resource allocation applications. CPLEX is at the core of countless cutting-edge commercial products worldwide.

ILLUSTRATIVE EXAMPLE FOR GADIM

The supply chain with 2 plants, 3 warehouses, and 3 customers for the two periods is taken into consideration. The production capacities for the two plants are 40 and 50. The inventory levels at the three warehouses are 30, 40 and 50. Total demands for the three customers for first period are 10, 20 and 30. For the second period are 20, 30 and 40. The production cost for the plant 1 and plant 2 are Rs 2 per unit and Rs 3 per unit respectively. The inventory carrying cost for the plant 1 and plant 2 are Rs 2 per unit per period and Rs 4 per unit period respectively. The inventory carrying cost for the warehouse 1, warehouse 2 and warehouse 3 are Rs 3 per unit per period, Rs 4 per unit per period, Rs 5 per unit per period respectively. Inventory capacity in plant 1 and plant 2 are 15 units and 20 units respectively. The backordering cost is Rs 5 per unit (See Tables 9 and 10).

This example problem is optimized using Genetic Algorithm by random generation of production rate in the plant, inventory level in the warehouse, unit of products transferred from plant to warehouse and then to customer with know demand for all periods.

Initially 10 chromosomes are randomly generated for the initialization process. The chromosomes are evaluated for roulette wheel selection process. And the selected chromosomes will undergo crossover and mutation operations. Finally mutated chromosomes are considered as a randomly initialized chromosome for the next generation. The process is repeated for 50 numbers (iteration or times) of generations.

RESULTS

Results Obtained Through GA

The optimal solution for the illustrative example through the optimization by GA given below.

Among the 50 iterations, at the end of 46 th Iteration nearer optimal result is achieved (See Tables 11 through 20).

The results obtained predict that Genetic algorithm is best suited for the real world application since it has nearest optimal solution and also less

Table 11. Optimal production rate

Production Rate	Plant 1	Plant 2
Period 1	17	47
Period 2	38	48

Table 12. Optimal warehouse stock

Warehouse Stock	Warehouse 1	Warehouse 2	Warehouse 3
Period 1	21	15	26
Period 2	20	20	50

Table 13. Optimal units transferred from plant to warehouse

UTPW for Period 1	Warehouse 1	Warehouse 2	Warehouse 3
Plant 1	0	2	13
Plant 2	21	13	13
UTPW for Period 2	Warehouse 1	Warehouse 2	Warehouse 3
Plant 1	11	1	28
Plant 2	8	19	21

Table 14. Optimal units transferred from warehouse to customer

UTPW for Period 1	Customer 1	Customer 2	Customer 3
Warehouse 1	9	7	4
Warehouse 2	0	9	6
Warehouse 3	1	4	20
UTPW for Period 2	Customer 1	Customer 2	Customer 3
Warehouse 1	7	2	11
Warehouse 2	0	14	6
Warehouse 3	13	14	23

Table 15. Total production cost

Optimal Cost	Period 1	Period 2	Total
Production Cost	171	224	395

Table 16. Optimal Transportation cost (plant to warehouse to customer)

Optimal Transportaion cost	Period 1	Period 2	Total
Plant to Warehouse	270	350	620
Warehouse to Customer	335	560	895
Total	605	910	1515

Table 17. Optimal total cost

Periods	Period 1	Period 2	Total
Total Cost	776	1146	1922

Table 18. The following table gives the results obtained through CPLEX and LINDO

Periods	Period 1	Period 2	Total
Total Cost	764	1153	1917

Table 19. Optimality test

Model/Matrix	Complete enumeration / LINDO (a)	Genetic Algorithm (b)	(b-a)/a*100
2*2*2	optimal	Near optimal	<10%
2*2*3	optimal	Near optimal	<10%
2*3*3	optimal	Near optimal	<10%
2*3*4	optimal	Near optimal	<10%
3*3*3	Not applicable	Optimal	
3*3*4	Not applicable	Optimal	

Table 20. Time complexity

Model/Matrix	Complete enumeration / CPEX (a)	Genetic Algorithm (b)
2*2*2	Min	less
2*2*3	Min	less
2*3*3	Average	less
2*3*4	High	less
3*3*3	High	less
3*3*4	Very high	less

time consuming when compared to other existing tools. Moreover the tools like CPLEX, LINDO cannot handle more than 150 constraints whereas Genetic Algorithm can handle very complex constraints.

Thus the optimization technique that has been evolved is more effective and the generalized tool developed provides flexible end user capabilities.

CONCLUSION AND FUTURE RESEARCH DIRECTIONS

This chapter has presented a near optimal type inventory-logistics cost minimizing model for a production/distribution network with multiple plants supplying to a multiple warehouses, which in turn distributes to a large number of customers.

The model was a synthesis of three components (1) the production rate at each plant for all periods (2) the inventory at each plant and warehouse for all periods. (3) route between entities where the product is transferred from plant to warehouse and then to customer for all periods. The decisions in the model were made through a comprehensive distribution-based cost framework that includes the inventory, transportation, and transit components of the supply chain.

The repeated trial runs on the test problem using GADIM show that the results obtained are encouraging. Early investigation shows that some modifications on the genetic search scheme such as using a newly modified crossover operator will further enhance its efficiency. One should see that the solution of Distribution Inventory is important for the supply chain industry. Thus our model demonstrates good ability in solving the Distribution Inventory problem can be applied to larger class of complex problem.

However, the model suffers from a number of limitations. Future research can address some of these issues:

- The model is limited to two-echelon. Extensions of the model need to be undertaken beyond two echelons.
- The model assumes a single product and stochastic demand for modeling simplicity. Better procedures need to be introduced to tackle the multi-product and probabilistic demand case.
- Time taken for transporting the product between entities has not taken into consideration.

REFERENCES

Abdul-Jalbar, B., Gutierrez, J., & Sicilia, J. (2005). Integer-ratio policies for distribution/inventory systems. *International Journal of Production Economics* (pp. 93–94, 407–415).

Badri, M. A. (1999). A simulation model for multi-product inventory control management. *Simulation, 72*(1), 20–32. doi:10.1177/003754979907200103

Beamon, B. M. (1999). Measuring supply chain performance. *International Journal of Operations & Production Management, 19*(3), 275–292. doi:10.1108/01443579910249714

Cakravastia, A. A. (2002). A two-stage model for the design of supply chain network. *International Journal of Production Economics, 80*(3), 231–248. doi:10.1016/S0925-5273(02)00260-8

Ghodsypour, S. H., & O'brien. C. (1998). A decision support system for supplier selection using an integrated analytic hierarchy process and linear programming. *International Journal of Production Economics* (pp. 56–57, 199–212).

Graves, S. C., & Schwarz, L. B. (1977). Single cycle continuous review policies for arborescent production/inventory systems. *Management Science, 23,* 529–540. doi:10.1287/mnsc.23.5.529

Gunasekaran, A., Patel, C., & McGaughey, R. E. (2004). A framework for supply chain performance measurement. *International Journal of Production Economics, 87,* 333–347. doi:10.1016/j.ijpe.2003.08.003

Holland, J. (1975). *Adaptation in natural and artificial systems.* Ann Arbor, Michigan: University of Michigan Press.

Muckstadt, J. A., & Roundy, R. O. (1987). Multi-item, one-warehouse, multi-retailer distribution systems. *Management Science*, 33(12), 1613–1621. doi:10.1287/mnsc.33.12.1613

Muckstadt, J. A., & Roundy, R. O. (1993). Analysis of multistage production systems. *Handbooks in OR & MS* (Vol. 4).

Vignaux, G. A., & Michalewicz, Z. (1991). A Specialized Genetic Algorithm for Numerical Optimization Problems. *IEEE International Conference on Evolutionary Computation* (pp. 266–278).

Yokoyama, M., & Lewis, H. W. III. (2003). Optimization of the stochastic dynamic production cycling problem by a genetic algorithm. *International Journal of Computers & Operations Research, 30*(12), 1831–1849. doi:10.1016/S0305-0548(02)00109-0

This work was previously published in International Journal of Information Systems and Supply Chain Management, Vol. 2, Issue 4, edited by J. Wang, pp. 24-42, copyright 2009 by IGI Publishing (an imprint of IGI Global).

Chapter 15
Data Mining in Decision Support for Bioenergy Production

Nasser Ayoub
Tokyo Institute of Technology, Japan & Helwan University, Egypt

Yuji Naka
Tokyo Institute of Technology, Japan

ABSTRACT

Deliberate exploitation of natural resources and excessive use of environmentally abhorrent materials have resulted in environmental disruptions threatening the life support systems. A human centric approach of development has already damaged nature to a large extent. This has attracted the attention of environmental specialists and policy makers. It has also led to discussions at various national and international conventions. The objective of protecting natural resources cannot be achieved without the involvement of professionals from multidisciplinary areas. This chapter recommends a model for the creation of knowledge-based systems for natural resources management. Further, it describes making use of unique capabilities of remote sensing satellites for conserving natural resources and managing natural disasters. It is exclusively for the people who are not familiar with the technology and who are given the task of framing policies.

INTRODUCTION

Vast quantities of data and information are captured and stored through the available information technologies. Moreover, the data available are changing rapidly that force the planners, in governmental and private sectors to think differently in order to face market challenges and predict future trends. The DM is one of the important tools that can master this data explosion and provide usable knowledge from them. The use of data mining in developing DSS is popular in medical science, information technology, (Chae, Ho, Cho, Lee, & Ji, 2001; Wang, 1997) and recently in environmental applications, (Gibert, Sanchez-Marre, & Rodriguez-Roda, 2006). However, in the field of biomass planning and utilization the work presented here is pioneer. To the best knowledge of the authors, no similar work is conducted in the field. Here DM is used to help in deciding the economical bioenergy production resources and the hot spots of energy production facilities by applying clustering methods to the available biomass resources data. Then Life Cycle Inventory, LCI, analysis is used to estimate the CO2 emission from bioenergy production. As biomass materials distribution is diverse, from one location to another, GIS is used to allocate biomass resources to its real location. The GA is used, as a powerful combinatorial algorithm, to solve the optimization problem in the local level of gBEDS. However, LCI and GA works are not explained here as they are beyond the topic of this chapter. However, for interested reader, a detailed explanation can be found in (Ayoub, 2007).

In this chapter, the multilevel decision problem for bioenergy production has been considered to help local and national planners in deciding about the economical bioenergy resources quantities, energy production methods, facilities locations, annual investments, environmental and social impacts. The solution methods and strategies for such complicated problem including the use of DM as the main part in solving the problem are also discussed in this work. Furthermore a DSS called general Bioenergy Decision System, gBEDS that is used in realizing the solution methods has been explained with a case study.

DECISION MAKING PROBLEMS

Planning and decision-making processes are in most, if not all, cases take a multidisciplinary course of actions; as the decisions have to address many involved stakeholders. Consequently, decisions are seen from multiple perspectives that mostly have tradeoff nature. In addition, the methods of applying the decisions are of prime importance to the overall project efficiency (Basson & Petrie, 2007; Hoskinson, Rope, & Fink, 2007; Marquez & Blanchar, 2006). This means that the success of the decisions made lies, primarily, on their follow up and evaluation rather than their appropriateness only. The decisions assessments, always, provide an aid in judging their suitability for specific actions and, mostly, help in stimulating better results by changing or modifying their drawbacks at early stages of bringing them into reality. The data being analyzed are often historical in nature: daily, weekly and yearly results (Chau, Cao, Anson, & Zhang, 2003) that is a fertile land for DM to be applied.

Although it is a sound solution, environmentally, for energy production, bioenergy is still less attractive, from economical point of view. Therefore, most of the bioenergy production projects are completely or, at least, partly funded by the government in all levels. That makes the communication between different governmental levels, i.e., national, regional, local, and operative; through electronic media (DSS) is a typical example of the E-Government problems where decision making and follow up need to be flexible and quick. All communications between different planners can be performed through automated plans (files with different formats) that can be discussed simultaneously using any communication media, such as video conferencing. That may clarify many misunderstanding between planners in each level. The misinterpretation exist because national level plans are strategic for the whole country and considered as a challenge for the following levels planners to understand strategic decisions and adjust them to the local factors. Some of these factors are regulation / deregulation, the competition in different energy and emissions markets or the competition between different energy carriers in satisfying the local energy demand (Catrinu, 2006; Rozakis, Soldatos, Kallivroussis, & Nicolaou, 2001).

Furthermore, bioenergy local systems planning should operate with measures and designs similar to those applied nationally in order to harmonize the local and national strategies. Although vertical linkages among planning levels are very important, there is rarity of integrated economic and environmental plans at the different levels (King, Annandale, & Bailey, 2000). Special attention is given to the analysis and development of a multilevel decision making life cycle strategy to support the E-Government promotion where different levels of decision making and analysis are exist i.e., decisions have to be made in multilevel.

MULTILEVEL DECISION MAKING LIFE CYCLE (MLDMLC)

When designing a DSS for a multilevel decision making problem in E-Government, such as bioenergy planning problem, the system developer has to keep in mind the possible interactions between the different decision levels through the analysis of Decision Making Life Cycle, DMLC and Decisions Life Cycle, DLC. Life cycle as defined in the dictionary is "the progression through a series of differing stages of development". From this prospective, the DLC in bioenergy multilevel decision making framework is defined as:

The track of developmental changes or approval of decision(s) throughout a hierarchy of subsequent bioenergy decision-making levels with increasing intensity of details starting from the top national level, aiming at determining the best way to solve a prespecified bioenergy problem.

On the other hand, the DMLC as defined by (Ayoub, 2007, p.14) is:

The trip taken by the system designer and bioenergy decision makers until reaching a satisfactory Decision Support Tool(s) that help in the quick making of optimal or near optimal decisions, about establishing the bioenergy facilities or systems, based on the current constraint, starting from problem definition and ending at providing the results to the decision makers, in all levels, in an easy and understandable format.

Here, the MLDMLC shown as Figure 1 is proposed, for handling the bioenergy planning

Figure 1. The multilevel decision making life cycle

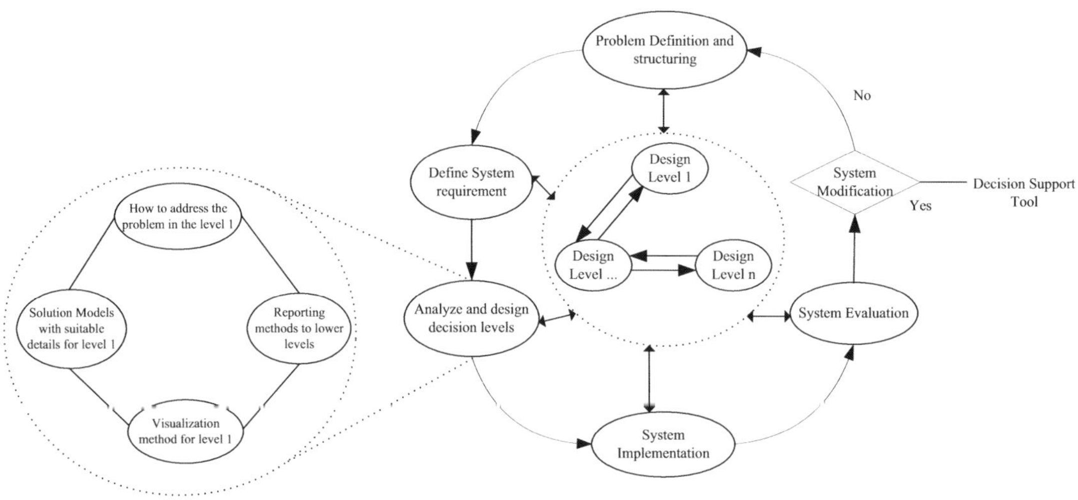

problem, by integrating the formal DMLC defined elsewhere (Borges, Pino, & Valle, 2005; Delen & Pratt, 2006) with the DLC and can be applied in the following steps:

- Problem definition and structuring based on a given set of indications or wish for change by an individual or group of responsible people in a firm or a country;
- Define the prospective decision levels i.e., national, regional, local, etc;
- Determine the best way to address the problem under study for each decision level;
- Develop the solution models with diverse details and appropriate to carry out the analysis at each level;
- Provide the results to the decision maker in an easily understandable format to allow exchange of decisions information and evaluation between different decision makers' levels as well as other concerned stakeholders.

Considering the MLDMLC while designing and developing the bioenergy DSS together with the appropriate Data Mining and Visualization techniques, will surely increase the cost and time of system development as the number of levels increases due to the task complexity of data collection and analysis, (Ayoub, 2007). However, this raise in cost will be recovered soon, as the system reinforces the robustness and effectiveness of decisions by overcoming several decision making weaknesses by:

- The ability to provide the decision makers in different levels with the suitable level of details that will save time and money for decision making procedure and their follow up, especially in the upper level.
- Ensure high communication performance through the mutual advantage of decisions visualization and semantics standardization.
- Increasing the possibilities of supporting future decisions of the same type through minor changes to system structure and data, that may reduce future costs.

By applying the bioenergy MLDMLC, it is recognized that the scope of a national planner, who proposes a biomass exploitation strategy for the whole country, is different from those of regional planners and executors who are applying policies in a small scale with specific requirements and limitations, e.g., resources availability, topological restrictions, energy needs, as shown in Figure 2. Such differences in scope must be taken into account when developing systems for decisions and planning support for the exploitation of biomass as a bioenergy resource.

For a macro-level of planning, such as that of national planners, a rough cost and simulation models are enough, since their main concerns are defining: the available biomass materials in the whole country; the available classes of energy conversion technologies for converting different biomass materials to different bioenergy; approximate economic effects, such as the cost of energy production or the potential legislative benefits; and the benefit in applying emission prevention legislation, e.g., Emission Taxation, Renewable Portfolio Standard and Fossil Fuel Taxation (Hakkila, 2006; Rozakis et al., 2001). The planning process in the two levels can be performed in four steps as given below:

- National level planning session is performed as shown by numbers 1 to 4 in Figure 2. The result of the national plan is the localities that considered as hot areas for energy production, type of institution, and specific supply chains.
- The national plans for the local areas are passed to the local levels in a form of proposal supply chains that have definite resources inputs, possible unit processes, potential products and the institution to be applied

Figure 2. Two levels with different objective

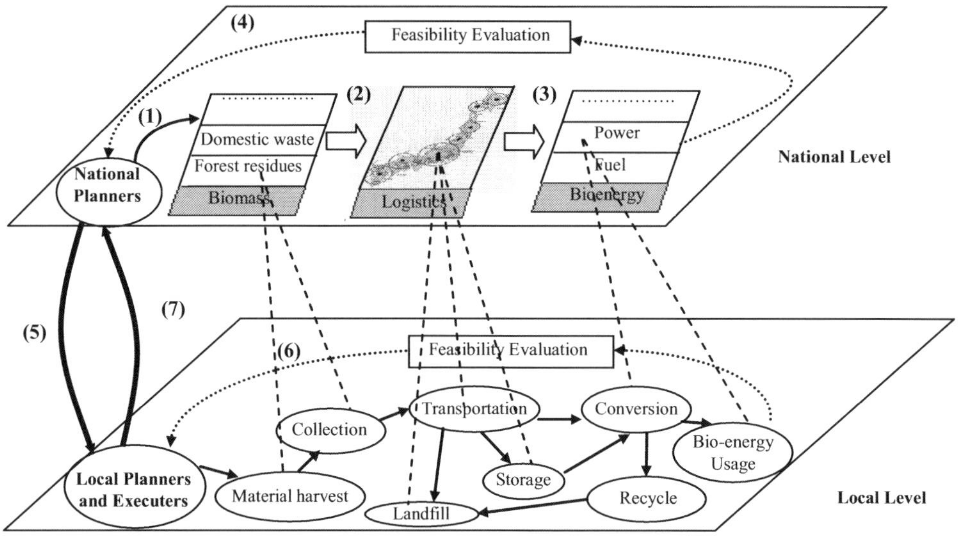

e.g., Renewable Portfolio Standard, number 5 in Figure 2.

- The local level planners evaluate the proposed plan by the national level, number 6 in Figure 2.
- The evaluation results, at the local level, can be one of three cases, number 7 in Figure 2;
    - Agree and apply: the application results are reported back to the national level with the same level of details.
    - Modify and apply: define the modifications, the reason for making the modifications, the methods of applying the modifications.
    - Reject: the reason(s) for rejection, suggestions for new plan(s) to be approved by the upper level and the advantages of the proposed plan are reported by the local level.

The results of one planning cycle can be applied, with moderate modifications, to the local areas with similar situation.

DATA MANIPULATION AND CLUSTERING

The data of biomass resources are collected to the city level in the national planning session and to the county level in local planning session. This means that in case of the national level, the exact locations for the resources production points are not known in certain. Furthermore, to define the collection places and the production facilities certain type of data clustering is needed to allocate the resources production points to bioenergy facility.

Clustering is a data mining technique and can be defined as "the process of grouping a collection of N patterns into distinct segments or clusters based on a suitable notion of closeness or similarity among these patterns". Good clusters show high similarity within a group and low similarity between patterns belonging to two different groups (Ye, 2003). Now a method on bioenergy resources data preparation is discussed next.

Figure 3. Converting the available bioenergy resources quantities into points.

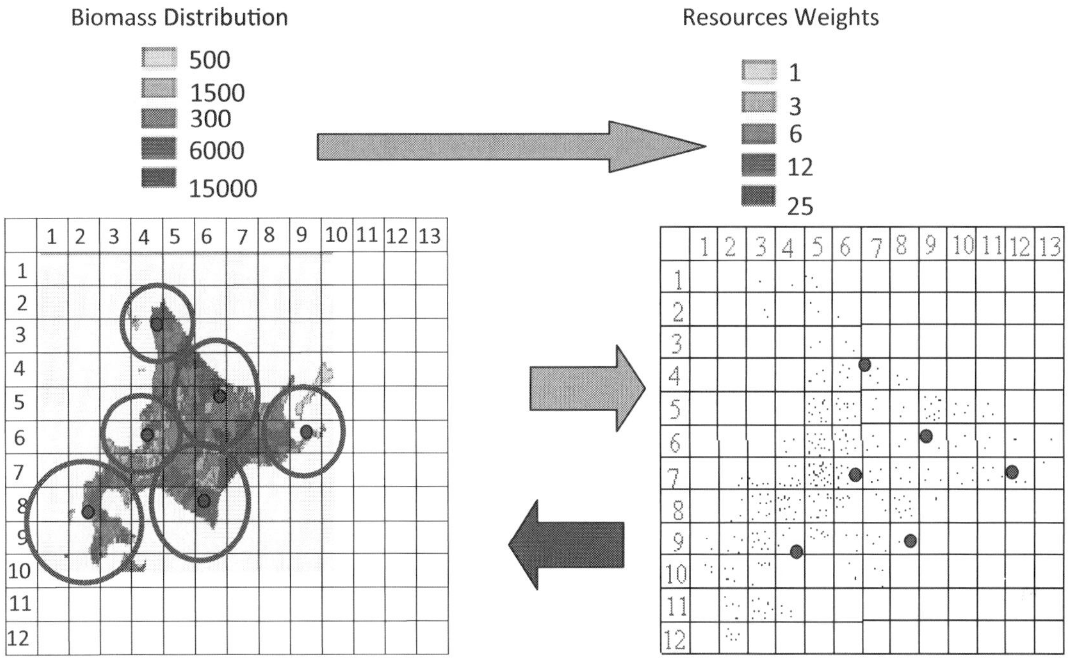

Bioenergy Resources Data Preparation and Clustering

A method for converting the available bioenergy resources quantities into resources production points is shown in Figure 3. The biomass resources quantities distribution are converted to weight points, or collection points, that are spread randomly around the centroid of the production area as seen in the right part of the figure. The resulted collection points are projected to the road network of the country or the region to define the economical collection points as seen in Figure 4. Then the clustering technique is used to cluster the collection points based on their attributes to define the location of storage place or bioenergy plants. In case of using the bioenergy plants locations, the electricity grid is used to define the best choice of the provided alternatives. Once the locations of storage places and production plants are defined the procurement area for each cluster is defined and the supply chain optimization part can start as will be explained later.

To integrate all the above mentioned techniques and methods, and in order to achieve the optimal solution for the decision problem a DSS is developed and introduced in the following sections including the use of the system in planning for real situation.

GENERAL BIOENERGY DECISION SYSTEM (gBEDS)

The gBEDS is a DSS for planning and implementing bioenergy systems in national and local decision levels. It uses the functions of data management, processing, and visualization through combining a group of methods and tools in one platform to present meaningful results that support the decision makers in planning for bioenergy projects. The planning session involves, clustering, managing and optimizing the flow of biomass materials and properties from supply sources to the energy network. This system is used to design the Bioenergy Supply Chain defined as "the series

Figure 4. Schematic representation for estimating the locations of bioenergy facilities

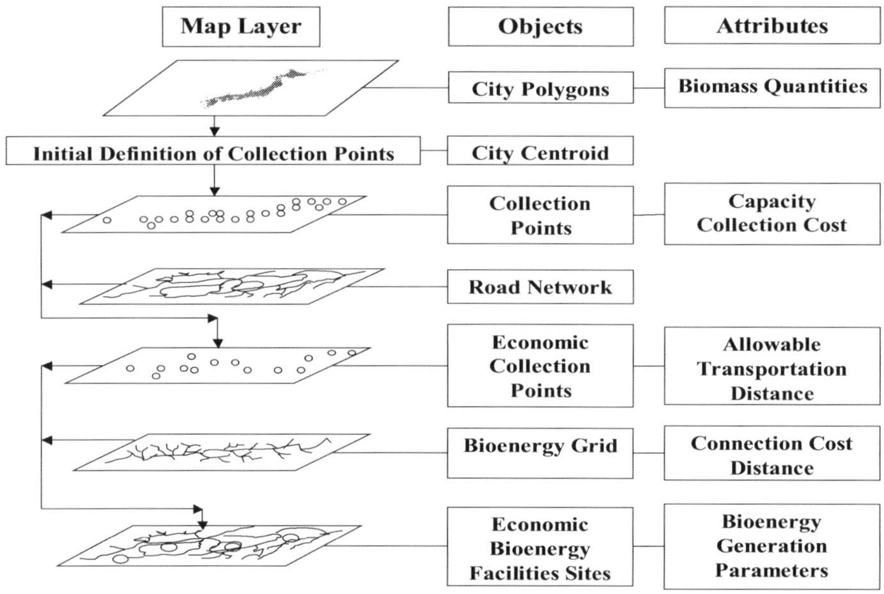

of physically interconnected handling processes of biomass from production until energy production (generation), transmission and distribution facilities". The gBEDS is used as a tool to guide the decision maker through the process of choosing the sources and technologies needed for different processes to realize the supply chain and their locations to satisfy community need.

The gBEDS Conceptual Structure

The gBEDS, as shown in Figure 5, consists of five parts; a common database; a Unit Process simulation or optimization module; the Data Mining engine; an in-line domain knowledge evaluation module; and a user interface. For more detailed explanation about the system components, the reader is directed to (Ayoub, Martins, Wang, Seki, & Naka, 2007; Ayoub, Seki, & Naka, 2007; Ayoub, Wang, Kagiyama, Seki, & Naka, 2006).

How gBEDS Works?

A general description of how DM Engine works with the modules of gBEDS is shown in Figure 6. It is clear that DM together with GIS are the heart of the gBEDS, as they are used to supply the subsequent modules with the necessary data for optimization calculations, e.g. locations and distances.

The GIS Database is prepared for the country or the region giving the data of biomass resources. By data clustering, of the GIS Database, we determine the feasible bioenergy resources quantities and their allocation to their calculated locations of bioenergy facilities and storage places, based on the spatial distribution of resources. In the Unit Process Simulator, UPS, the user defines a bioenergy supply chain. Depending on the planning level, the simulation or optimization model is used to calculate the different parameters of storage places and bioenergy facilities. The calculated parameters and impacts are, then, shown as

Figure 5. The conceptual structure of the gBEDS

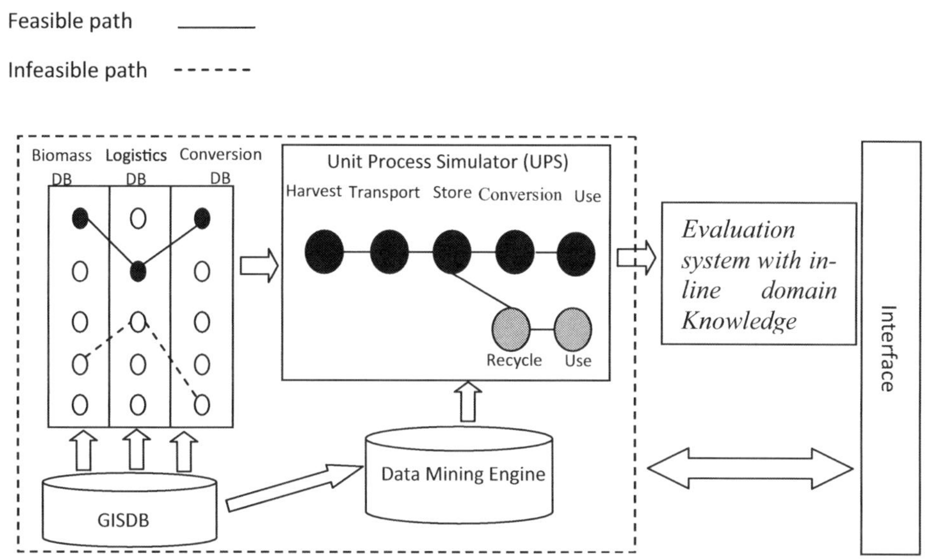

maps on the GIS screen, figures, or spread sheet data for further analysis (Ayoub, Martins, Wang, Seki, & Naka, 2007).

HOW DATA MINING IS USED IN gBEDS

Fuzzy *C*-means Clustering

In gBEDS the fuzzy *C*-means clustering, which is a non-hierarchical clustering method, is used

Figure 6. How gBEDS works?

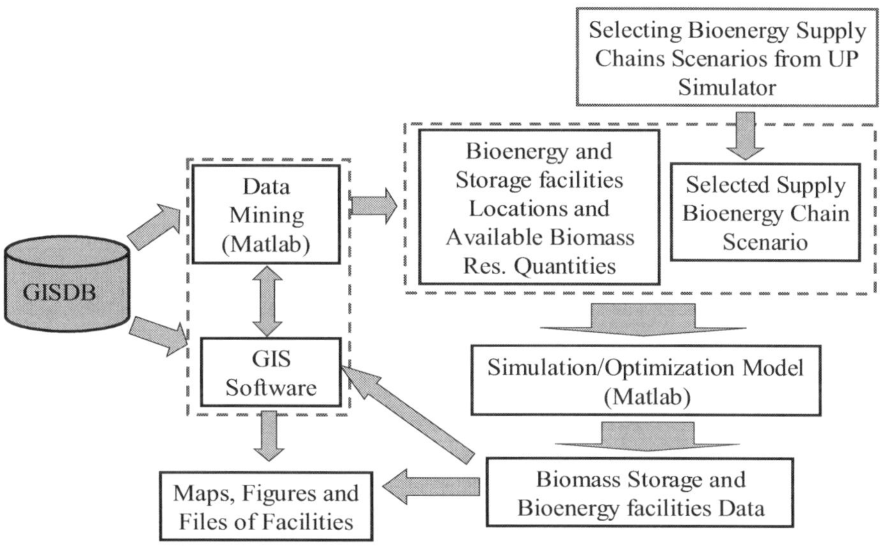

and designed to cluster data points into C clusters by minimizing the distance function. It is defined by Equation-1.

Equation-1:

$$E_c = \sum_{p=1}^{P_c} \sum_{i=1}^{n} \left(x_{cpi} - \bar{x}_{ci}\right)^2$$

where

c = index of the cluster, $c=1, 2, ..., C$.
P_c = collection points belong to cluster c
x_{cpi} = location of the p_{th} point of Pc data points on the i_{th} dimension of n-dimension variables for in the c_{th} cluster in standard dimension units (m)
x_{ci} = location of the centroid on the i_{th}-dimension in the c_{th} cluster in standard dimension units (m)
E_c = error sum of squares of distances for cluster c.

The data clustering methods may be applied for one or two times in the system depending on the supply chain defined by the user. That means if the supply chain includes storage, the clustering method will be applied first to define the potential location of the storage points by clustering the collection points considering the storage places as clusters centroids.

To locate the bioenergy conversion plants the resulted storage points, centroids of the first clustering iteration, are clustered to determine each bioenergy conversion plant location, assuming that it will be located at the centroid of the C-cluster. That means the transportation costs between the collection points to storage and from storage to conversion plants are optimized by the fuzzy C-means clustering at the same time.

Number of Clusters

In local planning level the number of clusters, storage and facilities, is small and defined by the decision makers based on the local inputs. On the other hand, at the national level, the numbers of storage places and bioenergy production facilities are defined automatically by a criterion known as cluster Validity Index, VI, that is measured for the input data set, (Shen, Chang, Lee, Deng, & Brown, 2005). The *VI* defines how well the clustering algorithm can partition the given data sets. In gBEDS the validity index to obtain the optimal number of clusters is represented by Equation-2 (Shen., 2005):

Equation-2, 3 and 4:

$$VI = \frac{Intra}{Inter}$$

$$Intra = \frac{1}{N} \sum_{i=1}^{C} \sum_{x \in U_i} \left\| x - z_i \right\|^2$$

$$Inter = \min_{i,j} \left(\left\| z_i - z_j \right\| \right)^2$$

Where

VI = Validity Index
C = number of clusters,
i = Cluster index, $i = 1, 2, \ldots, C-1$
j = Cluster index, $j = i + 1, i + 2, \ldots, C$
N = the number of collection points
z_i = spatial location of the i_{th} cluster center
z_j = spatial location of the j_{th} cluster center
x = spatial location of the storage or Plant U that belong to cluster i

The objective is to minimize the *intra*-value (as defined by Equation-3), which is average distance between the data points and the center of cluster and at the same time maximize the *inter*-value (see Equation-4) which is the inter distance between the clusters centers. The *VI* for C is obtained by minimizing the *intra–inter* ratio. The Validity Index table prepared for the specified data set as a pre-knowledge in the gBEDS and provided to

Table 1. Validity Index table provided to the user (storage locations)

Number of clusters	*VI*	*Intra* (m2)	*Inter* (m2)	Feasible biomass quantity (Ton)
10	0.086482	1.67E+09	1.93E+10	86568
50	0.51151	9.60E+08	1.88E+09	236250
100	0.76462	5.77E+08	7.55E+08	245740
150	1.3007	4.07E+08	3.13E+08	247200
200	1.5765	3.11E+08	1.97E+08	247770
250	2.4332	2.64E+08	1.08E+08	247610
300	4.1897	2.17E+08	5.18E+07	247690
350	3.6349	2.12E+08	5.83E+07	247570
450	4.3634	1.64E+08	3.75E+07	247770
500	10.178	1.57E+08	1.54E+07	247610

the user. A sample of *VI* table prepared by the system for the Japanese forestry residues storage is shown, in Table 1, for the national level.

The gBEDS also provides an option to allow the calculation of the *VI* by the user, however, the computation cost of the algorithm is high, 48 hours to establish *VI* table of 500 clusters for 6195 collection points in a Pentium 3- 1500 MHz CPU. A sample of *VI* table prepared by the system for the Japanese forestry residues storage is shown, in Table 1, for the national level.

Using the *VI* table, the numbers of clusters, e.g., storages and bioenergy conversion plants that meet users requirements are defined. It is important to refer that the number of clusters and the maximum feasible amount of utilized biomass are not inter-related as shown in Table 1. If 450 storage places are considered, the feasible quantity is more than that of 500 storage places. Based on the optimal results by the fuzzy *C*-means clustering of the collection points all bioenergy conversion plants attributes such as locations, capacities, number of workers, etc. can be evaluated easily.

Collection Points Clustering and Impacts Estimation

The desired Validity Index is defined from the table or according to user requirements and the system is ready to estimate the biomass utilization impacts along the selected supply chains by applying the clustering method to the collection points data. In case of forest and others, the system asks the user to define some site-specific inputs required for the impacts calculations such as, type of bioenergy required; efficiencies of available bioenergy conversion plants, average access distance from transportations routs, etc. Default values are provided for system users who do not have sufficient information about the required inputs.

The data related with the required storages, locations, areas, costs, etc., are calculated and saved in a matrix. In order to establish the storage layer in the GIS interface, database is stored in dbf files. Furthermore, the system defines the different impacts, environmental, economical and social, of bioenergy production from the input type biomass, costs, CO2 emissions per amount of

Figure 7. Supply chains for integrated biomass resources

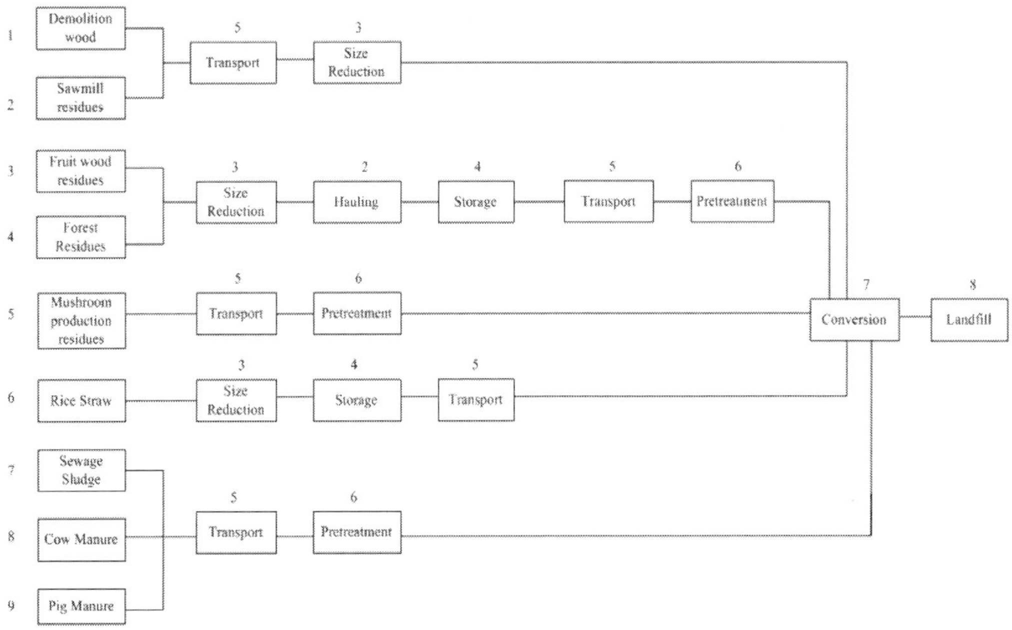

Resources Matrix= [1 2 3 4 5 6 7 8 9]
Matrix of Supply Chains Length = [4 4 7 7 4 5 4 4 4]
Chromosome = [5 3 7 8 5 3 7 8 3 2 4 5 6 7 8 3 2 4 5 6 7 8 3 2 4 5 6 7 8 5 6 7 8 3 4 5 7 8 5 6 7 8 5 6 7 8 5 6 7 8]

electricity produced, number of men-labor per the supply chain, total investments required for establishing the business, and others, for constructing a matrix of all data related to bioenergy production for each plant as a database file. In addition to these, at the national level, the optimization is performed by calculating the minimum distance only and the different impacts, from the selected bioenergy supply chain, are calculated as the sum of impacts produced from their Unit Process, UP, models defined for this level. However, at the local level the optimization model is used for further impacts reduction through investigating the different technologies defined in the UP models (Ayoub, 2007).

CASE STUDY

Now, a case study is being described by illustrating the above mentioned methods. This case study is performed for an integrated system of bioenergy resources with different supply chains to produce electricity in a Japanese city called Iida. Nine bioenergy resources have been chosen with their supply chains as shown in Figure 7. The quantities used in are given in Table 2. The bioenergy production plan includes; defining the locations of the new facilities; biomass resources quantities allocated for each bioenergy plant and its production capacities; annual investments, costs, emissions, energy and number of labor hours needed per year for bioenergy production. The output of applying this scenario may be considered as a future blueprint for the possible locations, capacities, types, etc. of bioenergy conversion plants.

Study Area

The planning session for Iida city, in Nagano prefecture, the Middle part of Japan, is selected here as a model for local level planning. The gross area of the city is 325.68 km^2 and a total population of 107,384 citizens. The city area is divided into 265 counties with population density

Figure 8. The potential location of bioenergy production facilities in Iida city.

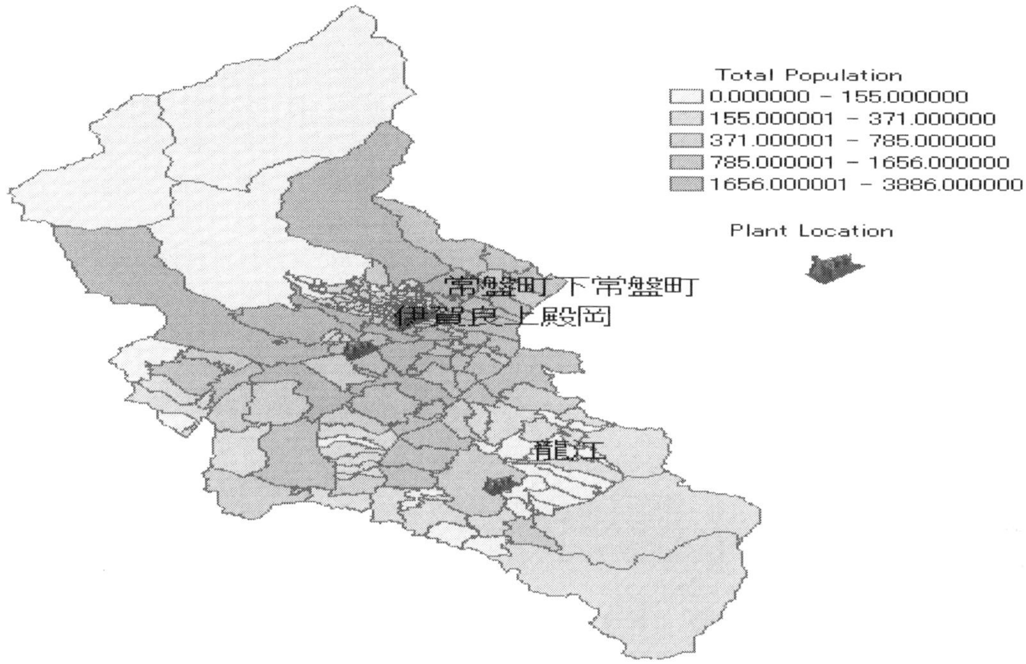

ranges from 0 citizens/km^2 where the county area is all forests to 9941.22 citizens/km^2 in the center of city. The city is served by a water treatment facility, an animal manure facility, 25 sawmills, and 3 wood chipping facilities.

Scenario Results

The planning for integrated bioenergy system tested here starts in gBEDS by first clustering the bioenergy resources to the number of facilities and defining the desired supply chain (see Figure 7).

Clustering Session

The output of the data clustering session, of the nine-biomass resources, is the location of potential facilities with the county name where they are located, and the quantities of bioenergy resources allocated to each bioenergy production plant are shown in Figure 8. It has been observed that, the locations of conversion facilities are found close to the areas of average and high population. This result is expected as the production of resources utilized depends on the number of population except the forestry residues that found to be of small amounts as seen in the resulted resources allocation quantities to each facility of Table 3.

Data Mining and Optimization

Once the locations of the bioenergy production facilities is decided the different impacts of establishing the facilities are calculated and visualized on the country map that enables the user to navigate the different impacts of establishing the bioenergy production facilities. For example, gBEDS, in the data mining session, calculates optimal sizes, capacities of the storage place (if it is included), the bioenergy conversion plants capacity, the minimum traveling distance between collection points and the storage, minimum distance from the storage to bioenergy conversion plant for minimizing the transportation costs, CO2 emissions and number of workers. In the following sub-sections, an introduction about data mining and clustering techniques is given and followed

Table 2.The bioenergy resources data used in solving the optimization problem

Biomass resources data											
Optimization Parameters	Forest Residues ton/year	Rice Straw ton/year	Mushroom Residues ton/year	Fruit Tree residues ton/year	Damaged Wood ton/ year	Pig Manure ton/year	Cow Manure ton/year	Sewage Sludge ton/ year	Demolition Wood ton/ year	Bioenergy plant operating hours ton/year	Distance to electricity network km
Bioenergy production plant 1	3129	647.35	3393.8	622.26	553.68	40241	0	9329.4	1805	4000	5
Bioenergy production plant 2	9105.8	2083.1	491.8	122.26	51.181	10238	0	0	156.75	4000	5
Bioenergy production plant 3	5298.3	2134.1	1513.8	291.97	204.72	22516	1.27E+05	447.09	1008	4000	5

by explaining the clustering method applied to GIS data in this work.

Based on a selected scenario, the optimization session is conducted for the three facilities to minimize the cost of bioenergy production. As a result, the bioenergy plants capacities, cost of one MWh in each plant, and annual investments are calculated. The average data, of the optimization solution for the three plants are shown in Table 3.

The data shown in Table 3 are computed by optimizing the three supply chains based on the cost and resulted a specific UP models, where each model belongs to a specific technology type, that work as a blue print for the coming level. Table 4 shows an example of UP models suggested by the system for three bioenergy production plants to produce bioenergy from the specified resources.

CONCLUSION

In this chapter a description for Data Mining as an effective tool for decision support is presented. A two level system called, general BioEnergy Decision System, gBEDS, as an effective tool in planning for bioenergy supply chains is introduced with the core part of Data Mining techniques, i.e. clustering. The gBEDS is integration of ideas and methods used for the first time in the research field of bioenergy planning and decision making. The Multilevel Decision Making Life Cycle concept has been applied, both, in the analysis of data required for the specified decision level as well as explaining the differences in prospective between the two decision levels. A general explanation about the proposed DM method used in this work is presented and followed by detailed explanation of the specific use of DM in the gBEDS prototype.

Table 3. The bioenergy optimization problem results based on cost criteria

	Bioenergy supply chains		
Bioenergy planning parameters	Plant 1	Plant 2	Plant 3
Plant Capacity MW	4.85	2.46	11.11
Annual Investment (Myen/year)	290.39	245.97	547.19
Supply Chain Cost Per MWh (yen/MWh)	14962	25040	12308
Supply Chain Emission Per MWh (KgCO2eq/MWh)	37.39	43.15	37.50
Supply Chain Energy Per MWh (MJ/MWh)	1212.49	1264.23	1245.84
Supply Chain Labor (L.h/year)	46632.93	47139.32	88530.09
I/O Energy Ratio	0.105	0.076	0.161
Forest Residues (ton/year)	2254.28	8570.29	4071.62
Rice Straw (ton/year)	553.72	1742.21	1622.60
Mushroom Residues (ton/year)	3326.26	486.88	1483.68
Fruit Tree residues (ton/year)	537.97	105.70	210.35
Damaged Wood (ton/year)	542.66	50.16	200.65
Pig Manure (ton/year)	39440.20	10034.26	22290.84
Cow Manure (ton/year)	0	0	124423.70
Sewage Sludge (ton/year)	9143.75	0	438.19
Demolition Wood (ton/year)	1770.87	153.63	987.94

Special attention was given to the clustering techniques used to tackle the problem of distributed nature of biomass and identifying locations of the biomass utilization facilities that is considered a preliminary optimization based on distance and the integration between GIS and DM in defining the feasible biomass quantity is also explained. It is also, demonstrated by a case study how gBEDS helps the decision makers to establish, design, and evaluate a system of bioenergy production, starting from location determination via DM and ending to suggesting specific technologies for the specified Unit Process.

The authors, believe that this research results, which for the first time approach the biomass planning process via non-conventional optimization techniques, e.g., Data Mining, GA and GIS, may provide vital footing for the development of DSS in the bioenergy planning field and other similar fields.

FUTURE RESEARCH DIRECTIONS

The results of this work open several opportunities for further research and improvement via different Data Mining techniques and can be summa-

Table 4. The UP models recommended by the optimization model based on cost criteria

Biomass resources	Bioenergy supply chain UPs technologies: Plant 1	Plant 2	Plant 3
Forest residues			
	Road side Hammer mill (12 ton/h)	Road Side Disk Drum Chipper (9 ton/h)	Road Side Disk Drum Chipper (9 ton/h)
	Self propelled forwarder (1.688 ton/h)	Self propelled forwarder (1.688 ton/h)	Self propelled forwarder (1.688 ton/h)
	Unprotected on crushed rock	Pole framed Structure on crushed rock	Poly Tarp Cover on crushed rock
	Dump Truck Diesel (15 ton)	Dump Truck Bio-Diesel (10 ton)	Dump Truck Diesel (10 ton)
	Tube bundle dryer (B) - Saturated steam	Tube bundle dryer (A) - Saturated steam	Tube bundle dryer (B) - Saturated steam
	160 KW, (3 t.ev.w/h)[a]	94 KW, (3 t.ev.w/h)	160 KW, (3 t.ev.w/h)
	Coal Co-firing	Coal Co-firing	Coal Co-firing
	Landfill without gas recovery system	Landfill without gas recovery system	Landfill with gas recovery system
Rice straw			
	Hesston 5800 Round Baler (3.3 ton/h)	John Deer Model 346 Square Baler (6 ton/h)	Hesston 5800 Round Baler (10 ton/h)
	Poly Tarp Cover on crushed rock	Pole framed Structure on crushed rock	Poly Tarp Cover on crushed rock
	Dump Truck Bio-Diesel (15 ton)	Dump Truck Bio-Diesel (15 ton)	Dump Truck Diesel (4 ton)
	Coal Co-firing	Coal Co-firing	Coal Co-firing
	Landfill without gas recovery system	Landfill without gas recovery system	Landfill with gas recovery system
Sewage sludge			
	Dump Truck Bio-Diesel (4 ton)	NA[b]	Dump Truck Diesel (4 ton)

continued on following page

Table 4. continued

Tube bundle dryer (B) - Saturated steam 160 KW, (3 t.ev.w/h)	NA	Tube bundle dryer (A) - Saturated steam 94 KW, (3 t.ev.w/h)
Landfill without gas recovery system	NA	Landfill with gas recovery system
Coal Co-firing	NA	Coal Co-firing

a) t.ev.w : ton evaporated water per hour
b) NA: not applicable

Figure 9. The structure of data manipulation for the two levels of decision making

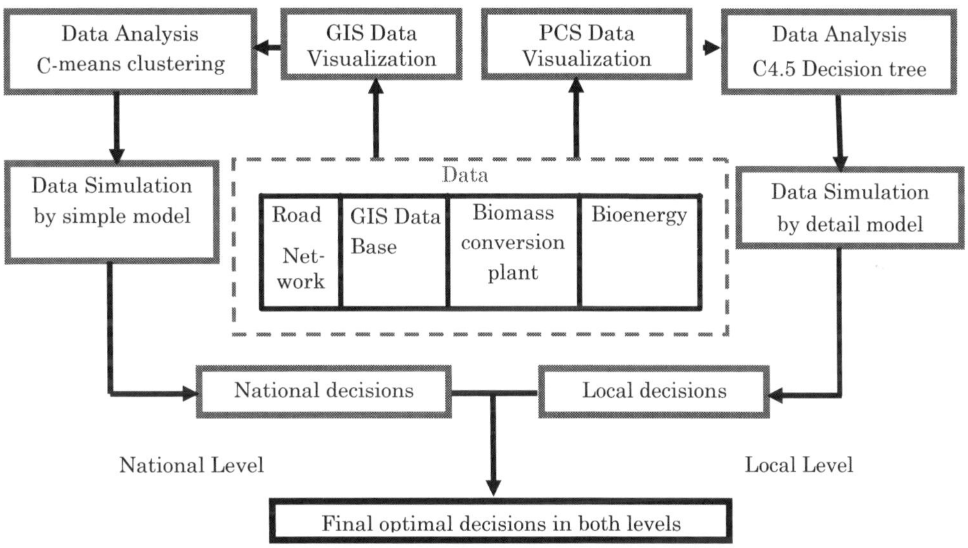

rized in the following paragraphs. To overcome the differences in nature between localities it is important to establish standard classes of localities that share similar characteristics, i.e., land, weather, locations, etc., which is a typical DM task. In this case, applying an institution or plan on one instance of specific class can be applied to other instances with moderate effort, time and resources.

The possible collaboration between the localities at the same level, such as material transportation and supply, is not considered in the gBEDS where planning is made for only one local area at a time. Therefore, this point, as well as the resources exchange between succeeding localities, is considered challenging tasks to be added in the local level planning. For example, if resources shortage has been occurred, there should be a procurement strategy from the nearest local area within certain distance for safe feed supply and this can be supported by the popular DM technique Decision Trees.

In the current DSS authors used one DM technique and the same data visualization method for both decision making levels which needs to be revisited and modified in the future versions of gBEDS. As seen in Figure 9, the structure of data manipulation in both DM and data visualization can be modified to include more detailed techniques, i.e., C4.5 Decision tree and PCS Data visualization techniques in the local level planning.

NOTE

Two figures (2, 5), one table (1) and some text are reproduced from our previous publication "Ayoub, N., Martins, R., Wang, K., Seki, H., & Naka, Y. (2007). Two levels decision system for efficient planning and implementation of bioenergy production. *Energy Conversion and Management, 48*(3), 709-723" under the permission of Elsevier Limited}.

REFERENCES

Ayoub, N. (2007). *A Multilevel Decision Making Strategy for Designing and Evaluating Sustainable Bioenergy Supply Chains.* Unpublished doctoral dissertation, Tokyo Institute of Technology, Yokohama, Japan.

Ayoub, N., Martins, R., Wang, K., Seki, H., & Naka, Y. (2007). Two levels decision system for efficient planning and implementation of bioenergy production. *Energy Conversion and Management, 48*(3), 709-723.

Ayoub, N., Wang, K., Kagiyama, T., Seki, H., & Naka, Y. (2006). A Planning Support System for Biomass-based Power Generation. In W. Marquardt & C.Pantelides (Eds.), *16th European Symposium on Computer Aided Process Engineering and 9th International Symposium on Process Systems Engineering Vol. 21B.* (pp. 1899-1904). Garmisch-Partenkirchen, Germany.

Basson, L., & Petrie, J. G. (2007). An integrated approach for the consideration of uncertainty in decision making supported by Life Cycle Assessment. *Environmental Modelling & Software Environmental Decision Support Systems, 22*(2), 167-176.

Borges, M. R. S., Pino, J. A., & Valle, C. (2005). Support for decision implementation and follow-up. *European Journal of Operational Research Decision Support Systems in the Internet Age, 160*(2), 336-352.

Catrinu, M. D. (2006). *Decision Aid for Planning Local Energy Systems: Application of Multi-Criteria Decision Analysis.* Unpublished doctoral dissertation, Norwegian University of Science and Technology, Trondheim, Norway.

Chae, Y. M., Ho, S. H., Cho, K. W., Lee, D. H., & Ji, S. H. (2001). Data mining approach to policy analysis in a health insurance domain. *International Journal of Medical Informatics, 62*(2-3), 103-111.

Chau, K. W., Cao, Y., Anson, M., & Zhang, J. (2003). Application of data warehouse and Decision Support System in construction management. *Automation in Construction, 12*(2), 213-224.

Delen, D., & Pratt, D. B. (2006). An integrated and intelligent DSS for manufacturing systems. *Expert Systems with Applications, 30*(2), 325-336.

Gibert, K., Sanchez-Marre, M., & Rodriguez-Roda, I. (2006). GESCONDA: An intelligent data analysis system for knowledge discovery and management in environmental databases. *Environmental Modelling & Software, 21*(1), 115-120.

Hakkila, P. (2006). Factors driving the development of forest energy in Finland. *Biomass and Bioenergy*: Proceedings of the third annual workshop of Task 31 'Sustainable production systems for bioenergy: Impacts on forest resources and utilization of wood for energy' October 2003, Flagstaff, Arizona, USA, *30*(4), 281-288.

Hoskinson, R. L., Rope, R. C., & Fink, R. K. (2007). Using a decision support system to optimize production of agricultural crop residue Biofeedstock. *Biomass and Bioenergy, 31*(4), 186-194.

King, P., Annandale, D., & Bailey, J. (2000). A Conceptual Framework for Integrated Economic and Environmental Planning in Asia - A Literature Review. *Journal of Environmental Assessment Policy and Management, 2*(3), 279-315.

Marquez, A. C., & Blanchar, C. (2006). A Decision Support System for evaluating operations investments in high-technology business. *Decision Support Systems, 41*(2), 472-487.

Rozakis, S., Soldatos, P. G., Kallivroussis, L., & Nicolaou, I. (2001). Multiple Criteria Analysis of Bio-energy Projects: Evaluation of Bio-Electricity Production in Farsala Plain, Greece. *Journal of Geographic Information and Decision Analysis, 5*(1), 49-64.

Shen, J., Chang, S. I., Lee, E. S., Deng, Y., & Brown, S. J. (2005). Determination of cluster number in clustering microarray data. *Applied Mathematics and Computation, 169*(2), 1172-1185.

Wang, H. (1997). Intelligent agent-assisted decision support systems: Integration of knowledge discovery, knowledge analysis, and group decision support. *Expert Systems with Applications, 12*(3), 323-335.

Ye, N. (2003). *The Handbook of Data Mining.* London: Lawrence Erlbaum Associates, Publishers.

ADDITIONAL READING

Anderson, B. F. (2002). *The Three Secrets of Wise Decision Making.* Portland, Oregon: Single Reef Press.

Anttiroiko, A. V., & Malkia, M. (Eds.). (2007). *Encyclopedia of digital government.* Hershey. London. Melbourne. Singapore: Idea Group Publishing.

Bigus, J. P. (1996). *Data mining with neural networks: Solving business problems:from application development to decision support.* McGraw-Hill.

Chinese, D., & Meneghetti, A. (2005). Optimisation models for decision support in the development of biomass-based industrial district-heating networks in Italy. *Applied Energy, 82*(3), 228-254.

Cios, K. J., Pedrycz, W., Swiniarski, R. W., & Kurgan, L. A. (2007). *Data Mining: A Knowledge Discovery Approach. New York,: Springer Science+Business Media, LLC.*

Courtney, J. F. (2001). Decision making and knowledge management in inquiring organizations: toward a new decision-making paradigm for DSS. *Decision Support Systems, 31*(1), 17-38.

Dornburg, V., Faaij, A. P. C., & Meuleman, B. (2006). Optimising waste treatment systems: Part A: Methodology and technological data for optimising energy production and economic performance. *Resources, Conservation and Recycling, 49*(1), 68-88.

Freppaz, D., Minciardi, R., Robba, M., Rovatti, M., Sacile, R., & Taramasso, A. (2004). Optimizing forest biomass exploitation for energy supply at a regional level. *Biomass and Bioenergy, 26*(1), 15-25.

Gibert, K., Sanchez-Marre, M., & Rodriguez-Roda, I. (2006). GESCONDA: An intelligent data analysis system for knowledge discovery and management in environmental databases. *Environmental Modelling & Software, 21*(1), 115-120.

Larichev, O. (1999). Normative and descriptive aspects of decision making. In T. Gal, T. Stewart & T. Hanne (Eds.), *Multicriteria decision making: Advances in MCDM models, algorithms, theory and applications* (pp. 5.1-5.24). Boston: Kluwer Academic Publishers.

Larose, D. T. (2006). *Data mining methods and models.* Hoboken, New Jersey: John Wiley & Sons, Inc.

Lin, F. r., Huang, K. j., & Chen, N. (2006). Integrating information retrieval and data mining to discover project team coordination patterns. *Decision Support Systems, 42*(2), 745-758.

Makaila, M., Anttiroiko, A. V., & Savolaine, R. (2004). *E-Transformation in Governance: New Directions in Government and Politics.* Hershey: Idea Group Publishing.

Min, H., Ko, C. S., & Ko, H. J. (2006). The spatial and temporal consolidation of returned products in a closed-loop supply chain network. *Computers & Industrial Engineering: Special Issue: Logistics and Supply Chain Management, Selected Papers from The 33rd. ICC&IE, 51*(2), 309-320.

Mitchell, C. P. (2000). Development of decision support systems for bioenergy applications. *Biomass and Bioenergy, 18*(4), 265-278.

Myatt, G. J. (2007). *A Practical Guide to Exploratory Data Analysis and Data Mining.* New Jersey: John Wiley & Sons, Inc.

Ramachandra, T. V., & Shruthi, B. V. (2007). Spatial mapping of renewable energy potential. *Renewable and Sustainable Energy Reviews, 11*(7), 1460-1480.

Taniar, D. (2007 .). *Research and trends in data mining technologies and applications.* Hershey: Idea Group Publishing.

Tonn, B., & Martin, M. (2000). Industrial energy efficiency decision making. *Energy Policy, 28*(12), 831-843.

Wang, J. (Ed.). (2003). *Data Mining: Opportunities and Challenges.* London: Idea Group Publishing.

Witten, I. H., & Frank, E. (2005). *Data Mining: Practical Machine Learning Tools and Techniques* (Second Edition ed.): Morgan Kaufmann.

This work was previously published in Social and Political Implications of Data Mining: Knowledge Management in E-Government, edited by MD H. Rahman, pp. 235-253, copyright 2009 by Information Science Reference (an imprint of IGI Global).

Chapter 16
Towards a Semiotic Metrics Suite for Product Ontology Evaluation

Joerg Leukel
University of Hohenheim, Germany

Vijayan Sugumaran
Oakland University, USA

ABSTRACT

In recent years, product ontology has been proposed for solving integration problems in product-related information systems such as e-commerce and supply chain management applications. A product ontology provides consensual definitions of concepts and inter-relationships being relevant in a product domain of interest. Adopting such an ontology requires means for assessing their suitability and selecting the "right" product ontology. In this article, the authors (1) propose a metrics suite for product ontology evaluation based on semiotic theory, and (2) demonstrate the feasibility and usefulness of the metrics suite using a supply chain model. The contribution of our research is the comprehensive metrics suite that takes into account the various quality dimensions of product ontology.

INTRODUCTION

Product-related information is of paramount importance in many interorganizational applications, since it concerns goods and services being procured, manufactured and sold to customers. Due to the involvement of multiple organizations, there is a need for integrating product-related information, e.g., by standardization or mediation. In the past years, product ontology has attracted both industry and academia because of its potential contribution to solving integration problems (Shim & Shim 2006). A product ontology provides, at least to some extent, consensual definitions of concepts and inter-relationships between these concepts in a product domain of interest. Most product ontologies define a hierarchy of product classes and respective properties for describing product instances. Such ontologies may support finding and comparing products being offered by multiple suppliers and described in distributed data sources, or allow for benchmarking the procure-

DOI: 10.4018/jiit.2009080701

ment activities of organizational units (Doring et al., 2006). Ontology users are required to annotate their product instance data accordingly.

Product ontologies have already emerged in diverse industries and for various tasks (Park et al., 2008). However, assessing the quality and suitability of a given product ontology, i.e., to what degree it actually meets user requirements, remains a critical question for potential ontology adopters. This question is the focus of ontology evaluation, which aims at providing metrics reflecting the ontology's quality and suitability. There is great difficulty in determining what elements of quality to evaluate. In other words, what factors should be considered in evaluating product ontology quality? Current research yields a number of approaches, metrics, and tools for automatically evaluating ontologies (Garcia-Castro et al., 2007; Hartmann, 2005). However, most of this research originates from the Semantic Web arena, and therefore relies mainly on the expressiveness of ontology languages such as DAML (DARPA Agent Markup Language) and OWL (Ontology Web Language); hence their scope is constrained by these languages and does not take the specific setting of product ontology into account.

Very often, an ontology is regarded as an artifact used by a community as a common vocabulary without considering the organizational properties of the respective community and thus the interrelations within the community (Zhdanova et al., 2007). For example, a community that often uses product ontologies is made of entities belonging to a supply chain. A supply chain is a system of entities participating in producing, transforming, and distributing goods and services from supply to demand. A single product ontology is thus used within supply chains and determining its quality and suitability has to consider the supply chain characteristics, e.g., by distinguishing different roles such as manufacturer and distributor. A major trend affecting supply chains is individualization, caused by customers demanding individualized products, which are tailored to their specific needs (e.g., custom-made products) (Coates, 1995) (Kirn, 2008). For instance, enabling customers to order custom-made shoes via an e-commerce application does not only concern the e-commerce firm but also the stakeholders in the respective supply chain (e.g., manufacturer and its suppliers). Here, a product ontology may help provide a common terminology and means of describing products along the entire supply chain.

In the context of supply chain and individualization, a product ontology should emphasize the importance of quality metrics that allow the assessment of product complexity in terms of richness of product description and product structure, and how the final product is composed of individual parts. Current evaluation metrics do not take these factors into account: Domain-independent metrics are not able to exploit the domain characteristics (e.g., Yao et al. 2005), whereas domain-specific evaluation metrics regard products as single and atomic items without considering existing interrelations that arise due to supply chain structures and customer requirements (e.g., Hepp et al. 2007). To overcome this limitation, we address product ontology evaluation on a broader scale by taking a semiotic perspective. Semiotics studies the properties of signs; for our purposes, it can provide a theoretical basis for distinguishing generic categories of quality. We define evaluation metrics based on Stamper's et al. (2000) semiotic framework and adopt the domain-independent semiotic metrics suite proposed by Burton-Jones' et al. (2005).

The objectives of this research are to: (1) develop a semiotic set of metrics that allow for assessing the quality of product ontologies, and (2) apply the metrics to a commonly available product ontology to demonstrate the feasibility and usefulness of the metrics suite. The contribution of this research is the comprehensive metrics suite that takes into account the various quality dimensions of product ontologies. A preliminary study of semiotic metrics for product ontology evaluation can be found in (Leukel & Sugumaran,

Figure 1. Roles within a supply chain

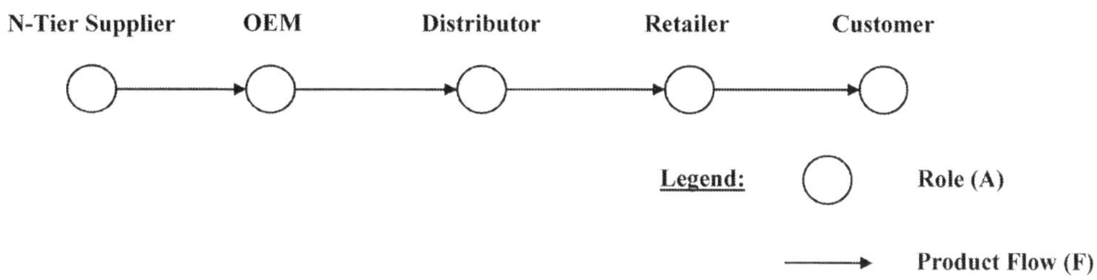

2007). The contribution of this research is that the current work adapts the metric suite developed by Burton-Jones et al. (2005) to the product domain in the context of supply chain management to determine which metrics are applicable and how they relate to the existing work in the product domain ontology. We map the metrics developed in both streams of research and develop a unified set of metrics for the product ontology domain.

The rest of the article is organized as follows. The next section defines the basic model of supply chain and product ontology. After that, we present our semiotic metrics suite. The subsequent section provides a preliminary validation of the metrics. In the section that follows, we review the related work. Finally, we draw some conclusions and outline avenues of future research.

BASIC MODEL OF SUPPLY CHAIN AND PRODUCT ONTOLOGY

A supply chain model is a representation of entities participating in producing, transforming, and distributing goods and services from supply to demand (Supply-Chain Council, 2007). Such a model thus concerns both products and organizations which we call actors. The inter-relations between actors are constituted by the flow of products. We define the supply chain model as a directed graph S = (A, F) using the following notions.

- A: set of actors
- a: element of A
- F: set of flows of products with $F \subseteq A \times A$-1
- f: element of F connecting two actors, f = (a1, a2)

To allow for distinguishing the specific role of each actor within a supply chain, we define five generic roles: 1) OEM, 2) N-tier supplier, 3) distributor, 4) retailer, and 5) customer. These roles represent the generic types of participants in a supply chain. A specific supply chain may include several instances of each of these roles. Typically, a supply chain is made of two major parts being separated by the original equipment manufacturer (OEM). The left hand side role of n-tier suppliers describe how the product is made from parts (thus it focuses on procurement and manufacturing) with n denoting the number of supply chain stages. For instance, by assigning the 1st tier supplier role to different actors each supplying a different part of the final product, one can represent the product structure in the supply chain model. 2nd tier suppliers provide product to the 1st tier supplier etc. The right hand side roles participate in distributing the product from the OEM across one or more stages which are: distributor and retailer. Respective actors do not apply manufacturing technology but change the product regarding location, time, and quantity (by means of logistics, e.g., bundling) until the product reaches the final customer. The sequence of roles can be regarded as the reference structure of many real-world supply chains, as shown in Figure 1. Note that this simple model contains only one ac-

Table 1. Elements of product ontology

Element	Definition	
PC	set of product classes	A product class pc represents the product concept and often consists of, besides the class name, natural language definition and some data management information (e.g., identifiers, version etc.).
RC	set of relations between product classes	Is-a relations build a hierarchy of product classes with RC $\subseteq$ PC × PC; often a taxonomy with top-level classes CT$\subseteq$ PC for separating domains (e.g., automotive, chemical, textile etc.).
PP	set of product properties	A product property pp is a template for describing product instances; consists of property name, a natural language definition, data type, and unit of measurement.
RP	set of class-property relations	Maps properties of PP to classes PC with RP $\subseteq$ PC × PP; the semantics of such relations can range from loose recommendation to mandatory.
PV	set of property values	Values v for properties besides standard data types such as integer, float, and string. Used for expressing a narrower domain, e.g., for colors, shapes, materials etc.
RV	set of property-value relations	Maps values of PV to properties PP with RV $\subseteq$ PV × PP.

tor for each role whereas in reality multiple actors exist for almost any supply chain stage.

Product ontology evaluation focuses on the ontology constructs that are used in this type of ontology. Product ontology relies essentially on providing an often broad and deep hierarchy of product classes based on is-a relationships, while other relationship types play a minor role. Having reviewed current product ontologies and respective conceptual models, we define product ontology PRO as a 6-tuple PRO = (PC, RC, PP, RP, PV, RV) as defined in Table 1.

The relationship between supply chain and product ontology model is two-fold. First, since the arrows (F) connecting two roles in the supply chain model represent flows of one or more products, each such arrow (f) has to be mapped to at least one respective product class "pc" in the ontology; otherwise the ontology would not be able to cover the respective part of the supply chain model. Figure 2 shows an example how product flows can be annotated with references to product classes (pc) to describe the product structure along the supply chain stages.

However, this type of relationship does not reflect how well the product class represents the semantics of the product flow. For instance, choosing rather abstract product classes would allow for describing many different product flows, but lose the actual semantics. Therefore, a second

Figure 2. Example: Relationship between supply chain and product ontology

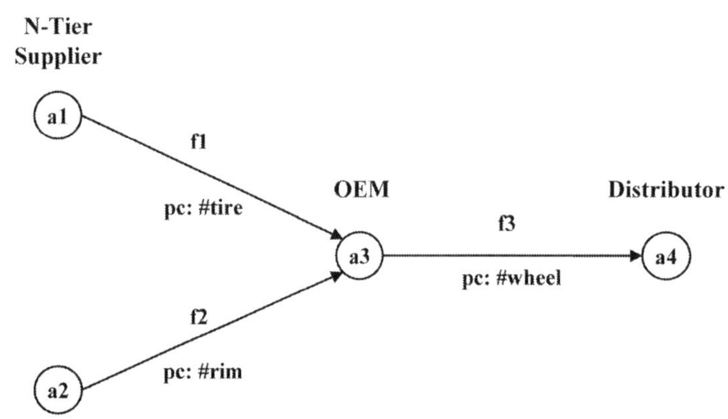

relationship has to be considered, which represents the richness of the product class. Thus, additional information from the product ontology is used to specify a product flow in greater detail. In the example shown in Figure 2, it is not specified whether the wheels are for cars, trucks, or both. If the product ontology contains more specific classes, one could, for instance, describe the supply chain of a specialized OEM which produces wheels for trucks only more precisely.

PROPOSED METRICS SUITE

In this section, we propose a metrics suite for product ontology evaluation based on semiotic theory (Stamper et al. 2000) and earlier work by Burton-Jones et al. (2005). First, we introduce the semiotic approach to ontology evaluation and define the basic metrics. Then we define five categories of semiotic metrics.

Semiotics

Stamper et al. (2000) present a general theoretical semiotic framework derived from linguistics that includes general elements of quality for evaluating signs. It includes pragmatic issues to develop a metrics suite that is widely applicable yet can be tailored to the needs of specific applications. They provide a 6-level semiotic framework to support the analysis of signs: a) physical, b) empirical, c) syntactic, d) semantic, e) pragmatic, and f) social. There is a strong relationship between the successive levels, i.e., each level contributes to the next level. The physical level deals with representation of signs in hardware, components, etc. and the empirical level is concerned with communication properties of signs including channel capacity, noise, and entropy. Since these levels are very implementation specific, they may not be highly relevant to quality assessment (Burton-Jones et al. 2005). The remaining levels are considered in developing the metrics suite for quality assessment.

Based on these levels, Burton-Jones et al. (2005) have developed a metric suite that consists of syntactic, semantic, pragmatic, and social qualities. As mentioned above, they do not include physical and empirical quality since they correspond to implementation aspects. With regard to product ontology, these two stages can also be omitted because signs do exist (physical in terms of explicit concepts) and can be seen (empirical in terms of a formal representation of the concepts). For our purposes, their metric suite can provide a theoretical basis for developing metrics, because (1) the general semiotic framework takes the various dimensions of the meaning of signs into account, and (2) it has been proven to be applicable and valuable for ontology evaluation.

Metrics

The metrics suite originally proposed by Burton-Jones et al. (2005) is adapted for the product ontology domain. While the metrics themselves and their constituent parts are applicable to product ontologies, additional constructs specific to the product ontology domain are identified and added to the overall metric suite. In particular, we consider the work of Hepp et al. (2007) in identifying additional constructs and create a unified set of metrics by explicating the similarities and differences between the two sets of metrics discussed in the literature.

The overall quality (Q) of an ontology is computed using a weighted function of its syntactic (S), semantic (E), pragmatic (P), and social (O) qualities (Burton-Jones et al. 2005):

$$Q = b1\times S + b2\times E + b3\times P + b4\times O$$

The weights sum to unity. In the absence of prespecified weights, the weights are assumed to be equal. Since the numerical values of these relative scores could exceed one for any given ontology, the scores for these metrics are normalized so that the values of all metrics vary between zero and one prior to calculating the overall ontological quality. According to Burton-Jones et al. (2005), the values for a given ontology will depend on external benchmarks such as the metric's average value across all ontologies in the ontology library. With respect to product ontology, the small size of the respective ontology library may prevent using such a benchmark.

Syntactic Quality

Syntactic Quality (S) measures the quality of the product ontology according to the way it is written. It consists of Lawfulness, the degree to which an ontology language's rules have been complied, and Richness, the proportion of features in the ontology language that have been used in an ontology. Lawfulness is entirely domain-independent because one would require that any given product ontology complies with the syntax of the ontology language used. Otherwise the ontology cannot be processed. The interpretation of richness, however, must consider the requirements of the product domain with respect to expressiveness. For instance, if a highly expressive language based on description logic is used but the domain does not require much reasoning support, the richness metric would have quite a low value.

Semantic Quality

Semantic Quality (E) evaluates the meaning of terms in the product ontology. Three attributes are used in this metric: interpretability, consistency, and clarity.

Interpretability refers to the meaning of terms in the ontology (e.g., names of product classes and product properties). Meaning could be determined by checking whether the terms used can be found in the product domain of interest, for example, by searching the standards and references. Consistency is whether terms have a consistent meaning in the ontology. The appearance of the same term in more than one concept could indicate inconsistency. Clarity is whether the context of terms is clear. For example, if a product ontology claims that class "Chair" has the property "Material", an application must know that this describes furniture, and not academics.

Pragmatic Quality

Pragmatic Quality (P) refers to the product ontology's usefulness for users, irrespective of syntax or semantics. Three criteria are used for this metric, namely, accuracy, comprehensiveness, and relevance.

Accuracy is whether the claims about products an ontology makes are "true." In general, it can only be determined by domain experts, or by reasoning if the product ontology defines respective relationships between concepts. Comprehensiveness is a measure of the size of the ontology. Larger product ontologies are more likely to be complete representations of their domains, and provide more knowledge to the user. Size could indicate both how well the product domain is covered and to which degree of detail it is represented by explicit concepts. However, one has to be careful since sometimes, simply considering the total number of concepts and relationships may lead to false assessments. Relevance is whether the ontology satisfies the user's specific requirements. It could be calculated by checking against a set of explicit requirements articulated for a particular scenario.

Social Quality

Social quality (O) reflects the fact that users and product ontologies exist in communities. It consists of two attributes, namely, authority and history. The authority is the number of other ontologies that link to it. More authoritative product ontologies signal that the knowledge they provide is accurate

Table 2. Determination of pragmatic metrics of Hepp et al. (2007)

Aspect	Metric	Determination
Amount of content	M11: Number of classes	Number of elements in PC
	M12: Number of properties	Number of elements in PP
	M13: Number of enumerative data types	Number of elements in PV
Hierarchical order and balance of scope	M21: Number of classes per top-level class CTi⊆ PC	Number of elements in PC which are subclass of CTi
	M22: Services ratio	(Number of elements in PC representing services) / (number of elements in PC)
	M23: Distribution properties of metric M21	Minimal value, maximal value, mean, median, first quartile, third quartile, interquartile range, standard deviation, and coefficient of variation of M21
	M24: Percentage of content in the three biggest top-level classes CT	(Number of subclasses of the three biggest classes CT) / M11
	M25: Size of the biggest top level class vs. median of M21	(Number of subclasses of the biggest class CT) / median of M21
	M26: Number of descendents per subordinate class COn	Number of elements in RC with COn superclass of COm
Class-specific property lists	M31: Specific property lists ratio	(Number of elements in PC with a specific property list in RP) / M11
	M32: Distribution of specific property lists per top-level class CTi	(Number of elements in PC which are subclass of CTi and have a specific property list in RP) / M21
	M33: Property usage in property lists	Minimal value, maximal value, mean, median, standard deviation, and coefficient of variation of {number of elements in RP with PPi}
	M34: Semantic weight of property PPi	1 / (number of property lists in RP including PPi)
	M35: Semantic value of property lists	Sum of M34 for each property PP in property list of class PCi
Growth and maintenance	M41: Number of new classes per month	((Number of elements in PC in the current version of PRO) – (number of elements in PC in the former version of PRO)) / (number of months between publication of the current and former version of PRO)
	M42: number of new classes per top-level class CTi	M41 for CTi

or useful. The history is the number of times the ontology is accessed, and more precisely how often its concepts are used in actual product-related information systems. It is assumed that ontologies with longer histories are more dependable.

For the purpose of evaluating product ontologies in the context of supply chain management, comprehensiveness can be used as an indicator for both coverage of supply chains and richness of the product concepts (as defined in the basic model section). Therefore, we extend the comprehensiveness metric at this level by integrating the work of Hepp el al. (2007), which proposes an elaborate set of pragmatic metrics for product ontology evaluation. We select the relevant metrics and transform their definition into our notation of product ontology. The result is shown in Table 2, which depicts the following four aspects of product ontology quality: (1) amount of ontology content, (2) hierarchical order and balance of scope, (3) class-specific property sets, and (4) ontology growth and maintenance.

Based on our analysis of the product categorization standards work by Hepp et al. (2007), it is evident that there is some commonality between the metrics identified by Hepp et al. (2007) and

Table 3. Product ontology metrics [adopted from Burton-Jones et al. (2005)]

Metric	Definition
Overall Quality (Q)	Q = b1·S + b2·E + b3·P + b4·O where: b1,..,b4 = weights
Syntactic Quality (S)	S = bs1·SL + bs2·SR where: SL = Lawfulness SR = Richness
Semantic Quality (E)	E = be1·EI + be2·EC + be3·EA where: EI = Interpretability EC = Consistency EA = Clarity
Pragmatic Quality (P)	P = bp1·PO + bp2·PU + bp3·PR where: PO = Comprehensiveness PU = Accuracy PR = Relevance
Social Quality (S)	O = bo1·OT + bo2·OH where: OT = Authority OH = History

Burton-Jones et al. (2005). Table 3 summarizes the integrated product ontology metrics The initial set of metrics identified by Burton-Jones et al. (2005) are still applicable for the product domain ontology, however, the pragmatic quality metric is extended with some additional metrics.

PRELIMINARY VALIDATION

In this section, we provide a preliminary validation of the proposed semiotic metrics suite. We apply the metrics suite to a supply chain scenario and demonstrate its relevance and usefulness. We describe the experimental design, report the results, and discuss the findings.

Validation Scenario

The purpose of the experiment is to demonstrate the feasibility and usefulness of the metrics suite in the specific setting of product ontology. In the context of individualization or mass customization in supply chains, we consider the following 5-tier supply chain model in the IT industry (Figure 3): The ultimate goal of the supply chain is to deliver custom-made desktop computers to end customers. The supply chain consists of a retailer, distributor, OEM, and 1st tier suppliers with S = (A, F) = ({a1, a2, a3, a4, a5, a6, a7}, {f1, f2, f3, f4, f5, f6}).

Product flows f1, f2, and f3 represent parts, whereas f4 and f5 represent both parts and computers; f6 is computers only. This supply chain model enables different options for individualization. For instance, both the distributor and retailer can create a new bundle of computers and parts and offer it as individualized computers.

In the above supply chain scenario set up for individualization, we study the use of eCl@ss product ontology which is available in both an OWL representation called eClassOWL (Hepp, 2006) and in simple comma-separated value files. The eCl@ss is originally an international

Figure 3. Supply chain model used in validation

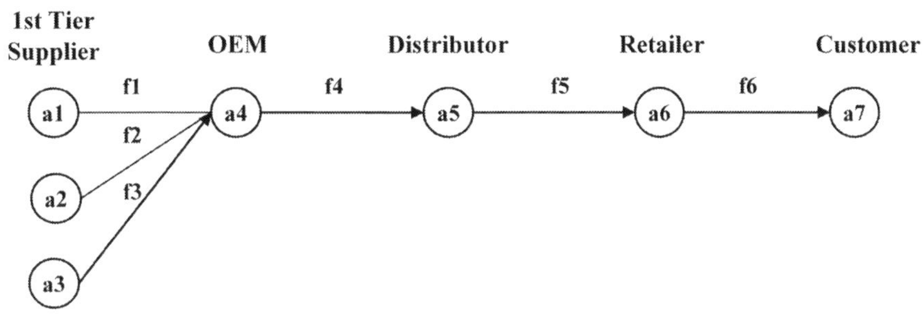

classification scheme for goods and services, and thus is of practical importance in many industries and countries (eCl@ss, 2008).

We conduct different levels of validation. First, we pre-check each metric whether it can be applied and what results can be expected. Then, we apply those metrics to the eCl@ss ontology and evaluate how well the product ontology can be used to support individualization in the supply chain scenario. In particular, we focus on the metrics that are the most relevant for our purposes, namely, Interpretability, Consistency, Clarity, and Relevance, since they refer to semantics and pragmatics. The results of our analysis are discussed in the next section.

Results

Table 4 summarizes the results of pre-checking of each metric. This assessment of eCl@ss ontology shown in Table 4 does not relate to the supply chain model yet, but provides the general findings.

Next, we report on applying the individual metrics to the supply chain scenario. Specifically, we highlight our findings with respect to Interpretability, Consistency, Clarity, and Relevance

Table 4. Applicability and results of metrics for eClassOWL

Metric	Applicability and Results
Lawfulness (SL)	eClassOWL is syntactically correct and has been checked formally by respective OWL tools. Its spreadsheet version contains several data errors though; see Hepp et al. (2007).
Richness (SR)	eClassOWL uses rather few OWL language features (missing, e.g., unionOf, intersectionOf, inverseOf).
Interpretability (EI)	The meaning of terms used in eClassOWL can be checked by searching for entries in a reference dictionary
Consistency (EC)	Inconsistencies can occur if is-a relationships are used falsely; here, cross-checks with reference dictionaries could help identify those.
Clarity (EA)	Interpreting terms at the lowest level of the taxonomy should consider the scope or context defined by its super classes.
Comprehensiveness (PO)	All detailed metrics can be applied; see Hepp et al. (2007) for results.
Accuracy (PU)	Whether a claim made is true or false can only be determined by domain experts. Reasoning cannot help, because there are is-a relationships only.
Relevance (PR)	Whether eClassOWL provides the knowledge required by an application could at least partly be answered by querying the ontology based on explicit requirements.
Authority (OT)	No ontology in the public ontology repositories links to eClassOWL yet.
History (OH)	Accesses to eClassOWL can be recorded by the ontology provider only. eCl@ss is mostly used in non-public applications, thus acceptance can hardly be determined.

metrics using a snapshot of eCl@ss.We choose a sample of 100 class names from the relevant segment 19 which is entitled 'Information, communication and media technology'.

Lawfulness and Richness In a supply chain context, participants exchange product information. .If the participants use different descriptions, ontologies can be used to resolve differences. To automate this process, ontologies should be machine processable. The eClassOWL representation has been checked by OWL tools and it is syntactically correct, however, the spreadsheet version contains several data errors. With respect to Richness, the eClassOWL uses only a few OWL language features (missing, e.g., unionOf, intersectionOf, inverseOf).

Interpretability of the terms used in eCl@ss is checked by searching for respective entries in WordNet. WordNet returns respective entries for 97 out of 100 terms (which correctly refer to IT). The three missing terms can be explained by typos and not separating words correctly (i.e., 'computersystem', 'mainmemory', and 'harddrive'); these probably could have been retrieved by using a more capable stemmer.

This result shows quite a high interpretability of the terms used in eCl@ss; we did not search for too technical or new terms (such as acronyms of interface standards) because WordNet has not been designed for technical domains. We thereby acknowledge that the results of the interpretability metric also depend on the quality of WordNet, which is used as the reference dictionary. We assume that the share of equal terms is lower in specific product domains (e.g., chemical domain) than in broader ones.

Consistency relates to the correct usage of is-a relationships. We examined the same sample of 100 product classes and made the following observations:

- The general layout of the eCl@ss taxonomy is consistent with separating computer systems from parts. The former is classified into notebook, PDA, server, PC etc.; the latter into graphic card, cooler, drive etc. Thus, the rationale is to arrive at increasingly specialized product classes.
- This rationale is, however, violated by inserting six generic product classes for both computer systems and parts (12 inconsistent classes in total). These classes represent 'other' goods (that do not belong to another class), 'parts' (in the most generic sense), 'accessories', 'assembly', 'maintenance service' and 'repair'.

The usage of such generic product classes cannot only be observed in the sample, but in the entire eCl@ss ontology.

Clarity of the terms used in eCl@ss relates to both product classes and properties. The classes are exclusively part of a four-level taxonomy. Therefore, the context of each term in the sample is already defined by the name of segment 19, and on the lower levels by names of classes on levels 2 and 3. Some terms at lower levels make the context explicit by extending the class name. For instance, the class 'Mouse (computer input device)' is a sub class of 'Input device for computer'; such an extension is made only for improving interpretation by human readers and not machines. With regard to product properties, there are some generic properties (e.g., 'manufacturer name', 'supplier product number') which are assigned to every single class, whereas other properties are specific and used in few or even one class only. We can state that clarity is high for all properties in the sample due to how properties are used in eCl@ss in general.

Comprehensiveness and Accuracy metrics are typically computed based on domain expert's feedback. For the subset of eCl@ss ontology under consideration, all detailed metrics can be applied and comprehensiveness and Accuracy rank high. This is demonstrated by Hepp et al. (2007) in their analysis of eCl@ss ontology.

Table 5. Example queries for assessing the relevance of eCl@ss

Query	Assessment of eCl@ss
Q1: Customer a7 searches for a thin client with specific technical features (i.e., dimensions).	'thin client' is linked to the 'personal computer' class by a keyword; requirements can be described using 25 class-specific properties.
Q2: Retailer a6 asks distributors for laptop computers with no software installed in order to create a bundle by himself.	'laptop computer' class allows specifying whether software is included or not by the Boolean property 'software included'.
Q3: Retailer a6 searches for similar offerings from multiple distributors using the product number of the OEM a4.	All product classes possess, besides the supplier product number, manufacturer name and manufacturer product number, thus enabling respective search beyond the previous stage in the supply chain.
Q4: Distributor a5 searches for a replacement of TFT monitor by plasma monitor.	The classes for 'TFT monitor' and 'plasma monitor' share the same super-class 'monitor'. However, the two lists of properties are completely different, which prevents supporting the replacement decision.
Q5: OEM a4 searches for suppliers a1-a3 which can deliver modem cards for all desktop computers.	The product class for PCs contains a suitable property indicating whether a modem card is present or not; however, this property is not linked to the corresponding class 'telecommunication card' because of limitations of the eCl@ss ontology model which defines is-a relationships only.

Relevance of the domain knowledge in eCl@ss can be studied by example queries. This metric also ties the ontology to the supply chain model. We define 5 queries each testing whether the eCl@ss provides a respective concept, thus being able to represent the required product. The queries and results are shown in Table 5.

Authority and History metrics rank very low for the eClassOWL ontology because no ontology in the public ontology repositories links to eClassOWL yet. Similarly, accesses to eClassOWL can be recorded by the ontology provider only. eCl@ss is mostly used in non-public applications, thus acceptance cannot be easily determined.

Implications

The semiotic metrics suite has several implications for users of product ontologies as well as its creators. In particular, applications such as agent-based systems can use the metrics suite to compare and select the most appropriate ontology to be used in a particular task. Ontology evaluation generally ignores specific requirements of a particular organization. Our metrics suite allows the ontology evaluation in the context of a set of products available from an organization in a supply chain. In other words, our framework helps us take into account the context in which the ontology is used in an organization. Since the metrics go beyond pragmatic dimensions such as comprehensiveness, they can help detect weak points in a given product ontology (e.g., lack of interpretability, inconsistencies, and irrelevant concepts).

Existing tools do not adequately support ontology searching and selection. Our proposed metrics suite provides a framework for evaluation of product ontologies and possibly selecting appropriate ontologies for supporting seamless e-commerce activities. The metrics can also be used by product ontology creators such as industry associations or standardization bodies as a means for analyzing the ontology and generating quantitative information about the various quality dimensions, as defined by the metrics suite.

RELATED WORK

The related work can be grouped into two major areas: product ontologies and ontology evaluation. Despite the former's importance, it is rather a specialized field which attracts interest from communities such as knowledge engineering (e.g., Fensel et al., 2001), data management (e.g., Beneventano et al., 2004), e-commerce (e.g., Leukel, 2004), and certainly product data management. While the phrase 'product ontology' is often used to stress the formal specification aspect, other widely used terms, though not equal in meaning, are 'product classification standard' or 'product classification system'. The quality and suitability of such artifacts, however, have rarely been the focus of dedicated research. Many researchers take these ontologies for granted and do not further investigate their structure and content.

To the best of our knowledge, only the work by Hepp et al. provides product ontology metrics as well as results of extensive quantitative evaluation (Hepp et al., 2007). While these metrics analyze product ontologies to a great extent and their rationale reflects a lot of domain expertise, they are confined only to the product ontology. Thus, they do not investigate the relationship between the ontology and, for instance, its users, the ontology language, or other ontologies. With regard to the proposed semiotic metrics suite, Hepp et al.'s metrics concern the pragmatic dimension only.

Ontology evaluation in general aims at assessing the relevance of diverse types of ontologies. There is a growing research community which develops methodologies, models, and tools for ontology evaluation, e.g., EON Workshop Series (Garcia-Castro et al., 2007). Studying the ontological quality is made difficult by a number of factors. Contrary to Information Retrieval, for instance, one cannot easily define the metrics 'precision' and 'recall', since these require a clear set of items – here concepts, inter-relations, and properties – being relevant in the respective domain of interest (Brewster et al., 2004). Ontology evaluation can be classified using attributes, for instance as described in (Hartmann, 2004) and (Brank et al., 2005). For the purpose of our work, we focus on one attribute which distinguishes functionality and structure.

The functionality of an ontology describes how suitable and appropriate it is for its intended usage in an information system. There are two major approaches. First, one could relate the ontology directly to requirements of the respective task. In this case, such requirements need to be elicited, formalized, and then mapped to elements of the ontology. Second, one could select a particular ontology and compare it to a reference ontology. The shortcoming of both approaches is that both requirements and the reference ontology for the domain of interest can be incomplete, wrong, lacking, and if available, subjective. This is in particular true for broad product ontologies such as UNSPSC1, eOTD2 and eCl@ss3, which all aim at becoming the first reference and global standard. Thus comparing them to another reference ontology is not feasible.

The structure of an ontology is formed by its elements and inter-relations. A major stream of research is rooted in the Semantic Web arena and its approaches rely on ontology languages such as OWL and its predecessor DAML+OIL. By systematically checking the actual usage of language features such as classes, properties, axioms, instances etc., one can determine the structural characteristics. For instance, complexity metrics are defined in (Yang et al., 2006); they include number of concepts, relations and paths, and the mean of relations and paths per concept. Similar metrics can be found in (Huang & Diao 2006), which defines metrics for assessing how balanced a taxonomy is. However, both proposals represent only a limited subset of the entire ontology language features and respective structural aspects.

A more elaborate set of metrics describing cohesion can be found in (Yao et al., 2005). Based on graph theory, these metrics determine the degree of relatedness of concepts in an ontology. Though mathematically sound, the results of such metrics cannot easily be interpreted in terms of quality. The same is true for many structural metrics. For instance, whether a big, nested ontology is better than a smaller one depends primarily on the domain of interest. This point of criticism complements the fact that structural metrics in general rely on the expressiveness of ontology languages (thus on what can be described formally). Consequently, these metrics should be regarded as a component of a broader evaluation framework.

The review of related work points out that (1) generic ontology evaluation limits its scope by respective ontology languages and thus cannot fully exploit the quality of domain ontologies, in particular product ontologies which rather rely on size and deepness than on formal complexity, and (2) domain ontology evaluation requires not only extensive domain expertise, but also an ontological foundation to arrive at both suitable and well-defined metrics. Most current research exploits the pragmatic quality dimension only without taking into account the setting of an ontology. By employing semiotics theory, the proposed metrics suite aims at overcoming these deficits.

CONCLUSION

This article proposed a metrics suite for product ontology evaluation based on semiotic theory and demonstrated the feasibility and usefulness of the metrics suite using a supply chain model. The contribution of our research is the comprehensive metrics suite that takes into account the various quality dimensions of product ontology.

Our approach also incorporates the specific setting of product ontologies. This setting is mainly determined by supply chains in which such ontologies are used. Thus, our work has studied supply chains based on a general supply chain model, which allows us to distinguish different roles of actors. In particular, we have addressed the individualization of supply chains which no longer makes it feasible to consider only a small snapshot of a supply chain. For instance, reducing the problem to the relationship between supplier and customer arrives at two-tier supply chain models.

However, finding a suitable product ontology for a particular industry is non-trivial because these ontologies are implemented using a variety of languages, methodologies, and platforms. Assessing their quality and selecting a particular ontology is difficult because of the heterogeneity. We have presented a semiotic set of product ontology metrics that allow for assessing the syntactic, semantic, pragmatic, and social quality, and showed how these metrics could be used for evaluating a real-world product ontology. This metric suite can be used by applications and decision makers to assess the quality of available ontologies in a particular domain. The semiotics based approach sheds light on creating various categories of quality and provides a systematic way to develop quality metrics.

Our proposed product ontology evaluation approach has some limitations. First, it does not provide explicit guidelines for determining the optimal weighting scheme of the various quality dimensions. Second, product ontologies evolve over time and the metric suite does not yet take into account this dynamic nature (Hepp, 2007). Third, it does not include any learning mechanism to update the evaluations based upon feedback from external users. Finally, empirical testing of the approach is needed to validate the relationship between an ontology's internal attributes reflected in its metrics and its external attributes such as its usefulness for supporting an application.

Our future work involves: (1) testing our metrics suite in more realistic and comprehensive usage scenarios, (2) developing a tool for automated ontology quality assessment, and (3) applying it to ontologies in other domains.

REFERENCES

Beneventano, D., Guerra, F., Magnani, S., & Vincini, M. (2004). A web service based framework for the semantic mapping amongst product classification schemas. *Journal of Electronic Commerce Research, 5*(2), 114–127.

Brank, J., Grobelnik, M., & Mladenic, D. (2005). A Survey of Ontology Evaluation Techniques. *Proceedings of the Conference on Data Mining and Data Warehouses (SiKDD 2005).*

Brewster, C., Alani, H., Dasmahapatra, S., & Wilks, Y. (2004). Data Driven Ontology Evaluation. *Proceedings of the International Conference on Language Resources and Evaluation (LREC-2004).*

Burton-Jones, A., Storey, V. C., Sugumaran, V., & Ahluwalia, P. (2005). A semiotic metrics suite for assessing the quality of ontologies. *Data & Knowledge Engineering, 55*(1), 84–102. doi:10.1016/j.datak.2004.11.010

Coates, J. (1995). Customization promises sharp competitive edge. *Research in Technology Management, 38*(1), 6–7.

Doring, S., Kiebling, W., Preisinger, T., & Fischer, S. (2006). Evaluation and optimization of the catalog search process of e-procurement platforms. *Electronic Commerce Research and Applications, 5*(1), 44–56. doi:10.1016/j.elerap.2005.08.004

eCl@ss (2008). *International Standard for the Classification and Description of Products and Service.* V8.0, http://www.eclass-online.com.

Fensel, D., McGuinness, D. L., Schulten, E., Ng, W. K., Lim, E.-P., & Yan, G. (2001). *Ontologies and Electronic Commerce. IEEE Intelligent Systems, 16*(1), 8–14. doi:10.1109/MIS.2001.1183337

Garcia-Castro, R., Vrandecic, D., Gomez-Perez, A., Sure, Y., & Huang, Z. (2007). *Evaluation of Ontologies and Ontology-based tools.*http://km.aifb.uni-karlsruhe.de/ws/eon2007.

Hartmann, J. (2005). Methods for ontology evaluation. *Knowledge Web Deliverable D1.2.3.* http://knowledgeweb.semanticweb.org.

Hepp, M. (2006). eClassOWL 5.1. Products and Services Ontology for e-Business. *User's Guide.* http://www.heppnetz.de/eclassowl.

Hepp, M. (2007). Possible Ontologies: How Reality Constrains the Development of Relevant Ontologies. *IEEE Internet Computing, 11*(1), 90–96. doi:10.1109/MIC.2007.20

Hepp, M., Leukel, J., & Schmitz, V. (2007). A Quantitative Analysis of Product Categorization Standards: Content, Coverage, and Maintenance of eCl@ss, UNSPSC, eOTD, and the RosettaNet Technical Dictionary. *Knowledge and Information Systems, 13*(1), 77–114. doi:10.1007/s10115-006-0054-2

Huang, N., & Diao, S. (2006). Structure-Based Ontology Evaluation. *Proceedings of the IEEE International Conference on e-Business Engineering (ICEBE 2006)* (pp. 132-137).

Kirn, S. (2008). *Individualization Engineering.* Cuvillier: Goettingen.

Leukel, J. (2004). Standardization of Product Ontologies in B2B Relationships – On the Role of ISO 13584. *Proceedings of the 10th Americas Conference on Information Systems (AMCIS 2004)* (pp. 4084-4091).

Leukel, J., & Sugumaran, V. (2007). Evaluating Product Ontologies in E-Commerce: A Semiotic Approach. *Proceedings of the 6th Workshop on e-Business (WeB 2007)* (pp. 240-249).

Park, J. M., Nam, J. H., Hu, Q. P., & Suh, H. W. (2008). Product Ontology Construction from Engineering Documents. *Proceedings of the International Conference on Smart Manufacturing Application (ICSMA 2008)* (pp. 305-310).

Shim, J., & Shim, S. S. Y. (2006). Ontology-based e-Catalog in e-commerce: Special Section. *Electronic Commerce Research and Applications, 5*(1), 1. doi:10.1016/j.elerap.2006.01.001

Stamper, R., Liu, K., Hafkamp, M., & Ades, Y. (2000). Understanding the role of signs and norms in organisations – a semiotic approach to information systems design. *Behaviour & Information Technology, 19*(1), 15–27. doi:10.1080/014492900118768

Supply-Chain Council. (2006). *Supply Chain Operations Reference Model (SCOR®) Version 8.0.*http://www.supply-chain.org.

Yang, Z., Zhang, D., & Ye, C. (2006). Evaluation Metrics for Ontology Complexity and Evolution Analysis. *Proceedings of the IEEE International Conference on e-Business Engineering (ICEBE 2006)* (pp. 162-169).

Yao, H., Orme, A. M., & Etzkorn, L. (2005). Cohesion Metrics for Ontology Design and Application. *Journal of Computer Science, 1*(1), 117–113.

Zhdanova, A. V., Krummenacher, R., Henke, J., & Fensel, D. (2005). Community-driven ontology management: DERI case study. *Proceedings of the 2005 IEEE/WIC/ACM International Conference on Web Intelligence* (pp. 73-79).

ENDNOTES

[1] http://www.unspsc.org

[2] http://www.eccma.org

[3] http://www.eclass-online.com

This work was previously published in International Journal of Intelligent Information Technologies, Vol. 5, Issue 4, edited by V. Sugumaran, pp. 1-15, copyright 2009 by IGI Publishing (an imprint of IGI Global).

Compilation of References

Abdul-Jalbar, B., Gutierrez, J., & Sicilia, J. (2005). Integer-ratio policies for distribution/inventory systems. International Journal of Production Economics (pp. 93–94, 407–415).

Aberdeen Group. (August 2006). The E-Procurement Benchmark report.

Abidi, S.S.R. (2001). Knowledge management in healthcare: Towards knowledgedriven decision-support services. International Journal of Medical Informatics, 63, 5-18.

Acir, N., & Guzelis, C. (2004). Automatic spike detection in EEG by a two-stageprocedure based on support vector machines. Computers in Biology and Medicine,34(7), 561-575.

Agarwal, S., Handschuh, S., & Staab, S. (2005). Annotation, Composition and Invocation of Semantic Web Services. Journal of Web Semantics, 2(1), 1–24.

Agrawal, R., Gunopulos, D., & Leymann, F. (1998). Mining process models from workflow logs. In Sixth International Conference on Extended Database Technology (pp. 469-483).

Agrawal, R., Mannila, H., Srikant, R., Toivonen, H., & Inkeri, A. (1996). Fast discovery of association rules, advances in knowledge discovery and data mining. The Association for the Advancement of Artificial Intelligence and the MIT Press (pp. 307-328).

Ahuja, R. K., Liu, J., Orlin, J. B., Sharma, D., & Shughart, L. (2002). Solving real-life locomotive scheduling problems (Tech. Rep. 4389-02). Cambridge, MA: Massachusetts Institute of Technology, Sloan School of Management.

Alarie, S., & Gamache, M. (2002). Overview of Solution Strategies Used in Truck Dispatching Systems for Open Pit Mines. International Journal of Mining . Reclamation and Environment, 16(1), 59–76. doi:10.1076/ijsm.16.1.59.3408

Alberts, D. S., & Hayes, R. E. (2003). Power to the Edge, Command and Control in the Information Age. Department of Defense Command and Control Program; Information Age Transformation Series. Washington, DC.

Alesso, H. P., & Smith, C. F. (2005). Developing Semantic Web Services. Wellesley, MA: A.K. Peters, Ltd.

Allotta, B., Colombo, C., Bertocci, F., Scarselli, L., Sgarbi, M., Bioli, G., et al. (2003). A system based on stereo vision for unpacking pieces of wood with an industrial robot. Proceedings of the 1st International Workshop on Advances in Service Robotics, Italy, 3, 1448-1457.

Alshawish, H. M. M., & Allen, C. R. (1995). 3D object recognition using coded light projection for robot assembly applications. In Proceedings of the 21st International Conference on Industrial Electronics, Control and Instrumentation, Orlando, FL, 2(6), 1240-1247.

Alvarez, R., & Tamarit, J. M. (1989). Algoritmos heurísticos deterministas y aleatorios en secuenciación de proyectos con recursos limitados . Qüestiió, 13(1), 173–191.

Andrew, J. C., & Baker, A. B. (1994). Experimental results on the application of satisfiability algorithms to scheduling problems. In Proc. of the 12th National Conference on Artificial Intelligence (pp. 1092–1097).

Ansar, A., & Daniilidis, K. (2003). Linear pose estimation from points or lines. IEEE Transactions on Pattern Analysis and Machine Intelligence, 25(5), 578–589. doi:10.1109/TPAMI.2003.1195992

APICS – The Association for Operations Management. (2005). The Certified Supply Chain Professional Learning Systems. APICS – Advancing Productivity, Innovation, and Competitive Success, Chicago, IL. Retrieved from http://www.apics.org/

Applegate, L. M., Austin, R. D., & McFarlan, F. W. (2003). Corporate Information Strategy and Management Text and Cases: Sixth Edition. New Delhi, India: Tata McGraw-Hill Publishing Company Limited.

Arasu, A., Babu, S., & Widom, J. (2003). CQL A language for continous queries over streams and relations. In 9th Intern. Conf. on Data Base Programming Languages (DBPL) (pp. 1-19).

ARGO. (2009). Global Ocean Sensor Network. Retrieved January 23, 2009, from www.argo.ucsd.edu

Au, W., Chan, K., & Yao, X. (2003). A novel evolutionary data mining algorithm with applications to churn prediction. IEEE Transactions on Evolutionary Computation, 7(6), 532–545. doi:10.1109/TEVC.2003.819264

Ayoub, N. (2007). A Multilevel Decision Making Strategy for Designing and Evaluating Sustainable Bioenergy Supply Chains. Unpublished doctoral dissertation, Tokyo Institute of Technology, Yokohama, Japan.

Ayoub, N., Martins, R., Wang, K., Seki, H., & Naka, Y. (2007). Two levels decision system for efficient planning and implementation of bioenergy production. Energy Conversion and Management, 48(3), 709-723.

Ayoub, N., Wang, K., Kagiyama, T., Seki, H., & Naka, Y. (2006). A Planning Support System for Biomass-based Power Generation. In W. Marquardt & C.Pantelides (Eds.), 16th European Symposium on Computer Aided Process Engineering and 9th International Symposium on Process Systems Engineering Vol. 21B. (pp. 1899-1904). Garmisch-Partenkirchen, Germany.

Baader, F. (1996). A formal definition for the expressive power of terminological knowledge representation languages. Journal of Logic and Computation, 6(1), 33–54. doi:10.1093/logcom/6.1.33

Babcock, B., Babu, S., Datar, M., Motwani, R., & Widom, J. J. (2002). Models and issues in data stream systems. In Proc. of the 21st Symp. on Principles of Database Systems (pp- 1-16).

Babu, S., & Widom, J. (2001). Continous queries over Streams. SIGMOD Record.

Badri, M. A. (1999). A simulation model for multi-product inventory control management. Simulation, 72(1), 20–32. doi:10.1177/003754979907200103

Balci, O. (1998). Verification, Validation, and Testing. In J. Banks, (Ed) Handbook of Simulation. New York: John Wiley & Sons.

Baldus, H., Klabunde, K., & Muesch, G. (2004). Reliable Set-Up of Medical Body-Sensor Networks. In Proc. EWSN 2004, Berlin, Germany.

Ballard, G. (1997). Lookahead planning: the missing link in production control. In proceedings 5th Annual Conference of the International Group for Lean Construction, Griffith University, Gold Coast, Australia.

Bandyopadhyay, S., Rees, J., & Barron, J. M. (2008). Reverse Auctions with Multiple Reinforcement Learning Agents. Decision Sciences, 39(1), 33–63. doi:10.1111/j.1540-5915.2008.00181.x

Barber, K. D., Dewhurst, F. W., Burns, R. L. D. H., & Rogers, J. B. B. (2003). Business-process modeling and simulation for manufacturing management: A practical way forward. Business Process Management Journal, 9(4), 527–542. doi:10.1108/14637150310484544

Barnes, A., & Hammell, R. J., II. (2008). Determining Information Technology Project Status using R-CAST. In the Proceedings of CONISAR 2008 (v1), Phoenix, AZ.

Barratt, M., & Rosdahl, K. (2002). Exploring business-to-business marketsites. European Journal of Purchasing & Supply Management, 8, 111–122. doi:10.1016/S0969-7012(01)00010-7

Barros, A., Decker, G., & Grosskopf, A. (2007). Complex Events in Business Processes. In W. Abramowicz (Ed.), Proceedings of the 10th International Conference on Business Information Systems, Poznan, Poland, (pp. 464-477). Berlin: Springer.

Bartolini, C., Preist, C., & Jennings, N. R. (2002). A Software Framework for Automated Negotiation. In Proceedings of AOSE 2002, International Workshop on Agent-Oriented Software Engineering, Bologna, Italy, (LNCS 2585, pp. 88-100). Berlin: Springer Verlag.

Basson, L., & Petrie, J. G. (2007). An integrated approach for the consideration of uncertainty in decision making supported by Life Cycle Assessment. Environmental Modelling & Software Environmental Decision Support Systems, 22(2), 167-176.

Bauer, P., Nouak, S., & Winkler, R. (1996). Retrieved April 11, 2009, from http://www.esru.strath.ac.uk/Reference/concepts/fuzzy/fuzzy_appl.10.htm

Bazaraa, M. S., Jarvis, J. J., & Sherali, H. D. (2005). Linear programming and network flows. Hoboken, NJ: John Wiley & Sons, Inc.

Beamon, B. (1998). Supply chain design and analysis: Models and methods. International Journal of Production Economics, 55, 281–294. doi:10.1016/S0925-5273(98)00079-6

Beamon, B. M. (1999). Measuring supply chain performance. International Journal of Operations & Production Management, 19(3), 275–292. doi:10.1108/01443579910249714

Beneventano, D., Guerra, F., Magnani, S., & Vincini, M. (2004). A web service based framework for the semantic mapping amongst product classification schemas. Journal of Electronic Commerce Research, 5(2), 114–127.

Benyoucef, M. Alj, H., Levy, K., & Keller, R. K., (2002) A Rule-Driven Approach for Defining the Behaviour of Negotiation Software Agents. In J. Plaice, et al, (Eds.), Proceedings of DCW 2002, (LNCS 2468, pp. 165-181). Berlin: Springer Verlag.

Beraldi, P., Ghiani, G., Greco, A., & Guerriero, E. (2006). Fix and Relax Heuristic for a Stochastic Lot-Sizing Problem. Computational Optimization and Applications, 33, 303–318. doi:10.1007/s10589-005-3055-2

Bergan, T., & Hirsch, G.B. (1976). Analysis and Planning for improved distribution of Nursing personnel and services. A national model of supply, demand, and distribution. Cambridge, MA: Pugh-Roberts Associates, Inc.

Berndt, D., & Watkins, A. (2005). High Volume Software Testing using Genetic Algorithms. In Proceedings of the 38th Hawaii International Conference on System Sciences (HICSS'05), (pp. 318b-318b), Big Island, HI.

Berner, E. S., & Ball, M. J. (1998). Clinical Decision Support Systems: Theory and Practice. Berlin: Springer.

Berstel, B. (2002). Extending the RETE Algorithm for Event Management. In Proc. of the 9th International Symposium on Temporal Representation and Reasoning, (TIME'02).

Bezdek, J. C. (1993). Fuzzy Models – What are they, and why? IEEE transactions on Fuzzy Systems, 1(1), 1–6.

Blum, A. L., & Furst, M. L. (1995). Fast planning through planning graph analysis. Artificial Intelligence, 90, 1636–1642.

Böcker, J., Lind, J., & Zirkler, B. (2001). Using a multiagent approach to optimise the train coupling and sharing system. European Journal of Operational Research, 131, 242–252. doi:10.1016/S0377-2217(00)00124-7

Bodendorf, F., & Zimmermann, R. (2005). Proactive Supply Chain Event Management with Agent Technology. International Journal of Electronic Commerce, 9(4), 57–89.

Bojovic, N., & Milenkovic, M. (2008). The best rail fleet mix problem. Operations Research, 8(1), 77–87.

Boland, N., Dumitrescu, I., Froyland, G., & Gleixner, A. M. (2009). LP-based disaggregation approaches to solving the open pit mining production scheduling problem with block processing selectivity. Computers & Operations Research, 36(4), 1064–1089. doi:10.1016/j.cor.2007.12.006

Bonabeau, E., Dorigo, M., & Theraulaz, G. (1999). Swarm Intelligence: From Natural to Artificial Systems. New York: Oxford University Press.

Bonet, B., & Geffner, H. (2001). Planning as heuristic search. Artificial Intelligence, 129, 5–33. doi:10.1016/S0004-3702(01)00108-4

Borangiu, T. (2002). Visual conveyor tracking for "pick-on-the-fly" robot motion control. In Proceedings of the 7th IEEE International Workshop on Advanced Motion Control, Maribor, Slovenia, (pp. 353-358).

Borges, M. R. S., Pino, J. A., & Valle, C. (2005). Support for decision implementation and follow-up. European Journal of Operational Research Decision Support Systems in the Internet Age, 160(2), 336-352.

Boutilier, C., Dean, T., & Hanks, S. (1999). Decision-theoretic planning: Structural assumptions and computational leverage. Journal of Artificial Intelligence Research, 11(1), 1–94.

Brachman, R. J., & Levesque, H. J. (2004). Knowledge Representation and Reasoning. San Francisco: Morgan Kaufmann.

Brade, D. (2000). Enhancing M&S Accreditation by Structuring V&V Results. Proceedings of the Winter Simulation Conference, (pp. 840-848).

Bradley, S. P., Hax, A. C., & Magnanti, T. L. (1977). Applied Mathematical Programming. Reading, MA: Addison-Wesley Publishing Company.

Brank, J., Grobelnik, M., & Mladenic, D. (2005). A Survey of Ontology Evaluation Techniques. Proceedings of the Conference on Data Mining and Data Warehouses (SiKDD 2005).

Bretscher, O. (1995). Linear Algebra With Applications, (3rd ed.). Upper Saddle River NJ: Prentice Hall.

Brewster, C., Alani, H., Dasmahapatra, S., & Wilks, Y. (2004). Data Driven Ontology Evaluation. Proceedings of the International Conference on Language Resources and Evaluation (LREC-2004).

Brimberg, J., Hansen, P., Laporte, G., Mladenovic, N., & Urosevic, D. (2008). The maximum return-on-investment plant location problem with market share. The Journal of the Operational Research Society, 59(3), 399–406. doi:10.1057/palgrave.jors.2602349

Brown, H. M. M., Burschka, D., & Hager, G. D. (2003). Advances in computational stereo. IEEE Transactions on Pattern Analysis and Machine Intelligence, 25(8), 993–1008. doi:10.1109/TPAMI.2003.1217603

Bruccoleri, M., Perrone, G., & Renna, P. (2005). Design and Evaluating Value Added Services in Manufacturing E-Marketplace. Amsterdam: Springer.

Brucker, P. (2001). Scheduling Algorithms. Secaucus, NJ: Springer-Verlag New York, Inc.

Bruns, R., & Dunkel, J. (2010). Event-Driven Architecture - Softwarearchitektur für ereignisgesteuerte Geschäftsprozesse. Berlin: Springer.

Burgess, T., Byrne, K., & Kidd, C. (2003). Making Project Status Visible in Complex Aerospace Projects. International Journal of Project Management, 21(4), 251–259. doi:10.1016/S0263-7863(02)00022-4

Burton-Jones, A., Storey, V. C., Sugumaran, V., & Ahluwalia, P. (2005). A semiotic metrics suite for assessing the quality of ontologies. Data & Knowledge Engineering, 55(1), 84–102. doi:10.1016/j.datak.2004.11.010

Caccetta, L. (2007). Application of Optimisation Techniques in Open Pit Mining. In A. Weintraub, C. Romero, T. Bjørndal, R. Epstein & J. Miranda (Eds.), Handbook Of Operations Research In Natural Resources, (pp. 547-559). New York: Springer Science+Business Media, Inc.

Caccetta, L., & Hill, S. P. (2003). An Application of Branch and Cut to Open Pit Mine Scheduling. Journal of Global Optimization, 27, 349–365. doi:10.1023/A:1024835022186

Caccetta, L., & Kelsey, P. J. (2001). Models for Mine Site Rehabilitation. In A. Jakeman (Ed.), Proceedings of the Australian National University International Congress on Modelling and Simulation, (pp. 2155-2160). Canberra, Australia: Modelling and Simulation Society of Australia and New Zealand Inc.

Cachon, G., & Terwiesch, C. (2009). Matching Supply with Demand: An Introduction to Operations Mangement, (2nd Ed.). New York: McGraw-Hill.

Cadoli, M., & Schaerf, A. (2005). Compiling problem specifications into SAT. Artificial Intelligence, 162(1-2), 89–120. doi:10.1016/j.artint.2004.01.006

Cakravastia, A. A. (2002). A two-stage model for the design of supply chain network. International Journal of Production Economics, 80(3), 231–248. doi:10.1016/S0925-5273(02)00260-8

Cannon-Bowers, J., & Salas, E. (1998). Team performance and training in complex environments: recent findings from applied research. Journal of Current Directions in Psychological Science, 7, 83–87. doi:10.1111/1467-8721.ep10773005

Canny, J. (1986). A computational approach to edge detection. IEEE Transactions on Pattern Analysis and Machine Intelligence, 8(6), 679–714. doi:10.1109/TPAMI.1986.4767851

Carlsson, C., & Turban, E. (2002). Introduction: DSS: Directions For The Next Decade. Decision Support Systems, 33(2), 105–110. doi:10.1016/S0167-9236(01)00137-3

Catrinu, M. D. (2006). Decision Aid for Planning Local Energy Systems: Application of Multi-Criteria Decision Analysis. Unpublished doctoral dissertation, Norwegian University of Science and Technology, Trondheim, Norway.

Çetin, N. (2004). Open-pit truck-shovel haulage system simulation. Unpublished Ph.D. Dissertation, School of Natural and Applied Sciences, Ataturk University, Turkey.

Chae, Y. M., Ho, S. H., Cho, K. W., Lee, D. H., & Ji, S. H. (2001). Data mining approach to policy analysis in a health insurance domain. International Journal of Medical Informatics, 62(2-3), 103-111.

Chaki, S., et al. (2003). Modular verification of software components in C. In Proc. of the 25th ACM/IEEE International Conference on Software Engineering, (pp. 385–395).

Chanda, E. K., & Dagdelen, K. (1995). Optimal blending of mine production using goal programming and interactive graphics systems. International Journal of Surface Mining . Reclamation and Environment, 9, 203–208. doi:10.1080/09208119508964748

Chandra, P., & Fischer, M. L. (1994). Coordination of production and distribution planning. European Journal of Operational Research, 72, 503–5717. doi:10.1016/0377-2217(94)90419-7

Charbonneau, P. (1995). Genetic algorithms in astronomy and astrophysics. The Astrophysical Journal. Supplement Series, 101, 309–334. doi:10.1086/192242

Chau, K. W., Cao, Y., Anson, M., & Zhang, J. (2003). Application of data warehouse and Decision Support System in construction management. Automation in Construction, 12(2), 213-224.

Chellapilla, K., & Fogel, D. (2001). Evolving an expert checkers playing program without using human expertise. IEEE Transactions on Evolutionary Computation, 5(4), 422–428. doi:10.1109/4235.942536

Chen, B., Ye, Y., & Zhang, J. (2006). Lot-sizing Scheduling with Batch Setup Times. Journal of Scheduling, 9(3), 299–310. doi:10.1007/s10951-006-8265-7

Chen, C., Zhao, Z., & Ball, M. O. (2002). A model for batch advanced available-to-promise. Production and Operations Management, 11(4), 424–440.

Chen, E., Kersten, G. E., & Vahidov, R. (2005). Agent-supported Negotiations in the E-marketplace. International Journal of Electronic Business, 3(1), 28–49. doi:10.1504/IJEB.2005.006387

Chen, J.-F., & Wu, T.-H. (2006). Total tardiness minimization on unrelated parallel machine scheduling with auxiliary equipment constraints. Omega The International Journal of Management Science, 34(1), 81–89. doi:10.1016/j.omega.2004.07.023

Chen, W. (2006). Designing an Extended Set of Coordination Mechanisms for Multi-Agent Systems. Doctoral dissertation, Computer and Information Sciences, University of Delaware, Newark, DE.

Chen, W., & Decker, K. (2004). Managing multi-agent coordination, planning, and scheduling. In Proceedings of the 3rd International Conference on Autonomous Agent and Multi-Agent Systems, New York.

Chen, W., & Decker, K. (2006). Analyzing characteristics of task structures to develop GPGP coordination mechanisms. In Proceedings of the 3rd International Conference on Autonomous Agent and Multi-Agent Systems (pp. 662–669), Hakodate, Japan.

Chen, W., Durfee, E., & Dumas, M. (2009). Human Agent Collaboration in a Simulated Combat Medical Scenario. In Proceedings of the International Symposium on Collaborative Technologies and Systems, Baltimore, Maryland.

Chen, Z.-L., & Powell, W. B. (2003). Exact algorithms for scheduling multiple families of jobs on parallel machines. Naval Research Logistic, 50(7), 823–840. doi:10.1002/nav.10091

Cheng, C., Chan, C. H., & Lin, K. (2006). Intelligent agents for e-marketplace: Negotiation with issue trade-offs by fuzzy inference systems. Decision Support Systems, 42, 626–638. doi:10.1016/j.dss.2005.02.009

Cheng, F. S., & Denman, A. (2005). A study of using 2D vision system for enhanced industrial robot intelligence. Proceedings of the IEEE International Conference on Mechatronics and Automation, Ontario, Canada, 3, 1185-1189.

Cheung, D., Han, J., Ng, V.T., & Fu, A.W. (1996). Fast distributed algorithm for mining association rules. In International Conference on Parallel and Distributed Information Systems, Tokyo, Japan (pp. 31-42).

Chin, R. T., & Harlow, C. A. (1982). Automated visual inspection: A survey. IEEE Transactions on Pattern Analysis and Machine Intelligence, 4(6), 557–573. doi:10.1109/TPAMI.1982.4767309

Choi, S. P. M., Liu, J., & Chan, S. P. (2001). A genetic agent based negotiation system. Computer Networks, 37, 195–204. doi:10.1016/S1389-1286(01)00215-8

Clarke, T. A., & Fryer, J. G. (1998). The development of camera calibration methods and models. The Photogrammetric Record, 16(91), 51–66. doi:10.1111/0031-868X.00113

Clarke, T. A., Fryer, J. F., & Wang, X. (1998). The principal point and CCD cameras. The Photogrammetric Record, 16(92), 293–312. doi:10.1111/0031-868X.00127

Coates, J. (1995). Customization promises sharp competitive edge. Research in Technology Management, 38(1), 6–7.

Cohen, M. A., & Lee, H. L. (1988). Strategic analysis of integrated production-distribution systems: Models and methods. Operations Research, 36(2), 216–228. doi:10.1287/opre.36.2.216

Cook, J.E., & Wolf, A.L. (1998, November). Event-based detection of concurrency. In Proceedings of the Sixth International Symposium on the Foundations of Software Engineering (FSE-6), Orlando, FL (pp. 35-45).

Cook, S. A. (1971). The complexity of theorem-proving procedures. In Proc. of the 3rd Annual ACM Symposium on the Theory of Computing, (pp. 151–158).

Coral8. (2009). Retrieved January 23, 2009, from http://www.coral8.com/

Cordeau, J.-F., Pasin, F., & Solomon, M. M. (2006). An integrated model for logistics network design. Annals of Operations Research, 144(1), 59–82. doi:10.1007/s10479-006-0001-3

Cordeau, J.-F., Toth, P., & Vigo, D. (1998). A Survey of Optimization Models for Train Routing and Scheduling. Transportation Science, 32(4), 380–404. doi:10.1287/trsc.32.4.380

Cormen, T. H., Leiserson, C. E., Rivest, R. L., & Stein, C. (2001). Introduction to Algorithms (Second ed.). Cambridge, MA: MIT Press.

Cox, E. (1995). Fuzzy Logic for Business and Industry. Rockland, MA: Charles River Media, Inc.

Coyle, R.G. (1996). System dynamics modelling: A practical approach. London: Chapman and Hall.

Crama, Y., Pochet, Y., & Wera, Y. (2001). A discussion of production planning approaches in the process industry, (Tech. Rep. GEMME-0102). Liège, Belgique: Université de Liège.

Dahiya, S., Chhabra, J., & Kumar, S. (2007). Use of Genetic Algorithm for Software Maintainability Metrics' Conditioning. In Proceedings of the 15th International Conference on Advanced Computing and Communication (ADCOM'07), (pp. 87-92), Guwahati, India.

Daniels, C., & LaMarsh, W. (2007). Complexity as a Cause of Failure in Information Technology Project Management. In Proceedings of the IEEE International Conference on System of Systems Engineering (SoSE '07). San Antonio, TX: IEEE.

Dauzére-Péres, S., & Lasserre, J.-B. (1994). An integrated approach in production planning and scheduling. Lecture Notes in Economics and Mathematical Systems, 411.

David, M. (1988). Handbook of Applied Advanced Geostatistical Ore Reserve Estimation. Amsterdam: Elsevier.

David, P., DeMenthon, D., Duraiswami, R., & Samet, H. (2003). Simultaneous pose and correspondence determination using line features. Proceedings on the IEEE Computer Society Conference on Computer Vision and Pattern Recognition, Madison, WI, 2, 18-20.

Davidson, C. (1997). Creatures from primordial silicon. New Scientist, 156(2108), 30–35.

Davis, M., & Putnam, H. (1960). A Computing Procedure for Quantification Theory. Journal of the ACM, 7(3), 201–215. doi:10.1145/321033.321034

Davis, M., Logemann, G., & Loveland, D. (1962). A machine program for theorem-proving. Communications of the ACM, 5(7), 394–397. doi:10.1145/368273.368557

Dawes, M., & Sampson, U. (2003). Knowledge management in clinical practice: A systematic review of information seeking behavior in physicians. International Journal of Medical Informatics, 71, 9-15.

de Matta, R., & Miller, T. (2004). Production and interfacility transportation scheduling for a process industry. European Journal of Operational Research, 158, 72–88. doi:10.1016/S0377-2217(03)00358-8

Dechter, R., & Frost, D. (1999). Backtracking algorithms for constraint satisfaction problems (Tech. Rep.). Irvine, CA: Univ. of California, Irvine.

Decker, K., & Li, J. (2000). Coordinating mutually exclusive resources using GPGP. Journal of Autonomous Agent and Multi-Agent Systems, 3(2), 133. doi:10.1023/A:1010074611407

Delen, D., & Pratt, D. B. (2006). An integrated and intelligent DSS for manufacturing systems. Expert Systems with Applications, 30(2), 325-336.

Delp, E. J. (1997). Image processing software and hardware. IEEE Signal Processing Magazine, 15, 8–11.

Dementhon, D. F., & Davis, L. S. (1995). Model-based object pose in 25 lines of code. International Journal of Computer Vision, 15(1-2), 123–141. doi:10.1007/BF01450852

Department of Defense. (1996). Modeling & Simulation Verification, Validation, and Accreditation (US DoD Instruction 5000.61). Washington, DC: Author.

Dessureault, S., Scoble, M., & Dunbar, S. (1999). Intelligent Production Management in Mining Systems. In J. A. Meech, M. M. Veiga, M. H. Smith, & S. R. LeClair (Eds.), Proceedings of The Second International Conference on Intelligent Processing and Manufacturing of Materials (pp. 145-150), British Columbia University.

Dhome, M., Richetin, M., Lapreste, J. T., & Rives, G. (1989). Determination of the attitude of 3D objects from a singleperspective view. IEEE Transactions on Pattern Analysis and Machine Intelligence, 11(12), 1265–1278. doi:10.1109/34.41365

Dias, J., Captivo, M. E., & Clímaco, J. (2006). Capacitated dynamic location problems with opening, closure and reopening of facilities. IMA Journal of Management Mathematics, 17(4), 317–348. doi:10.1093/imaman/dpl003

Diaz, J. C., & Abderrahim, M. (2007). Modified SoftPOSIT algorithm for 3D visual tracking. In Proceedings of the

IEEE International Symposium on Intelligent Signal Processing, Madrid, Spain, (pp. 1-6).

Dillenberger, C., Escudero, L. F., Wollensak, A., & Zhang, W. (1994). On practical resource allocation for production planning and scheduling with period overlapping setups. European Journal of Operational Research, 75, 275–286. doi:10.1016/0377-2217(94)90074-4

Dimitrakopoulos, R., & Ramazan, S. (2004). Uncertainty-based production scheduling in open pit mining. Transactions of the Society for Mining, Metallurgy, and Exploration, 316, 106–112.

Dimitrakopoulos, R., Farrelly, C. T., & Godoy, M. (2002). Moving forward from traditional optimization: grade uncertainty and risk effects in open-pit design. Mining Technology: Institute of Mining and Metallurgy Transactions section A, 1(111), 82-88.

Dobrev, P., Kalaydjiev, O., & Angelova, G. (2007). From Conceptual Structures to Semantic Interoperability of Content. (LNCS Vol. 4604, pp. 192-205). Berlin: Springer-Verlag.

Dogan, K., & Goetschalckx, M. (1999). A primal decomposition method for the integrated design of multi-period production-distribution systems. IIE Transactions, 31, 1027–1036.

Doring, S., Kiebling, W., Preisinger, T., & Fischer, S. (2006). Evaluation and optimization of the catalog search process of e-procurement platforms. Electronic Commerce Research and Applications, 5(1), 44–56. doi:10.1016/j.elerap.2005.08.004

Dreiseitl, S., Ohno-Machado, L., Kittler, H., Vinterbo, S., Billhardt, H., & Binder, M. (2001). A comparison of machine learning methods for the diagnosis of pigmented skin lesions. Journal of Biomedical Informatics, 34, 28-36.

Drumond, F. P., & Mateus, G. R. (2000). Efficient Production Planning and Scheduling Integrated Application. In F. Carter (Ed.) Proceedings of the World Conference on Production and Operations Management (pp. 165-174), Sevilla, Spain.

Dumas, M., Aldred, L. J., Governatori, G., & ter Hofstede, A. H. M. (2005). Probabilistic Automated Bidding in Multiple Auctions. Electronic Commerce Research, 5(1), 25–49. doi:10.1023/B:ELEC.0000045972.75533.90

Dunkel, J., & Bruns, R. (2008). Reference Architecture for event-driven RFID applications. In 2nd Intern. Workshop on RFID Technology (IWRT) (pp. 129-135), Barcelona.

Dunkel, J., Fernández, A., Ortiz, R., & Ossowski, S. (2008). Event-Driven Architecture for Decision Support in Traffic Management Systems. In 11th Intern. IEEE Conf. on Intelligent Transportation Systems, (pp. 7-13).

East, T. (2000). Return on Investment (ROI) Analysis in the Face of Uncertainty. Retrieved March, 15, 2009, from http://www2.amia.org/pubs/symposia/D200476.PDF

Ebert, C., & Baisch, A. (1998). Industrial Applications of Criticality Predictions in Software Development. In Proceedings of the Ninth International Symposium on Software Reliability Engineering, (pp. 80-89), Paderborn, Germany.

eCl@ss (2008). International Standard for the Classification and Description of Products and Service. V8.0, http://www.eclass-online.com.

Eclipse. (2009). Retrieved from http://www.eclipse.org/

Eiben, A. E., & Smith, J. E. (2003). Introduction to Evolutionary Computing. Heidelberg, Germany: Springer-Verlag.

El Sheikh, A. A. R., Paul, R. J., Harding, A. S., & Balmer, D. W. (1987). A Microcomputer-Based Simulation Study of a Port. The Journal of the Operational Research Society, 38, 673–681.

En, N., & Sörensson, N. (2003). An Extensible SAT-solver. In Sat (Vol. 2919, pp. 502–518). Berlin: Springer.

Engelbrecht, A. (2007). Computational Intelligence: An Introduction (2nd Ed.). New York: John Wiley & Sons.

Eren, T. (2007). A Multicriteria Scheduling with Sequence-Dependent Setup Times. Applied Mathematical Sciences, 1(58), 2883–2894.

Eren, T., & Güner, E. (2005). A bicriteria scheduling with sequence-dependent setup times. Applied Mathematics and Computation, 179(1), 378–385. doi:10.1016/j.amc.2005.11.112

Ernst, M., Millstein, T., & Weld, D. (1997). Automatic SAT-Compilation of Planning Problems.In Proc. of the 15th International Joint Conference on Artificial Intelligence (IJCAI-97), (pp.1169–1176).

Erol, K., Hendler, J., & Nau, D. (1994). HTN planning: Complexity and expressivity. In Proceedings of the 12th National Conference on Artificial Intelligence (pp. 1123-1128), Seattle, WA.

Escudero, L. F., & Salmeron, J. (2005). On a Fix-and-Relax Framework for a Class of Project Scheduling Problems. Annals of Operations Research, 140, 163–188. doi:10.1007/s10479-005-3970-8

Esmaeili, M., Aryanezhad, M., & Zeephongsekul, P. (2009). A game theory approach in seller-buyer supply chain. European Journal of Operational Research, 195, 442–448. doi:10.1016/j.ejor.2008.02.026

Esper. ESPERTECH (2009). Esper Reference Documentation, Version 2.0.0. Tech. Rep. Retrieved January 23, 2009, from http://esper.codehaus.org

Estrin, D. Girod. L., Pottie, G. & Srivastava, M. (2001). Instrumenting the World with Wireless Sensor Networks, In Intern. Conf. on Acoustics, Speech, and Signal Processing (ICASSP 2001), Salt Lake City.

Everett, J. E. (1997). Simulation to reduce variability in iron ore stockpiles. Mathematics and Computers in Simulation, 43, 563–568. doi:10.1016/S0378-4754(97)00045-1

Everett, J. E. (2001). Iron ore production scheduling to improve product quality. European Journal of Operational Research, 129(2), 355–361. doi:10.1016/S0377-2217(00)00233-2

Everett, J. E. (2002). A decision support simulation model for the management of an elective surgery waiting system. Journal for Health Care Management Science, 5(2), 89–95. doi:10.1023/A:1014468613635

Everett, J. E., Howard, T., & Kamperman, M. (2001). The Development of a MIS for Iron Ore Mining Operations. In A. Harriger, (Ed.), Proceedings of the Informing Science Conference, Krakow, Poland.

Family of Standards for Modeling and Simulation (M&S) High Level Architecture (HLA) – (a) IEEE 1516-2000 Framework and Rules; (b) IEEE 1516.1-2000 Federate Interface Specification; (c) IEEE 1516.2-2000 Object Model Template (OMT) Specification IEEE 1516-2000, IEEE Press.

Fan, X. Sun, B, Sun, S., McNeese, M., Yen, J., & et al. (2006) RPD-Enabled Agents Teaming with Humans for Multi-Context Decision Making. In Proceedings of the Fifth International Joint Conference on Autonomous Agents and Multi-Agent Systems (AAMAS'06), (pp. 34-41). Hokkaido, Japan: ACM.

Fan, X., & Yen, J. (2007). R-CAST: Integrating Team Intelligence for Human-Centered Teamwork. In Proceedings of the Twenty-Second AAAI Conference on Artificial Intelligence (AAAI-07). Menlo Park, CA: Association for the Advancement of Artificial Intelligence (www.aai.org).

Fan, X., Shuang, S., & Yen, J. (2005a). On Shared Situation Awareness for Supporting Human Decision-Making Teams. In Proceedings of 2005 AAAI Spring Symposium on AI Technologies for Homeland Security, (pp. 17-24). Stanford, CA: American Association for Artificial Intelligence.

Fan, X., Shuang, S., & Yen, J. (2005a). On Shared Situation Awareness for Supporting Human Decision-Making Teams. In Proceedings of 2005 AAAI Spring Symposium on AI Technologies for Homeland Security, (pp. 17-24). Stanford, CA: American Association for Artificial Intelligence.

Fan, X., Shuang, S., McNeese, M., & Yen, J. (2005b). Extending Recognition-Primed Decision Model For Human-Agent Collaboration. In Proceedings of the Fourth International Joint Conference on Autonomous Agents and Multi Agent Systems. (pp. 945-952).The Netherlands: ACM.

Fan, X., Shuang, S., Sun, B., Airy, G., McNeese, M., & Yen, J. (2005c). Collaborative RPD-Enabled Agents Assisting the Three-Block Challenge in Command and Control in Complex and Urban Terrain. In Proceedings of 2005 Conference on Behavior Representation in Modeling and Simulation (BRIMS), (pp. 113-123), Universal City, CA.

Faugeras, O. (1993). Three-dimensional computer vision: a geometric viewpoint. Cambridge, MA: MIT Press.

Felsen, J. (1976). Decision Making Under Uncertainty: An Artificial Intelligence Approach. New York: CDS Publishing Company.

Fensel, D., McGuinness, D. L., Schulten, E., Ng, W. K., Lim, E.-P., & Yan, G. (2001). Ontologies and Electronic Commerce. IEEE Intelligent Systems, 16(1), 8–14. doi:10.1109/MIS.2001.1183337

Fernandez, J., Benchetrit, D., & Gachet, D. (2001). Automated visual inspection to assembly of frontal airbag sensors of automobiles. In Proceedings on the 8th IEEE International Conference on Emerging Technologies and Factory Automation, Antibes Juan-Les-Pins, France, 2, 631-634.

Ferreira, D., Morabito, R., & Rangel, S. (2009). Solution approaches for the soft drink integrated production lot sizing and scheduling problem. European Journal of Operational Research, 196, 697 706. doi:10.1016/j.ejor.2008.03.035

Flood, R. (1999). Rethinking the fifth discipline. Learning within the unknowable. London: Ruthledge.

Fogel, D. (2008). Welcome from the President. IEEE Computational Intelligence Society. Retrieved April 2, 2009, from http://www.ieee-cis.org/about_cis/

Forrest, S., Hofmeyr, S.A., Somayaji, A., & Longstaff, T.A. (1996). A sense of self for unix processes. In 1996 IEEE Symposium on Security and Privacy (pp. 120-128). IEEE Computer Society. Building the virtual state. Information technology and institutional change. Washington, DC: Brookings Institution Press.

Forrester, J.W. (1961, July-August). Industrial Dynamics: A Major Breakthrough for Decision Makers. Harvard Business Review, 37-66.

Fowler, M. (2002). Enterprise Application Architecture. Reading, MA: Addison-Wesley.

French, S. (2000). Decision Analysis and Decision Support. Hoboken, NJ: John Wiley & Sons.

Friedman, C., Hripcsak, G., Johnson, S.B., Cimino, J.J., & Clayton, P.D. (1990). A generalized relational schema for an integrated clinical patient database. In Proceedings of the 14th Annual Symposium on Computer Applications in Medical Care (pp. 335-339).

Froyland, G., Boland, N., & Dumitrescu, I. (2008). A multistage stochastic programming approach to open pit mine production scheduling with uncertain geology. In R. Beatson & R. Murray (Eds.), VII Joint Australia-New Zealand Mathematics Convention. Christchurch, New Zealand: University of Canterbury.

Fu, K. S., & Mui, J. K. (1981). A survey on image segmentation. Pattern Recognition, 13(1), 3–16. doi:10.1016/0031-3203(81)90028-5

Fujitec, F. (1988). FLEX-8000 series elevator group control system. Osada, Japan: Fujitec Co., Ltd.

Fukusawa, R., Aragão, M. V. P., Porto, O., & Uchoa, E. (2002). Solving the freight car flow problem to optimality. Electronic Notes in Theoretical Computer Science, 66(6), 1–11. doi:10.1016/S1571-0661(04)80524-3

Garcia-Castro, R., Vrandecic, D., Gomez-Perez, A., Sure, Y., & Huang, Z. (2007). Evaluation of Ontologies and Ontology-based tools.http://km.aifb.uni-karlsruhe.de/ws/eon2007.

Garey, M., & Johnson, D. (1979). Computers and Intractability: A Guide to the Theory of NP- Completeness. San Francisco: Freeman.

Garvey, A., Humphrey, M., & Lesser, V. (1993). Task interdependencies in design-to-time real-time scheduling. In Proceedings of the 11th National Conference on Artificial Intelligence (pp. 580–585), Washington, DC.

Ge, Y., & Chang, C. (2006). Capability-based Project Scheduling with Genetic Algorithms. In Proceedings of the International Conference on Computational Intelligence for Modelling Control and Automation, and International Conference on Intelligent Agents, Web Technologies, and Internet Commerce (CIMCA-IAWTIC'06), (pp. 161), Sydney, Australia.

Gerald, B., King, N., & Natchek, D. (2002). Oracle E-Business Suite Manufacturing and Supply Chain management. New York: McGraw-Hill/Osborne.

Germain, R., Claycomb, C., & Dröge, C. (2008). Supply chain variability, organizational structure, and performance: The moderating effect of demand unpredictability. Journal of Operations Management, 26, 557–570. doi:10.1016/j.jom.2007.10.002

Ghodsypour, S. H., & O'brien. C. (1998). A decision support system for supplier selection using an integrated analytic hierarchy process and linear programming. International Journal of Production Economics (pp. 56–57, 199–212).

Gholamnejad, J., & Osanloo, M. (2007). Incorporation of ore grade uncertainty into the push back design process. Journal of the South African Institute of Mining and Metallurgy, 107, 177–185.

Gibbs, W. (1996). Programming with primordial ooze. Scientific American, (October): 48–50. doi:10.1038/scientificamerican1096-48

Gibert, K., Sanchez-Marre, M., & Rodriguez-Roda, I. (2006). GESCONDA: An intelligent data analysis system for knowledge discovery and management in environmental databases. Environmental Modelling & Software, 21(1), 115-120.

Gido, J., & Clements, J. (2009). Successful Project Management (4th Ed.). Mason, OH: South-Western.

Gido, J., & Clements, J. P. (2006). Successful Project Management, (3rd Ed.). Mason, OH: Thomson Higher Education.

Godoy, M., & Dimitrakopoulos, R. (2004). Managing risk and waste mining in long-term production scheduling. Transactions of the Society for Mining, Metallurgy, and Exploration, 316, 43–50.

Goldberg, E., & Novikov, Y. (2007). Berkmin: A fast and robust sat-solver. Discrete Applied Mathematics, 155(12), 1549–1561. doi:10.1016/j.dam.2006.10.007

Gomes, C., Kautz, H., Sabharwal, A., & Selman, B. (2008). Satisfiability Solvers. In Handbook of Knowledge Representation (Foundations of Artificial Intelligence), (pp.89–134). New York: Elsevier B.V.

Gomes, C., Selman, B., & Kautz, H. (1998). Boosting combinatorial search through randomization. In Proc. of the 15th National Conference on Artificial Intelligence (AAAI-98) (pp. 431–437).

Graham, J., & Decker, K. (2000). Towards a distributed, environment-centered agent framework. In J. G. Carbonell & J. Siekmann (Ed.), Intelligent Agents VI, (LNAI-1757, pp. 290–304). Berlin: Springer Verlag.

Graves, S. C., & Schwarz, L. B. (1977). Single cycle continuous review policies for arborescent production/inventory systems. Management Science, 23, 529–540. doi:10.1287/mnsc.23.5.529

Greasley, A. (2003). Using business-process simulation within a business-process reengineering approach. Business Process Management Journal, 9(4), 408–420. doi:10.1108/14637150310484481

Grieger, M. (2003). Electronic marketplaces: A literature review and a call for supply chain management research. European Journal of Operational Research, 144, 280–294. doi:10.1016/S0377-2217(02)00394-6

Griffin, A., Lacetera, J., Sharp, R., & Tolk, A. (Eds.). (2002). C4ISR/Sim Technical Reference Model Study Group Final Report (C4ISR/Sim TRM), (SISO-Reference Document 008-2002-R2). Simulation Interoperability Standards Organization, Orlando, FL.

Grosz, B., & Kraus, S. (1996). Collaborative plans for complex group action. Journal of Artificial Intelligence, 86(2), 269–357. doi:10.1016/0004-3702(95)00103-4

Gu, X., Song, G., & Xiao, L. (2006). Design of a Fuzzy Decision-making Model and Its Application to Software

Functional Size Measurement. In Proceedings of the International Conference on Computational Intelligence for Modelling Control and Automation, and International Conference on Intelligent Agents, Web Technologies, and Internet Commerce (CIMCA-IAWTIC'06), (pp. 199), Sydney, Australia.

Gunasekaran, A., Patel, C., & McGaughey, R. E. (2004). A framework for supply chain performance measurement. International Journal of Production Economics, 87, 333–347. doi:10.1016/j.ijpe.2003.08.003

Haase, K., & Kimms, A. (2000). Lot Sizing and Scheduling with Sequence Dependent Setup Costs and Times and Efficient Rescheduling Opportunities. International Journal of Production Economics, 66(2), 159–169. doi:10.1016/S0925-5273(99)00119-X

Haase, V. (1998). Software Process Improvement Planning with Neural Networks. In Proceedings of the 24th EUROMICRO Conference (EUROMICRO'98), (pp. 20808-20815), Vesteras, Sweden.

Hakkila, P. (2006). Factors driving the development of forest energy in Finland. Biomass and Bioenergy: Proceedings of the third annual workshop of Task 31 'Systainable production systems for bioenergy: Impacts on forest resources and utilization of wood for energy' October 2003, Flagstaff, Arizona, USA, 30(4), 281-288.

Hald, A. (1998). A History of Mathematical Statistics from 1750 – 1930. New York: Wiley.

Hall, M. B., & Harris, C. A. (1993). Fuzzy logic expert system for iron ore processing. In Conference Record of the IEEE Industry Applications Society Annual Meeting (vol. 3, pp. 2190-2199), Toronto, Canada.

Hammell, R. J. II, & Sudkamp, T. (1996). An Adaptive Hierarchical Fuzzy Model. Expert Systems with Applications, 11(2), 125–136. doi:10.1016/0957-4174(96)00040-1

Hammell, R. J., II, & Schafer, R. (2007). A Fuzzy Associative Memory for the Classification of Chemical Warfare Agent Simulants. In Proceedings of the Annual Conference of the North American Fuzzy Information Processing Society (NAFIPS 2007), San Diego, CA.

Hanratty, T., Fan, X., & Hammell, R. J., II. (2009). Leveraging Visualization to Improve Sensemaking with a Computational RPD Model: A Military Perspective. In Proceedings of the 9th Bi-Annual International Conference on Naturalistic Decision Making, Covent Garden, London, accepted.

Hanratty, T., Hammell, R. J., II, Yen, J., & Fan, X. (2007). Utilizing Concept Maps to Improve Human-Agent Collaboration Within a Recognition-Primed Decision Model. In Proceedings of the 2007/IEEE/WIC/ACM International Conference on Intelligent Agent Technology (IAT 2007). Silicon Valley, CA: IEEE.

Haralick, R. M., Lee, D., Ottenburg, K., & Nolle, M. (1995). Analysis and solutions of the three point perspective pose estimation problem. In Proceedings of the IEEE Computer Society Conference on Computer Vision and Pattern Recognition, Maui, Hawaii, (pp. 592-598).

Hartmann, J. (2005). Methods for ontology evaluation. Knowledge Web Deliverable D1.2.3. http://knowledgeweb.semanticweb.org.

Hausman, W. H. (2002). Supply Chain Performance Metrics: The Practices of Supply Chain Management. Boston: Kluwer Academic Publishers.

Hax, A. C., & Meal, H. C. (1975). Hierarchical integration of production planning and scheduling. In M. A. Geisler (Ed.), Studies in Management Science (pp. 53-69). Amsterdam, The Netherlands.

Haykin, S. (1999). Neural networks: A comprehensive foundation (2nd ed.). Upper Saddle River, NJ: Prentice Hall.

Heathfield, H., & Louw, G. (1999). New challenges for clinical informatics: Knowledge management tools. Health Informatics Journal, 5(2), 67-73.

Heckerman, D. (1999). A Tutorial on Learning with Bayesian Networks. In M. Jordan (ed.), Learning in Graphical Models (pp. 301-354). Cambridge, MA: MIT Press.

Heikkila, J., & Silven, O. (1997). A four-step camera calibration procedure with implicit image correction. In Proceedings on the IEEE Computer Society Conference on Computer Vision and Pattern Recognition, San Juan, Puerto Rico, (pp. 1106-1112).

Heinrich, C., & Simchi-Levi, D. (2005). Do IT Investments Really Pay Off. Supply Chain Management Review, 9(4), 22–28.

Hepp, M. (2006). eClassOWL 5.1. Products and Services Ontology for e-Business. User's Guide. http://www.heppnetz.de/eclassowl.

Hepp, M. (2007). Possible Ontologies: How Reality Constrains the Development of Relevant Ontologies. IEEE Internet Computing, 11(1), 90–96. doi:10.1109/MIC.2007.20

Hepp, M., Leukel, J., & Schmitz, V. (2007). A Quantitative Analysis of Product Categorization Standards: Content, Coverage, and Maintenance of eCl@ss, UNSPSC, eOTD, and the RosettaNet Technical Dictionary. Knowledge and Information Systems, 13(1), 77–114. doi:10.1007/s10115-006-0054-2

Herbst, J. (2000). A machine learning approach to workflow management. In 11th European Conference on Machine Learning. Berlin, Germany: Springer (LNCS 1810, pp. 183-194).

Herbst, J., & Karagiannis, D. (1999, August), An inductive approach to the acquisition and adaptation of workflow models. In M. Ibrahim & B. Drabble (Eds.), Proceedings of the IJCAI'99 Workshop on Intelligent Workflow and Process Management: The New Frontier for AI in Business (pp. 52-57). Stockholm, Sweden.

Herbst, J., & Karagiannis, D. (2000). Integrating machine learning and workflow management to support acquisition and adaptation of workflow models. International Journal of Intelligent Systems in Accounting, Finance and Management, 9, 67-92.

Hersh, W., Mailhot, M., Arnott-Smith, C., & Lowe, H. (2002). Selective automated indexing of findings and diagnoses in radiology reports. Journal of Biomedical Informatics, 34, 262-273.

Hinchey, M., Sterritt, R., & Rouff, C. (2007). Swarms and Swarm Intelligence. IEEE Computer, 40(3), 111–113.

Hirsch, G.B., & Bergan, T.A. (1973). Simulating ambulatory care systems: The relationship between structure and behaviour. In Proceedings of the Summer Computer Simulation Conference, Montreal, Quebec, Canada (pp. 877-884).

Hirsch, G.B., & Henderson, J.D. (1977). A simulation model for "what if?" questions in plan development. American Journal of Health Planning, 2(1), 40-46.

Hirsch, G.B., & Killingsworth, W.R. (1975). A new framework for projecting dental manpower requirements. Inquiry, 12(2), 126-142.

Hirsch, G.B., & Miller, S. (1974). Evaluating HMO policies with a computer simulation model. Medical Care, 12(8), 668-681.

Hirsch, G.B., Bergan, T.A., & Frohman, A.L. (1974). Organizational change in the provision of human services. In Proceedings of the Summer Computer Simulation Conference, Houston, Texas.

Hirsch, G.B., Forsyth, B.R., Bergan, T.A., & Goodman, M.R. (1976). Projecting a teaching hospital's future utilization: A dynamic simulation approach. Journal of Medical Education, 51(11), 937-939.

Hochbaum, D. S., & Chen, A. (2000). Performance analysis and best implementations of old and new algorithms for the open-pit mining problem. Operations Research, 48(6), 894–914. doi:10.1287/opre.48.6.894.12392

Hoffmann, J., & Nebel, B. (2001). The FF Planning System: Fast Plan Generation Through Heuristic Search. Journal of Artificial Intelligence Research, 14, 253–302.

Holland, J. (1975). Adaptation in natural and artificial systems. Ann Arbor, Michigan: University of Michigan Press.

Hoos, H. (2002). An adaptive noise mechanism for WalkSAT. In Proc. of the 18th National Conference on Artificial Intelligence (AAAI-2002), (pp. 655–660).

Hoos, H., & Stützle, T. (2000). SATLIB: An Online Resource for Research on SAT. In Proc. Of SAT 2000, (pp. 283–292). Retrieved from http:/www.satlib.org

Horaud, R., Conio, B., Leboulleux, O., & Lacolle, B. (1989). An analytic solution for the perspective 4-point problem. Computer Vision Graphics and Image Processing, 47, 33–44. doi:10.1016/0734-189X(89)90052-2

Horaud, R., Dornaika, F., Lamiroy, B., & Christy, S. (1997). Object pose: The link between weak perspective, paraperspective, and full perspective. International Journal of Computer Vision, 22(2), 173–189. doi:10.1023/A:1007940112931

Hoskinson, R. L., Rope, R. C., & Fink, R. K. (2007). Using a decision support system to optimize production of agricultural crop residue Biofeedstock. Biomass and Bioenergy, 31(4), 186-194.

Howard, T. J., Carson, M. W., & Everett, J. E. (2005). Simulation modeling of iron ore product quality for process and infrastructure development. In A. Zerger & R. M. Argent (Eds.), Proceedings of the International Congress on Modeling and Simulation, (pp. 1251-1257), Melbourne, Australia.

Howson, C., & Urbach, P. (1993). Scientific Reasoning: The Bayesian Approach. Peru, IL: Open Court.

Hripcsak, G. (1993). Monitoring the monitor: Automated statistical tracking of a clinical event monitor. Computers and Biomedical Research, 26(5), 449-466.

Hripcsak, G., Austin, J.H., Alderson, P.O., & Friedman, C. (2002). Use of natural language processing to translate clinical information from a database of 889,921 chest radiographic reports. Radiology, 224(1), 157-163.

Hu, Y., Huang, J., Chen, J., Liu, M., & Xie, K. (2007). Software Project Risk Management Modeling with Neural Network and Support Vector Machine Approaches. In Proceedings of the Third International Conference on Natural Computation (ICNC'07), (pp. 358-362), Haikou, China.

Huang, J. (2007). A Case for Simple SAT Solvers. In Proc. 13th international conf. on principles and practice of constraint programming (cp'07), lncs 4741 (pp. 839–846). Berlin: Springer.

Huang, N., & Diao, S. (2006). Structure-Based Ontology Evaluation. Proceedings of the IEEE International Conference on e-Business Engineering (ICEBE 2006) (pp. 132-137).

Hung, W. Y., Kucherenko, S., Samsatli, N. J., & Shan, N. (2004). A Flexible and Generic Approach to Dynamic Modeling of Supply Chains. The Journal of the Operational Research Society, 55, 801–813. doi:10.1057/palgrave.jors.2601740

Hutchinson, S., Hager, G. D., & Corke, P. I. (1996). A tutorial on visual servo control. IEEE Transactions on Robotics and Automation, 12(5), 651–670. doi:10.1109/70.538972

Idri, A., & Abran, A. (2001). A Fuzzy Logic Based Set of Measures for Software Project Similarity: Validation and Possible Improvements. In Proceedings of the Seventh International Software Metrics Symposium (METRICS'01), (pp. 85-97), London.

Iwata, K., Nakashima, T., Anan, Y., & Ishii, N. (2008). Error Estimation Models Integrating Previous Models and Using Artificial Neural Networks for Embedded Software Development Projects. In Proceedings of the 20th IEEE International Conference on Tools with Artificial Intelligence (ICTAI'08), (pp. 371-378). Dayton, OH.

Jackson, J.R. (2000). The urgent call for knowledge management in medicine. The Physician Executive, 26(1), 28-31.

Jacobson, S., Johnson, A., Sullivan, K., Fleischer, M., & Kumar, A. (1997). Metaheuristic for a Flexible Assembly System Design Problems. Journal of Heuristics, 3(2), 139–159. doi:10.1023/A:1009679413948

Janetzki, V., Allen, M., & Cimino, J.J. (2004). Using natural language processing to link from medical text to online information resources. In Proceedings of Medinfo (pp. 1665).

Jassbi, J., Hafari, H., & Khanmohammadi, S. (2008). Fuzzy Expert System for Determining the Criticality of Activities in Mega Projects. In Proceedings of the Eighth international Conference on Hybrid Intelligent Systems, (pp. 246-251), Barcelona, Spain.

JBoss. (2009). JBoss Rules. Retrieved January 23, 2009, from http://www.jboss.com/products/rules

Jeffrey, S., Alonso, G., Franklin, M., Hong, W., & Widom, J. (2006). A pipelined framework for online cleaning of sensor data streams. In ICDE, (pp. 140-142).

Jennings, N. (1996). Coordination techniques for distributed artificial intelligence. In G. M. P. O'Hare & N. R. Jennings (Ed.), Foundations of Distributed Artificial Intelligence (pp. 187–210). Hoboken, NJ: Wiley.

Jeong, B., Sim, S., Jeong, H., & Kim, S. (2002). An available-to-promise system for TFT LCD manufacturing in the supply chain. Computers & Industrial Engineering, 43(1,2), 191-212.

Jess. (2009). Jess the Rule Engine for the Java Platform. Retrieved January 23, 2009, from http://herzberg.ca.sandia.gov/

Juang, P., Oki, H., Wang, Y., Martonosi, M., Peh, L. S., & Rubenstein, D. (2002). Energy-Efficient Computing for Wildlife Tracking: Design Tradeoffs and Early Experiences with ZebraNet. In Proc. ASPLOS X, San Jose, CA.

Kamrad, B., & Ernst, R. (2001). An economic model for evaluating mining and manufacturing ventures with output yield uncertainty. Operations Research, 49(5), 690–699. doi:10.1287/opre.49.5.690.10610

Kandaswamy, A., Kumar, C.S., Ramanathan, R.P., Jayaraman, R., & Malmurugan, N. (2004). Neural classification of lung sounds using wavelet coefficients. Computers in Biology and Medicine, 34, 523-537.

Karadogan, A., Kahriman, A., & Ozer, U. (2008). Application of fuzzy set theory in the selection of underground mining method. Journal of the South African Institute of Mining and Metallurgy, 108, 73–79.

Karp, R. M. (1972). Reducibility Among Combinatorial Problems. In Complexity of Computer Computations, (85-103). New York: Plenum.

Kasai, Y., & Morimoto, Y. (1988). Electronically controlled continuously variable transmission. In Proceedings of the International Congress on Transportation Electronics, Dearborn, MI.

Kautz, H., & Selman, B. (1992). Planning as satisfiability. In Proc. of the 10th European Conference on Artificial intelligence (ECAI'92), (pp. 359–363). Hoboken, NJ: John Wiley & Sons, Inc.

Kautz, H., & Selman, B. (1996). Pushing the envelope: Planning, propositional logic, and stochastic search. In Proc. of 13th the National Conference on Artificial Intelligence (AAAI'96), (pp. 1194–1201).

Kautz, H., McAllester, D., & Selman, B. (1996). Encoding Plans in Propositional Logic. In Proc. of the 5th International Conference on Principles of Knowledge Representation and Reasoning (KR'96) (pp. 374-384).

Kautz, H., Sabharwal, A., & Selman, B. (2008). Incomplete algorithms. In Handbook of Satisfiability, (pp. 185–203). Amsterdam: IOS Press.

Keane, A. J., & Brown, S. M. (1996). The design of a satellite boom with enhanced vibration performance using genetic algorithm techniques. In I.C. Parmee, (Ed.), Adaptive Computing in Engineering Design and Control '96 - Proceedings of the Second International Conference, (pp. 107-113), University of Plymouth.

Kennedy, J., & Eberhart, R. (with Shi, Y.) (2001). Swarm Intelligence. San Diego, CA: Academic Press.

Kewley, R., & Embrechts, M. (2002). Computational military tactical planning system. IEEE Transactions on Systems, Man and Cybernetics. Part C, Applications and Reviews, 32(2), 161–171. doi:10.1109/TSMCC.2002.801352

Kim, W., & Lee, J. K. (2002). Agent and e-business models. Decision Support Systems, (Special Issue), 34.

King, P., Annandale, D., & Bailey, J. (2000). A Conceptual Framework for Integrated Economic and Environmental Planning in Asia - A Literature Review. Journal of Environmental Assessment Policy and Management, 2(3), 279-315.

Kirn, S. (2008). Individualization Engineering. Cuvillier: Goettingen.

Kleijnen, J. P. C., & Smits, M. T. (2003). Performance Metrics in Supply Chain Management. The Journal of the

Operational Research Society, 54, 507–514. doi:10.1057/palgrave.jors.2601539

Klette, R., Schluns, K., & Koschan, A. (1998). Computer vision. Singapore: Springer.

Kline, M. (1972). Mathematical thought from ancient to modern times. New York: Oxford University Press.

Kolonja, B., Kalasky, D. R., & Mutmansky, J. M. (1993). Optimization and dispatching criteria for open-pit truck haulage system design using multiple comparisons with the best and common random numbers. In G. W. Evans, M. Mollaghasemi, E. C. Russell, & W. E. Biles (Eds.), Proceedings of the Winter Simulation Conference (pp. 393-401), Los Angeles, CA.

Kovalerchuk, B., Vityaev, E., & Ruiz, J.F. (2001). Consistent and complete data and expert mining in medicine. In K.J. Cios (Ed.), Medical data mining and knowledge discovery. New York: Physica-Verlag.

Kuchta, M., Newman, A., & Topal, E. (2004). Implementing a Production Schedule at LKAB's Kiruna Mine. Interfaces, 34(2), 124–134. doi:10.1287/inte.1030.0059

Kuperman, G.J., Gardner, R.M., & Pryor, T.A. (1991). The HELP system. New York: Springer-Verlag.

Kurbel, K., & Loutchkor, I. (2005). A model for multilateral negotiations on an agent-based job marketplace. Electronic Commerce Research and Applications, 4(3), 187–203. doi:10.1016/j.elerap.2005.01.002

Kvaale, H. (1988). A decision support simulation model for design of fast patrol boats. European Journal of Operational Research, 37(1), 92–99. doi:10.1016/0377-2217(88)90283-4

Kwon, O. K., Martland, C. D., & Sussman, J. M. (1998). Routing and scheduling temporal and heterogeneous freight car traffic on rail networks. Transportation Research Part E (Logistics and Transpn. Rev.), 34(2), 101-115.

Ladd, S. (1996). Genetic Algorithms in C++. New York: M & T Books.

Landt, J. (2005). The history of RFID. IEEE Potentials, 24(4), 8–11. doi:10.1109/MP.2005.1549751

Larkin, L. I. (1985). A fuzzy logic controller for aircraft flight control. In M. Sugeno (ed.), Industrial Applications of Fuzzy Control (pp.87-104). Amsterdam: North Holland.

Larsen, P. M. (1980). Industrial application of fuzzy logic control. International Journal of Man-Machine Studies, 12(1), 3–10. doi:10.1016/S0020-7373(80)80050-2

Lee, S. K., & Hwang, C. S. (2003). Architecture layers and engineering approach for agent-based system. Information and Software Technology, 45(13), 889–898. doi:10.1016/S0950-5849(03)00057-0

Lee, Y., & Chen, C.-Y. (2009). An optimization heuristic for the berth scheduling problem. European Journal of Operational Research, 196, 500–508. doi:10.1016/j.ejor.2008.03.021

Lerchs, H., & Grossmann, I. (1965). Optimum design of open-pit mines. Canadian Institute of Mining Bulletin, 68, 17–24.

Lesser, V., Decker, K., & Wagner, T. (2004). Evolution of the GPGP/TÆMS Domain-Independent Coordination Framework. Journal of Autonomous Agents and Multi-Agent Systems, 9(1-2), 87–143. doi:10.1023/B:AGNT.0000019690.28073.04

Lesser, V., Wagner, T., & Raja, A. (2006). Modeling uncertainty and its implications to sophisticated control in TAEMS agents. Journal of Autonomous Agent and Multi-Agent Systems, 13(3), 235–292. doi:10.1007/s10458-006-7669-2

Leukel, J. (2004). Standardization of Product Ontologies in B2B Relationships – On the Role of ISO 13584. Proceedings of the 10th Americas Conference on Information Systems (AMCIS 2004) (pp. 4084-4091).

Leukel, J., & Sugumaran, V. (2007). Evaluating Product Ontologies in E-Commerce: A Semiotic Approach. Proceedings of the 6th Workshop on e-Business (WeB 2007) (pp. 240-249).

Levin, G., & Roberts, E.B. (1976). The dynamics of human service delivery. Cambridge, MA: Ballinger Publishing Company.

Levin, G., Hirsch, G.B., & Roberts, E. (1972). Narcotics and the community: A system simulation. American Journal of Public Health, 62, 8-6.

Lewis, B. (2007). The 70-percent failure. Retrieved August 15, 2008, from http://www.infoworld.com/articles/op/xml/01/10/29/011029opsurvival.html

Lewis, B. (2008). The 70-percent failure. InfoWord. Retrieved August 2008, from http://www.infoworld.com/articles/op/xml/01/10/29/011029opsurvival.html

Li, H., Durfee, E., & Shin, K. (2003). Multiagent planning with internal execution resource constraints. In Proceedings of the 2nd International Conference on Autonomous Agent and Multi-Agent Systems (pp. 56-0567), Melbourne, Australia.

Li, S., Li, G., & Zhang, S. (2004). Minimizing makespan with release times on identical parallel batching machines. Discrete Applied Mathematics, 148(1), 127–134. doi:10.1016/j.dam.2004.11.004

Liebowitz, J. (Ed.). (1998). The Handbook of Applied Expert Systems. Salem, MA: CRC Press.

Lifschitz, V., Morgenstern, L., & Plaisted, D. (2008). Knowledge Representation and Classical Logic. In Handbook of Knowledge Representation (Foundations of Artificial Intelligence), (pp. 3–88). New York: Elsevier B.V.

Lin, C., & Hsieh, P. (2004). A Fuzzy Decision Support System for Strategic Portfolio Management. Decision Support Systems, 38(3), 383–398. doi:10.1016/S0167-9236(03)00118-0

Lin, Y. (2009). Two-commodity reliability evaluation of a stochastic-flow network with varying capacity weight in terms of minimal paths. Computers & Operations Research, 36, 1050–1063. doi:10.1016/j.cor.2007.12.005

Liu, X., Kane, G., & Bambroo, M. (2003). An Intelligent Early Warning System for Software Quality Improvement and Project Management. In Proceedings of the 15th IEEE International Conference on Tools with Artificial Intelligence (ICTAI '03), (pp. 32-38), Sacremento, CA.

Lopez-Martin, C., Yanex-Marquez, C., & Guitierrez-Tornes, A. (2006). Fuzzy Logic Systems for Software Development Effort Estimation Based Upon Clustering of Programs Segmented by Personal Practices. In Proceedings of the Electronics, Robotics, and Automotive Mechanics Conference (CERMA'06), (pp. 367-372), Cuernavaca, Morelos, Mexico.

Lou, X., & Chu, F. (2006). A branch and bound algorithm of the single machine schedule with sequence dependent setup times for minimizing total tardiness. Applied Mathematics and Computation, 138(1), 575–588.

Louta, M., Roussaki, I., & Pechlivanos, L. (2008). An intelligent agent negotiation strategy in the electronic marketplace environment. European Journal of Operational Research, 187(3), 1327–1345. doi:10.1016/j.ejor.2006.09.016

Lu, J., Fu, M., & Sha, J. (2005). Research of import iron ore logistics system based on the minimum cost theory. In Proceedings of the Second IEEE International Conference on Services Systems and Services Management (pp. 391-396), Chongqing, China.

Luckham, D. (2002). Power of Events. Reading, MA: Addison-Wesley.

Luckham, D. (2006). A View of Current State of Event Processing. In First Workshop on Event Processing, New York. Retrieved January 23, 2009, http://complex-events.com/wp-content/uploads/2006/03/dluckham-workshop-03-2006.pdf

Maes, P., Guttman, R. H., & Moukas, A. G. (1999). Agent that buy and sell: Transforming Commerce as we know it. Communications of the ACM, 42(3), 81–91. doi:10.1145/295685.295716

Magdalena, L. (2009, February). Soft Computing for Students and Society. IEEE Computational Intelligence, 47-50.

Mahfoud, S., & Mani, G. (1996). Financial forecasting using genetic algorithms. Applied Artificial Intelligence, 10(6), 543–565. doi:10.1080/088395196118425

Mainwaring, A., Polastre, J., Szewczyk, R., Culler, D., & Anderson, J. (2002). Wireless Sensor Networks for Habitat Monitoring. In WSNA, Atlanta, GA.

Malamas, E. N., Petrakis, E. G. M., Zervakis, M. E., Petit, L., & Legat, J. D. (2003). A survey on industrial vision systems, applications and tools. Image and Vision Computing, 21(2), 171–188. doi:10.1016/S0262-8856(02)00152-X

Malaviya, A., & Peters, L. Extracting Meaningful Handwriting Features with Fuzzy Aggregation Method. In Proceedings of the Third International Conference on Document Analysis and Recognition (ICDAR), (pp 841-844).

Malone, T., & Crowston, K. (1994). The interdisciplinary study of coordination. Journal of ACM Computing Surveys, 26(1), 87–119. doi:10.1145/174666.174668

Man, Z., Wei, T., Xiang, L., & Lishan, K. (2008). Research on Multi-project Scheduling Problem Based on Hybrid Genetic Algorithm. In Proceedings of the International Conference on Computer Science and Software Engineering (CSSE'08), (pp. 390-394), Wuhan, China.

Marczyk, A. (2004). Genetic Algorithms and Evolutionary Computing. Retrieved March 30, 2009, from http://www.talkorigins.org/faqs/genalg/genalg.html

Marquez, A. C., & Blanchar, C. (2006). A Decision Support System for evaluating operations investments in high-technology business. Decision Support Systems, 41(2), 472-487.

Martin, C. (2005). Logistics and Supply Chain Management, (3rd Ed.). Englewood Cliffs, NJ: Prentice Hall.

Martin, J., & Oxman, S. (1988) Building Expert Systems: A Tutorial. Englewood Cliffs, NJ: Prentice Hall.

Martinez, K., Ong, R., Hart, J. K., & Stefanov, J. (2004). GLACSWEB: A Sensor Web for Glaciers. In Proc. EWSN, Berlin, Germany.

Maruster, L., van der Aalst, W.M.P., Weijters, T., van den Bosch, A., & Daelemans, W. (2001, October 25-26). Automated discovery of workflow models from hospital data. In B. Kröse et al. (Eds.), Proceedings 13th Belgium-Netherlands Conference on Artificial Intelligence (BNAIC'01), Amsterdam, The Netherlands (pp. 183-190).

Mateus, G. R., Luna, H. P. L., Junior, J. F., Vieira, R. L. F., Costa, R. A. V., & Pratocinio, Z. K. G. (1994). Iron ore transportation at MBR: characterization and modeling (Tech. Rep. RT-009/94). Belo Horizonte, Brazil: Computer Science Department, Federal University of Minas Gerais.

McNeill, F. M., & Thro, E. (1994). Fuzzy Logic: A Practical Approach. London: Academic Press, Inc.

Melo, M. T., Nickel, S., & Saldanha-da-Gama, F. (2009). Facility location and supply chain management – A review. European Journal of Operational Research, 196, 401–412. doi:10.1016/j.ejor.2008.05.007

Menabde, M., Froyland, G., Stone, P., & Yeates, G. (2004). Mining Schedule Optimisation for Conditionally Simulated Orebodies. In R. Dimitrakopoulos & S. Ramazan (Eds.), Proceedings of Orebody Modeling and Strategic Mine Planning: Uncertainty and Risk Management, (pp. 357-342), Perth, Australia.

Mitchell, D. (2005). A SAT Solver Primer. EATCS Bulletin, 85, 112–133.

Mitchell, D. G., Selman, B., & Levesque, H. (1992). Hard and easy distributions of SAT solvers. In Proc. of the 10th National Conference on Artificial Intelligence (AAAI-92) (pp. 459–465).

Mitchell, D., Selman, B., & Levesque, H. (1992). Hard and Easy Distributions of SAT Problems. In Proc. of aaai (association for the advancement of artificial intelligence), 92, 459–465.

Mitchell, M. (1998). An Introduction to Genetic Algorithms. Cambridge, MA: MIT Press.

Montani, S., & Bellazzi, R. (2002). Supporting decisions in medical applications: The knowledge management perspective. International Journal of Medical Informatics, 68, 79-90.

Moskewicz, M., Madigan, C., Zhao, Y., Zhang, L., & Malik, S. (2001). Chaff: Engineering an Efficient SAT Solver. In Proc. Design Automation Conference (pp. 530–535).

Moya, L., & Weisel, E. (2008) The Difficulties with Validating Agent Based Simulations of Social Systems.

Proceedings Spring Multi-Simulation Conference, Agent-Directed Simulation, Ottawa, Canada.

Muckstadt, J. A., & Roundy, R. O. (1987). Multi-item, one-warehouse, multi-retailer distribution systems. Management Science, 33(12), 1613–1621. doi:10.1287/mnsc.33.12.1613

Muckstadt, J. A., & Roundy, R. O. (1993). Analysis of multistage production systems. Handbooks in OR & MS (Vol. 4).

Mueller, J. P. (1996). The Design of Intelligent Agents. Berlin: Springer.

Muguira, J. A., & Tolk, A. (2006). Applying a Methodology to Identify Structural Variances in Interoperations. Journal for Defense Modeling and Simulation, 3(2), 77–93. doi:10.1177/875647930600300203

Muhdi, R. A. (2006). Evaluation Modeling: Development, Characteristics and Limitations. Proceedings of the National Occupational Research Agenda (NORA), (pp. 87-92).

Nachiappan, S. P., & Jawahar, N. (2007). A genetic algorithm for optimal operating parameters of VMI system in a two-echelon supply chain. European Journal of Operational Research, 182, 1433–1452. doi:10.1016/j.ejor.2006.09.040

Nadel, A. (2002). Backtrack Search Algorithms for Propositional Logic Satisfiability: Review and Innovations. Unpublished master's thesis, Univ. of Tel-Aviv, Israel.

Naik, G. (1996, January 16). Back to Darwin: In sunlight and cells, science seeks answers to high-tech puzzles. The Wall Street Journal, p. A1.

Naimark, L., & Foxlin, E. (2002). Circular data matrix fiducial system and robust image processing for a wearable vision inertial self-tracker. In Proceedings of ACM/IEEE International Symposium on Mixed and Augmented Reality, Darmstadt, Germany, (pp. 27-36).

Nair, R., Pynadath, D., Yokoo, M., Tambe, M., & Marsella, S. (2003). Taming Decentralized POMDPs: Towards Efficient Policy Computation for Multiagent Settings. In Proceedings of the 18th International Joint Conference on Artificial Intelligence (pp. 705-711), Acapulco, Mexico.

National Research Council. (2002). Modeling and Simulation in Manufacturing and Defense Acquisition: Pathways to Success. Washington, DC: Committee on Modeling and Simulation Enhancements for 21st Century Manufacturing and Defense Acquisition, National Academies Press.

National Research Council. (2006). Defense Modeling, Simulation, and Analysis: Meeting the Challenge. Washington, DC: Committee on Modeling and Simulation for Defense Transformation, National Academies Press.

Navarro, J. A. (2007). Encoding and Solving Problems in Effectively Propositional Logic. Unpublished doctoral dissertation, University of Manchester, UK.

Nelson, B. J., Papanikolopoulos, N. P., & Khosla, P. K. (1996). Robotic visual servoing and robotic assembly tasks. IEEE Robotics & Automation Magazine, 3(2), 23–31. doi:10.1109/100.511777

Neumann, D. (2002). An Enhanced Neural Network Technique for Software Risk Analysis. IEEE Transactions on Software Engineering, 28(9), 904–912. doi:10.1109/TSE.2002.1033229

Newman, A., Martinez, M., & Kuchta, M. (2006). A review of long- and short-term production scheduling at LKAB's Kiruna Mine. In J. W. Herrmann (Ed.), Handbook of Production Scheduling, (pp. 265-286). New York: Springer Science+Business Media, Inc.

Ngo, L., Haddawy, P., & Nguyen, H. (1998). A modular structured approach to conditional decision-theoretic planning. In Proceedings of Artificial Intelligence Planning Systems, (pp. 111–118), Menlo Park, CA.

Nguyen, H. T., & Walker, E. (2000). A First Course in Fuzzy Logic - Second Editon. Boca Raton, FL: Chapman & Hall/CRC Press LLC.

North Atlantic Treaty Organization. (2002). NATO Code of Best Practice for Command and Control Assessment. Revised ed. Washington, DC: CCRP Press

Obayashi, S., Sasaki, D., Takeguchi, Y., & Hirose, N. (2000). Multiobjective evolutionary computation for supersonic wing-shape optimization. IEEE Transactions on Evolutionary Computation, 4(2), 182–187. doi:10.1109/4235.850658

Oh, J. K., & Lee, C. H. (2007). Development of a stereo vision system for industrial robots. In Proceedings of the International Conference on Control, Automation and Systems, Seoul, Korea, (pp. 659-663).

Oliver, J. R. (1997). A machine learning approach to automated negotiation and prospects for electronic commerce. Journal of Management Information Systems, 13(3), 83–112.

OpenForecast. (2009). Retrieved from http://openforecast.sourceforge.net/

Ören, T. I., & Numrich, S. K. S.K., Uhrmacher, A.M., Wilson, L. F., & Gelenbe, E. (2000). Agent-directed simulation: challenges to meet defense and civilian requirements. In Proceedings of the 32nd Conference on Winter Simulation, Orlando, Florida, December 10 – 13. San Diego, CA: Society for Computer Simulation International.

Osanloo, M., Gholamnejad, J., & Karimi, B. (2007). Long-term open pit mine production planning: a review of models and algorithms. International Journal of Mining . Reclamation and Environment, 22(1), 3–35. doi:10.1080/17480930601118947

Otto, A. (2003). Supply Chain Event Management: Three Perspectives. The International Journal of Logistics Management, 14(2), 1–13. doi:10.1108/09574090310806567

Page, E. H., Briggs, R., & Tufarolo, J. A. (2004). Toward a Family of Maturity Models for the Simulation Interconnection Problem. In Proceedings of the Spring Simulation Interoperability Workshop. New York: IEEE CS Press.

Palmer, R. C. (1991). The bar code book. Peterborough, NH: Helmers Publishing.

Park, J. M., Nam, J. H., Hu, Q. P., & Suh, H. W. (2008). Product Ontology Construction from Engineering Documents. Proceedings of the International Conference on Smart Manufacturing Application (ICSMA 2008) (pp. 305-310).

Paton, N., & Díaz, O. (1999). Active Database Systems. ACM Computing Surveys, 31(1), 63–103. doi:10.1145/311531.311623

Pearl, J. (1985). Bayesian Networks: A Model of Self-Activated Memory for Evidential Reasoning. In Proceedings of the 7th Conference of the Cognitive Science Society, (pp. 329-334), Irvine, CA.

Pendharkar, P. C., & Rodger, J. A. (2000). Nonlinear programming and genetic search application for production scheduling in coal mines. Annals of Operations Research, 95, 251–267. doi:10.1023/A:1018958209290

Perrone, G., Renna, P., Cantamessa, M., Gualano, M., Bruccoleri, M., & Lo Nigro, G. (2003). An Agent Based Architecture for production planning and negotiation in catalogue based e-marketplace. In Proceedings of the 36th CIRP-International Seminar on Manufacturing Systems, 03-05 June 2003, Saarbruecken, Germany

Pham, D.T., Wang, Y., & Dimov, S. (2005). Intelligfent manufacturing strategy selection. In Proc 1st International Virtual Conference on Intelligent Production Machines and Systems (pp. 312-318). Oxford: Elsevier.

Pham, D.T., Wang, Y., & Dimov, S. (2006). Incorporating delay and feedback in intelligent manufacturing strategy selection. In Proc 2nd International Virtual Conference on Intelligent Production Machines and Systems (pp. 246-252). Oxford: Elsevier.

Phillips-Wren, G., & Jain, L. C. (Eds.). (2005). Intelligent Decision Support Systems in Agent-Mediated Environments. The Netherlands: IOS Press.

Phillips-Wren, G., Ichalkaranje, N., & Jain, L. C. (Eds.). (2008). Intelligent Decision Making –An AI-Based Approach. Berlin: Springer-Verlag.

Phuong, N. (1995). Fuzzy Set Theory and Medical Expert Systems: Survey and Model. In G. Goos, J. Hartmanis, & J. Leeuwen (eds.), Lecture Notes in Computer Science: SOFSEM'95: Theory and Practice of Informatics (pp. 431-436). Berlin/Heidelberg: Springer.

Picard, J. C. (1976). Maximum closure of a graph and its applications to combinatorial problems. Management Science, 22, 1268–1272. doi:10.1287/mnsc.22.11.1268

Pochet, Y., & Wolsey, L. A. (2006). Production planning by mixed integer programming. New York: Springer Science+Business Media, Inc.

Pohl, J. G., Wood, A. A., Pohl, K. J., & Chapman, A. J. (1999). IMMACCS: A Military Decision-Support System. DARPA-JFACC 1999 Symposium on Advances in Enterprise Control, San Diego, CA.

Pomerleau, D. (1993). Knowledge-based Training of Artificial Neural Networks for Autonomous Robot Driving. In J. Connell & S. Mahadevan (Eds.), Robot Learning (pp. 19-43). Boston: Kluwer Academic Publishing.

Potter, S. S., Elm, W. C., & Gualtieri, J. W. (2006). Making Sense of Sensemaking: Requirements of a Cognitive Analysis to Support C2 Decision Support System Design. Proceedings of the Command and Control Research and Technology Symposium. Washington, DC: CCRP Press.

Power, D. J., & Sharda, R. (2007). Model-driven decision support systems: Concepts and research directions. Journal for Decision Support Systems, 43(3), 1044–1061. doi:10.1016/j.dss.2005.05.030

Prather, J.C., Lobach, D.F., Goodwin, L.K., Hales, J.W., Hage, M.L., & Hammond, W.E. (1997). Medical data mining: Knowledge discovery in a clinical data warehouse. In Proceedings of the AMIA Annual Symposium Fall 1997 (pp. 101-105).

Presley, A., Sarkis, J., Barnett, W., & Liles, D. (2001). Engineering the Virtual Enterprise: An Architecture-Driven Modeling Approach. International Journal of Flexible Manufacturing Systems, 13, 145–162. doi:10.1023/A:1011131417956

Prestwich, S. (2008). CNF Encodings. In Handbook of Satisfiability, (pp. 75–97). Amsterdam: IOS Press.

Prindible, M. (2006). An Introduction to R-CAST: Developing Simple Agents using Decision-making modeling and the R-CAST Architecture. The Laboratory for Intelligent Agents, College of Information Science and Technology, The Pennsylvania State University. Retrieved March 1, 2007, from http://agentlab.psu.edu/rcast/rcast_tutorial.pdf

Project Management Institute (PMI). (2004). Project Management Body of Knowledge. Newtown Square, PA: Project Management Institute, Incorporated.

Project Management Institute (PMI). (2008). The Project Management Professional (PMP) Credential. Project Management Institute, Inc. Retrieved August 2008, from http://www.pmi.org/CareerDevelopment/Pages/Obtaining-Credential.aspx

PRTM – Pittiglio. Rabin, Todd, & McGrath. (2005). Managing Complexity in Global Supply Chain. PRTM's 2005 Supply Chain Complexity Survey. Retrieved from http://www.prtm.com/

Pullen, J. M., Brunton, R., Brutzman, D., Drake, D., Hieb, M. R., Morse, K. L., & Tolk, A. (2005). Using Web Services to Integrate Heterogeneous Simulations in a Grid Environment. Journal on Future Generation Computer Systems, 21, 97–106. doi:10.1016/j.future.2004.09.031

Rahardja, K., & Kosaka, A. (1996). Vision-based bin-picking: Recognition and localization of multiplecomplex objects using simple visual cues. In Proceedings of the IEEE/RSJ International Conference on Intelligent Robots and Systems, Osaka, Japan, 3, 1448-1457.

Rahwan, I., Sonenberg, L., Jennings, N. R., & McBurney, P. (2007). STRATUM: A methodology for designing heuristic agent negotiation strategies. Applied Artificial Intelligence, 21(6), 489–527. doi:10.1080/08839510701408971

Raman, A., & Ton, Z. (2005). Operational Execution at Arrow Electronics. Harvard Business Case, Harvard University, Cambridge, MA.

Ramazan, S. (2007). The new Fundamental Tree Algorithm for production scheduling of open pit mines. European Journal of Operational Research, 177, 1153–1166. doi:10.1016/j.ejor.2005.12.035

Ramazan, S., & Dimitrakopoulos, R. (2003). Production Scheduling Optimisation in a Nickel Laterite Deposit: MIP and LP Applications and Infeasibility in the Presence of Orebody Variability. In Carlton, (Ed.), Proceedings of the Australasian Institute of Mining and Metallurgy International Symposium on Mine Planning and Equipment Selection (pp. 1-5). Kalgoorlie, Wichita.

Ramazan, S., & Dimitrakopoulos, R. (2004). Recent applications of operations research and efficient MIP formulations in open pit mining. Transactions of the Society for Mining, Metallurgy, and Exploration, 316, 73–78.

Rangone, A., & Bertelè, U. (2004). Il B2B in Italia: Finalmente parlano i dati. Rapporto dell'osservatorio B2B, Marzo 2004.

Richardson, G.P. (1996). System dynamics. In S.I. Gass & C.M. Harris (Eds.), Encyclopedia of operations research and management science (pp. 656-660). Boston, MA: Kluwer Academic Publishers.

Richardson, G.P. (1999). Citation for winner of the 1999 Jay W. Forrester Award: Jac Vennix. System Dynamics Review, 15 4, 375-377.

Richardson, G.P., & Andersen, D.F. (1995). Teamwork in group model building. System Dynamics Review, 11(2), 113-137.

Rintanen, J. (2009). Planning as SAT. In Handbook of Satisfiability, (Vol. 185, Frontiers in Artificial Intelligence and Applications, pp. 483–504). Amsterdam: IOS Press.

Rizki, M., Zmuda, M., & Tamburino, L. (2002). Evolving pattern recognition systems. IEEE Transactions on Evolutionary Computation, 6(6), 594–609. doi:10.1109/TEVC.2002.806167

Roberts, E.B., & Hirsch, G.B. (1976, Winter). Strategic modelling for health care managers. Health Care Management Review, 69-77.

Robinson, A., & Voronkov, A. (Eds.). (2001). Handbook of Automated Reasoning (Vols. I & II). Amsterdam: The MIT Press.

Robinson, N., Gretton, C., Pham, D. N., & Sattar, A. (2008). A compact and efficient sat encoding for planning. Icaps (pp. 296-303).

Robinson, S. (2008). Conceptual modelling for simulation Part I: definition and requirements. The Journal of the Operational Research Society, 59, 278–290. doi:10.1057/palgrave.jors.2602368

Römer, K., & Matttern, F. (2004). The Design Space of Wireless Sensor Networks. IEEE Wireless Communications, 11(6), 54–61. doi:10.1109/MWC.2004.1368897

Ross, T. J. (1995). Fuzzy Logic with Engineering Applications. Boston: McGraw-Hill, Incorporated.

Rozakis, S., Soldatos, P. G., Kallivroussis, L., & Nicolaou, I. (2001). Multiple Criteria Analysis of Bio-energy Projects: Evaluation of Bio-Electricity Production in Farsala Plain, Greece. Journal of Geographic Information and Decision Analysis, 5(1), 49-64.

Runge, I. (1998). Mining Economics and Strategy. Littleton, CO: Society for Mining, Metallurgy and Exploration, Inc.

Russell, S., & Norvig, P. (2003), Artificial Intelligence: A Modern Approach (2nd Ed.). Upper Saddle River, NJ: Prentice Hall.

Ruzon, M., & Tomasi, C. (2001). Edge, junction, and corner detection using color distributions. IEEE Transactions on Pattern Analysis and Machine Intelligence, 23(11), 1281–1295. doi:10.1109/34.969118

Saaty, T. (1977). Scenarios and priorities in transport planning: application to the Sudan. Transportation Research, 11, 346–351. doi:10.1016/0041-1647(77)90044-2

Sabino, J. A., Stützle, T., Birattari, M., & Leal, J. E. (2006). ACO Applied to Switch Engine Scheduling in a Railroad Yard (extended abastract). In Proceedings of the V International Symposium on Ant Colony Optimization and Swarm Intelligence (pp. 502-503). Heidelberg: Springer Berlin.

Saha, S. (2006). Improving Agreements In Multi-Issue Negotiation. Journal of Electronic Commerce Research, 7(1), 41–49.

Saker, B. R., & Diponegoro, A. (2009). Optimal production plans and shipment schedules in a supply-chain system with multiple suppliers and multiple buyers. European Journal of Operational Research, 194, 753–773. doi:10.1016/j.ejor.2008.01.025

Salim, V., & Cai, X. (1995). Scheduling cargo trains using genetic algorithms. In Proceedings of the IEEE International Conference on Evolutionary Computation (pp. 224-227). Perth, WA.

Samarasinghe, S. (2006). Neural Networks for Applied Sciences & Engineering. New York: Auerbach.

Sambridge, M., & Gallagher, K. (1993). Earthquake hypocenter location using genetic algorithms. Bulletin of the Seismological Society of America, 83(5), 1467–1491.

Sandia. (2009). Jess, the Rule Engine for the Java Platform. Retrieved January 23, 2009, from http://herzberg.ca.sandia.gov/

Santos, R. R., Meech, J. A., & Ramos, L. T. S. (1995). Thickener operations at Carajas using a fuzzy logic controller Systems. IEEE International Conference on Man and Cybernetics: Intelligent Systems for the 21st Century, (vol. 2., pp. 1636-1639).

Sargent, R. G. (1999). Validation and verification of simulation models. Proceedings of the Winter Simulation Conference, 1, 39–48.

Sargent, R. G. (2000). Verification, Validation, and Accreditation of Simulation Models. Proceedings of the Winter Simulation Conference, (pp. 50-59).

Sargent, R. G. (2007). Verification and validation of simulation models. Proceedings of the Winter Simulation Conference, (pp. 124-137).

Sarimveis, H., Patrinos, P., Tarantilis, C. D., & Kiranoudis, C. T. (2008). Dynamic modeling and control of supply chain systems: a review. Computers & Operations Research, 35, 3530–3561. doi:10.1016/j.cor.2007.01.017

Sarin, S. C., & West-Hansen, J. (2005). The long-term mine production scheduling problem. Institute of Industrial Engineers Transactions, 37, 109–121.

Sarter, N. B., & Woods, D. D. (1991). Situation Awareness: A Critical But Ill-Defined Phenomenon. The International Journal of Aviation Psychology, 1(1), 45–57. doi:10.1207/s15327108ijap0101_4

Sattarvand, J., & Niemann-Delius, C. (2008). Perspective of metaheuristic optimization methods in open pit production planning. In J. Dubiński (Ed.), XXI World Mining Congress (pp. 143-155). London: Taylor & Francis Group.

Sayarshad, H. R., & Ghoseiri, K. (2009). A simulated annealing approach for the multi-periodic rail-car fleet sizing problem. Computers & Operations Research, 36(6), 1789–1799. doi:10.1016/j.cor.2008.05.004

Schenker, D., & Khoshgoftaar, R. (1998). The Application of Fuzzy Enhanced Case-Based Reasoning For Identifying Fault-Prone Modules. In Proceedings of the Third International Symposium on High-Assurance Systems Engineering, (pp. 90-97), Washington, DC.

Schiefer, J., Rozsnyai, S., Rauscher, C., & Saurer, G. (2007). Event-driven rules for sensing and responding to business situations. In Proceedings of the 2007 Inaugural International Conference on Distributed Event-Based Systems, (pp. 198-205).

Schofield, C. G. (1980). Homogenization and Blending Systems. Clausthal, Germany: TransTech.

Schubart, J.R., & Einbinder, J.S. (2000). Evaluation of a data warehouse in an academic health sciences center. International Journal of Medical Informatics, 60, 319-333.

Schwartz, D.G., & Klir, G.J. (1993). Fuzzy Logic: Concepts to constructs. AI Expert, 26-33.

Scott, M., & Dimitrakopoulos, R. (2001). Quantitative analysis of mineral resources for strategic planning: implications for Australian geological surveys. Natural Resources Research, 10(3), 159–177. doi:10.1023/A:1012536823294

Searle, J., & Brennan, J. (2006). General Interoperability Concepts. In Integration of Modelling and Simulation (pp. 3-1 – 3-8), (Educational Notes RTO-EN-MSG-043). Neuilly-sur-Seine, France

Selman, B., Kautz, H., & Cohen, B. (1996). Local Search Strategies for Satisfiability Testing. In Cliques, Coloring, and Satisfiability: Second DIMACS Implementation Challenge, DIMACS Series in Discrete Mathematics and Theoretical Computer Science, (Vol. 26, pp. 521–532).

Selman, B., Levesque, H., & Mitchell, D. (1992). A new method for solving hard satisfiability problems. In Proc. of aaai 92 (pp. 440–446).

Senge, P., Kleiner, A., Roberts, C., Ross, R., & Smith, B. (1994). The fifth discipline fieldbook. New York: Doubleday

Seyyedi, M., Teshnehlab, M., & Shams, F. (2005). Measuring Software Processes Performance on the Fuzzy Multi Agent Measurements. In Proceedings of the International Conference on Information Technology: Coding and Computing (ITCC'05), (pp. 410-415), Las Vegas, NV.

Shapiro, L. G., & Stockman, G. C. (2001). Computer vision. London: Prentice Hall.

Sharon, G., & Etzion, O. (2008). Event-processing network model and implementation. IBM Systems Journal, 47(2), 321–334.

Shen, J., Chang, S. I., Lee, E. S., Deng, Y., & Brown, S. J. (2005). Determination of cluster number in clustering microarray data. Applied Mathematics and Computation, 169(2), 1172-1185.

Shim, J. P., Merrill, W., & Courtney, J. (2002). Past, Present and Future of Decision Support Technology. Decision Support Systems, 33(2), 111–126. doi:10.1016/S0167-9236(01)00139-7

Shim, J., & Shim, S. S. Y. (2006). Ontology-based e-Catalog in e-commerce: Special Section. Electronic Commerce Research and Applications, 5(1), 1. doi:10.1016/j.elerap.2006.01.001

Shim, S.-O., & Kim, Y.-D. (2006). A branch and bound algorithm for an identical parallel machine scheduling problem with a job splitting property. Computers & Operations Research, 35(3), 863–875. doi:10.1016/j.cor.2006.04.006

Shipley, M., & Johnson, M. (2009). A Fuzzy Approach for Selecting Project Membership to Achieve Cognitive Style Goals. European Journal of Operational Research, 192(3), 918–928. doi:10.1016/j.ejor.2007.10.006

Silva, J. P. M., & Sakallah, K. (1996). Grasp — a new search algorithm for satisfiability. In ICCAD '96: Proc. of the 1996 IEEE/ACM international conference on Computer-aided design, (pp. 220–227).

Simulation Interoperability Standards Organization (SISO). (2006). The Base Object Model Standard; SISO-STD-003-2006: Base Object Model (BOM) Template Standard; SISO-STD-003.1-2006: Guide for Base Object Model (BOM) Use and Implementation. Orlando, FL: SISO Documents.

Singh, V., & Mohan, R. S. (2005). Application of image processing and radial basis neural network techniques for ore sorting and ore classification. Minerals Engineering, 18(15), 1412–1420. doi:10.1016/j.mineng.2005.03.003

Sipser, M. (1997). Introduction to the Theory of Computation. Boston: PWS Publishing Company.

Sisk, T. (1998). The History of Project Management. Retrieved March 15, 2009, from https://www.microsoft.com/downloads/details.aspx?FamilyID=c1f9b881-d879-4b54-b07b-55041685f15f&displaylang=en

Smith, M. L., Sheppard, I., & Karunatillake, G. (2003). Using MIP for strategic life-of-mine planning of the lead/zinc stream at Mount Isa Mines. In Application of Computers and Operations Research in the Minerals Industries, (pp. 465-473).

Soroor, J., Tarokh, M. J., & Shemshadi, A. (2009). Initiating a state of the art system for real-time supply chain coordination. European Journal of Operational Research, 196, 635–650. doi:10.1016/j.ejor.2008.03.008

Sourd, F. (2006). Dynasearch neighborhood for the earliness-tardiness scheduling problem with release dates and setup constraints. Operations Research Letters, 34(5), 591–598. doi:10.1016/j.orl.2005.06.005

Srivastava, B., & Koehler, J. (2003). Web Service Composition - Current Solutions and Open Problems.

Proceedings ICAPS 2003 Workshop on Planning for Web Services.

Stamper, R., Liu, K., Hafkamp, M., & Ades, Y. (2000). Understanding the role of signs and norms in organisations – a semiotic approach to information systems design. Behaviour & Information Technology, 19(1), 15–27. doi:10.1080/014492900118768

Stanford Encyclopedia of Philosophy. (2008). Search: Fuzzy Logic. Metaphysics Research Lab, CSLI, Stanford University. Retrieved August 22, 2008, from http://plato.stanford.edu/entries/logic-fuzzy

Stearns, N.S, Bergan, T.A., Roberts E.B., & Quigley, L. (1976, Fall). Systems intervention: New help for hospitals. Health Care Management Review, 1, 9-18.

Stearns, N.S., Bergan, T.A., Roberts, E.B., & Cavazos, L.F. (1978). A systems intervention for improving medical school-hospital interrelationships. Journal of Medical Education, 53, 464-472.

Stecco, G., Cordeau, J.-F., & Moretti, E. (2008). A branch-and-cut algorithm for a production scheduling problem with sequence-dependent and time-dependent setup times. Computers & Operations Research, 35(8), 2635–2655.

Steedman, M. (1995). Improvements to Propositional Satisfiability Search Algorithms. Unpublished doctoral dissertation, Univ. of Pennsylvania.

Steffen, O. K. H. (1997). Planning of open pit mines on a risk basis. Journal of the Southern African Institute of Mining and Metallurgy, March/April, 47-56.

Stergiou, C., & Siganos, D. (1996). Neural Networks. SURPRISE 96 Journal. Retrieved March 30, 2009, from http://www.doc.ic.ac.uk/~nd/surprise_96/journal/vol4/cs11/report.html

Sterman, J. D. (2000). Business dynamics: Systems thinking and modelling for a complex world. Boston, MA: Irwin McGraw-Hill.

Sterman, J.D. (1994). Learning in and about complex systems. System Dynamics Review, 10(2-3), 291-330.

Stockmeyer, L. J., & Meyer, A. R. (1973). Word problems requiring exponential time. In Proc. of the 5th Annual ACM Symposium on the Theory of Computing (pp. 1–9).

Streambase. (2009). Streambase event processing engine. Retrieved January 23, 2009, from http://www.streambase.com/

Sudkamp, T., & Hammell, R. J. II. (1994). Interpolation, Completion, and Learning Fuzzy Rules. IEEE Transactions on Systems, Man, and Cybernetics, 24(2), 332–342. doi:10.1109/21.281432

Sudkamp, T., & Hammell, R. J. II. (1998). Scalability in Fuzzy Rule-based Learning. Journal of Information Science, 109, 135–147. doi:10.1016/S0020-0255(98)00014-0

Supply Chain Council. (2009). SCOR Model (Version 9.0) of Supply Chain Council (SCC). Retrieved from http://www.supply-chain.org/

Supply-Chain Council. (2006). Supply Chain Operations Reference Model (SCOR®) Version 8.0.http://www.supply-chain.org.

Ta, C. H., Kresta, J. V., Forbes, J. F., & Marquez, H. J. (2005). A Stochastic Optimization Approach to Mine Truck Allocation. International Journal of Surface Mining . Reclamation and Environment, 19(3), 162–175. doi:10.1080/13895260500128914

Tadayon, N. (2005). Neural Network Approach for Software Cost Estimation. In Proceedings of the International Conference on Information Technology: Coding and Computing (ITCC'05), (pp. 815-818), Las Vegas, NV.

Tang, K. S., Man, K. F., Kwong, S., & He, Q. (1996). Genetic algorithms and their applications. IEEE Signal Processing Magazine, 13(6), 22–37. doi:10.1109/79.543973

Tang, L., & Liu, G. (2007). A mathematical programming model and solution for scheduling production orders in Shanghai Baoshan Iron and Steel Complex. European Journal of Operational Research, 182, 1453–1468. doi:10.1016/j.ejor.2006.09.090

Tasan, O. S., & Semra, T. (2008). A review of the current applications of genetic algorithms in assembly line balancing. Journal of Intelligent Manufacturing, 19(1), 49–69. doi:10.1007/s10845-007-0045-5

The 29 Palms Experiment (2009) Tracking vehicles with a UAV-delivered sensor network. Retrieved January 23, 2009, from http://robotics.eecs.berkeley.edu/~pister/29Palms0103/

Thomas, D. J., & Griffin, P. M. (1996). Coordinated supply chain management. European Journal of Operational Research, 94, 1–15. doi:10.1016/0377-2217(96)00098-7

Tian, F., Zhang, H., & Lu, Y. (2003). Research on Modeling with Dynamic Bayesian Networks. In Proceedings of the IEEE International Conference on Web Intelligence (WI'03), (pp. 606-609), Halifax, Canada.

Tibco. (2009). Tibco BusinessEvents. Retrieved January 23, 2009, from http://www.tibco.com/software/complex_event_processing/

Tolk, A. (1999). Requirements for Simulation Systems when being used as Decision Support Systems. [), IEEE CS press]. Proceedings Fall Simulation Interoperability Workshop, I, 29–35.

Tolk, A., & Diallo, S. Y. (2008) Model-Based Data Engineering for Web Services. In Nayak R. et al. (eds) Evolution of the Web in Artificial Intelligence Environments, (pp. 137-161). Berlin: Springer.

Tolk, A., & Jain, L. C. (Eds.). (2008). Complex Systems in the Knowledge-based Environment. Berlin: Springer Verlag.

Tolk, A., Diallo, S. Y., & Turnitsa, C. D. (2008). Implied Ontological Representation within the Levels of Conceptual Interoperability Model. [IDT]. International Journal of Intelligent Decision Technologies, 2(1), 3–19.

Tolk, A., Diallo, S. Y., Turnitsa, C. D., & Winters, L. S. (2006). Composable M&S Web Services for Net-centric Applications. Journal for Defense Modeling and Simulation, 3(1), 27–44. doi:10.1177/875647930600300104

Tolwinski, B., & Underwood, R. (1996). A scheduling algorithm for open pit mines. IMA Journal of Mathematics Applied in Business & Industry, 7, 247–270.

Tonfoni, G., & Jain, L. C. (Eds.). (2003). Innovations in Decision Support Systems. Australia: Advanced Knowledge International.

Topal, E. (2008). Early start and late start algorithms to improve the solution time for long-term underground mine production scheduling. Journal of the South African Institute of Mining and Metallurgy, 108, 99–107.

Tosic, V., Pagurek, B., Esfandiari, B., & Patel, K. (2001). On the Management of Compositions of Web Services. In Proceedings Object-Oriented Web Services (OOPSLA).

Trucco, E., & Verri, A. (1998). Introductory techniques for 3D computer vision. Upper Saddle River, NJ: Prentice Hall.

Truong, T. H., Rothschild, B. J., & Azadivar, F. (2005) Decision support system for fisheries management. In Proceedings of the 37th Conference on Winter Simulation, (pp. 2107-2111).

Tsai, R. Y. (1987). A versatile camera calibration technique for high-accuracy 3D machine vision metrology using off-the-shelf TV cameras and lenses. IEEE Journal on Robotics and Automation, 3(4), 323–343. doi:10.1109/JRA.1987.1087109

Tsiakis, P., & Papageorgiou, L. G. (2008). Optimal production allocation and distribution supply chain networks. International Journal of Production Economics, 111, 468–483. doi:10.1016/j.ijpe.2007.02.035

Turowski, K. (2002). Agent-based e-commerce in case of mass customization. International Journal of Production Economics, 75, 69–81. doi:10.1016/S0925-5273(01)00182-7

Umbers, I. G., & King, P. J. (1980). An analysis of human decision-making in cement kiln control and the implications for automation. International Journal of Man-Machine Studies, 12(1), 11–23. doi:10.1016/S0020-7373(80)80051-4

Underwood, R., & Tolwinski, B. (1998). A mathematical programming viewpoint for solving the ultimate pit problem. European Journal of Operational Research, 107, 96–107. doi:10.1016/S0377-2217(97)00141-0

Van Asperen, E., Dekker, R., Polman, M., & Arons, H. S. (2003). Modeling ship arrivals in ports. In Proceedings of the 2003 IEEE Winter Simulation Conference, (pp. 1737-1744).

van Dam, A. (1999). Education: the unfinished revolution. [CSUR]. ACM Computing Surveys, 31(4). doi:10.1145/345966.346038

van Harmelen, F., Lifschitz, V., & Porter, B. (Eds.). (2008). Handbook of Knowledge Representation. New York: Elsevier.

Varma, V. K., Ferguson, I., & Wild, I. (2000). Decision Support System for Sustainable Forest Management. Forest Ecology and Management, 128(1-2), 49–55. doi:10.1016/S0378-1127(99)00271-6

Velev, M. N., & Bryant, R. E. (2001). Effective use of boolean satisfiability procedures in the formal verification of superscalar and VLIW microprocessors. In Proc. of the 38th Conference on Design Automation, (pp. 226–231).

Venayagamoorthy, G. (2009, February). A Successful Interdisciplinary Course on Computational Intelligence. IEEE Computational Intelligence, 14-23.

Vieira, P., Neto, E., Bessa, E., & Gomide, F. (1999). Railway dispatch planning and control. In Proceedings of the 18th International Conference of the North American Fuzzy Information Processing Society, (pp. 134-138), New York.

Vignaux, G. A., & Michalewicz, Z. (1991). A Specialized Genetic Algorithm for Numerical Optimization Problems. IEEE International Conference on Evolutionary Computation (pp. 266–278).

Vilím, P. (2002). Batch Processing with Sequence Dependent Setup Times: New Results. Lecture Notes in Computer Science, 2470, 153–167.

Vincenti, G., Hammell, R. J., II, & Trajkovski, G. (2005). Scouting for Imprecise Temporal Associations to Support Effectiveness of Drugs During Clinical Trials. In Proceedings of the Annual Conference of the North American Fuzzy Information Processing Society (NAFIPS 2005), (pp. 22-25), Ann Arbor, MI.

Vivanco, R. (2007). Use of a Genetic Algorithm to Identify Source Code Metrics Which Improves Cognitive Complexity Predictive Models. In Proceedings of the 15th International Conference on Program Comprehension (ICPC'07), (pp. 297-300), Banff, Alberta, Canada.

W3C-The World Wide Web Consortium. (2004). OWL Web Ontology Language Reference. Retrieved on January 23, 2009 from www.w3.org/TR/owl-ref

Wagner, T. (2005). DARPA IPTO/IXO. We are not alone. In Proceedings of the DARPATech (pp. 75-80), Anaheim, CA.

Wagner, T., Garvey, A., & Lesser, V. (1997). Complex goal criteria and its application in design-to-criteria scheduling. In Proceedings of the 4th National Conference on Artificial Intelligence (pp. 294–301), Providence, Rhode Island.

Wang, F. & Liu, P. (2005). Temporal management of RFID data. Very Large Data Bases (VLDB), 1128-1139.

Wang, H. (1988). A Layered Architecture for Manufacturing Operations Planning. Computers & Industrial Engineering, 14(2), 201–210. doi:10.1016/0360-8352(88)90029-0

Wang, H. (1997). Intelligent agent-assisted decision support systems: Integration of knowledge discovery, knowledge analysis, and group decision support. Expert Systems with Applications, 12(3), 323-335.

Wang, L., & Mendel, J. M. (1991). Analysis and Design of Fuzzy Logic Controller. Technical Report USC-SIPI-184, Signal and Image Processing Institute, University of Southern California, Los Angeles.

Wang, Q., & Sevim, H. (1995). Alternative to parameterization in finding a series of maximum-metal pits for production planning. SME Transactions, 298, 178–182.

Wang, Y. (2007). Integration of data mining with game theory. Journal of International Federation for Information Processing, 207/2006, 275-280. Boston, MA: Springer

Watson, I. (1997). Applying Case-based Reasoning: Techniques for Enterprise Systems. San Francisco, CA: Morgan Kaufmann Publishers, Inc.

Weijters, T., & van der Aalst, W.M.P. (2001, October 25-26). Process mining: Discovering workflow models from event-based data. In B. Kröse et al. (Eds.), Proceedings 13th Belgium-Netherlands Conference on Artificial Intelligence (BNAIC'01), Amsterdam, The Netherlands (pp. 283-290).

Weill, P., & Broadbent, M. (1998). Leveraging the New Infrastructure: How Market Leaders Capitalize on Information Technology. Boston: Harvard Business School Press.

Weiss, G. (1999). Multiagent Systems: A Modern Approach to DAI, (pp. 151–157). Cambridge, MA: MIT Press.

Wen, Y., & Oakley, J. P. (1995). Implementation of oriented filters for arc detection. In Proceedings of the 6th British conference on Machine vision, Birmingham, UK, 2, 483-492.

Weng, J., Cohen, P., & Herniou, M. (1992). Camera calibration with distortion models and accuracy evaluation. IEEE Transactions on Pattern Analysis and Machine Intelligence, 14(10), 965–980. doi:10.1109/34.159901

Weng, M. X., Lu, J., & Ren, H. (2001). Unrelated parallel machine scheduling with setup consideration and a total weighted completion time objective. International Journal of Production Economics, 70(3), 215–226. doi:10.1016/S0925-5273(00)00066 9

Westfechtel, B., Munch, B. P., & Conradi, R. (2001). A layered architecture for uniform version management. IEEE Transactions on Software Engineering, 27(12), 1111–1133. doi:10.1109/32.988710

White, J. W., & Olson, J. P. (1986). Computer-based dispatching in mines with concurrent operating objectives. AIME Transactions, 280.

Williamson, M., Decker, K., & Sycara, K. (1996). Unified information and control flow in hierarchical task networks. In Proceedings of AAAI-96 workshop on Theories of Planning, Action, and Control, Portland, Oregon.

Wilton, D. (2001). The Application Of Simulation Technology To Military Command And Control Decision Support. In Proceedings Simulation Technology and Training Conference (SimTecT), Canberra, Australia.

Wos, L., Overbeek, R., Lusk, E., & Boyle, J. (1984). Automated reasoning: Introduction and applications. Englewood Cliffs, NJ: Prentice Hall.

Wu, E., Diao, Y., & Rizvi, S. (2006). High-performance complex event processing over streams. In Proceedings of the 2006 ACM SIGMOD Intern. Conf. on Management of Data, (pp. 407-418).

Wurman, P. R., Wellman, M. P., & Walsh, W. E. (2002). Specifyng Rules for Electronic Aucions. AI Magazine, 23(3), 15–23.

Xiang, Z., & Navab, N. (2000). Tracking and pose estimation for computer assisted localization in industrial environments. In Proceedings on the 5th IEEE Workshop on Applications of Computer Vision, Palm Springs, CA, (pp. 214-221).

Xing, Z., Chen, Y., & Zhang, W. (2006). MaxPlan: Optimal planning by decomposed satisfiability and backward reduction. In Proc. of the international planning competition (ipc), (pp. 53–56).

Xiong, M. H., Tor, S. B., Bhatnagar, R., Khoo, L. P., & Venkat, S. (2006). A DSS approach to managing customer enquiries for SMEs at the customer enquiry stage. International Journal of Production Economics, 103(1), 316–331. doi:10.1016/j.ijpe.2005.08.008

Xiong, Y., & Shafer, S. A. (1993). Depth from focusing and defocusing. In Computer Vision and Pattern Recognition, (pp. 68-73).

Xu, K., Liu, L., & Wu, C. (2007). A three-layered method for business processes discovery and its application in manufacturing industry. Computers in Industry, 58(3), 265–278. doi:10.1016/j.compind.2006.06.001

Xu, W., & Zhang, L. (1993). Pose estimation problem in computer vision. In Proceedings of the IEEE Region 10 Conference on International Computer, Communication, Control and Power Engineering, Beijing, China, 2, 1138-1141.

Yang, B., Yao, L., & Huang, H. (2007). Early Software Quality Prediction Based on a Fuzzy Neural Network Model. In Proceedings of the Third International Conference on Natural Computation (ICNC'07), (pp. 760-764), Haikou, China.

Yang, Y., & Liu, X. (1999). A re-examination of text categorization methods. In Proceedings of the 22nd Annual International ACM Conference on Research andDevelopment in Information Retrieval (SIGIR'99) (pp. 42-49).

Yang, Z., Zhang, D., & Ye, C. (2006). Evaluation Metrics for Ontology Complexity and Evolution Analysis. Proceedings of the IEEE International Conference on e-Business Engineering (ICEBE 2006) (pp. 162-169).

Yao, H., Orme, A. M., & Etzkorn, L. (2005). Cohesion Metrics for Ontology Design and Application. Journal of Computer Science, 1(1), 117–113.

Ye, N. (2003). The Handbook of Data Mining. London: Lawrence Erlbaum Associates, Publishers.

Yen, J., & Langari, R. (1999). Fuzzy Logic: Intelligence, Control, and Information. Upper Saddle River, NJ: Prentice Hall.

Yilmaz, L. (2007). Using meta-level ontology relations to measure conceptual alignment and interoperability of simulation models. In Proceedings of the Winter Simulation Conference, (pp. 1090-1099).

Yilmaz, L., Ören, T., & Aghaee, N. (2006). Intelligent agents, simulation, and gaming. Journal for Simulation and Gaming, 37(3), 339–349. doi:10.1177/1046878106289089

Yokoyama, M., & Lewis, H. W. III. (2003). Optimization of the stochastic dynamic production cycling problem by a genetic algorithm. International Journal of Computers & Operations Research, 30(12), 1831–1849. doi:10.1016/S0305-0548(02)00109-0

Yong, H., Juhua, C., Zhenbang, R., Liu, M., & Kang, X. (2006). A Neural Networks Approach for Software Risk Analysis. In Proceedings of the Sixth IEEE International Conference on Data Mining – Workshops (ICDMW'06), (pp. 722-725), Hong Kong.

Yu, B., Harding, J. A., & Popplewell, K. (2000). A Reusable Enterprise Model. International Journal of Operations & Production Management, 20(1), 50–69. doi:10.1108/01443570010301083

Zadeh, L. A. (1965). Fuzzy sets. Information and Control, 8, 338–353. doi:10.1016/S0019-9958(65)90241-X

Zadeh, L. A. (1973). Outline of a new approach to the analysis of complex systems and decision processes. IEEE Transactions on Systems, Man, and Cybernetics, 3, 28–44.

Zadeh, L. A. (1975). The Concept of a Linguistic Variable - I. Information Sciences, 8, 199–249. doi:10.1016/0020-0255(75)90036-5

Zadeh, L. A. 1987). A theory of approximate reasoning. In R. Yager, S. Orchinnikov, R. Tong, H. Nguyen (Eds.), Fuzzy Sets and Applications (pp 367-412). New York: John Wiley & Sons.

Zaki, M.J., Parthasarathy, S., Ogihara, M., & Li, W. (1997). New parallel algorithms for fast discovery of association rules. Data Mining and Knowledge Discovery, 1(4), 343-37.

Zeigler, B. P. (1986) Toward a Simulation Methodology for Variable Structure Modeling. In Elzas, Oren, Zeigler (Eds.) Modeling and Simulation Methodology in the Artificial Intelligence Era.

Zeigler, B. P., & Hammonds, P. E. (2007) Modeling & Simulation-Based Data Engineering: Introducing Pragmatics into Ontologies for Net-Centric Information Exchange. New York: Academic Press.

Zeng, H., & Rine, D. (2004). Estimation of Software Defects Fix Effort Using Neural Networks. In Proceedings of the 28th Annual International Computer Software and Applications Conference (COMPSAC'04), (pp. 20-21). Hong Kong.

Zhang, L., Madigan, C., Moskewicz, M., & Malik, S. (2001). Efficient conflict driven learning in a boolean satisfiability solver. In Proc. IEEE/ACM International Conference on Computer- Aided Design, (pp. 279–285).

Zhang, Z. (1999). Flexible camera calibration by viewing a plane from unknown orientations. In Proceedings of the 7th IEEE International Conference on Computer Vision, Kerkyra, Greece, 1, 666-673.

Zhao, Z., Ball, M. O., & Kotake, M. (2005). Optimization-Based Available-To-Promise with Multi-Stage Resource Availability. Annals of Operations Research, 135(1), 65–85. doi:10.1007/s10479-005-6235-7

Zhao, Z., You, W., & Lv, Q. (2008). Applications of Fuzzy Critical Chain Method in Project Scheduling. In Proceedings of the Fourth International Conference on Natural Computation (ICNC'08), (pp. 473-477). Jinan, China.

Zhdanova, A. V., Krummenacher, R., Henke, J., & Fensel, D. (2005). Community-driven ontology management: DERI case study. Proceedings of the 2005 IEEE/WIC/ACM International Conference on Web Intelligence (pp. 73-79).

Zilouchian, A. (2001). Intelligent Control Systems Using Soft Computing Methodologies. Boca Raton, FL: CRC Press

Zimmer, D., & Unland, R. (1999). On the semantics of complex events in active database management systems. In ICDE, (pp. 392-399).

Zimmerman, H. J. (1985). Fuzzy Set Theory and Its Applications. Boston: Kluwer-Nijhoff.

Ziou, D., & Tabbone, S. (1998). Edge detection techniques: An overview. International Journal of Pattern Recognition and Image Analysis, 8(4), 537–559.

Zlotkin, G., & Rosenschein, J. (1996). Mechanisms for Automated Negotiation in State Oriented Domains. Journal of Artificial Intelligence Research, 5, 163–238.

About the Contributors

Barin Nag is a Professor in the College of Business & Economics of Towson University. He has a PhD in Management Science and Computer Science from the University of Maryland at College Park. He has Bachelor's and Master's degree in Electrical Engineering from the University of Calcutta, India. He has over forty publications in artificial intelligence applications in logistics, production scheduling, financial management, and supply chain management.

* **

Youssef Achbany received his PhD thesis from the Université catholique de Louvain, Belgium in 2009. He also obtained, in 2004, a MSc degree in information systems in 2004 from Facultés Universitaires Notre-Dame de la Paix in Namur, Belgium. He now works as a functional analyst in the finance industry. His research interests include multi-agents systems, reinforcement learning and probabilistic reputation models.

Franklin Assunção Almeida is an undergraduate Computer Science student at Federal University of Minas Gerais, Belo Horizonte, Brazil. Since 2007 he has been working on the development of mathematical models and algorithms to the integrated planning problem in the mining industry.

Anthony Paul Barnes is from Glenarden, Maryland. He earned a B.S in Computer Science from Claflin University in Orangeburg, South Carolina; and a Master of Computer Science from North Carolina State University in Raleigh, North Carolina. He is currently pursuing a Doctorate of Science in Applied Information Technology at Towson University in Towson, Maryland. His area of research is focused on applying decision support systems to the field of project management. In the year 2000, Anthony was inducted into the inaugural class of the Gates Millennium Scholars Program that is funded by the Bill & Melinda Gates Foundation.

Gianluca Bioli was born in Pisa in 1969. He obtained his Master Degree In Engineering from the University of Pisa in November 1995. In 1999 he received a two year long grant from Politecnico di Torino for the development of research work concerning the industrial application of neuro-fuzzy control techniques. He is currently Junior Researcher at Scuola Superiore Sant'Anna, where he joined the Steel and Industrial Automation Division (SIAD) of PERCRO laboratory in 2001. His research interests mainly concern the control of industrial robots and machine tools and the design and development of software tools for simulation of manufacturing processes.

Ralf Bruns is a professor of computer science at the Fachhochschule Hannover (University of Applied Sciences and Arts). His research interests include software architecture, event-driven architecture, and internet technologies. He received a Diploma degree and a Doctoral degree in computer science from University of Oldenburg (Germany). He is a member if the German Computer Science Society.

Valentina Colla was born in La Spezia in 1970. She obtained her Master Degree In Engineering from the University of Pisa in June 1994 and her PhD in Robotics from Scuola Superiore Sant'Anna of Pisa in January 1998. In the same institution from 1998 until 2000 She was post-doctoral research fellow and She developed research work in the application of neuro-fuzzy control techniques to robotic and mechatronic systems. From 2000 until 2008 She was researcher at Scuola Superiore Sant'Anna and in this institution She is currently a Technical Research Manager. Since 2003 she coordinates the Steel and Industrial Automation Division (SIAD) of PERCRO laboratory. Her research interests include mathematical modeling of industrial processes, modeling and control of industrial robots and machine tools and application of soft-computing techniques for optimization and control of manufacturing procedures.

Oscar Leonel Chacon-Mondragon, received the B.Sc. degree in chemical engineering from Universidad Autonoma de Nuevo Leon, Monterrey, Mexico in 1968, the M.Sc. degree from the University of Houston, Houston Tx in 1976, and the Ph.D. degree from the University of Texas at Austin in 1987. Currently he is Full Time Professor and Chairman of the Graduate Program in Systems engineering in the Universidad Autonoma de Nuevo Leon. His research interest is in the area of fault pattern recognition methods in power systems transmission lines and the heuristic methods applied to assignment problems.

Mark Christensen is from Reisterstown, Maryland. He earned a B.A. in Economics and Business Administration with a minor in Communication Arts from Ursinus College in Collegeville, PA; and an M.S. in Applied Information Technology with a certificate in Information Systems Management from Towson University in Towson, Maryland. He is an employee of the Social Security Administration in Woodlawn, Maryland.

Keith Decker is an Associate Professor in the Department of Computer and Information Sciences at the University of Delaware. His research interests include multi-agent systems, computational organization design, distributed planning and scheduling, and bioinformatics. He received his BS in applied math from Carnegie Mellon University in 1984, his MS in computer science from Rensselaer Polytechnic Institute in 1987, and his Ph.D. in computer science from the University of Massachusetts in 1995. He received a NSF CAREER Award in 1998 and a DARPA recognition award in 2006. Decker is on the editorial board for the Autonomous Agents and Multi-Agent Systems Journal and is program co-chair for AAMAS 2009.

Jürgen Dunkel is a professor of computer science at the Fachhochschule Hannover (University of Applied Sciences and Arts). He received a Diploma degree in computer science from University of Hagen and a Doctoral degree from University of Dortmund (Germany). His research interests include software architecture, event-driven architecture, and model-driven software development. He is a member if the German Computer Science Society.

Carlos Camarão de Figueiredo has a PhD in Computer Science by the Victoria University of Manchester, England (1994), where he developed a proof system for the formal verification of correctness of object-based programs. He is an Associate Professor at the Department of Computer Science of the Federal University of Minas Gerais, Brazil, where he has been working since 1996. His research interests include propositional satisfiability and applications, programming languages and functional programming.

Geraldo Robson Mateus is Full Professor in Computer Science at Federal University of Minas Gerais, Belo Horizonte, Brazil. He received his PhD and MS in computer science from Federal University of Rio de Janeiro, Brazil, in 1980 and 1986, respectively. He spent 1991 and 1992 at the University of Ottawa, Canada, as a visiting researcher. His research interests span network optimization, combinatorial optimization, algorithms, logistics, transport and telecommunication. He is a member of INFORMS, IFORS, SBC, SIAM and SOBRAPO. He has published over 200 scientific papers, 40 journal papers and book chapters and 2 books, and is a leader of several national and international projects. He has worked as a consultant for some companies such as Usiminas, Vale, MBR, Telemig, Telemar, France Telecom, and Embratel.

Mateus Galvão de Melo is a undergraduate Computer Science student at the Federal University of Minas Gerais, Brazil, where he's been doing research around the satisfiability problem since 2007. His research interests include propositional satisfiability, artificial intelligence and information recovery.

Robert J. Hammell II is an Associate Professor in the Department of Computer and Information Sciences at Towson University; he has been with the university since the fall of 2001. He teaches graduate and undergraduate courses for the Department as well as graduate courses in the Center for Applied Information Technology. Prior to joining Towson University, Dr. Hammell served for 22 years in the U.S. Army, retiring as a Lieutenant Colonel. Most of his career was spent in the Acquisition Corps; he is a graduate of the Advanced Program Management Course taught at the Defense Systems Management College. For the last six years of his career he served as a Senior Computer Scientist for the U.S. Army Research Laboratory, focusing on the area of tactical intelligent systems. His current research interests include applications of soft computing techniques for machine learning, decision support systems, pattern recognition, and chemical hazard prediction.

Julie Hoksbergen is from Pasadena, Maryland. She earned a B.S in Computer Information Systems with a Minor in Business Administration and an M.S. in Applied Information Technology with a certificate in Information Systems Management, all from Towson University in Towson, Maryland. She is an employee of the Social Security Administration in Woodlawn, Maryland.

Jean-Charles Lange is a researcher in Operational Research at the Université Catholique de Louvain, Belgium. He holds a bachelor and a master degree in Philosophy and a bachelor and a master degree in management sciences. His research is focusing on production planning and on the management of returns flows of logistic items. Prior to joining the UCL, he has worked as a scheduling manager in a pharmaceutical company.

Oliver Pawlowski is a Ph.D. student at Fachhochschule Hannover (University of Applied Sciences and Arts). He received a B.Sc. and a M.Sc. in computer science from Fachhochaschule Hannover in 2004 and 2006, resp. His research interests include event-based systems and parallel computing.

Bruno Santos Pimentel is a mechanical engineer and a PhD student in Computer Science at Federal University of Minas Gerais, Belo Horizonte, Brazil. His research is focused on the development of mathematical models and algorithms to optimize integrated production and transportation decisions on mining operations. He received his MS degree in Computer Science also from Federal University of Minas Gerais in 2002. He is a member of INFORMS and has worked as an engineer and consultant for Brazilian government agencies and for private companies such as Vallourec & Mannesmann, Arcelor Mittal and Vale.

Haiying Qiao is a Project Manager in SmartOps, applying OR models to solve real world supply chain problems. He received his BS in management science from Wuhan Institute of Technology and his MS in systems engineering from Shanghai Jiao Tong University. He received his PhD in operations research from the University of Maryland, College Park.

Paolo Renna is an Assistant Professor at Department of Environmental Engineering and Physics in the Engineering Faculty of Basilicata University (Italy). He took Ph.D. degree at Polytechnic of Bari in Advanced Production Systems. His academic researches principally deal with the development of innovative negotiation and production planning in distributed environments and manufacturing scheduling in dynamic environment. Several contributions have been presented on design Multi Agent Architecture and test by discrete event simulation in Business to Business environment. Among the contributions, he is co-author of two research books about e-marketplaces and production planning in production networks. Moreover, he has developed coordination approaches in multi-plant production planning environment and innovative scheduling approaches in flexible and reconfigurable manufacturing systems.

Yasmin Rios-Solis is an associated professor of the Graduate Program in Systems Engineering (PISIS) at the Autonomous University of Nuevo Leon (UANL), Mexico. She is interested in operations research, optimization and more particularly in machine scheduling. She obtained the PhD degree from the University Pierre et Marie Curie in France and did a postdoc at the University of Bielefeld in Germany.

Omar Jorge Ibarra-Rojas, received the Degree in mathematics from Faculty of Physic-Mathematics (FCFM) of UANL, Mexico, in 2005. Since 2007 he Works on the research field of optimization in the Graduate Program in Systems Engineering (PISIS), currently he is an aspirant for the program of PhD degree in PISIS, and he is interested in heuristic methodologies and optimization problems such as parallel machines scheduling, assignment, and routing.

Mirko Sgarbi was born in Grosseto in 1972. He obtained his Master Degree In Engineering from the University of Pisa in July 2000. He is currently Junior Researcher at Scuola Superiore Sant'Anna, where he joined the Steel and Industrial Automation Division (SIAD) of PERCRO laboratory in 2002. His research interests mainly concern the application of soft computing algorithms in steel industry, the design and development of software tools for simulation of manufacturing processes with machine tools and industrial robots and the application of vision systems in manufacturing processes.

Vi Tran is a PhD researcher in Information Systems at the Université catholique de Louvain, Belgium. She is completing a PhD thesis on agent based approach for model driven engineering of user interface. Before that, she was a software engineer at FCG, working in web services technologies.

Newton José Vieira has a DSc degree in Computer Science by the Catholic University of Rio de Janeiro (PUC-Rio), Brazil (1987), where he developed an automated reasoning model for knowledge based systems. He is an Associate Professor at the Department of Computer Science of the Federal University of Minas Gerais, Brazil, where he has been working since 1980. His research interests include automated reasoning and knowledge based systems.

Yves Wautelet is Assistant Professor in Information Systems Management at Hogeschool-Universiteit Brussel, Belgium. He has formerly been IT project manager and postdoc fellow at the Université catholique de Louvain, Belgium. He completed a PhD thesis focusing on project and risk management issues in large enterprise software design. Dr. Wautelet also holds a bachelor and master in management sciences as well as a master in Information Systems. His research interests include aspects of software engineering such as requirements engineering, project management, software development life cycles, component-based development, epistemological foundations of SE and CASE-Tools but also information systems strategy.

James Wood is from Pasadena, Maryland. He earned a B.S in Computer Information Systems with a minor in Business and an M.S. in Applied Information Technology with a certificate in Information Systems Management, all from Towson University in Towson, Maryland. He is an employee of the Social Security Administration in Woodlawn, Maryland.

Dong-Qing Yao is an Associate Professor at the College of Business and Economics of Towson University, USA. His research interests are in supply chain management. He has published in journals such as IIE Transactions, the European Journal of Operational Research, the International Journal of Production Economics, Computers and Industrial Engineering, the Journal of Business Research, and OMEGA. He serves on the Editorial Board of the International Journal of Applied Management Science.

Index

F

G

H

I

J

K

L

M

N

O

T

U

V

W

X

Y